Canadian Mathematical Society
Société mathématique du Canada

More information about this series at http://www.springer.com/series/4318

Constantin P. Niculescu · Lars-Erik Persson

Convex Functions and Their Applications

A Contemporary Approach

Second Edition

Constantin P. Niculescu
Department of Mathematics
University of Craiova
Craiova
Romania

and

Academy of Romanian Scientists
Bucharest
Romania

Lars-Erik Persson
UiT, The Artic University of Norway
Campus Narvik
Norway

and

Luleå University of Technology
Luleå
Sweden

ISSN 1613-5237 ISSN 2197-4152 (electronic)
CMS Books in Mathematics
ISBN 978-3-030-08679-4 ISBN 978-3-319-78337-6 (eBook)
https://doi.org/10.1007/978-3-319-78337-6

Mathematics Subject Classification (2010): 26B25, 26D15, 46A55, 46B20, 46B40, 52A01, 52A40, 90C25

Printed on acid-free paper

This Springer imprint is published by the registered company Springer International Publishing AG part of Springer Nature
The registered company address is: Gewerbestrasse 11, 6330 Cham, Switzerland

Contents

Preface ix

List of symbols xiii

1 Convex Functions on Intervals 1
- 1.1 Convex Functions at First Glance . . . 1
- 1.2 Young's Inequality and Its Consequences . . . 11
- 1.3 Log-convex Functions . . . 20
- 1.4 Smoothness Properties of Convex Functions . . . 24
- 1.5 Absolute Continuity of Convex Functions . . . 33
- 1.6 The Subdifferential . . . 36
- 1.7 The Integral Form of Jensen's Inequality . . . 41
- 1.8 Two More Applications of Jensen's Inequality . . . 51
- 1.9 Applications of Abel's Partial Summation Formula . . . 55
- 1.10 The Hermite–Hadamard Inequality . . . 59
- 1.11 Comments . . . 64

2 Convex Sets in Real Linear Spaces 71
- 2.1 Convex Sets . . . 71
- 2.2 The Orthogonal Projection . . . 83
- 2.3 Hyperplanes and Separation Theorems in Euclidean Spaces . . . 89
- 2.4 Ordered Linear Spaces . . . 95
- 2.5 $\mathrm{Sym}(N, \mathbb{R})$ as a Regularly Ordered Banach Space . . . 97
- 2.6 Comments . . . 103

3 Convex Functions on a Normed Linear Space 107
- 3.1 Generalities and Basic Examples . . . 107
- 3.2 Convex Functions and Convex Sets . . . 115
- 3.3 The Subdifferential . . . 122
- 3.4 Positively Homogeneous Convex Functions . . . 130
- 3.5 Inequalities Associated to Perspective Functions . . . 135
- 3.6 Directional Derivatives . . . 141
- 3.7 Differentiability of Convex Functions . . . 147

3.8 Differential Criteria of Convexity 153
3.9 Jensen's Integral Inequality in the Context of Several Variables . 161
3.10 Extrema of Convex Functions 165
3.11 The Prékopa–Leindler Inequality 173
3.12 Comments 179

4 Convexity and Majorization **185**
4.1 The Hardy–Littlewood–Pólya Theory of Majorization 185
4.2 The Schur–Horn Theorem 195
4.3 Schur-Convexity 197
4.4 Eigenvalue Inequalities 202
4.5 Horn's Inequalities 209
4.6 The Way of Hyperbolic Polynomials 212
4.7 Vector Majorization in $\mathbb{R}^N$ 218
4.8 Comments 224

5 Convexity in Spaces of Matrices **227**
5.1 Convex Spectral Functions 227
5.2 Matrix Convexity 234
5.3 The Trace Metric of $\mathrm{Sym}^{++}(n, \mathbb{R})$ 241
5.4 Geodesic Convexity in Global NPC Spaces 248
5.5 Comments 252

6 Duality and Convex Optimization **255**
6.1 Legendre–Fenchel Duality 255
6.2 The Correspondence of Properties under Duality 263
6.3 The Convex Programming Problem 272
6.4 Ky Fan Minimax Inequality 279
6.5 Moreau–Yosida Approximation 283
6.6 The Hopf–Lax Formula 290
6.7 Comments 297

7 Special Topics in Majorization Theory **301**
7.1 Steffensen–Popoviciu Measures 301
7.2 The Barycenter of a Steffensen–Popoviciu Measure 308
7.3 Majorization via Choquet Order 314
7.4 Choquet's Theorem 317
7.5 The Hermite–Hadamard Inequality for Signed Measures 322
7.6 Comments 324

A Generalized Convexity on Intervals **327**
A.1 Means 328
A.2 Convexity According to a Pair of Means 330
A.3 A Case Study: Convexity According to the Geometric Mean . . . 333

B Background on Convex Sets **339**
B.1 The Hahn–Banach Extension Theorem 339
B.2 Separation of Convex Sets 343
B.3 The Krein–Milman Theorem 346

C Elementary Symmetric Functions **349**
C.1 Newton's Inequalities 350
C.2 More Newton Inequalities 354
C.3 Some Results of Bohnenblust, Marcus, and Lopes 356
C.4 Symmetric Polynomial Majorization 359

D Second-Order Differentiability of Convex Functions **361**
D.1 Rademacher's Theorem 361
D.2 Alexandrov's Theorem 364

E The Variational Approach of PDE **367**
E.1 The Minimum of Convex Functionals 367
E.2 Preliminaries on Sobolev Spaces 370
E.3 Applications to Elliptic Boundary-Value Problems 372
E.4 The Galerkin Method 375

References **377**

Index **411**

Preface

Convexity is a simple and natural notion which can be traced back to Archimedes (circa 250 B.C.), in connection with his famous estimate of the value of π (by using inscribed and circumscribed regular polygons). He noticed the important fact that the perimeter of a convex figure is smaller than the perimeter of any other convex figure surrounding it.

As a matter of fact, we experience convexity all the time and in many ways. The most prosaic example is our upright position, which is secured as long as the vertical projection of our center of gravity lies inside the convex envelope of our feet. Also, convexity has a great impact on our everyday life through numerous applications in industry, business, medicine, and art. So do the problems of optimum allocation of resources, estimation and signal processing, statistics, and finance, to name just a few.

The recognition of the subject of convex functions as one that deserves to be studied in its own right is generally ascribed to J. L. W. V. Jensen [230], [231]. However he was not the first to deal with such functions. Among his predecessors we should recall here Ch. Hermite [213], O. Hölder [225], and O. Stolz [463]. During the whole twentieth century, there was intense research activity and many significant results were obtained in geometric functional analysis, mathematical economics, convex analysis, and nonlinear optimization. The classic book by G. H. Hardy, J. E. Littlewood, and G. Pólya [209] played a prominent role in the popularization of the subject of convex functions.

What motivates the constant interest for this subject?

First, its elegance and the possibility to prove deep results even with simple mathematical tools.

Second, many problems raised by science, engineering, economics, informatics, etc. fall in the area of convex analysis. More and more mathematicians are seeking the hidden convexity, a way to unveil the true nature of certain intricate problems. There are two basic properties of convex functions that make them so widely used in theoretical and applied mathematics:

The maximum is attained at a boundary point.

Any local minimum is a global one. Moreover, a strictly convex function admits at most one minimum.

The modern viewpoint on convex functions entails a powerful and elegant interaction between analysis and geometry. In a memorable paper dedicated to the Brunn–Minkowski inequality, R. J. Gardner [176, p. 358], described this reality in beautiful phrases: [convexity] "appears like an octopus, tentacles reaching far and wide, its shape and color changing as it roams from one area to the next. It is quite clear that research opportunities abound."

Over the years a number of notable books dedicated to the theory and applications of convex functions appeared. We mention here: H. H. Bauschke and P. L. Combettes [34], J. M. Borwein and J. Vanderwerff [73], S. Boyd and L. Vandenberghe [75], J.-B. Hiriart-Urruty and C. Lemaréchal [218], L. Hörmander [227], M. A. Krasnosel'skii and Ya. B. Rutickii [259], J. E. Pečarić, F. Proschan and Y. C. Tong [388], R. R. Phelps [397], [398], A. W. Roberts and D. E. Varberg [420], R. T. Rockafellar [421], and B. Simon [450]. The references at the end of this book include many other fine books dedicated to one aspect or another of the theory.

The title of the book by L. Hörmander, *Notions of Convexity,* is very suggestive for the present state of art. In fact, nowadays the study of convex functions has evolved into a larger theory about functions which are adapted to other geometries of the domain and/or obey other laws of comparison of means. Examples are log-convex functions, multiplicatively convex functions, subharmonic functions, and functions which are convex with respect to a subgroup of the linear group.

Our book aims to be a thorough introduction to contemporary convex function theory. It covers a large variety of subjects, from the one real variable case to the infinite-dimensional case, including Jensen's inequality and its ramifications, the Hardy–Littlewood–Pólya theory of majorization, the Borell–Brascamp–Lieb form of the Prékopa–Leindler inequality (as well as its connection with isoperimetric inequalities), the Legendre–Fenchel duality, Alexandrov's result on the second differentiability of convex functions, the highlights of Choquet's theory, and many more. It is certainly a book where inequalities play a central role but in no case a book on inequalities. Many results are new, and the whole book reflects our own experiences, both in teaching and research.

The necessary background is advanced calculus, linear algebra, and some elements of real analysis. When necessary, the reader will be guided to the most pertinent sources, where the quoted results are presented as transparent as possible.

This book may serve many purposes, ranging from honors options for undergraduate students to one-semester graduate course on Convex Functions and Applications.

For example, Chapter 1 and Appendix A offer a quick introduction to generalized convexity, a subject which became very popular during the last decades. The same combination works nicely as supplementary material for a seminar debating heuristics of mathematical research. Chapters 1–6 together with Appendix B could be used as a reference text for a graduate course. And the options can continue.

In order to avoid any confusion relative to our notation, a symbol index was added for the convenience of the reader.

A word of caution is necessary. In this book,

$$\mathbb{N} = \{0, 1, 2, ...\} \quad \text{and} \quad \mathbb{N}^* = \{1, 2, 3, ...\}.$$

According to a recent tendency in mathematical terminology we call a number x *positive* if $x \geq 0$ and *strictly positive* if $x > 0$. A function f is called *increasing* if $x \leq y$ implies $f(x) \leq f(y)$ and *strictly increasing* if $x < y$ implies $f(x) < f(y)$.

Notice also that our book deals only with *real* linear spaces and *all* Borel measures under attention are assumed to be *regular*.

This Second Edition corrects a few errors and typos in the original and includes considerably more material emphasizing the rich applicability of convex analysis to concrete examples.

Chapter 2, on *Convex sets in real linear spaces*, is entirely new and together with Appendix B (mostly dedicated to the Hahn–Banach separation theorems) assures all necessary background for a thorough presentation of the theory and applications of convex functions defined on linear normed spaces. The traditional section devoted to the existence of orthogonal projections in Hilbert spaces is supplemented here with the extension of all basic results to the case of uniformly convex spaces and followed by comments on Clarkson's inequalities. Our discussion on cones in Chapter 2 includes a special section devoted to the order properties of the space $\mathrm{Sym}(n, \mathbb{R})$, of all $n \times n$-dimensional symmetric matrices, an important example of a regularly ordered Banach space which is not a Banach lattice.

Chapter 3, devoted to *Convex functions on a linear normed space*, appears here in a new form, including special sections on the interplay between convex functions and convex sets, the inequalities associated to perspective functions, the subdifferential calculus and extrema of convex functions.

Chapters 4–6 are entirely new. Chapter 4, devoted to the connection between *Convexity and majorization*, includes not only the classical theory of majorization but also its connection with the Horn inequalities and the theory of hyperbolic polynomials (developed by L. Gårding). For the first time in a book, a detailed presentation of Sherman's theorem of majorization (as well as some of its further generalizations) is included. Chapter 5 deals with *Convexity in spaces of matrices.* We focus here on three important topics: the theory of convex spectral functions (treating a special case of convexity under the presence of a symmetry group), the matrix convexity (à la Löwner), and the geodesic convexity in the space $\mathrm{Sym}^{++}(n, \mathbb{R})$, of all $n \times n$-dimensional positively definite matrices endowed with the trace metric. Chapter 6, on *Duality and convex optimization*, illustrates the power of convex analysis to provide useful tools in handling practical problems posed by science, engineering, economics, informatics, etc. Special attention is paid to the Legendre–Fenchel–Moreau duality theory and its applications. We also discuss the convex programming problem, the von Neumann and Ky Fan minimax theorems, the Moreau–Yosida regularization and the implications of the theory of convex functions in deriving the Hopf-Lax formula for the Hamilton Jacobi equation.

Chapter 7 replaces Chapter 4 in the first edition, but the main subject remains the Choquet theory and its extension to a special class of signed measures called Steffensen–Popoviciu measures. Except for Choquet's main theorem, all other results are original and were published by the first named author in collaboration with his students.

In this second edition, Appendix A deals with *Generalized convexity on intervals*, reporting a number of basic facts related to convexity with respect to a pair of means. This appendix summarizes the contents of the former Chapter 2 in the first edition.

The list of references, though extensive, is not intended to be complete, covering only the sources which we used or have a historical importance.

The draft of this new edition was prepared by the first named author at Craiova. The final form was agreed by both authors.

We wish to thank our families for their continuous support and encouragements.

We also wish to thank all our colleagues and friends who read and commented on various versions and parts of the manuscript: Nicolae Cîndea, Frank Hansen, Alois Kufner, Sorin Micu, Gabriel Prăjitură, Ionel Rovenţa, Natasha Samko, Eleftherios Symeonidis, Flavia-Corina Mitroi-Symeonidis, Laurenţiu Temereancă, and Peter Wall.

Craiova, Romania — *Constantin P. Niculescu*
Luleå, Sweden — *Lars-Erik Persson*
May 2018

List of symbols

$\mathbb{N}$, $\mathbb{Z}$, $\mathbb{Q}$, $\mathbb{R}$, $\mathbb{C}$: the classical numerical sets (naturals, integers, etc.)

$\mathbb{N}^\star$: the set of strictly positive integers

$\mathbb{R}_+$: the set of positive real numbers

$\mathbb{R}_+^\star$: the set of strictly positive real numbers

$\overline{\mathbb{R}}$: the set of extended real numbers

$\emptyset$: empty set

∂A: boundary of A

$\overline{A}$: closure of A

$\operatorname{int} A$: interior of A

A°: polar of A

C^*: dual cone

E^*: dual space

$\operatorname{rbd}(A)$: relative boundary of A

$\operatorname{ri}(A)$: relative interior of A

$B_r(a)$: open ball with center a and radius r

$\overline{B}_r(a)$: closed ball with center a and radius r

$[x, y]$: line segment

$\operatorname{aff}(A)$: affine hull of A

$\text{conv}(A)$: convex hull of A

$\dim E$: dimension of E

$\overline{\text{conv}}(A)$: closed convex hull of A

$\text{ext}\, K$: set of extreme points of K

$\text{span}(A)$: linear span of A

$\overline{\text{span}}(A)$: closed linear span of A

$\lambda A + \mu B = \{\lambda x + \mu y : x \in A,\ y \in B\}$

$|A|$: cardinality of A

$\text{diam}(A)$: diameter of A

$\text{Vol}_N(K)$: N-dimensional volume of K

χ_A: characteristic function of A

$\text{cl}\, f$: closure of f

$\text{dom}\, f$: domain of f

I_E: identity of the space E

ι_C: indicator function of C

$f|_K$: restriction of f to K

f^*: Legendre–Fenchel transform of f (convex conjugate of f)

$d_U(x) = d(x, U)$: distance from x to U

$d_H(A, B)$: Pompeiu–Hausdorff distance

$P_K(x)$: orthogonal projection

$\text{dom}(f)$: effective domain of f

$\text{epi}(f)$: epigraph of f

$\text{graph}(f)$: graph of f

$\partial f(a)$: subdifferential of f at a

$\operatorname{supp}(f)$: support of f

$f * g$: convolution

$f \Box g$: infimal convolution

$\mathbb{R}^N$: Euclidean N-space

$\mathbb{R}^N_+ = \{(x_1, \ldots, x_N) \in \mathbb{R}^N : x_1, \ldots, x_N \geq 0\}$, the positive orthant

$\mathbb{R}^N_{++} = \{(x_1, \ldots, x_N) \in \mathbb{R}^N : x_1, \ldots, x_N > 0\}$

$\mathbb{R}^N_{\geq} = \{(x_1, \ldots, x_N) \in \mathbb{R}^N : x_1 \geq \cdots \geq x_N\}$

$\langle x, y \rangle$, $x \cdot y$: inner product

$\mathrm{M}_N(\mathbb{R})$, $\mathrm{M}_N(\mathbb{C})$: spaces of $N \times N$-dimensional matrices

$\mathrm{O}(N)$: orthogonal group

$\Pi(N)$: permutation group

$\operatorname{Sym}(N, \mathbb{R})$: the space of all symmetric matrices of $\mathrm{M}_N(\mathbb{R})$

$\operatorname{Sym}^+(N, \mathbb{R})$: the subset of all positive matrices of $\operatorname{Sym}(N, \mathbb{R})$

$\operatorname{Sym}^{++}(N, \mathbb{R})$: the subset of all strictly positive matrices of $\operatorname{Sym}(N, \mathbb{R})$

I: identity matrix

A^*: adjoint matrix

$\det A$: determinant of A

$\ker A$: kernel (null space) of A

$\operatorname{rng} A$: range of A

$\operatorname{trace} A$: trace of A

$\mathcal{O}(A)$: orbit of A

$A \#_{1/2} B$: geometric mean of the matrices A and B

$\liminf_{x \to a} f(x) = \lim_{r \to 0} \inf\{f(x) : x \in \operatorname{dom}(f) \cap B_r(a)\}$: lower limit

$\limsup_{x\to a} f(x) = \lim_{r\to 0} \sup\{f(x) : x \in \mathrm{dom}(f) \cap B_r(a)\}$: upper limit

$f'_+(a;v)$ and $f'_-(a;v)$: sided directional derivatives

$f'(a;v)$: first Gâteaux differential

$f''(a;v,w)$: second Gâteaux differential

$\mathrm{d}f$: First-order Fréchet differential

$\mathrm{d}^2 f$: Second-order Fréchet differential

∇: gradient

Hess, ∇^2: Hessian

$A(K)$: space of real-valued continuous and affine functions

$\mathrm{Conv}(K)$: space of real-valued continuous and convex functions

$C(K)$: space of real-valued continuous functions

$C^m(\Omega) = \{f : D^\alpha f \in C(\Omega) \text{ for all } |\alpha| \leq m\}$

$C^m(\overline{\Omega}) = \{f : D^\alpha f \text{ uniformly continuous on } \Omega \text{ for all } |\alpha| \leq m\}$

$C_c^\infty(\Omega)$: space of functions of class C^∞ with compact support

$L^p(\Omega)$: space of p-th-power Lebesgue integrable functions on Ω

$L^p(\mu)$: space of p-th-power μ-integrable functions

$\|f\|_{L^p}$: L^p-norm

ℓ^p: space of p-th-power absolutely summing real sequences

$\mathrm{Lip}(f)$: Lipschitz constant

$W^{m,p}(\Omega)$: Sobolev space on Ω

$\|f\|_{W^{m,p}}$: Sobolev norm

$W_0^{m,p}(\Omega)$: norm closure of $C_c^\infty(\Omega)$ in $W^{m,p}(\Omega)$

$\Gamma(E)$: the set of lower semicontinuous proper convex functions defined on E

$\mathcal{B}(X)$: Borel σ-algebra associated to X

Prob(X): set of Borel probability measures on X

δ_a: Dirac measure concentrated at a

$\mathcal{L}$: Lebesgue measure

bar(μ): barycenter of μ

$E(f)$: conditional expectation (mean value) of f

var(f): variance of f

$\prec_{Ch}, \prec_{HLP}, \prec_{Sch}, \prec_{SPM}, \prec_{\mathrm{ds}}, \prec_{\mathrm{uds}}$: various types of majorization

$\blacksquare$: end of the proof

Chapter 1

Convex Functions on Intervals

The study of convex functions of one real variable offers an excellent glimpse of the beauty and fascination of advanced mathematics. The reader will find here a large variety of results based on simple and intuitive arguments that have remarkable applications. At the same time they provide the starting point of deep generalizations in the setting of several variables, that will be discussed in the next chapters.

1.1 Convex Functions at First Glance

Throughout this book the letter I will denote a nondegenerate interval (that is, an interval containing an infinity of points).

1.1.1 Definition *A function* $f\colon I \to \mathbb{R}$ *is called* convex *if*

$$f((1-\lambda)x + \lambda y) \leq (1-\lambda)f(x) + \lambda f(y) \tag{1.1}$$

for all points x *and* y *in* I *and all* $\lambda \in [0,1]$. *It is called* strictly convex *if the inequality* (1.1) *holds strictly whenever* x *and* y *are distinct points and* $\lambda \in (0,1)$. *If* $-f$ *is convex (respectively, strictly convex), then we say that* f *is* concave *(respectively,* strictly concave*). If* f *is both convex and concave, then* f *is said to be* affine.

The affine functions are precisely the functions of the form $mx+n$, for suitable constants m and n. One can easily prove that the following three functions are convex (though not strictly convex): the positive part $x^+ = \max\{x,0\}$, the negative part $x^- = \max\{-x,0\}$, and the absolute value $|x| = \max\{-x,x\}$. Together with the affine functions they provide the building blocks for the entire class of convex functions on intervals. See Lemma 1.9.1 and Lemma 1.9.2 below.

C. P. Niculescu and L.-E. Persson, *Convex Functions and Their Applications*, CMS Books in Mathematics, https://doi.org/10.1007/978-3-319-78337-6_1

Simple computations show that the square function x^2 is strictly convex on $\mathbb{R}$ and the square root function $\sqrt{x}$ is strictly concave on $\mathbb{R}_+$. In many cases of interest the convexity is established via the second derivative test. See Corollary 1.4.8. Some other criteria of convexity related to basic theory of convex functions will be presented in what follows.

The convexity of a function $f\colon I \to \mathbb{R}$ means geometrically that the points of the graph of $f|_{[u,v]}$ are under (or on) the chord joining the endpoints $(u, f(u))$ and $(v, f(v))$, for all $u, v \in I$, $u < v$; see Figure 1.1. Thus the inequality (1.1) is equivalent to

$$f(x) \leq f(u) + \frac{f(v) - f(u)}{v - u}(x - u) \tag{1.2}$$

for all $x \in [u, v]$, and all $u, v \in I$, $u < v$.

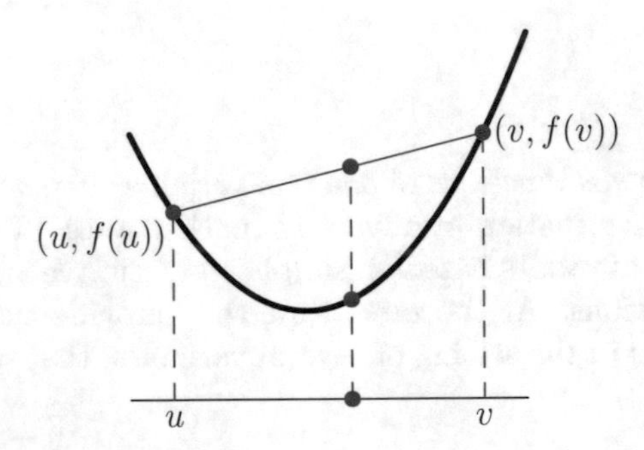

Figure 1.1: Convex function: the graph is under the chord.

This remark shows that the convex functions are majorized by affine functions on any compact subinterval. The existence of affine minorants will be discussed in Section 1.5.

Every convex function f is bounded on each compact subinterval $[u, v]$ of its interval of definition. In fact, $f(x) \leq M = \max\{f(u), f(v)\}$ on $[u, v]$ and writing an arbitrary point $x \in [u, v]$ as $x = (u+v)/2 + t$ for some t with $|t| \leq (v-u)/2$, we easily infer that

$$\begin{aligned} f(x) = f\Big(\frac{u+v}{2} + t\Big) &\geq 2f\Big(\frac{u+v}{2}\Big) - f\Big(\frac{u+v}{2} - t\Big) \\ &\geq 2f\Big(\frac{u+v}{2}\Big) - M. \end{aligned}$$

1.1.2 Theorem *A convex function $f\colon I \to \mathbb{R}$ is continuous at each interior point of I.*

Proof. Suppose that $a \in \operatorname{int} I$ and choose $\varepsilon > 0$ such that $[a - \varepsilon, a + \varepsilon] \subset I$. Then

$$f(a) \leq \frac{1}{2} f(a - \varepsilon) + \frac{1}{2} f(a + \varepsilon),$$

and

$$f(a \pm t\varepsilon) = f((1-t)a + t(a \pm \varepsilon)) \leq (1-t) f(a) + tf(a \pm \varepsilon)$$

for every $t \in [0,1]$. Therefore

$$t\,(f(a \pm \varepsilon) - f(a)) \geq f(a \pm t\varepsilon) - f(a) \geq -t\,(f(a \mp \varepsilon) - f(a)),$$

which yields

$$|f(a \pm t\varepsilon) - f(a)| \leq t \max\{|f(a-\varepsilon) - f(a)|, |f(a+\varepsilon) - f(a)|\},$$

for every $t \in [0,1]$. The continuity of f at a is now clear. ∎

Simple examples such as $f(x) = 0$ if $x \in (0,1)$, and $f(0) = f(1) = 1$, show that upward jumps can appear at the endpoints of the interval of definition of a convex function. Fortunately, these possible discontinuities are removable:

1.1.3 Proposition *If $f\colon [a,b] \to \mathbb{R}$ is a convex function, then the limits $f(a+) = \lim_{x\searrow a} f(x)$ and $f(b-) = \lim_{x\nearrow b} f(x)$ exist in $\mathbb{R}$ and*

$$\widetilde{f}(x) = \begin{cases} f(a+) & \text{if } x = a \\ f(x) & \text{if } x \in (a,b) \\ f(b-) & \text{if } x = b \end{cases}$$

is a continuous convex function.

This result is a consequence of the following:

1.1.4 Lemma *If $f\colon I \to \mathbb{R}$ is convex, then either f is monotonic on $\operatorname{int} I$, or there exists a point $\xi \in \operatorname{int} I$ such that f is decreasing on the interval $(-\infty,\xi] \cap I$ and increasing on the interval $[\xi,\infty) \cap I$.*

Proof. Choose $a < b$ arbitrarily among the interior points of I and put $m = \inf\{f(x) : x \in [a,b]\}$. Since f is continuous on $[a,b]$, this infimum is attained at a point $\xi \in [a,b]$, that is,

$$m = f(\xi).$$

If $a \leq x < y < \xi$, then y is a convex combination of x and ξ, precisely, $y = \frac{\xi - y}{\xi - x}x + \frac{y-x}{\xi-x}\xi$. Since f is convex,

$$f(y) \leq \frac{\xi - y}{\xi - x} f(x) + \frac{y - x}{\xi - x} f(\xi) \leq f(x)$$

and thus f is decreasing on the interval $[a,\xi]$. If $\xi < b$, a similar argument shows that f is increasing on $[\xi,b]$.

The proof ends with a gluing process (to the left of a and the right of b), observing that the property of convexity makes impossible the existence of three numbers $u < v < w$ in I such that $f(u) < f(v) > f(w)$. ∎

1.1.5 Corollary (a) *Every convex function $f\colon I \to \mathbb{R}$ which is not monotonic on $\operatorname{int} I$ has an interior global minimum.*

(b) *If a convex function $f\colon \mathbb{R} \to \mathbb{R}$ is bounded from above, then it is constant.*

Attaining supremum at endpoints is not a property characteristic of convex functions, but a variant of it does the job.

1.1.6 Theorem (S. Saks [432]) *Let f be a real-valued function defined on an interval I. Then f is (strictly) convex if and only if for every compact subinterval J of I, and every affine function L, the supremum of $f+L$ on J is attained at an endpoint (and only there).*

This statement remains valid if the perturbations L are supposed to be linear (that is, of the form $L(x) = mx$ for suitable $m \in \mathbb{R}$). See Exercise 1 for examples of convex functions derived from Theorem 1.1.6.

Proof. We will restrict ourselves to the case of convex functions. The case of strictly convex functions can be treated in the same manner.

Necessity: If f is convex, so is the sum $F = f + L$. Since every point of a subinterval $J = [x, y]$ is a convex combination $z = (1-\lambda)x + \lambda y$ of x and y, we have

$$\begin{aligned}\sup_{z\in J} F(z) &= \sup_{\lambda\in[0,1]} F((1-\lambda)x + \lambda y) \\ &\leq \sup_{\lambda\in[0,1]} [(1-\lambda)F(x) + \lambda F(y)] = \max\{F(x), F(y)\}.\end{aligned}$$

Sufficiency: Given a compact subinterval $J = [x, y]$ of I, there exists an affine function $L(x) = mx + n$ which agrees with f at the two endpoints x and y. Then

$$\sup_{\lambda\in[0,1]} [(f-L)((1-\lambda)x + \lambda y)] = 0,$$

which yields

$$\begin{aligned}0 &\geq f((1-\lambda)x + \lambda y) - L((1-\lambda)x + \lambda y) \\ &= f((1-\lambda)x + \lambda y) - (1-\lambda)L(x) - \lambda L(y) \\ &= f((1-\lambda)x + \lambda y) - (1-\lambda)f(x) - \lambda f(y)\end{aligned}$$

for every $\lambda \in [0, 1]$. ■

1.1.7 Remark *Call a function $f : I \to \mathbb{R}$* quasiconvex *if*

$$f((1-\lambda)x + \lambda y) \leq \max\{f(x), f(y)\}$$

for all $x, y \in I$ and $\lambda \in [0, 1]$. For example, so are the monotonic functions and the functions that admit interior points $c \in I$ such that they are decreasing on $(-\infty, c] \cap I$ and increasing on $[c, \infty) \cap I$. Unlike the case of convex functions, the sum of a quasiconvex function and a linear function may not be quasiconvex. See the case of the sum $x^3 - 3x$. Related to quasiconvex functions are the quasiconcave *functions, characterized by the property*

$$f((1-\lambda)x + \lambda y) \geq \min\{f(x), f(y)\}$$

for all $x, y \in I$ and $\lambda \in [0, 1]$. See Exercises 11 and 12 for more information.

The following characterization of convexity within the class of continuous functions also proves useful in checking convexity.

1.1.8 Theorem (J. L. W. V. Jensen [231]) *A function $f\colon I \to \mathbb{R}$ is convex if and only if it verifies the following two conditions:*
(a) *f is continuous at each interior point of I; and*
(b) *f is midpoint convex, that is,*

$$f\Big(\frac{x+y}{2}\Big) \leq \frac{f(x)+f(y)}{2} \quad \text{for all } x, y \in I.$$

Proof. The necessity follows from Theorem 1.1.2. The sufficiency is proved by *reductio ad absurdum.* If f were not convex, then there would exist a subinterval $[a, b]$ such that the graph of $f|_{[a,b]}$ is not under the chord joining $(a, f(a))$ and $(b, f(b))$; that is, the function

$$\varphi(x) = -f(x) + f(a) + \frac{f(b)-f(a)}{b-a}(x-a), \quad x \in [a, b]$$

verifies $\gamma = \inf\{\varphi(x) : x \in [a, b]\} < 0$. Notice that $-\varphi$ is midpoint convex, continuous and $\varphi(a) = \varphi(b) = 0$. Put $c = \inf\{x \in [a, b] : \varphi(x) = \gamma\}$; then necessarily $\varphi(c) = \gamma$ and $c \in (a, b)$. By the definition of c, for every $h > 0$ for which $c \pm h \in (a, b)$ we have

$$\varphi(c-h) > \varphi(c) \quad \text{and} \quad \varphi(c+h) \geq \varphi(c)$$

so that

$$-\varphi(c) > \frac{-\varphi(c-h) - \varphi(c+h)}{2}$$

in contradiction with the fact that $-\varphi$ is midpoint convex. ∎

1.1.9 Corollary *Let $f\colon I \to \mathbb{R}$ be a continuous function. Then f is convex if and only if*

$$f(x+h) + f(x-h) - 2f(x) \geq 0$$

for all $x \in I$ and all $h > 0$ such that both $x+h$ and $x-h$ are in I.

Notice that both Theorem 1.1.8 and its Corollary 1.1.9 above have straightforward variants in the case of strictly convex functions.

Corollary 1.1.9 allows us to check immediately the strict convexity/concavity of some very common functions, such as the exponential function, the logarithmic function, and the restriction of the sine function to $[0, \pi]$. Indeed, in the first case, the fact that

$$a, b > 0, \; a \neq b, \quad \text{implies} \quad \frac{a+b}{2} > \sqrt{ab}$$

is equivalent to

$$\mathrm{e}^{x+h} + \mathrm{e}^{x-h} - 2\mathrm{e}^{x} > 0$$

for all $x \in \mathbb{R}$ and all $h > 0$.

Many other examples can be deduced using the closure under functional operations with convex/concave functions.

1.1.10 Proposition (The operations with convex functions)

(a) *Adding two convex functions (defined on the same interval) we obtain a convex function; if one of them is strictly convex, then the sum is also strictly convex.*

(b) *Multiplying a (strictly) convex function by a (strictly) positive scalar we obtain also a (strictly) convex function.*

(c) *Suppose that f and g are two positive convex functions defined on an interval I. Then their product is convex provided that they are synchronous in the sense that*

$$(f(x) - f(y))\,(g(x) - g(y)) \geq 0$$

for all $x, y \in I$; for example, this condition occurs if f and g are both decreasing or both increasing.

(d) *The restriction of every (strictly) convex function to a subinterval of its domain is also a (strictly) convex function.*

(e) *A convex and increasing (decreasing) function of a convex (concave) function is convex. The variant where convexity is strict and monotonicity is also strict works as well. Moreover, the role of convexity and concavity can be interchanged.*

(f) *Suppose that f is a bijection between two intervals I and J. If f is strictly increasing, then f is (strictly) convex if and only if f^{-1} is (strictly) concave. If f is a decreasing bijection, then f and f^{-1} are of the same type of convexity.*

(g) *If f is a strictly positive concave function, then $1/f$ is a convex function. Here the role of concavity and convexity cannot be changed to each other.*

(h) *The maximum of two (strictly) convex functions $f, g : I \to \mathbb{R}$,*

$$\max\{f, g\}\,(x) = \max\{f(x), g(x)\}$$

is also a (strictly) convex function.

(i) *The superposition $f\,(ax + b)$, of a convex function f and an affine function $ax + b$, is a convex function.*

The details are straightforward.

We next discuss the extension of the inequality of convexity (1.1) to arbitrarily long finite families of points. The basic remark in this respect is the fact that the intervals are closed under arbitrary convex combinations, that is,

$$\sum_{k=1}^{n} \lambda_k x_k \in I$$

for all $x_1, \ldots, x_n \in I$ and all $\lambda_1, \ldots, \lambda_n \in [0, 1]$ with $\sum_{k=1}^{n} \lambda_k = 1$. This can be proved by induction on the number n of points involved in the convex combinations. The case $n = 1$ is trivial, while for $n = 2$ it follows from the definition of a convex set. Assuming the result is true for all convex combinations with at most $n \geq 2$ points, let us pass to the case of combinations with $n + 1$ points, $x = \sum_{k=1}^{n+1} \lambda_k x_k$. The nontrivial case is when all coefficients λ_k lie in $(0, 1)$.

But in this case, due to our induction hypothesis, x can be represented as a convex combination of two elements of I,

$$x = (1 - \lambda_{n+1})\Big(\sum_{k=1}^{n} \frac{\lambda_k}{1-\lambda_{n+1}} x_k\Big) + \lambda_{n+1} x_{n+1},$$

hence x belongs to I.

The above remark on intervals has a notable counterpart for convex functions:

1.1.11 Lemma (The discrete case of Jensen's inequality) *A real-valued function f defined on an interval I is convex if and only if for all points $x_1, \ldots, x_n$ in I and all scalars $\lambda_1, \ldots, \lambda_n$ in $[0,1]$ with $\sum_{k=1}^{n} \lambda_k = 1$ we have*

$$f\Big(\sum_{k=1}^{n} \lambda_k x_k\Big) \le \sum_{k=1}^{n} \lambda_k f(x_k).$$

If f is strictly convex, the above inequality is strict if the points x_k are not all equal and the scalars λ_k are positive.

Proof. The first assertion follows by mathematical induction. As concerns the second assertion, suppose that the function f is strictly convex and

$$f\left(\sum_{k=1}^{n} \lambda_k x_k\right) = \sum_{k=1}^{n} \lambda_k f(x_k) \tag{1.3}$$

for some points $x_1, \ldots, x_n \in I$ and some scalars $\lambda_1, \ldots, \lambda_n \in (0,1)$ that sum up to 1. If $x_1, \ldots, x_n$ are not all equal, the set $S = \{k : x_k < \max\{x_1, \ldots, x_n\}\}$ will be a proper subset of $\{1, ..., n\}$ and $\lambda_S = \sum_{k \in S} \lambda_k \in (0,1)$. Since f is strictly convex, we get

$$\begin{aligned}
f\left(\sum_{k=1}^{n} \lambda_k x_k\right) &= f\left(\lambda_S \left(\sum_{k \in S} \frac{\lambda_k}{\lambda_S} x_k\right) + (1-\lambda_S)\left(\sum_{k \notin S} \frac{\lambda_k}{1-\lambda_S} x_k\right)\right) \\
&< \lambda_S f\left(\sum_{k \in S} \frac{\lambda_k}{\lambda_S} x_k\right) + (1-\lambda_S) f\left(\sum_{k \notin S} \frac{\lambda_k}{1-\lambda_S} x_k\right) \\
&< \lambda_S \sum_{k \in S} \frac{\lambda_k}{\lambda_S} f(x_k) + (1-\lambda_S) \sum_{k \notin S} \frac{\lambda_k}{1-\lambda_S} f(x_k) = \sum_{k=1}^{n} \lambda_k f(x_k),
\end{aligned}$$

which contradicts our hypothesis (1.3). Therefore all points x_k should coincide. ■

The extension of Lemma 1.1.11 to the case of arbitrary finite measure spaces (as well as an estimate of its precision) will be discussed in Section 1.7.

An immediate consequence of Lemma 1.1.11 (when applied to the exponential function) is the following result which extends the well-known *AM–GM* inequality (that is, the inequality between the arithmetic mean and the geometric mean):

1.1.12 Theorem (The weighted form of the *AM–GM* inequality; L. J. Rogers [424]) *If $x_1, \ldots, x_n \in (0, \infty)$ and $\lambda_1, \ldots, \lambda_n \in (0, 1)$, $\sum_{k=1}^n \lambda_k = 1$, then*

$$\sum_{k=1}^{n} \lambda_k x_k > x_1^{\lambda_1} \cdots x_n^{\lambda_n}$$

unless $x_1 = \cdots = x_n$.

Replacing x_k by $1/x_k$ in the last inequality we get

$$x_1^{\lambda_1} \cdots x_n^{\lambda_n} > 1 / \sum_{k=1}^{n} \frac{\lambda_k}{x_k}$$

unless $x_1 = \cdots = x_n$. This represents the weighted form of the *geometric mean–harmonic mean inequality* (that is, of *GM–HM* inequality).

For $\lambda_1 = \cdots = \lambda_n = 1/n$ we recover the usual *AM–GM–HM inequality, which asserts that for every family $x_1, \ldots, x_n$ of positive numbers, not all equal to each other, we have*

$$\frac{x_1 + \cdots + x_n}{n} > \sqrt[n]{x_1 \cdots x_n} > \frac{n}{\left(\frac{1}{x_1} + \cdots + \frac{1}{x_n}\right)}.$$

A stronger concept of convexity, called logarithmic convexity (or log-convexity), will be discussed in Section 1.3.

Exercises

1. Infer from Theorem 1.1.6 that the following differentiable functions are strictly convex:

   (a) $-\sin x$ on $[0, \pi]$ and $-\cos x$ on $[-\pi/2, \pi/2]$;

   (b) x^p on $[0, \infty)$ if $p > 1$; x^p on $(0, \infty)$ if $p < 0$; $-x^p$ on $[0, \infty)$ if $p \in (0, 1)$;

   (c) $(1 + x^p)^{1/p}$ on $[0, \infty)$ if $p > 1$.

2. (New from old) Assume that $f(x)$ is a (strictly) convex function on an interval $I \subset (0, \infty)$. Prove that $xf(1/x)$ is (strictly) convex on any interval J for which $x \in J$ implies $1/x \in I$. Then infer that $x \log x$ is a strictly convex function on $[0, \infty)$ and $x \sin(1/x)$ is strictly concave on $[1/\pi, \infty)$.

3. Suppose that $f : \mathbb{R} \to \mathbb{R}$ is a convex function and P and Q are distinct points of its graph. Prove that:

   (a) if we modify the graph of f by replacing the portion joining P and Q with the corresponding linear segment, the result is still the graph of a convex function g;

   (b) all points of the graph of g are on or above the secant line joining P and Q.

4. Suppose that $f_1, \ldots, f_n$ are positive concave functions with the same domain of definition. Prove that $(f_1 \cdots f_n)^{1/n}$ is also a concave function.

5. (From discrete to continuous) Prove that the discrete form of Jensen's inequality implies (and is implied by) the following integral form of this inequality: If $f : [c,d] \to \mathbb{R}$ is a convex function and $g : [a,b] \to [c,d]$ is a Riemann integrable function, then

$$f\left(\frac{1}{b-a}\int_a^b g(x)\mathrm{d}x\right) \leq \frac{1}{b-a}\int_a^b f\left(g(x)\right)\mathrm{d}x.$$

6. (Optimization without calculus) The *AM–GM* inequality offers a very convenient way to minimize a sum when the product of its terms is constant (or to maximize a product whose factors sum to a constant). Infer from Theorem 1.1.12 that

$$\min_{x,y>0}\left(x+y+\frac{1}{x^2y}\right) = 4/\sqrt{2} \quad \text{and} \quad \max_{6|x|+5|y|+|z|=4} x^2yz = \frac{1}{45}.$$

7. (a) Prove that Theorem 1.1.8 remains true if the condition of midpoint convexity is replaced by the following one: $f((1-\alpha)x+\alpha y) \leq (1-\alpha)f(x)+\alpha f(y)$ for some fixed parameter $\alpha \in (0,1)$, and for all $x, y \in I$.

   (b) Prove that Theorem 1.1.8 remains true if the condition of continuity is replaced by boundedness from above on every compact subinterval.

8. Let $f\colon I \to \mathbb{R}$ be a convex function and let $x_1, \ldots, x_n \in I$ $(n \geq 2)$. Prove that

$$(n-1)\Big[\frac{f(x_1)+\cdots+f(x_{n-1})}{n-1} - f\Big(\frac{x_1+\cdots+x_{n-1}}{n-1}\Big)\Big]$$

cannot exceed

$$n\Big[\frac{f(x_1)+\cdots+f(x_n)}{n} - f\Big(\frac{x_1+\cdots+x_n}{n}\Big)\Big].$$

9. (Two more proofs of the *AM–GM* inequality) Let $x_1, \ldots, x_n > 0$ $(n \geq 2)$ and for each $1 \leq k \leq n$ put

$$A_k = \frac{x_1+\cdots+x_k}{k} \quad \text{and} \quad G_k = (x_1 \cdots x_k)^{1/k}.$$

   (a) (T. Popoviciu) Prove that

$$\Big(\frac{A_n}{G_n}\Big)^n \geq \Big(\frac{A_{n-1}}{G_{n-1}}\Big)^{n-1} \geq \cdots \geq \Big(\frac{A_1}{G_1}\Big)^1 = 1.$$

   (b) (R. Rado) Prove that

$$n(A_n - G_n) \geq (n-1)(A_{n-1} - G_{n-1}) \geq \cdots \geq 1 \cdot (A_1 - G_1) = 0.$$

   [*Hint*: Apply the result of Exercise 8 respectively to $f = -\log$ and $f = \exp$.]

10. (The power means in the discrete case) Let $x = (x_1, \ldots, x_n)$ and $\lambda = (\lambda_1, \ldots, \lambda_n)$ be two n-tuples of strictly positive numbers such that $\sum_{k=1}^{n} \lambda_k = 1$. The (weighted) *power mean* of order t is defined as

$$M_t(x; \lambda) = \Big(\sum_{k=1}^{n} \lambda_k x_k^t \Big)^{1/t} \quad \text{for } t \in \mathbb{R} \setminus \{0\}$$

and

$$M_0(x; \lambda) = \prod_{k=1}^{n} x_k^{\lambda_k}. \tag{1.4}$$

We also define $M_{-\infty}(x; \lambda) = \min\{x_k : k = 1, ..., n\}$ and $M_{\infty}(x, \lambda) = \max\{x_k : k = 1, ..., n\}$. Notice that M_1 is the arithmetic mean, M_0 is the geometric mean, and M_{-1} is the harmonic mean. Moreover, $M_{-t}(x; \lambda) = M_t(x^{-1}; \lambda)^{-1}$. Prove that:

(a) $M_s(x; \lambda) \leq M_t(x; \lambda)$ whenever $s \leq t$ in $\mathbb{R}$;

(b) $\lim_{t \to 0} M_t(x, \lambda) = \prod_{k=1}^{n} x_k^{\lambda_k}$;

(c) $\lim_{t \to -\infty} M_t(x; \lambda) = M_{-\infty}(x; \lambda)$ and $\lim_{t \to \infty} M_t(x; \lambda) = M_{\infty}(x; \lambda)$.

[*Hint:* (a) According to the weighted form of the *AM–GM* inequality, $M_0(x; \lambda) \leq M_t(x; \lambda)$ for all $t \geq 0$. Then apply Jensen's inequality to the function $x^{t/s}$ to infer that $M_s(x; \lambda) \leq M_t(x; \lambda)$ whenever $0 < s \leq t$. To end the proof of (a) use the formula $M_{-t}(x; \lambda) = M_t(x^{-1}; \lambda)^{-1}$.]

Remark. The power means discussed above are a special case of integral power means, that will be presented in Section 1.7, Exercise 1. Indeed, an n-tuple $\lambda = (\lambda_1, \ldots, \lambda_n)$ of strictly positive numbers that sum to unity can be identified with the discrete probability measure on the set $\Omega = \{1, 2, ..., n\}$ defined by $\lambda(A) = \sum_{k \in A} \lambda_k$ for $A \in \mathcal{P}(\Omega)$. The λ-integrable functions $x : \Omega \to \mathbb{R}$, $x(k) = x_k$, are nothing but the strings $x = (x_1, \ldots, x_n)$ of real numbers, so that

$$M_t(x; \lambda) = \int_{\Omega} x^t(k) \mathrm{d}\lambda(k) \quad \text{for } t \in \mathbb{R} \setminus \{0\}$$

and $M_0(x; \lambda) = \exp\left(\int_{\Omega} \log x(k) \mathrm{d}\lambda(k) \right)$.

11. Prove that:

(a) a function $f : I \to \mathbb{R}$ is quasiconvex if and only if all sublevel sets $L_{\lambda} = \{x : f(x) \leq \lambda\}$ are intervals;

(b) the product of two positive concave functions is a quasiconcave function;

(c) if f is positive and convex, then $-1/f$ is quasiconvex;

(d) if f is negative and quasiconvex, then $1/f$ is quasiconcave;

(e) If f is a positive convex function and g is a positive concave function, then f/g is quasiconvex.

12. Suppose that f is a continuous real-valued function defined on an interval I. Prove that f is quasiconvex if and only if it is either monotonic or there exists an interior point $c \in I$ such that f is decreasing on $(-\infty, c] \cap I$ and increasing on $[c, \infty) \cap I$.

13. A sequence of real numbers $a_0, a_1, \ldots, a_n$ (with $n \geq 2$) is said to be *convex* provided that

$$\Delta^2 a_k = a_k - 2a_{k+1} + a_{k+2} \geq 0$$

for all $k = 0, \ldots, n-2$; it is said to be *concave* provided $\Delta^2 a_k \leq 0$ for all k.

(a) Solve the system

$$\Delta^2 a_k = b_k \quad \text{for } k = 0, \ldots, n-2$$

(in the unknowns a_k) to prove that the general form of a convex sequence $\mathbf{a} = (a_0, a_1, \ldots, a_n)$ with $a_0 = a_n = 0$ is given by the formula

$$\mathbf{a} = \sum_{j=1}^{n-1} c_j \mathbf{w}^j,$$

where $c_j = 2a_j - a_{j-1} - a_{j+1}$ and $\mathbf{w}^j$ has the components

$$w_k^j = \begin{cases} k(n-j)/n, & \text{for } k = 0, \ldots, j \\ j(n-k)/n, & \text{for } k = j, \ldots, n. \end{cases}$$

(b) Prove that the general form of a convex sequence $\mathbf{a} = (a_0, a_1, \ldots, a_n)$ is $\mathbf{a} = \sum_{j=0}^{n} c_j \mathbf{w}^j$, where c_j and $\mathbf{w}^j$ are as in the case (a) for $j = 1, \ldots, n-1$. The other coefficients and components are:

$$c_0 = a_0, c_n = a_n, w_k^0 = (n-k)/n \text{ and } w_k^n = k/n \text{ for } k = 0, \ldots, n).$$

Remark. The theory of convex sequences can be subordinated to that of convex functions. If $f\colon [0, n] \to \mathbb{R}$ is a convex function, then $f(0), ..., f(n)$ is a convex sequence; conversely, if $a_0, ..., a_n$ is a convex sequence, then the piecewise linear function $f\colon [0, n] \to \mathbb{R}$ obtained by joining the points (k, a_k) is convex.

1.2 Young's Inequality and Its Consequences

The following special case of the weighted form of the *AM–GM* inequality is known as *Young's inequality*:

$$ab \leq \frac{a^p}{p} + \frac{b^q}{q} \quad \text{for all } a, b \geq 0, \tag{1.5}$$

whenever $p, q \in (1, \infty)$ and $1/p + 1/q = 1$; the equality holds if and only if $a^p = b^q$. Young's inequality can be also obtained as a consequence of strict convexity of the exponential function. In fact,

$$ab = \mathrm{e}^{\log ab} = \mathrm{e}^{(1/p)\log a^p + (1/q)\log b^q}$$
$$< \frac{1}{p}\,\mathrm{e}^{\log a^p} + \frac{1}{q}\,\mathrm{e}^{\log b^q} = \frac{a^p}{p} + \frac{b^q}{q}$$

for all $a, b > 0$ with $a^p \neq b^q$. Yet another argument is provided by the study of variation of the differentiable function

$$F(a) = \frac{a^p}{p} + \frac{b^q}{q} - ab, \quad a \geq 0,$$

where $b \geq 0$ is a parameter. This function attains at $a = b^{q/p}$ its strict global minimum, which yields $F(a) > F(b^{q/p}) = 0$ for all $a \geq 0$, $a \neq b^{q/p}$. A refinement of the inequality (1.5) is presented in Exercise 2.

W. H. Young [493] actually proved a much more general inequality which covers (1.5) for $f(x) = x^{p-1}$:

1.2.1 Theorem (Young's inequality) *Suppose that $f\colon [0,\infty) \to [0,\infty)$ is a strictly increasing continuous function such that $f(0) = 0$ and $\lim_{x\to\infty} f(x) = \infty$. Then*

$$uv \leq \int_0^u f(x)\,\mathrm{d}x + \int_0^v f^{-1}(y)\,\mathrm{d}y$$

for all $u, v \geq 0$, and equality occurs if and only if $v = f(u)$.

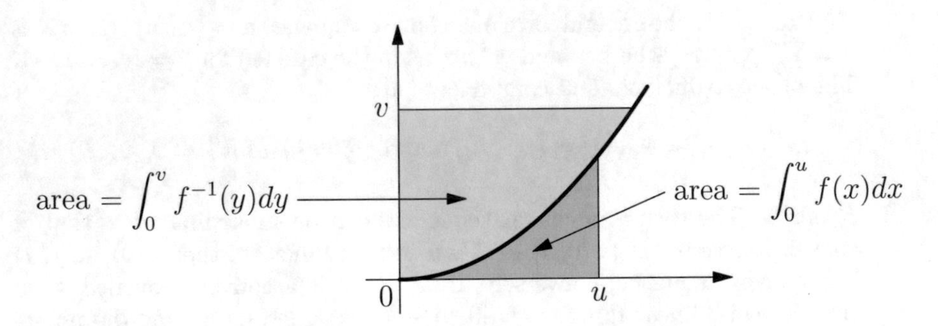

Figure 1.2: The area of the union of the two curvilinear triangles exceeds the area of the rectangle with sides u and v.

Proof. Using the definition of the derivative one can easily prove that the function

$$F(u) = \int_0^u f(x)\,\mathrm{d}x + \int_0^{f(u)} f^{-1}(y)\,\mathrm{d}y - uf(u), \quad u \in [0, \infty),$$

is differentiable, with F' identically 0. Thus $F(u) = F(0) = 0$ for all $u \geq 0$.

If $u, v \geq 0$ and $v \geq f(u)$, then

$$\begin{aligned}
uv = uf(u) + u(v - f(u)) &= \int_0^u f(x)\,\mathrm{d}x + \int_0^{f(u)} f^{-1}(y)\,\mathrm{d}y + u(v - f(u)) \\
&= \int_0^u f(x)\,\mathrm{d}x + \int_0^v f^{-1}(y)\,\mathrm{d}y + \left[u(v - f(u)) - \int_{f(u)}^v f^{-1}(y)\,\mathrm{d}y\right] \\
&\leq \int_0^u f(x)\,\mathrm{d}x + \int_0^v f^{-1}(y)\,\mathrm{d}y.
\end{aligned}$$

The other case where $v \leq f(u)$ can be treated in a similar way. ∎

A pictorial proof of Theorem 1.2.1 is shown in Figure 1.2.

Applications of this result to Legendre duality will be presented in Exercise 7, Section 1.6.

Young's inequality (1.5) is the source of many important integral inequalities, most of them valid in the context of arbitrary measure spaces. The interested reader may find a thorough presentation of the basic elements of measure theory in the monograph of E. Hewitt and K. Stromberg [214], Sections 12 and 13.

1.2.2 Theorem (The Rogers–Hölder inequality for $p > 1$) *Let $p, q \in (1, \infty)$ with $1/p + 1/q = 1$, and let $f \in L^p(\mu)$ and $g \in L^q(\mu)$. Then fg is in $L^1(\mu)$ and we have*

$$\left|\int_\Omega fg\,\mathrm{d}\mu\right| \leq \int_\Omega |fg|\,\mathrm{d}\mu \tag{1.6}$$

and

$$\int_\Omega |fg|\,\mathrm{d}\mu \leq \|f\|_{L^p}\,\|g\|_{L^q}. \tag{1.7}$$

As a consequence,

$$\left|\int_\Omega fg\,\mathrm{d}\mu\right| \leq \|f\|_{L^p}\,\|g\|_{L^q}. \tag{1.8}$$

The above result extends in a straightforward manner to the pairs $p = 1$, $q = \infty$ and $p = \infty$, $q = 1$. In the complementary domain, $p \in (-\infty, 1)\backslash\{0\}$ and $1/p + 1/q = 1$, the inequality sign in (1.6)–(1.8) should be reversed. See Exercise 3.

From the Rogers–Hölder inequality it follows that for all $p, q, r \in (1, \infty)$ with $1/p + 1/q = 1/r$ and all $f \in L^p(\mu)$ and $g \in L^q(\mu)$ we have $fg \in L^r(\mu)$ and

$$\|fg\|_{L^r} \leq \|f\|_{L^p}\,\|g\|_{L^q}. \tag{1.9}$$

The special case of inequality (1.8), where $p = q = 2$, is known as the *Cauchy–Bunyakovsky–Schwarz inequality*. See Remark 1.2.6.

Proof. The first inequality is trivial. If f or g is zero μ-almost everywhere, then the second inequality is trivial. Otherwise, using Young's inequality, we have

$$\frac{|f(x)|}{\|f\|_{L^p}} \cdot \frac{|g(x)|}{\|g\|_{L^q}} \leq \frac{1}{p} \cdot \frac{|f(x)|^p}{\|f\|_{L^p}^p} + \frac{1}{q} \cdot \frac{|g(x)|^q}{\|g\|_{L^q}^q}$$

for all x in Ω, such that $fg \in L^1(\mu)$. Thus

$$\frac{1}{\|f\|_{L^p}\|g\|_{L^q}} \int_\Omega |fg|\, \mathrm{d}\mu \leq 1$$

and the proof of (1.7) is done. The inequality (1.8) is immediate. ■

1.2.3 Remark (Conditions for equality in Theorem 1.2.2) *The basic observation is the fact that*

$$f \geq 0 \quad \text{and} \quad \int_\Omega f\, \mathrm{d}\mu = 0 \quad \text{imply} \quad f = 0 \ \mu\text{-almost everywhere.}$$

Consequently we have equality in (1.6) *if and only if*

$$f(x)g(x) = \mathrm{e}^{\mathrm{i}\theta}|f(x)g(x)|$$

for some real constant θ and for μ-almost every x.

Suppose that $p, q \in (1, \infty)$ and f and g are not zero μ-almost everywhere. In order to get equality in (1.7) *it is necessary and sufficient to have*

$$\frac{|f(x)|}{\|f\|_{L^p}} \cdot \frac{|g(x)|}{\|g\|_{L^q}} = \frac{1}{p} \cdot \frac{|f(x)|^p}{\|f\|_{L^p}^p} + \frac{1}{q} \cdot \frac{|g(x)|^q}{\|g\|_{L^q}^q}$$

μ-almost everywhere. The equality case in Young's inequality shows that this is equivalent to $|f(x)|^p/\|f\|_{L^p}^p = |g(x)|^q/\|g\|_{L^q}^q$ μ-almost everywhere, that is,

$$A|f(x)|^p = B|g(x)|^q \quad \mu\text{-almost everywhere}$$

for some positive numbers A and B.

If $p = 1$ and $q = \infty$, we have equality in (1.7) *if and only if there is a constant $\lambda \geq 0$ such that $|g(x)| \leq \lambda$ μ-almost everywhere, and $|g(x)| = \lambda$ μ-almost everywhere in the set $\{x : f(x) \neq 0\}$.*

1.2.4 Theorem (Minkowski's inequality) *For $1 \leq p < \infty$ and $f, g \in L^p(\mu)$ we have*

$$\|f + g\|_{L^p} \leq \|f\|_{L^p} + \|g\|_{L^p}. \tag{1.10}$$

In the discrete case, using the notation of Exercise 10 in Section 1.1, this inequality reads

$$M_p(x + y, \lambda) \leq M_p(x, \lambda) + M_p(y, \lambda). \tag{1.11}$$

The integral analogue of Minkowski's inequality for $0 < p < 1$ holds in reversed direction. See Exercise 7. The case $p = 0$ is presented in Section 3.1, Example 3.1.5.

Proof. For $p = 1$, the inequality (1.10) follows immediately by integrating the inequality $|f + g| \leq |f| + |g|$. For $p \in (1, \infty)$ we have

$$\begin{aligned} |f + g|^p &\leq (|f| + |g|)^p \leq (2 \sup\{|f|, |g|\})^p \\ &\leq 2^p(|f|^p + |g|^p), \end{aligned}$$

which shows that $f+g \in L^p(\mu)$. Moreover, according to Theorem 1.2.2,

$$\begin{aligned}
\|f+g\|_{L^p}^p &= \int_\Omega |f+g|^p \,\mathrm{d}\mu \le \int_\Omega |f+g|^{p-1}\,|f|\,\mathrm{d}\mu + \int_\Omega |f+g|^{p-1}\,|g|\,\mathrm{d}\mu \\
&\le \Big(\int_\Omega |f|^p \,\mathrm{d}\mu\Big)^{1/p}\Big(\int_\Omega |f+g|^{(p-1)q}\,\mathrm{d}\mu\Big)^{1/q} \\
&\quad + \Big(\int_\Omega |g|^p \,\mathrm{d}\mu\Big)^{1/p}\Big(\int_\Omega |f+g|^{(p-1)q}\,\mathrm{d}\mu\Big)^{1/q} \\
&= (\|f\|_{L^p} + \|g\|_{L^p})\|f+g\|_{L^p}^{p/q},
\end{aligned}$$

where $1/p+1/q=1$, and it remains to observe that $p-p/q=1$. ∎

1.2.5 Remark *If $p=1$, we obtain equality in (1.10) if and only if there is a positive measurable function φ such that*

$$f(x)\varphi(x) = g(x)$$

μ-almost everywhere on the set $\{x : f(x)g(x) \ne 0\}$.

If $p \in (1,\infty)$ and f is not 0 almost everywhere, then we have equality in (1.10) if and only if $g=\lambda f$ almost everywhere, for some $\lambda \ge 0$.

In the particular case where (Ω,Σ,μ) is the measure space associated with the counting measure on a finite set,

$$\mu\colon \mathcal{P}(\{1,\dots,n\}) \to \mathbb{N}, \quad \mu(A) = |A|,$$

we retrieve the classical discrete forms of the above inequalities. For example, the discrete version of the Rogers–Hölder inequality can be read

$$\Big|\sum_{k=1}^n \xi_k\eta_k\Big| \le \Big(\sum_{k=1}^n |\xi_k|^p\Big)^{1/p}\Big(\sum_{k=1}^n |\eta_k|^q\Big)^{1/q}$$

for all $\xi_k,\ \eta_k \in \mathbb{C}$, $k \in \{1,\dots,n\}$. On the other hand, a moment's reflection shows that we can pass immediately from these discrete inequalities to their integral analogues, corresponding to finite measure spaces.

1.2.6 Remark (More on Cauchy–Bunyakovsky–Schwarz inequality) *A.-L. Cauchy, in his famous* Cours d'Analyse [99], *derived the discrete case of this inequality from Lagrange's algebraic identity,*

$$\Big(\sum_{k=1}^n a_k^2\Big)\Big(\sum_{k=1}^n b_k^2\Big) = \sum_{1\le j<k\le n} (a_jb_k - a_kb_j)^2 + \Big(\sum_{k=1}^n a_kb_k\Big)^2.$$

Thus

$$\Big|\sum_{k=1}^n a_kb_k\Big| \le \Big(\sum_{k=1}^n a_k^2\Big)^{1/2}\Big(\sum_{k=1}^n b_k^2\Big)^{1/2}$$

for all families of real numbers $a_1, \ldots, a_n, b_1, \ldots, b_n$. *The equality case is straightforward. The corresponding inequality for integrals was proved independently by V. Y. Bunyakovsky [84] and H. A. Schwarz [440].*

In 1890, H. Poincaré [400] noticed the integral version of Lagrange's algebraic identity (that yields the Cauchy–Bunyakovsky–Schwarz inequality in full generality): If μ *is a probability measure on a space* Ω *and* f *and* g *are two functions belonging to the space* $L^2(\mu)$, *then*

$$\left(\int_\Omega f^2 \mathrm{d}\mu\right)\left(\int_\Omega g^2 \mathrm{d}\mu\right) - \left(\int_\Omega fg \mathrm{d}\mu\right)^2$$
$$= \frac{1}{2}\int_\Omega \int_\Omega (f(x)g(y) - f(y)g(x))^2 \,\mathrm{d}\mu(x)\mathrm{d}\mu(y).$$

He used this integral identity to derive the one-dimensional case of an inequality bearing his name. See Exercise 8.

Another simple proof of the Cauchy–Bunyakovsky–Schwarz inequality is provided by an identity equivalent to the law of cosines: for every pair of nonzero vectors x *and* y *in a real inner vector space,*

$$\left\| \frac{x}{\|x\|} - \frac{y}{\|y\|} \right\|^2 = 2 - 2\frac{\langle x, y\rangle}{\|x\| \, \|y\|}.$$

The book of J. M. Steele [456] offers many other proofs and a bulk of interesting applications (including Heisenberg's Uncertainty Principle and van der Corput's inequality). Still another application (motivated by information theory) makes the object of Exercise 10.

Exercises

1. (The Bernoulli inequality) (a) Prove that for all $x > -1$ we have

$$(1 + x)^p \geq 1 + px \quad \text{if } p \in (-\infty, 0] \cup [1, \infty)$$

and

$$(1 + x)^p \leq 1 + px \quad \text{if } p \in [0, 1];$$

if $p \notin \{0, 1\}$, the equality occurs only for $x = 0$.

(b) Use the substitution $1 + x = a/b^{q-1}$ to infer Young's inequality for full range of parameters, that is, for all $x, y > 0$,

$$xy \leq \frac{x^p}{p} + \frac{y^q}{q} \quad \text{if } p \in (1, \infty) \text{ and } 1/p + 1/q = 1$$

and

$$xy \geq \frac{x^p}{p} + \frac{y^q}{q} \quad \text{if } p \in (0, 1) \text{ and } 1/p + 1/q = 1.$$

Remark. The last inequality easily yields the following version of Rogers–Hölder inequality for $p \in (0,1)$ and $1/p + 1/q = 1$: If $f \in L^p(\mu)$, $f \geq 0$, and $g \in L^q(\mu)$, $g \geq 0$, then

$$\int_\Omega fg \,\mathrm{d}\mu \geq \|f\|_{L^p} \|g\|_{L^q}. \tag{1.12}$$

Alternatively, this case follows from the usual one (corresponding to $p > 1$) via a change of function. See the monograph of E. Hewitt and K. Stromberg [214], Theorem 13.6, p. 191. For more general results, see L.-E. Persson [394].

2. (J. M. Aldaz [8]) Use calculus to prove the following refinement of the inequality (1.5): If $1 < p \leq 2$ and $1/p + 1/q = 1$, then for all $a, b \geq 0$

$$\frac{1}{q}\left(a^{p/2} - b^{q/2}\right)^2 \leq \frac{a^p}{p} + \frac{b^q}{q} - ab \leq \frac{1}{p}\left(a^{p/2} - b^{q/2}\right)^2.$$

This fact yields a refinement of the Rogers–Hölder inequality and also a new proof of the uniform convexity of real L^p-spaces for $1 < p < \infty$.

Remark. Another refinement of Young's inequality was proposed by Y. Al-Manasrah and F. Kittaneh [11]: $(1-\mu)a + \mu b \geq a^{1-\mu}b^\mu + \min\{\mu, 1-\mu\} \cdot \left(\sqrt{a} - \sqrt{b}\right)^2$ for all $a, b \in [0, \infty)$ and $\mu \in [0, 1]$.

3. (A symmetric form of the Rogers–Hölder inequality) Let p, q, r be nonzero real numbers such that $1/p + 1/q = 1/r$.

(a) Prove that the inequality

$$\|fg\|_{L^r} \leq \|f\|_{L^p} \|g\|_{L^q}.$$

holds in each of the following three cases:

$$p > 0,\ q > 0,\ r > 0 \quad \text{or} \quad p < 0,\ q > 0,\ r < 0 \quad \text{or} \quad p > 0,\ q < 0,\ r < 0.$$

(b) Prove that the preceding inequality holds in the reversed direction in each of the following cases:

$$p > 0,\ q < 0,\ r > 0 \quad \text{or} \quad p < 0,\ q > 0,\ r > 0 \quad \text{or} \quad p < 0,\ q < 0,\ r < 0.$$

Remark. For more general results, see L.-E. Persson [394] and the references therein.

4. (a) Young's inequality extends to more than two variables. Prove that if $p_1, \ldots, p_n \in (1, \infty)$ and $1/p_1 + \cdots + 1/p_n = 1$, then

$$\prod_{k=1}^n x_k \leq \sum_{k=1}^n \frac{x_k^{p_k}}{p_k}$$

for all $x_1, \ldots, x_n \geq 0$.

(b) Infer the following version of the Rogers–Hölder inequality : If (Ω, Σ, μ) is a measure space and $f_1, \ldots, f_n$ are functions such that $f_k \in L^{p_k}(\mu)$ for some $p_k \in [1, \infty]$, and $\sum_{k=1}^n 1/p_k = 1$, then

$$\left| \int_\Omega \left(\prod_{k=1}^n f_k \right) \mathrm{d}\mu \right| \leq \prod_{k=1}^n \|f_k\|_{L^{p_k}}.$$

5. (Hilbert's inequality) Suppose that $p, q \in (1, \infty)$, $1/p + 1/q = 1$ and f, g are two positive functions such $f \in L^p(0, \infty)$ and $g \in L^q(0, \infty)$. Infer from the Rogers–Hölder inequality that

$$\int_0^\infty \int_0^\infty \frac{f(x) g(y)}{x + y} \mathrm{d}x\mathrm{d}y \leq \frac{\pi}{\sin \frac{\pi}{p}} \left(\int_0^\infty f^p(x) \, \mathrm{d}x \right)^{1/p} \left(\int_0^\infty g^q(y) \, \mathrm{d}y \right)^{1/q},$$

the constant $\pi / \sin(\pi/p)$ being sharp.

[*Hint*: Notice that the left-hand side equals

$$\int_0^\infty \int_0^\infty \frac{f(x)}{(x+y)^{1/p}} \left(\frac{x}{y} \right)^{1/pq} \frac{g(y)}{(x+y)^{1/q}} \left(\frac{y}{x} \right)^{1/pq} \mathrm{d}x\mathrm{d}y.]$$

6. (A general form of Minkowski's inequality) Suppose that $(X, \mathcal{M}, \mu)$ and $(Y, \mathcal{N}, \nu)$ are two σ-finite measure spaces, f is a positive function on $X \times Y$ which is $\mu \times \nu$-measurable, and let $p \in [1, \infty)$. Then

$$\left(\int_X \left(\int_Y f(x,y) \, \mathrm{d}\nu(y) \right)^p \mathrm{d}\mu(x) \right)^{1/p} \leq \int_Y \left(\int_X f(x,y)^p \, \mathrm{d}\mu(x) \right)^{1/p} \mathrm{d}\nu(y).$$

7. (The Minkowski inequality for $0 < p < 1$) Suppose that $0 < p < 1$ and $f, g \in L^p(\mu)$ are two positive functions. Then

$$\|f\|_{L^p} + \|g\|_{L^p} \leq \|f + g\|_{L^p} \leq 2^{(1-p)/p} \left(\|f\|_{L^p} + \|g\|_{L^p} \right).$$

[*Hint:* The function x^p is concave on $[0,\infty)$ if $0 < p < 1$ and every concave function $f : [0, \infty) \to \mathbb{R}$ with $f(0) \geq 0$ verifies the inequality $f(x + y) \leq f(x) + f(y)$ for all $x, y \geq 0$.]

8. (The 1-dimensional case of Poincaré inequality [400]) Infer from the integral version of Lagrange's algebraic identity the following inequality: If $f : [0, 1] \to \mathbb{R}$ is a function of class C^1 that verifies the condition $\int_0^1 f \mathrm{d}x = 0$, then

$$\int_0^1 f^2 \mathrm{d}x \leq \frac{1}{2} \int_0^1 f'^2(s) \mathrm{d}s.$$

The n-dimensional variant of Poincaré inequality plays a major role in partial differential equations and their applications. See [151].

9. Let (Ω, Σ, μ) be a measure space and let $f\colon \Omega \to \mathbb{C}$ be a measurable function, which belongs to $L^t(\mu)$ for t in a subinterval I of $(0, \infty)$. Infer from the Cauchy–Bunyakovsky–Schwarz inequality that the function $t \to \log \int_\Omega |f|^t \, d\mu$ is convex on I.

   Remark. The result of this exercise is equivalent to *Lyapunov's inequality* [291]: *If $a \geq b \geq c$, then*

$$\left(\int_\Omega |f|^b \, d\mu\right)^{a-c} \leq \left(\int_\Omega |f|^c \, d\mu\right)^{a-b} \left(\int_\Omega |f|^a \, d\mu\right)^{b-c}$$

   (provided the integrability aspects are fixed). Equality holds if and only if one of the following four conditions holds:

   (1) f is constant on some subset of Ω and 0 elsewhere;

   (2) $a = b$;

   (3) $b = c$;

   (4) $c(2a - b) = ab$.

10. Denote by $\text{Prob}\,(n) = \{(\lambda_1, ..., \lambda_n) \in (0, \infty)^n : \sum_{k=1}^n \lambda_k = 1\}$ the set of all probability distributions of order n and suppose there are given two such distributions $Q = (q_1, ..., q_n)$ and $P = (p_1, ..., p_n)$. The *Kullback–Leibler divergence* from Q to P is defined by the formula

$$D_{KL}(P||Q) = \sum_{k=1}^n p_k \log \frac{p_k}{q_k}$$

   and represents a measure of how one probability distribution diverges from a second expected probability distribution.

   From Jensen's inequality it follows that $D_{KL}(P||Q) \geq 0$ (with equality if and only if $P = Q$). This is known as *Gibbs' inequality*. A better lower bound is provided by *Pinsker's inequality*, formulated by J. M. Borwein and A. S. Lewis [72], p. 63, as follows:

$$2\sum_{k=1}^n p_k \log \frac{p_k}{q_k} \geq 3\sum_{k=1}^n \frac{(p_k - q_k)^2}{p_k + 2q_k} \geq \left(\sum_{k=1}^n |p_k - q_k|\right)^2.$$

   (a) Derive the right-hand side inequality from the Cauchy–Bunyakovsky–Schwarz inequality.

   (b) Prove that $3(x-1)^2 \leq 2\,(x+2)\,(x \log x - x + 1)$ for all $x > 0$ and infer from it the inequality $3(u-v)^2 \leq 2\,(u + 2v)\left(u \log \frac{u}{v} - u + v\right)$ for all $u, v > 0$. Then put $u = p_k$ and $v = q_k$ and sum up for k from 1 to n to get the left hand side inequality.

11. (a) Prove the *discrete Berwald inequality* :

$$\frac{1}{n+1}\sum_{k=0}^n a_k \geq \left(\frac{3(n-1)}{4(n+1)}\right)^{1/2} \left(\frac{1}{n+1}\sum_{k=0}^n a_k^2\right)^{1/2}$$

for every concave sequence $a_0, a_1, \ldots, a_n$ of positive numbers.

(b) Derive from (a) the integral form of this inequality: if f is a concave positive function on $[a, b]$, then

$$\left(\frac{1}{b-a}\int_a^b f^2(x)\mathrm{d}x\right)^{1/2} \leq \frac{2\sqrt{3}}{3}\left(\frac{1}{b-a}\int_a^b f(x)\mathrm{d}x\right).$$

[*Hint*: By Minkowski's inequality (Theorem 1.2.4 above), if the Berwald inequality works for two concave sequences, then it also works for all linear combinations of them, with positive coefficients. Then apply the assertion (b) of Exercise 13, Section 1.1.]

1.3 Log-convex Functions

This section is aimed to a brief discussion on a stronger concept of convexity.

1.3.1 Definition *A strictly positive function $f : I \to (0, \infty)$ is called log-convex (respectively log-concave) if $\log f$ (respectively $-\log f$) is a convex function.*

Equivalently, the condition of log-convexity of f means

$$x, y \in I \text{ and } \lambda \in (0,1) \implies f((1-\lambda)x + \lambda y) \leq f(x)^{1-\lambda} f(y)^{\lambda}.$$

The following result collects some immediate consequences of the generalized *AM–GM* inequality:

1.3.2 Lemma (a) *The multiplicative inverse of any strictly positive concave function is a log-convex function.*

(b) *Every log-convex function is also convex.*

(c) *Every strictly positive concave function is also log-concave.*

According to this lemma, $1/x$ and $1/\sin x$ (as well as their exponentials) are log-convex, respectively, on $(0, \infty)$ and $(0, \pi)$.

Notice that the function e^{-x^2} is log-concave but not concave.

An important example of a log-convex function is the gamma function,

$$\Gamma\colon (0, \infty) \to \mathbb{R}, \quad \Gamma(x) = \int_0^\infty t^{x-1}\mathrm{e}^{-t}\,\mathrm{d}t \quad \text{for } x > 0.$$

See A. D. R. Choudary and C. P. Niculescu [107], pp. 352–360, for its basic theory.

The fact that Γ is log-convex is a consequence of Rogers–Hölder inequality. Indeed, for every $x, y > 0$ and $\lambda \in [0,1]$ we have

$$\Gamma((1-\lambda)x+\lambda y) = \int_0^\infty t^{(1-\lambda)x+\lambda y-1}\mathrm{e}^{-t}\mathrm{d}t = \int_0^\infty (t^{x-1}\mathrm{e}^{-t})^{1-\lambda}(t^{y-1}\mathrm{e}^{-t})^{\lambda}\,\mathrm{d}t$$
$$\leq \left(\int_0^\infty t^{x-1}\mathrm{e}^{-t}\,\mathrm{d}t\right)^{1-\lambda}\left(\int_0^\infty t^{y-1}\mathrm{e}^{-t}\,\mathrm{d}t\right)^{\lambda} = \Gamma^{1-\lambda}(x)\Gamma^{\lambda}(y).$$

Remarkably, Γ is the unique log-convex extension of the factorial function.

1.3.3 Theorem (H. Bohr and J. Mollerup [62], [20]) *The gamma function is the only function $f\colon (0,\infty) \to \mathbb{R}$ that satisfies the following three conditions:*
(a) $f(x+1) = xf(x)$ *for all* $x > 0$;
(b) $f(1) = 1$;
(c) f *is* log-convex.

An important class of log-convex functions is that of completely monotonic functions.

A function $f\colon (0,\infty) \to \mathbb{R}$ is called *completely monotonic* if f has derivatives of all orders and satisfies $(-1)^n f^{(n)}(x) \geq 0$ for all $x > 0$ and $n \in \mathbb{N}$.

Necessarily, every completely monotonic function is positive, decreasing, and convex.

One can prove easily that the functions

$$\mathrm{e}^{-x}, \mathrm{e}^{-1/x}, \frac{1}{(1+x)^2} \text{ and } \frac{\ln(1+x)}{x}$$

(as well as their sums, products, and derivatives of even order) are completely monotonic on $(0,\infty)$. More interesting examples can be found in the papers of M. Merkle [318] and K. S. Miller and S. G. Samko [324].

S. N. Bernstein has proved in 1928 that a necessary and sufficient condition that a function $f(x)$ be completely monotonic is the existence of a representation of the form

$$f(x) = \int_0^\infty \mathrm{e}^{-xt}\mathrm{d}\mu(t), \tag{1.13}$$

where μ is a positive Borel measure on $(0,\infty)$ and the integral converges for $0 < x < \infty$. A simple proof based on the Krein–Milman theorem (see Appendix B, Theorem B.3.2) can be found in B. Simon [450], pp. 143–152. His argument also covers the fact (known to Bernstein) that every completely monotonic function is the restriction to $(0,\infty)$ of an analytic function in the right semi-plane $\operatorname{Re} z > 0$.

1.3.4 Theorem (A. M. Fink [163]) *Every completely monotonic function is log-convex.*

Proof. It suffices to prove that every completely monotonic function f verifies the inequality $f(x)f''(x) \geq (f'(x))^2$. See Exercise 1. Taking into account the aforementioned integral representation due to Bernstein, this inequality is equivalent to

$$\int_0^\infty \mathrm{e}^{-xt_1}\mathrm{d}\mu(t_1)\int_0^\infty t_2^2\mathrm{e}^{-xt_2}\mathrm{d}\mu(t_2) \geq \int_0^\infty t_1\mathrm{e}^{-xt_1}\mathrm{d}\mu(t_1)\int_0^\infty t_2\mathrm{e}^{-xt_2}\mathrm{d}\mu(t_2)$$

and also to

$$\int_0^\infty\int_0^\infty t_2^2\mathrm{e}^{-x(t_1+t_2)}\mathrm{d}\mu(t_1)\mathrm{d}\mu(t_2) \geq \int_0^\infty\int_0^\infty t_1t_2\mathrm{e}^{-x(t_1+t_2)}\mathrm{d}\mu(t_1)\mathrm{d}\mu(t_2),$$

which is clear because by symmetry

$$\int_0^\infty\int_0^\infty t_2^2\mathrm{e}^{-x(t_1+t_2)}\mathrm{d}\mu(t_1)\mathrm{d}\mu(t_2) = \frac{1}{2}\int_0^\infty\int_0^\infty \left(t_1^2+t_2^2\right)\mathrm{e}^{-x(t_1+t_2)}\mathrm{d}\mu(t_1)\mathrm{d}\mu(t_2).$$

■

The convex functions, log-convex functions, and quasiconvex functions are all just special cases of a considerably more general concept, that of convex function with respect to a pair of means. This makes the subject of Appendix A.

Exercises

1. (The second derivative test of log-convexity) Prove that a twice differentiable function $f : I \to (0,\infty)$ is log-convex if and only if

$$f(x)f''(x) \geq \left(f'(x)\right)^2 \quad \text{for all } x \in I.$$

Infer from this result the log-convexity of the gamma function.

2. (The log-convex analogue of Theorem 1.1.8) Prove that a strictly positive continuous function f defined on an interval I is log-convex if and only if $f\left(\frac{x+y}{2}\right) \leq \sqrt{f(x)f(y)}$ for all $x, y \in I$.

3. (Operations with log-convex functions) (a) Clearly, a product of log-convex functions is also a log-convex function. Prove that the same happens for sums.

(b) Suppose that $x = (x_1, \ldots, x_n)$ and $\lambda = (\lambda_1, \ldots, \lambda_n)$ are n-tuples of strictly positive numbers such that $\sum_{k=1}^n \lambda_k = 1$). Infer from the assertion (a) the convexity of the function $t \to t\log M_t(x;\lambda)$ on $\mathbb{R}$ and of the function $t \to \log M_t(x;\lambda)$ on $(0,\infty)$. The last assertion yields

$$M_{\alpha s+\beta t}(x;\lambda) \leq M_s^\alpha(x;\lambda)M_t^\beta(x;\lambda)$$

whenever $\alpha, \beta \in [0,1]$, $\alpha+\beta = 1$ and $s, t > 0$.

(c) Suppose that f is a convex function and g is an increasing log-convex function. Prove that $g \circ f$ is a log-convex function.

(d) Prove that the harmonic mean of two log-concave functions is also log-concave.

[*Hint*: (a) Note that this assertion is equivalent to the following inequality for positive numbers: $a^\alpha b^\beta + c^\alpha d^\beta \leq (a+c)^\alpha (b+d)^\beta$.]

4. (Operations with log-concave functions) (a) Prove that the product of log-concave functions is also log-concave.

   (b) Prove that the convolution preserves log-concavity, that is, if f and g are two integrable log-concave functions defined on $\mathbb{R}$, then the function

$$(f * g)(x) = \int_{\mathbb{R}} f(x-y)g(y)\mathrm{d}y$$

   is log-concave too.

   Remark. The statement (b) of Exercise 4 is a particular case of Prékopa–Leindler inequality. See Section 3.10.

5. (P. Montel [335]) Let I be an interval. Prove that the following assertions are equivalent for every function $f\colon I \to (0,\infty)$:

   (a) f is log-convex;

   (b) the function $x \to \mathrm{e}^{\alpha x} f(x)$ is convex on I for all $\alpha \in \mathbb{R}$;

   (c) the function $x \to [f(x)]^\alpha$ is convex on I for all $\alpha > 0$.

   [*Hint*: For (c) $\Rightarrow$ (a), note that $([f(x)]^\alpha - 1)/\alpha$ is convex for all $\alpha > 0$ and $\log f(x) = \lim_{\alpha \to 0+}([f(x)]^\alpha - 1)/\alpha$. The limit of a sequence of convex functions is itself convex.]

6. Prove that the function $\frac{e^x \Gamma(x)}{x^x}$ is log-convex.

7. (Some geometric consequences of log-concavity)

   (a) A convex quadrilateral $ABCD$ is inscribed in the unit circle. Its sides satisfy the inequality $AB \cdot BC \cdot CD \cdot DA \geq 4$. Prove that $ABCD$ is a square.

   (b) Suppose that A, B, C are the angles of a triangle, expressed in radians. Prove that

$$\sin A \sin B \sin C < \left(\frac{3\sqrt{3}}{2\pi}\right)^3 ABC < \left(\frac{\sqrt{3}}{2}\right)^3,$$

   unless $A = B = C$.

   [*Hint*: Note that the sine function is log-concave, while $x/\sin x$ is log-convex on $(0, \pi)$.]

8. (E. Artin [20]) Let U be an open convex subset of $\mathbb{R}^n$ and let μ be a Borel measure on an interval I. Consider the integral transform

$$F(x) = \int_I K(x,t)\,\mathrm{d}\mu(t),$$

where the kernel $K(x,t)\colon U \times I \to [0,\infty)$ satisfies the following two conditions:

(a) $K(x,t)$ is μ-integrable in t for each fixed x;

(b) $K(x,t)$ is log-convex in x for each fixed t.

Prove that F is log-convex on U.

[*Hint*: Apply the Rogers–Hölder inequality, noticing that

$$K((1-\lambda)x + \lambda y, t) \leq (K(x,t))^{1-\lambda}(K(y,t))^{\lambda}.]$$

Remark. The *Laplace transform* of a function $f \in L^1(0,\infty)$ is given by the formula $(\mathcal{L}f)(x) = \int_0^\infty f(t)\mathrm{e}^{-tx}\,\mathrm{d}t$. By Exercise 8, the Laplace transform of any positive function is log-convex. In the same way one can show that the moment $\mu_\alpha = \int_0^\infty t^\alpha f(t)\,\mathrm{d}t$, of any random variable with probability density f, is a log-convex function in α on each subinterval of $[0,\infty)$ where it is finite.

9. (Combinatorial properties of sequences) Call a sequence $(a_n)_n$ of strictly positive numbers log-*concave* (respectively log-*convex*) if $a_{n-1}a_{n+1} \leq a_n^2$ for all $n \geq 1$ (respectively $a_{n-1}a_{n+1} \geq a_n^2$ for all $n \geq 1$).

(a) Let $(a_n)_n$ be a strictly positive and strictly decreasing sequence of numbers and let $A_n = \sum_{k=0}^n a_k$. Prove that $(A_n)_n$ is log-concave. Illustrate this result using the Maclaurin expansion of $\log(1-x)$ on $(0,1)$.

(b) Prove that the sequence of binomial coefficients $\binom{n}{k}$ is log-concave in k for fixed n.

(c) (I. Newton) Suppose that $P(x) = \sum_{k=0}^n \binom{n}{k} a_k x^k$ is a real polynomial with real zeros. Prove that $a_0, a_1, ..., a_n$ is a log-convex sequence.

(d) (H. Davenport–G.Pólya) If both $(x_n)_n$ and $(y_n)_n$ are log-convex, then so is the sequence of their binomial convolution,

$$z_n = \sum_{k=0}^n \binom{n}{k} x_k y_{n-k}.$$

Remark. More information on log-convex/log-concave sequences are available in the papers of L. L. Liu and Y. Wang [290] and R. P. Stanley [454]. See also Appendix C.

1.4 Smoothness Properties of Convex Functions

The starting point is the following restatement of Definition 1.1.1. A function $f\colon I \to \mathbb{R}$ is convex if and only if

$$f(x) \leq \frac{b-x}{b-a} \cdot f(a) + \frac{x-a}{b-a} \cdot f(b), \tag{1.14}$$

equivalently,

$$\begin{vmatrix} 1 & a & f(a) \\ 1 & x & f(x) \\ 1 & b & f(b) \end{vmatrix} \geq 0, \tag{1.15}$$

whenever $a < x < b$ in I. Indeed, every point x belonging to an interval $[a, b]$ can be written uniquely as a convex combination of a and b, more precisely,

$$x = \frac{b-x}{b-a} \cdot a + \frac{x-a}{b-a} \cdot b.$$

Subtracting $f(a)$ from both sides of the inequality (1.14) and repeating the operation with $f(b)$ instead of $f(a)$, we obtain that every convex function $f\colon I \to \mathbb{R}$ verifies the *three chords inequality*,

$$\frac{f(x)-f(a)}{x-a} \leq \frac{f(b)-f(a)}{b-a} \leq \frac{f(b)-f(x)}{b-x} \tag{1.16}$$

whenever $a < x < b$ in I. See Figure 1.3. Clearly, this inequality actually characterizes the convexity of f. Moreover, the three chords inequality with strict inequalities provides a characterization of strict convexity. Very close to these remarks is Galvani's characterization of convexity. See Exercise 3.

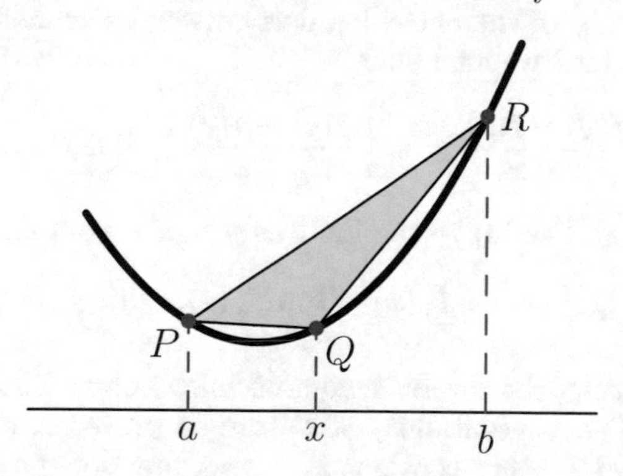

Figure 1.3: *slope* $PQ \leq$ *slope* $PR \leq$ *slope* QR

1.4.1 Remark *The three chords inequality can be strengthened as follows: If $f : I \to \mathbb{R}$ is a convex function and x, y, a, b are points in the interval I such that $x \leq a$, $y \leq b$, $x \neq y$ and $a \neq b$, then*

$$\frac{f(x)-f(y)}{x-y} \leq \frac{f(a)-f(b)}{a-b}.$$

1.4.2 Theorem (O. Stolz [463]) *Let $f\colon I \to \mathbb{R}$ be a convex function. Then f has finite left and right derivatives at each interior point of I and $x < y$ in* int I *implies*

$$f'_-(x) \leq f'_+(x) \leq \frac{f(y)-f(x)}{y-x} \leq f'_-(y) \leq f'_+(y).$$

Moreover, on int I, f'_- *is left continuous and* f'_+ *is right continuous.*

Therefore, if a convex function is differentiable on int I, *then it is also continuously differentiable.*

Proof. Indeed, according to the three chords inequality, we have

$$\frac{f(x)-f(a)}{x-a} \leq \frac{f(y)-f(a)}{y-a} \leq \frac{f(z)-f(a)}{z-a}$$

for all $x \leq y < a < z$ in I. This fact assures us that the left derivative at a exists and

$$f'_-(a) \leq \frac{f(z)-f(a)}{z-a}.$$

A symmetric argument will then yield the existence of $f'_+(a)$ and the availability of the relation $f'_-(a) \leq f'_+(a)$. On the other hand, starting with $x < u \leq v < y$ in int I, the same three chords inequality yields

$$\frac{f(u)-f(x)}{u-x} \leq \frac{f(v)-f(x)}{v-x} \leq \frac{f(v)-f(y)}{v-y},$$

so letting $u \to x+$ and $v \to y-$, we obtain that $f'_+(x) \leq f'_-(y)$.

For the continuity of the one-sided derivatives, let us notice that from the continuity of f on int I we infer that

$$\frac{f(y)-f(x)}{y-x} = \lim_{z \searrow x} \frac{f(y)-f(z)}{y-z} \geq \lim_{z \searrow x} f'_+(z)$$

whenever $x < z < y$. Passing to the limit as $y \searrow x$ we obtain

$$f'_+(x) \geq \lim_{z \searrow x} f'_+(z).$$

Since f'_+ is increasing, the reverse inequality also holds. Thus f'_+ is right continuous on int I. The left continuity of f'_- can be proved in a similar way. ■

By Theorem 1.4.2, every continuous convex function $f : [a,b] \to \mathbb{R}$ admits one-sided derivatives at the endpoints, but they can be infinite:

$$-\infty \leq f'_+(a) < \infty \quad \text{and} \quad -\infty < f'_-(b) \leq \infty.$$

This theorem also yields the fact that a continuous convex function $f\colon [a,b] \to \mathbb{R}$ can be extended to a convex function on $\mathbb{R}$ if and only if $f'_+(a)$ and $f'_-(b)$ exist and are finite.

How "nondifferentiable" can a convex function be?

Theorem 1.4.2 implies that every convex function $f\colon I \to \mathbb{R}$ is differentiable except for an enumerable subset. In fact, by considering the set of nondifferentiability,

$$I_{\text{nd}} = \{x : f'_-(x) < f'_+(x)\},$$

and choosing for each $x \in I_{\text{nd}}$ a rational point $r_x \in (f'_-(x), f'_+(x))$ we get a one-to-one function $\varphi\colon x \to r_x$ from I_{nd} into $\mathbb{Q}$. Consequently, I_{nd} is at most countable. Notice that this reasoning depends on the axiom of choice. An example of a convex function on $\mathbb{R}$ for which $I_{\rm nd}$ is infinitely countable is $f(x) = \sum_{n=0}^{\infty} |x-n|/2^n$.

Theorem 1.4.2 provides an alternative argument for the property of convex functions of being locally Lipschitz on open intervals. Indeed, if $f: I \to \mathbb{R}$ is a convex function and $[a,b]$ is a compact interval contained in the interior of I, then

$$f'_+(a) \leq f'_+(x) \leq \frac{f(y)-f(x)}{y-x} \leq f'_-(y) \leq f'_-(b)$$

for all $x, y \in [a,b]$ with $x < y$. Therefore $f|_{[a,b]}$ verifies the Lipschitz condition $|f(x) - f(y)| \leq L\,|x-y|$ with $L = \max\{|f'_+(a)|, |f'_-(b)|\}$. An immediate consequence is the following result:

1.4.3 Theorem *If $(f_n)_n$ is a pointwise converging sequence of convex functions defined on an open interval I, then its limit f is also convex. Moreover, the convergence is uniform on any compact subinterval and*

$$f'_-(a) \leq \liminf_{n\to\infty} (f_n)'_-(a) \leq \limsup_{n\to\infty} (f_n)'_+(a) \leq f'_+(a) \text{ for all } a \in I.$$

Proof. The convexity of f is trivial. By the three chords inequality (1.16), for every $h > 0$ with $a + h \in I$,

$$(f_n)'_+(a) \leq \frac{f_n(a+h) - f_n(a)}{h}$$

so that

$$\limsup_{n\to\infty} (f_n)'_+(a) \leq \limsup_{n\to\infty} \frac{f_n(a+h) - f_n(a)}{h} = \frac{f(a+h) - f(a)}{h}.$$

Taking $h \to 0$, we obtain the right-hand side inequality in the statement of Theorem 1.4.3. The left-hand side inequality can be proved in a similar way. Taking into account an above remark about local Lipschitzianity of convex functions, one can easily infer the uniform convergence on compact subintervals of I. ■

Equality may not hold in the above theorem. To see this, consider the sequence of convex functions $f_n(x) = |x|^{1+1/n}$ that converges on $\mathbb{R}$ to the function $f(x) = |x|$. Then $(f_n)'_+(0) = 0$ for all n while $f'_+(0) = 1$.

The next result (which exhibits the power of one-variable techniques in a several variables context) yields an upper estimate of Jensen's inequality:

1.4.4 Theorem *Let $f: [a,b] \to \mathbb{R}$ be a convex function and let*

$$[m_1, M_1], \ldots, [m_n, M_n]$$

be compact subintervals of $[a,b]$. Given $\lambda_1, \ldots, \lambda_n$ in $[0,1]$, with $\sum_{k=1}^n \lambda_k = 1$, the function

$$E(x_1, \ldots, x_n) = \sum_{k=1}^n \lambda_k f(x_k) - f\Big(\sum_{k=1}^n \lambda_k x_k\Big)$$

attains its maximum on $\Omega = [m_1, M_1] \times \cdots \times [m_n, M_n]$ *at a vertex, that is, at a point of* $\{m_1, M_1\} \times \cdots \times \{m_n, M_n\}$.

The proof depends upon the following refinement of Lagrange's mean value theorem:

1.4.5 Lemma *Let* $h\colon [a,b] \to \mathbb{R}$ *be a continuous function. Then there exists a point* $c \in (a,b)$ *such that*

$$\underline{D}h(c) \leq \frac{h(b) - h(a)}{b - a} \leq \overline{D}h(c).$$

Here

$$\underline{D}h(c) = \liminf_{x \to c} \frac{h(x) - h(c)}{x - c} \quad \text{and} \quad \overline{D}h(c) = \limsup_{x \to c} \frac{h(x) - h(c)}{x - c},$$

are, respectively, the *lower derivative* and the *upper derivative* of h at c. According to Theorem 1.4.2, in the case of convex functions, $\underline{D}h(c) = h'_-(c)$ and $\overline{D}h(c) = h'_+(c)$.

Proof. As in the smooth case, we consider the function

$$H(x) = h(x) - \frac{h(b) - h(a)}{b - a}(x - a), \quad x \in [a,b].$$

Clearly, H is continuous and $H(a) = H(b)$. If H attains its supremum at $c \in (a,b)$, then $\underline{D}H(c) \leq 0 \leq \overline{D}H(c)$ and the conclusion of Lemma 1.4.5 is immediate. The same is true when H attains its infimum at an interior point of $[a,b]$. If both extremes are attained at the endpoints, then H is constant and the conclusion of Lemma 1.4.5 works for all c in (a,b). ■

Proof of Theorem 1.4.4. Clearly, we may assume that f is also continuous. We shall show (by *reductio ad absurdum*) that

$$E(x_1, \ldots, x_k, \ldots, x_n) \leq \sup\{E(x_1, \ldots, m_k, \ldots, x_n), E(x_1, \ldots, M_k, \ldots, x_n)\}$$

for all $(x_1, x_2, \ldots, x_n) \in \Omega$ and all $k \in \{1, \ldots, n\}$. In fact, if

$$E(x_1, x_2, \ldots, x_n) > \sup\{E(m_1, x_2, \ldots, x_n), E(M_1, x_2, \ldots, x_n)\}$$

for some $(x_1, x_2, \ldots, x_n) \in \Omega$, we consider the function

$$h\colon [m_1, M_1] \to \mathbb{R}, \quad h(x) = E(x, x_2, \ldots, x_n).$$

According to Lemma 1.4.5, there exists $\xi \in (m_1, x_1)$ such that

$$h(x_1) - h(m_1) \leq (x_1 - m_1)\overline{D}h(\xi).$$

Since $h(x_1) > h(m_1)$, it follows that $\overline{D}h(\xi) > 0$, equivalently,

$$\overline{D}f(\xi) > \overline{D}f(\lambda_1\xi + \lambda_2 x_2 + \cdots + \lambda_n x_n).$$

Or, $\overline{D}f = f'_+$ is an increasing function on (a, b), which yields

$$\xi > \lambda_1\xi + \lambda_2 x_2 + \cdots + \lambda_n x_n,$$

and thus $\xi > (\lambda_2 x_2 + \cdots + \lambda_n x_n)/(\lambda_2 + \cdots + \lambda_n)$.

A new appeal to Lemma 1.4.5 (applied this time to $h|_{[x_1,M_1]}$) yields the existence of an $\eta \in (x_1, M_1)$ such that $\eta < (\lambda_2 x_2 + \cdots + \lambda_n x_n)/(\lambda_2 + \cdots + \lambda_n)$. But this contradicts the fact that $\xi < \eta$.

1.4.6 Corollary *Let $f\colon [a, b] \to \mathbb{R}$ be a convex function. Then*

$$(1-\lambda)f(a)+\lambda f(b)-f\Big((1-\lambda)a+\lambda b\Big) \geq (1-\lambda)f(c)+\lambda f(d)-f\Big((1-\lambda)c+\lambda d\Big)$$

for all $a \leq c \leq d \leq b$ and $\lambda \in [0, 1]$.

Corollary 1.4.6 is a first step toward the inequality of majorization. See Section 4.1. It can be easily turned into a characterization of convexity within the class of continuous functions. For this, use Theorem 1.1.8.

Since the first derivative of a convex function may not exist at a dense subset, a characterization of convexity in terms of second-order derivatives is not possible unless we relax the concept of twice differentiability.

The *upper* and the *lower second symmetric derivative* of f at x are, respectively, defined by the formulas

$$\overline{D}^2 f(x) = \limsup_{h\downarrow 0} \frac{f(x+h) + f(x-h) - 2f(x)}{h^2}$$

$$\underline{D}^2 f(x) = \liminf_{h\downarrow 0} \frac{f(x+h) + f(x-h) - 2f(x)}{h^2}.$$

It is not difficult to check that if f is twice differentiable at a point x, then

$$\overline{D}^2 f(x) = \underline{D}^2 f(x) = f''(x); \tag{1.17}$$

however $\overline{D}^2 f(x)$ and $\underline{D}^2 f(x)$ can exist even at points of discontinuity; for example, consider the case of the signum function and the point $x = 0$. The basic remark to prove the relations (1.17) is the formula

$$f(x+h) - f(x) - hf'(x) - \frac{h^2}{2} f''(x) = o(h^2),$$

that works at any point x at which f is twice differentiable. Indeed, by denoting the left-hand side as $g(h)$, we infer from the mean value theorem that $g(h) = g(h) - g(0) = hg'(k)$ and $g'(k) = g'(k) - g'(0) = kg''(0) + o(k) = o(k)$. Therefore $g(h) = ho(k) = o(h^2)$.

1.4.7 Theorem *Suppose that I is an open interval. A real-valued function f is convex on I if and only if f is continuous and $\overline{\mathcal{D}}^2 f \geq 0$.*

As a consequence, if a function $f : I \to \mathbb{R}$ is convex in the neighborhood of each point of I, then it is convex on the whole interval I.

Proof. If f is convex, then clearly $\overline{\mathcal{D}}^2 f \geq \underline{\mathcal{D}}^2 f \geq 0$. The continuity of f follows from Theorem 1.1.2.

Now, suppose that $\overline{\mathcal{D}}^2 f > 0$ on I. If f is not convex, then we can find a point x_0 such that $\overline{\mathcal{D}}^2 f(x_0) \leq 0$, which will be a contradiction. In fact, in this case there exists a subinterval $I_0 = [a_0, b_0]$ such that $f((a_0 + b_0)/2) > (f(a_0) + f(b_0))/2$. A moment's reflection shows that one of the intervals $[a_0, (a_0 + b_0)/2]$, $[(3a_0 + b_0)/4, (a_0 + 3b_0)/4]$, $[(a_0 + b_0)/2, b_0]$ can be chosen to replace I_0 by a smaller interval $I_1 = [a_1, b_1]$, with $b_1 - a_1 = (b_0 - a_0)/2$ and $f((a_1 + b_1)/2) > (f(a_1) + f(b_1))/2$. Proceeding by induction, we arrive at a situation where the principle of included intervals gives us the point x_0.

In the general case, consider the sequence of functions

$$f_n(x) = f(x) + \frac{1}{n} x^2.$$

Then $\overline{\mathcal{D}}^2 f_n > 0$, and the above reasoning shows us that f_n is convex. Clearly $f_n(x) \to f(x)$ for each $x \in I$, so that the convexity of f is a consequence of Theorem 1.4.3 above. ■

1.4.8 Corollary (The second derivative test) *Suppose that $f : I \to \mathbb{R}$ is a twice differentiable function. Then:*

(a) *f is convex if and only if $f'' \geq 0$;*

(b) *f is strictly convex if and only if $f'' \geq 0$ and the set of points where f'' vanishes does not include intervals of positive length.*

The following variant of the second derivative test plays an important role in optimization:

1.4.9 Theorem (A second-order condition of optimization) *If $f : I \to \mathbb{R}$ is a twice differentiable function and $f'' \geq C$ for a suitable constant $C > 0$, then every critical point is a global minimizer.*

Theorem 1.4.9 follows easily from Taylor's formula. See [107], p. 235.

The condition $f'' \geq C$ (for $C \in \mathbb{R}$) is equivalent to the convexity of the function $f - \frac{C}{2}x^2$, which in turn is equivalent to strong convexity of f if $C > 0$. See Exercise 5. The case $C < 0$ is that of *semiconvexity* and a first encounter with it will be provided by Proposition 1.7.7 (that estimates the gap in Jensen's inequality).

Exercises

1. (An application of the second derivative test of convexity). Prove that:

   (a) $\frac{x}{e^x-1}$ and $\frac{e^x-1}{x}$ are convex and positive on $\mathbb{R}$;

   (b) the functions $\log((e^{ax}-1)/(e^x-1))$ and $\log(\sinh ax/\sinh x)$ are convex on $\mathbb{R}$ if $a \geq 1$;

   (c) the function $b \log\cos(x/\sqrt{b}) - a\log\cos(x/\sqrt{a})$ is convex on $(0, \pi/2)$ if $b \geq a \geq 1$;

   (d) $\binom{x}{a} = \frac{x(x-1)\cdots(x-a+1)}{a!}$ is a convex function on $[a, \infty)$, whenever a is a positive integer;

   (e) the *running average* $F(x) = \frac{1}{x}\int_0^x f(t)dt$, of a differentiable convex function $f : (0, \infty) \to \mathbb{R}$ is convex too.

2. Suppose that $0 < a < b < c$ (or $0 < b < c < a$, or $0 < c < b < a$). Prove the following inequalities:

   (a) $ab^p + bc^p + ca^p > ac^p + ba^p + cb^p$ for $p \geq 1$;

   (b) $a^b b^c c^a > a^c c^b b^a$;

   (c) $$\frac{a(c-b)}{(c+b)(2a+b+c)} + \frac{b(a-c)}{(a+c)(a+2b+c)} + \frac{c(b-a)}{(b+a)(a+b+2c)} > 0.$$

   [*Hint*: Notice that the condition of convexity (1.14) can be reformulated as $(z-y)f(x) - (z-x)f(y) + (y-x)f(z) \geq 0$ for all $x < y < z$ in I. For concave functions, this inequality holds in reversed direction.]

3. (Galvani's characterization of convexity [173]) Given a function $f : I \to \mathbb{R}$ and a point $a \in I$, one can associate with them the *slope* function,

   $$s_a : I \backslash \{a\} \to \mathbb{R}, \quad s_a(x) = \frac{f(x) - f(a)}{x - a},$$

   whose value at x is the slope of the chord joining the points $(a, f(a))$ and $(x, f(x))$ of the graph of f. Prove that f is convex (respectively, strictly convex) if and only if its associated slope functions s_a are increasing (respectively, strictly increasing).

4. Infer from Exercise 3 that every positive concave function f defined on $[0, \infty)$ is increasing (and thus $f(x+y) \leq f(x) + f(y)$ for all $x, y \geq 0$).

5. According to B. T. Polyak [404], a function $f : I \to \mathbb{R}$ is called *strongly convex* if there exists a constant $C > 0$ such that

   $$f((1-\lambda)x + \lambda y) \leq (1-\lambda)f(x) + \lambda f(y) - \frac{C}{2}\lambda(1-\lambda)(x-y)^2 \quad (1.18)$$

   for all points x and y in I and all $\lambda \in [0,1]$. Every strongly convex function is strictly convex. The converse fails, as shows the case of the function $f(x) = x^4$. Prove that:

(a) f is strongly convex (with constant C) if and only if $g = f - Cx^2$ is convex.

(b) for a strongly convex function (with constant C), Jensen's inequality takes the form

$$f\Big(\sum_{k=1}^{n} \lambda_k x_k\Big) \leq \sum_{k=1}^{n} \lambda_k f(x_k) - \frac{C}{2}\sum_{k=1}^{n} \lambda_k (x_k - m)^2,$$

where $m = \sum_{k=1}^{n} \lambda_k x_k$.

(c) a twice differentiable function f is strongly convex (with constant $C > 0$) if and only if $f'' \geq C$.

6. Derive Corollary 1.4.6 directly from Remark 1.4.1.

7. (An integral analogue of Corollary 1.4.6) Suppose that $f : [a, b] \to \mathbb{R}$ is a convex function and $c, d \in [a, b]$. Prove that

$$\frac{1}{b-a}\int_a^b f(x)\mathrm{d}x - f\left(\frac{a+b}{2}\right) \geq \frac{1}{d-c}\int_c^d f(x)\mathrm{d}x - f\left(\frac{c+d}{2}\right)$$

and illustrate this result in the case of functions exp and log.

8. (G. Chiti [103]) Suppose that $f : [0, \infty) \to \mathbb{R}$ is an increasing convex function and $x_1 \geq x_2$, $y_1 \geq y_2$ are real points. Prove that

$$f(|x_1 - y_1|) + f(|x_2 - y_2|) \leq f(|x_1 - y_2|) + f(|x_2 - y_1|).$$

[*Hint:* Notice that if $f : \mathbb{R} \to \mathbb{R}$ is a convex function, then $f(x) - f(x-b) \leq f(x+a) - f(x+a-b)$ whenever $x \in \mathbb{R}$ and $a, b \geq 0$.]

9. (H. L. Montgomery) Prove that $1/\sin^2 x < 1/x^2 + 1$ for all $x \in (0, \pi/2]$. This inequality is not sharp. Try to find a better upper bound.

10. Suppose that $(f_n)_n$ is a sequence of differentiable convex functions defined on an open interval I, that converges pointwise to a differentiable function f. Prove that $(f_n')_n$ converges uniformly to f' on each compact subinterval of I.

11. (Kantorovich's inequality) Let $m, M, a_1, \ldots, a_n$ be positive numbers, with $m < M$. Prove that the maximum of

$$f(x_1, \ldots, x_n) = \Big(\sum_{k=1}^{n} a_k x_k\Big)\Big(\sum_{k=1}^{n} a_k / x_k\Big)$$

for $x_1, \ldots, x_n \in [m, M]$ is equal to

$$\frac{(M+m)^2}{4Mm}\Big(\sum_{k=1}^{n} a_k\Big)^2 - \frac{(M-m)^2}{4Mm}\min_{X \subset \{1,\ldots,n\}}\Big(\sum_{k \in X} a_k - \sum_{k \in \complement X} a_k\Big)^2.$$

Remark. The following particular case,

$$\Big(\frac{1}{n}\sum_{k=1}^{n} x_k\Big)\Big(\frac{1}{n}\sum_{k=1}^{n}\frac{1}{x_k}\Big) \le \frac{(M+m)^2}{4Mm} - \frac{(1+(-1)^{n+1})(M-m)^2}{8Mmn^2},$$

represents an improvement on Schweitzer's inequality for odd n.

12. Let $a_k, b_k, c_k, m_k, M_k, m_k', M_k'$ be positive numbers with $m_k < M_k$ and $m_k' < M_k'$ for $k \in \{1, \ldots, n\}$ and let $p > 1$. Prove that the maximum of

$$\Big(\sum_{k=1}^{n} a_k x_k^p\Big)\Big(\sum_{k=1}^{n} b_k y_k^p\Big)\Big/\Big(\sum_{k=1}^{n} c_k x_k y_k\Big)^p$$

for $x_k \in [m_k, M_k]$ and $y_k \in [m_k', M_k']$ ($k \in \{1, \ldots, n\}$) is attained at a $2n$-tuple whose components are endpoints.

13. Assume that $f : I \to \mathbb{R}$ is strictly convex and continuous and $g : I \to \mathbb{R}$ is continuous. For $a_1, \ldots, a_n > 0$ and $m_k, M_k \in I$, with $m_k < M_k$ for $k \in \{1, \ldots, n\}$, consider the function

$$h(x_1, \ldots, x_n) = \sum_{k=1}^{n} a_k f(x_k) + g\Big(\sum_{k=1}^{n} a_k x_k \Big/ \sum_{k=1}^{n} a_k\Big)$$

defined on $\prod_{k=1}^{n}[m_k, M_k]$. Prove that a necessary condition for a point $(y_1, \ldots, y_n)$ to be a point of maximum is that at most one component y_k is inside the corresponding interval $[m_k, M_k]$.

1.5 Absolute Continuity of Convex Functions

The fact that differentiation and integration are operations inverse to each other induces a duality between the class of continuous convex functions defined on an interval I and the class of increasing functions on that interval.

Given an increasing function $\varphi : I \to \mathbb{R}$ and a point $c \in I$ one can attach to them a new function f, given by

$$f(x) = \int_c^x \varphi(t)\,\mathrm{d}t.$$

As φ is bounded on bounded intervals, it follows that f is locally Lipschitz (and thus continuous). It is also a convex function. In fact, according to Theorem 1.1.4, it suffices to show that f is midpoint convex. In fact, for $x \le y$ in I we have

$$\frac{f(x)+f(y)}{2} - f\Big(\frac{x+y}{2}\Big) = \frac{1}{2}\Big(\int_{(x+y)/2}^{y} \varphi(t)\,\mathrm{d}t - \int_x^{(x+y)/2} \varphi(t)\,\mathrm{d}t\Big) \ge 0$$

since φ is increasing.

Notice that f is differentiable at all points of continuity of φ and $f' = \varphi$ at those points.

1.5.1 Remark *There exist convex functions whose first derivative fails to exist on a dense set. For this, let $r_1, r_2, r_3, \ldots$ be an enumeration of the rational numbers in $[0,1]$ and put*

$$\varphi(t) = \sum_{\{k: r_k \leq t\}} \frac{1}{2^k}.$$

Then

$$f(x) = \int_0^x \varphi(t)\, \mathrm{d}t$$

is a continuous convex function whose first derivative does not exist at the points r_k. F. Riesz exhibited an example of strictly increasing function φ with $\varphi' = 0$ almost everywhere. See [214], Example 18.8, pp. 278–282. The corresponding function f in his example is strictly convex though $f'' = 0$ almost everywhere. As we shall see in the Comments at the end of this chapter, Riesz's example is typical from the generic point of view.

Remarkably, every continuous convex function admits an integral representation as above.

1.5.2 Theorem *If $f: I \to \mathbb{R}$ is a continuous convex function, then for all $a < b$ in I we have*

$$f(b) - f(a) = \int_a^b f'_+(t)\, \mathrm{d}t.$$

This formula also works by replacing $f'_+(t)$ with $f'_-(t)$ (as well as with any function φ such that $\varphi(t) \in [f'_-(t), f'_+(t)]$ for $t \in (a,b)$).

Proof. Choose arbitrarily u and v such that $a < u < v < b$. If $u = t_0 < t_1 < \cdots < t_n = v$ is a division of $[u,v]$, then

$$f'_+(t_{k-1}) \leq \frac{f(t_k) - f(t_{k-1})}{t_k - t_{k-1}} \leq f'_+(t_k)$$

for all $k = 1, ..., n$. This yields

$$\begin{aligned}\sum_{k=1}^n f'_+(t_{k-1})\,(t_k - t_{k-1}) \leq f(v) - f(u) &= \sum_{k=1}^n [f(t_k) - f(t_{k-1})] \\ &\leq \sum_{k=1}^n f'_+(t_k)\,(t_k - t_{k-1}),\end{aligned}$$

and taking into account that f'_+ is an increasing function we infer that

$$f(v) - f(u) = \int_u^v f'_+(t)\, \mathrm{d}t.$$

Let u decrease to a. Since f is continuous at a and $f'_+(t) \leq f'_+(v)$ on $(a,v]$, an appeal to Beppo Levi's theorem (that is, the monotone convergence theorem for integrals, [107], Theorem 11.3.10, p. 378) gives us the Lebesgue integrability of f'_+ on $[a, v]$ and the formula

$$f(v) - f(a) = \int_a^v f'_+(t)\, \mathrm{d}t.$$

Then let v increase to b to deduce, via a similar argument, the asserted integral expressions for $f(b) - f(a)$, and the integrability of f'_+ on $[a, b]$. ■

In real analysis, a function $f : [a, b] \to \mathbb{R}$ is called *absolutely continuous* if it is differentiable almost everywhere and f' is Lebesgue integrable on $[a, b]$. Lebesgue has proved that this fact is equivalent to the existence of an integral representation of the form

$$f(x) = f(a) + \int_a^x g(t)\mathrm{d}t$$

for a suitable Lebesgue integrable function g. Thus, Theorem 1.5.2 shows that every continuous convex function is absolutely continuous. One can also prove that every Lipschitz continuous function is absolutely continuous.

The basic theory of absolutely continuous functions can be found, e.g., in the monographs [107], pp. 495–500 and [214], Chapter V, Section 18.

Exercises

1. Suppose that $f : [a, b] \to \mathbb{R}$ is an absolutely continuous function. Prove that for every $\varepsilon > 0$ there is a $\delta > 0$ such that for every finite family $([a_k, b_k])_{k=1}^n$ of pairwise disjoint subintervals of $[a, b]$ such that $\sum_{k=1}^n (b_k - a_k) < \delta$ we have

   $$\sum\nolimits_{k=1}^n (f(b_k) - f(a_k)) < \varepsilon.$$

   Remark. The property mentioned in Exercise 1 actually characterizes the absolute continuity of f. See [107].

2. Prove that every absolutely continuous function is uniformly continuous. The converse is not true. See the case of the function

   $$f(x) = \begin{cases} 0 & \text{if } x = 0 \\ x \sin(1/x) & \text{if } x \in (0, 1]. \end{cases}$$

3. (Integration by parts for absolutely continuous functions) Suppose that f and g are two absolutely continuous functions on $[a, b]$. Prove that

   $$\int_a^b f(t)g'(t)\mathrm{d}t = f(b)g(b) - f(a)g(a) - \int_a^b f'(t)g(t)\mathrm{d}t.$$

1.6 The Subdifferential

Our definition of a convex function took into consideration the position of the graph relative to the line segment between any two points on the graph. The aim of this section is to present a dual characterization, based on suitable substitutes of tangent lines.

1.6.1 Definition *A function f admits a supporting line at a point $a \in I$ if there exists a real number λ (called the* subgradient *of f at a) such that*

$$f(x) \geq f(a) + \lambda(x - a), \quad \text{for all } x \in I. \tag{1.19}$$

The set of subgradients of f at the point a is called the subdifferential of f at a, and is denoted $\partial f(a)$.

Geometrically, the subdifferential gives us the slopes of the supporting lines for the graph of f. See Figure 1.4. The subdifferential at a point is always a convex set, possibly empty.

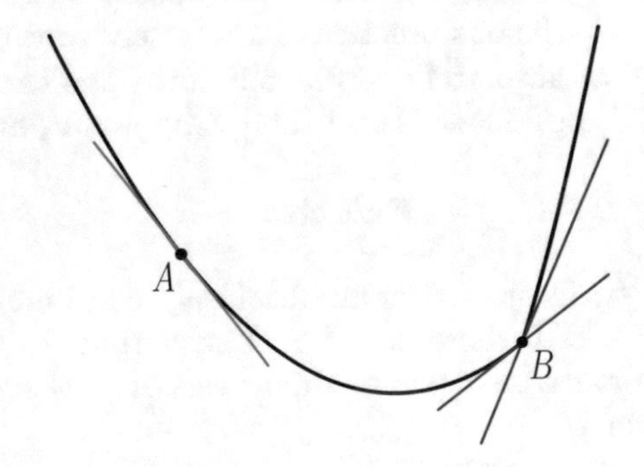

Figure 1.4: Convexity: the existence of supporting lines at interior points.

The convex functions have the remarkable property that $\partial f(x) \neq \emptyset$ at all interior points. However, even in their case, the subdifferential could be empty at the endpoints. An example is given by the continuous convex function $f(x) = 1 - \sqrt{1 - x^2}$, $x \in [-1, 1]$, which fails to have a support line at $x = \pm 1$.

1.6.2 Theorem *If f is a convex function on an interval I, then $\partial f(a) = [f'_-(a), f'_+(a)]$ at all interior points a of I.*

The conclusion above includes the endpoints of I provided that f is differentiable there. As a consequence, the differentiability of a convex function f at a point means that f admits a unique supporting line at that point.

Proof. A point λ belongs to $\partial f(a)$ if and only if

$$\frac{f(x) - f(a)}{x - a} \leq \lambda \leq \frac{f(y) - f(a)}{y - a}$$

whenever $x < a < y$ in I; we use the fact that $x - a < 0$. According to the inequality of three slopes, the first ratio increases to $f'_{-}(a)$ as x increases to a, while the second ratio decreases to $f'_{+}(a)$ as y decreases to a. Therefore λ belongs to $\partial f(a)$ if and only if $\lambda \in \left[f'_{-}(a), f'_{+}(a)\right]$. ■

Many elementary inequalities like $e^x \geq 1 + x$ (on $\mathbb{R}$), and $\sin x \leq x$ (on $\mathbb{R}_+$) are just illustrations of Theorem 1.6.2.

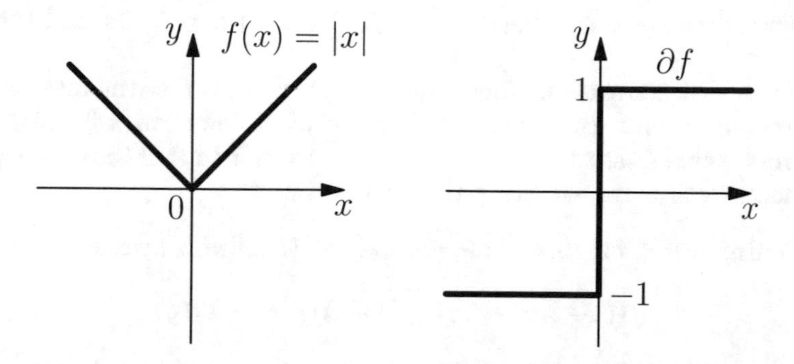

Figure 1.5: The absolute value function and its subdifferential.

Combining Theorem 1.4.2 and Theorem 1.6.2 we obtain the following result:

1.6.3 Corollary *If $f : I \to \mathbb{R}$ is a convex function, then every function φ from I to $\mathbb{R}$ that verifies the condition $\varphi(x) \in \partial f(x)$ whenever $x \in \operatorname{int} I$ is increasing on $\operatorname{int} I$.*

1.6.4 Remark *The convex functions are the only functions f that satisfy the condition $\partial f(x) \neq \emptyset$ at all interior points of the interval of definition.*

To prove this, let $u, v \in I$ with $u \neq v$ and $t \in (0,1)$. Then $(1-t)u+tv \in \operatorname{int} I$, so that for all $\lambda \in \partial f((1-t)u + tv)$ we get

$$f(u) \geq f((1-t)u + tv) + t(u - v) \cdot \lambda,$$
$$f(v) \geq f((1-t)u + tv) - (1-t)(u - v) \cdot \lambda.$$

By multiplying the first inequality by $1-t$, the second one by t and then adding them side by side, we get $(1-t)f(u) + tf(v) \geq f((1-t)u + tv)$, hence f is a convex function.

Every continuous convex function is the upper envelope of its supporting lines:

1.6.5 Theorem *Suppose that f is a continuous convex function on a nondegenerate interval I and $\varphi\colon I \to \mathbb{R}$ is a function such that $\varphi(x)$ belongs to $\partial f(x)$ for all $x \in \operatorname{int} I$. Then*

$$f(z) = \sup\{f(x) + (z - x)\varphi(x) : x \in \operatorname{int} I\} \quad \text{for all } z \in I.$$

Proof. The case of interior points follows from Theorem 1.6.2. Suppose now that z is an endpoint, say the left one. We already noticed that

$$f(z+t)-f(z)\leq t\varphi(z+t)\leq f(z+2t)-f(z+t)$$

for $t>0$ small enough, which yields $\lim_{t\to 0+} t\varphi(z+t)=0$. Given $\varepsilon>0$, there is $\delta>0$ such that $|f(z)-f(z+t)|<\varepsilon/2$ and $|t\varphi(z+t)|<\varepsilon/2$ for $0<t<\delta$. This shows that $f(z+t)-t\varphi(z+t)>f(z)+\varepsilon$ for $0<t<\delta$ and the result follows. ∎

The fact that convex functions admit supporting lines at the interior points has many important consequences. In what follows we will discuss the Legendre–Fenchel duality of convex functions from $\mathbb{R}$ to $\mathbb{R}\cup\{\infty\}$. Many other consequences will be presented in the next sections.

1.6.6 Definition *A function $f:\mathbb{R}\to\mathbb{R}\cup\{\infty\}$ is called convex if*

$$f((1-\lambda)x+\lambda y)\leq(1-\lambda)f(x)+\lambda f(y)$$

for all $x,y\in\mathbb{R}$ and all $\lambda\in(0,1)$. The function f is called strictly convex if the inequality is strict for $x\neq y$.

In what follows we will only deal with *proper* functions, that is, with functions f whose domains $\operatorname{dom} f=\{x:f(x)\in\mathbb{R}\}$ are nonempty.

Any function $f:I\to\mathbb{R}$ that is convex in the usual sense can be turned into a proper convex function in the sense of Definition 1.6.6 by extending it with ∞ outside I. The domain of every proper convex function $g:\mathbb{R}\to\mathbb{R}\cup\{\infty\}$ is a nonempty interval and the restriction g to this set is a usual convex function.

Given a proper convex function $f:\mathbb{R}\to\mathbb{R}\cup\{\infty\}$, one defines its *convex conjugate* (or *Legendre–Fenchel transform*) as

$$f^*:\mathbb{R}\to\mathbb{R}\cup\{\infty\},\quad f^*(y)=\sup\{xy-f(x):x\in\mathbb{R}\}.$$

Notice that f^* is a *proper convex function.* Indeed, if $\operatorname{dom} f$ is a singleton, say $\{x_0\}$, then $f^*(y)=x_0y-f(x_0)$ for every $y\in\mathbb{R}$. If $\operatorname{dom} f$ is a nondegenerate interval, then for each $a\in\operatorname{int}(\operatorname{dom} f)$ there is an $y\in\mathbb{R}$ such that $f(x)\geq f(a)+y(x-a)$, which yields

$$xy-f(x)\leq ay-f(a),$$

so, $y\in\operatorname{dom} f^*$. Thus, $\operatorname{dom} f^*$ includes $\bigcup_{x\in\mathbb{R}}\partial f(x)$.

If $\lambda\in(0,1)$ and $y,z\in\operatorname{dom} f^*$, then

$$\begin{aligned}f^*((1-\lambda)y+\lambda z)&=\sup\{x[(1-\lambda)y+\lambda z]-f(x):x\in I\}\\&\leq(1-\lambda)\sup\{xy-f(x):x\in I\}+\lambda\sup\{xz-f(x):x\in I\}\\&=(1-\lambda)f^*(y)+\lambda f^*(z)\end{aligned}$$

and the convexity of f^* follows.

The function f^* is also *lower semicontinuous* in the sense that all its *level sets* $L(\lambda) = \{y : f^*(y) \leq \lambda\}$ are closed sets. Since f^* is the upper envelope of a family of continuous functions,

$$L(\lambda) = \{y : f^*(y) \leq \lambda\} = \bigcap\nolimits_{x \in \mathbb{R}} \{y : xy - f(x) \leq \lambda\},$$

and an intersection of closed sets is also a closed set.

An immediate consequence of the definition of convex conjugation is *Young's inequality*,

$$xy \leq f(x) + f^*(y) \quad \text{for all } x, y \in \mathbb{R}.$$

This inequality extends the previous form of Young's inequality stated as equation (1.5). Indeed, the convex conjugate of the function $f(x) = |x|^p/p$, $x \in \mathbb{R}$ $(p > 1)$, is the function $f^*(y) = |y|^q/q$, $y \in \mathbb{R}$ $(1/p + 1/q = 1)$.

1.6.7 Lemma *Suppose that $f : \mathbb{R} \to \mathbb{R} \cup \{\infty\}$ is a lower semicontinuous proper convex function. Then, for all $x, y \in \mathbb{R}$,*

$$xy = f(x) + f^*(y) \text{ if, and only if, } y \in \partial f(x).$$

Proof. Indeed, $xy = f(x) + f^*(y)$ if and only if $xy \geq f(x) + f^*(y)$, which in turn is equivalent to $xy \geq f(x) + yz - f(z)$ for all $z \in \mathbb{R}$. The latter inequality can be rewritten as $f(z) \geq f(x) + y(z - x)$ for all $z \in \mathbb{R}$, which means that $y \in \partial f(x)$. ∎

In general

$$f \leq g \text{ implies } g^* \leq f^*$$

but for lower semicontinuous proper convex functions this becomes an equivalence.

1.6.8 Theorem (*Legendre–Fenchel duality for lower semicontinuous proper convex functions*) *Let $f : \mathbb{R} \to \mathbb{R} \cup \{\infty\}$ be a proper function. The following assertions are equivalent:*

(*a*) *f is convex and lower semicontinuous;*

(*b*) *$f = g^*$ for some proper function g;*

(*c*) *$f^{**} = f$.*

We will come back to this result in Section 6.1, where its several variables variant is presented with full details.

1.6.9 Remark (Orlicz spaces) *An* Orlicz function *is any convex function $\Phi : [0, \infty) \to \mathbb{R}$ such that:*

(Φ1) *$\Phi(0) = 0$, $\Phi(x) > 0$ for $x > 0$;*

(Φ2) *$\Phi(x)/x \to 0$ as $x \to 0$ and $\Phi(x)/x \to \infty$ as $x \to \infty$;*

(Φ3) *there exists a positive constant K such that $\Phi(2x) \leq K\Phi(x)$ for every $x \geq 0$.*

Let (X, Σ, μ) be a complete σ-finite measure space and let $S(\mu)$ be the vector space of all equivalence classes of μ-measurable real-valued functions defined on X. The Orlicz space $L^{\Phi}(X)$ is the subspace of all $f \in S(\mu)$ such that

$$I_{\Phi}(f/\lambda) = \int_X \Phi(|f(x)|/\lambda)\, \mathrm{d}\mu < \infty$$

for some $\lambda > 0$. *One can prove that $L^{\Phi}(X)$ is a linear space such that*

$$|f| \leq |g| \text{ and } g \in L^{\Phi}(X) \text{ implies } f \in L^{\Phi}(X).$$

Moreover, it constitutes a Banach space when endowed with the norm

$$\|f\|_{\Phi} = \inf\{\lambda > 0 \,|\, I_{\Phi}(f/\lambda) \leq 1\}$$

and the dual space of $L^{\Phi}(X)$ is $L^{\Psi}(X)$, where Ψ is the conjugate of Φ.

The Orlicz spaces extend the $L^p(\mu)$ spaces. Their theory is exposed in books like that of M. A. Krasnosel'skii and Ya. B. Rutickii [259]. The Orlicz space $\mathcal{L}\log^+\mathcal{L}$ (corresponding to the Lebesgue measure on $[0,\infty)$ and the function $\Phi(t) = t\,(\log t)^+$) plays an important role in Fourier Analysis. See, e.g., the book of A. Zygmund [500]. Applications to the interpolation theory are given by C. Bennett and R. Sharpley [42].

Exercises

1. Let $f\colon I \to \mathbb{R}$ be a convex function. Show that:

   (a) any local minimum of f is a global one;

   (b) f attains a global minimum at a if and only if $0 \in \partial f(a)$;

   (c) if f has a global maximum at an interior point of I, then f is constant.

2. (Convex mean value theorem) Consider a continuous convex function $f\colon [a,b] \to \mathbb{R}$. Prove that $(f(b) - f(a))/(b-a) \in \partial f(c)$ for some point $c \in (a,b)$.

3. Prove that a function $f : \mathbb{R} \to \mathbb{R} \cup \{\infty\}$ is lower semicontinuous if and only if $x_n \to x$ in $\mathbb{R}$ implies $f(x) \leq \liminf_{n\to\infty} f(x_n)$. Infer that the extension $\tilde{f}$ of a convex function $f : [a,b) \to \mathbb{R}$ with ∞ outside $[a,b)$ is lower semicontinuous if and only if f is continuous at a and $\lim_{x\to b} f(x) = \infty$.

4. Compute ∂f, ∂f^* and f^* for $f(x) = |x|$, $x \in \mathbb{R}$.

5. Prove that:

   (a) the conjugate of $f(x) = x^p/p$, $x \geq 0$, is $f^*(y) = (y^+)^q/q$, $y \in \mathbb{R}$ ($p > 1$, $1/p + 1/q = 1$);

   (b) the conjugate of $f(x) = \mathrm{e}^x$, $x \in \mathbb{R}$, is the function $f^*(y) = y\log y - y$ for $y > 0$ and $f^*(0) = 0$;

   (c) the conjugate of $f(x) = -\log x$, $x > 0$, is the function $f^*(y) = -1 - \log(-y)$, $y < 0$.

6. Prove that the function $f(x) = \frac{1}{2}x^2$ is the unique function $g : \mathbb{R} \to \mathbb{R}$ such that $g = g^*$.

   [*Hint:* Use Young's inequality.]

7. Suppose that $f\colon [0, \infty) \to [0, \infty)$ is a strictly increasing continuous function such that $f(0) = 0$ and $\lim_{x\to\infty} f(x) = \infty$. Prove that the function $\Phi(u) = \int_0^u f(x)\,\mathrm{d}x$, $u \geq 0$, is strictly convex and differentiable and the convex conjugate of Φ is the function $\Psi(v) = \int_0^v f^{-1}(x)\,\mathrm{d}x$, $v \geq 0$. Then, infer the conclusion of Theorem 1.2.1,

$$uv \leq \int_0^u f(x)\,\mathrm{d}x + \int_0^v f^{-1}(x)\,\mathrm{d}x$$

   for all $u, v \geq 0$, with equality if and only if $v = f(u)$.

1.7 The Integral Form of Jensen's Inequality

The aim of this section is to extend Jensen's discrete inequality to the general framework of finite measure spaces. Recall that a *finite measure space* is any triplet (Ω, Σ, μ) consisting of an abstract nonempty set Ω, a σ-algebra Σ of subsets of Ω, and a σ-additive measure $\mu : \Sigma \to \mathbb{R}_+$ such that $0 < \mu(\Omega) < \infty$. By replacing μ by $\mu/\mu(\Omega)$, one can reduce the study of finite measure spaces to that of *probability spaces*, characterized by the fact that $\mu(\Omega) = 1$.

1.7.1 Remark *The language of measure-theoretic probability theory is slightly different from that of traditional measure theory. Most books on probability theory use the notation (Ω, Σ, P) for a probability space consisting of a sample space Ω, a σ-algebra Σ of events (viewed as subsets of Ω) and a probability measure $P : \Sigma \to \mathbb{R}_+$. In the probabilistic context, the real measurable functions $X : \Omega \to \mathbb{R}$ are called* random variables. *Their definitory property is that $X^{-1}(A) \in \Sigma$ for every Borel subset A of $\mathbb{R}$. An important point in probability theory is the use of the notion of independence (both for events and random variables).*

Jensen's inequality relates two important concepts that can be attached to a finite measure space (Ω, Σ, μ): the integral arithmetic mean and the barycenter. As will be shown below, another probabilistic concept, that of variance, provides an efficient way for evaluating the gap in Jensen's inequality.

The *integral arithmetic mean* (or the *mean value*) of a μ-integrable function $f\colon \Omega \to \mathbb{R}$ is defined by the formula

$$M_1(f) = \frac{1}{\mu(\Omega)} \int_\Omega f(x)\,\mathrm{d}\mu(x).$$

In the context of probability theory, this number is also called the *expectation* (or *expected value*) of f and is denoted $E(f)$. The expectation generates a functional $E : L^1(\mu) \to \mathbb{R}$ having the following three properties:

- linearity, $E(\alpha f + \beta g) = \alpha E(f) + \beta E(g)$;
- positivity, $f \geq 0$ implies $E(f) \geq 0$;
- calibration, $E(1) = 1$.

A very convenient way to introduce the expectation of a real-valued function $f \in L^1(\mu)$ is provided by the Riemann–Stieltjes integral. The key ingredient is the *cumulative distribution function* of f, which is defined by the formula

$$F : \mathbb{R} \to [0,1], \quad F(x) = \mu\left(\{\omega : f(\omega) < x\}\right);$$

on several occasions we will use the notation F_f instead of F, to specify the integrable function under attention.

This function is increasing and its limits at infinity are $\lim_{x\to-\infty} F(x) = 0$ and $\lim_{x\to\infty} F(x) = 1$. Also, the cumulative distribution function is left continuous, that is,

$$\lim_{x\to x_0-} F(x) = F(x_0) \text{ at every } x_0 \in \mathbb{R}.$$

The distribution function allows us to introduce the expectation of a random variable by a formula that avoids the use of μ:

$$E(f) = \int_{-\infty}^{\infty} x \mathrm{d}F_f(x).$$

For convenience, the concept of barycenter will be introduced in the context of probability measures μ defined on the σ-algebra $\mathcal{B}(I)$ of Borel subsets of an interval I (the so-called Borel probability measures on the interval I). Precisely, we will consider the class

$$\mathcal{P}^1(I) = \left\{\mu : \mu \text{ Borel probability measure on } I \text{ and } \int_I |x| \,\mathrm{d}\mu(x) < \infty\right\}.$$

This class includes all Borel probability measures null outside a bounded subinterval.

1.7.2 Definition *The barycenter (or resultant) of a Borel probability measure $\mu \in \mathcal{P}^1(I)$ is the point*

$$\mathrm{bar}(\mu) = \int_I x \mathrm{d}\mu(x).$$

Necessarily, when the barycenter $\mathrm{bar}(\mu)$ exists, it must be in I. Indeed, if $\mathrm{bar}(\mu) \notin I$, then either x is an upper bound for I or is a lower bound. Since

$$\int_I [x - \mathrm{bar}(\mu)] \,\mathrm{d}\mu(x) = 0,$$

this situation will impose $x - \mathrm{bar}(\mu) = 0$ μ-almost everywhere, which is not possible because $\mu(I) = 1$.

A first example illustrating Definition 1.7.2 concerns the case of a discrete probability measure $\lambda = \sum_{k=1}^{n} \lambda_k \delta_{x_k}$ concentrated at the points $x_1, ..., x_n \in \mathbb{R}$. Here δ_z represents the Dirac measure concentrated at z, that is, the measure given by

$$\delta_z(A) = 1 \text{ if } z \in A \text{ and } \delta_z(A) = 0 \text{ otherwise.}$$

In this case,

$$E(f) = \int_{\mathbb{R}} f(x)\mathrm{d}\lambda(x) = \sum_{k=1}^{n} \lambda_k f(x_k)$$

for every function $f : \mathbb{R} \to \mathbb{R}$. Therefore

$$\operatorname{bar}(\lambda) = \int_{\mathbb{R}^N} x\mathrm{d}\lambda(x) = \sum_{k=1}^{n} \lambda_k x_k.$$

The barycenter of the restriction of Lebesgue measure to an interval $[a, b]$ is the middle point because

$$\frac{1}{b-a}\int_a^b x\mathrm{d}x = \frac{a+b}{2}.$$

The barycenter of the Gaussian probability measure $\frac{1}{\sqrt{2\pi}}e^{-x^2/2}\mathrm{d}x$ is the origin. Indeed,

$$\frac{1}{\sqrt{2\pi}}\int_{-\infty}^{\infty} xe^{-x^2/2}\mathrm{d}x = 0.$$

1.7.3 Theorem (The integral form of Jensen's inequality) *Let (Ω, Σ, μ) be a probability space and let $g\colon \Omega \to \mathbb{R}$ be a μ-integrable function. If f is a convex function defined on an interval I that includes the image of g, then $E(g) \in I$ and*

$$f(E(g)) \le \int_{\Omega} f(g(x))\mathrm{d}\mu(x).$$

Notice that the right-hand side integral always exists but it might be ∞ if the μ-integrability of $f \circ g$ is not expressly asked. When both functions g and $f \circ g$ are μ-integrable, then the above inequality becomes

$$f(E(g)) \le E(f \circ g).$$

Moreover, if in addition f is strictly convex, then this inequality becomes an equality if and only if g is constant μ-almost everywhere.

In order to avoid unnecessary repetitions, the details are postponed till Section 3.7, where the more general case of several variables will be presented.

1.7.4 Corollary *If $\mu \in \mathcal{P}^1(I)$ and $f : I \to \mathbb{R}$ is a μ-integrable convex function, then*

$$f(\operatorname{bar}(\mu)) \le \int_I f(x)\mathrm{d}\mu(x).$$

Proof. Apply Theorem 1.7.3 for g the identity of the interval I. ∎

1.7.5 Corollary (*Jensen's inequality for series*) *Suppose that $f : I \to \mathbb{R}$ is a convex function, $(x_n)_n$ is a sequence in I, and $(\lambda_n)_n$ is a sequence in $[0,1]$ such that $\sum_{n=1}^{\infty} \lambda_n = 1$. If the series $\sum_{n=1}^{\infty} \lambda_n x_n$ is absolutely convergent, then the series $\sum_{n=1}^{\infty} \lambda_n f(x_n)$ is either convergent or has infinite sum. In both cases,*

$$f\left(\sum_{n=1}^{\infty} \lambda_n x_n\right) \leq \sum_{n=1}^{\infty} \lambda_n f(x_n).$$

1.7.6 Remark (*Differential entropy*) *As above, we assume that (Ω, Σ, P) is a probability space. By applying Jensen's inequality to the convex function $f(x) = x \log x$ in $[0, \infty)$ we obtain the following upper estimate for the differential entropy $h(g) = -\int_{\Omega} g \log g \, \mathrm{d}\mu$ of a positive μ-integrable function g:*

$$h(g) \leq -\left(\int_{\Omega} g \, \mathrm{d}\mu\right) \log \left(\int_{\Omega} g \, \mathrm{d}\mu\right).$$

If $\int_{\Omega} g \, \mathrm{d}\mu = 1$, then $h(g) \leq 0$ even though the function f has a variable sign (it attains the minimum value $-1/\mathrm{e}$ at $z = 1/\mathrm{e}$).

A natural question is how large is the *Jensen gap*,

$$E(f \circ g) - f(E(g)).$$

The following result noticed by O. Hölder [225] provides the answer in an important special case.

1.7.7 Proposition *Suppose that $f\colon [a,b] \to \mathbb{R}$ is a twice differentiable function for which there exist real constants m and M such that $m \leq f'' \leq M$. Then*

$$\begin{aligned} \frac{m}{2} \sum_{1 \leq i < j \leq n} \lambda_i \lambda_j (x_i - x_j)^2 &\leq \sum_{k=1}^{n} \lambda_k f(x_k) - f\Big(\sum_{k=1}^{n} \lambda_k x_k\Big) \\ &\leq \frac{M}{2} \sum_{1 \leq i < j \leq n} \lambda_i \lambda_j (x_i - x_j)^2, \end{aligned} \tag{1.20}$$

whenever $x_1, ..., x_n \in [a,b]$, $\lambda_1, \ldots, \lambda_n \in [0,1]$ and $\sum_{k=1}^{n} \lambda_k = 1$.

Proof. This is a consequence of the discrete Jensen's Inequality when applied to the convex functions $f - mx^2/2$ and $Mx^2/2 - f$. ∎

1.8.8 Corollary (The gap in the *AM–GM* inequality) *If $0 < m \leq x_1, ..., x_n \leq M$, $\lambda_1, \ldots, \lambda_n \in [0,1]$ and $\sum_{k=1}^{n} \lambda_k = 1$, then*

$$\begin{aligned} \frac{m}{2n^2} \sum_{1 \leq j < k \leq n} (\log x_j - \log x_k)^2 &\leq \frac{1}{n} \sum_{k=1}^{n} x_k - \left(\prod_{k=1}^{n} x_k\right)^{1/n} \\ &\leq \frac{M}{2n^2} \sum_{1 \leq j < k \leq n} (\log x_j - \log x_k)^2. \end{aligned}$$

Proposition 1.7.7 exhibits the role of the variance in estimating the precision in Jensen's inequality.

If (Ω, Σ, μ) is a probability space, the *variance* of a function $g \in L^2(\mu)$ is defined by the formula

$$\begin{aligned} \operatorname{var}(g) &= E\left((g - E(g))^2\right) \\ &= E\left(g^2\right) - (E(g))^2. \end{aligned} \tag{1.21}$$

The variance is an indicator of how much the values of g are spread out. A variance of zero indicates that all the values of g are identical (except possibly for a subset of Ω of probability zero).

The square root of variance is called the *standard deviation.*

Since a probability measure is a finite measure, the space $L^2(\mu)$ is included in $L^1(\mu)$. Thus expectation and variance applies to every function that belongs to $L^2(\mu)$.

Let us consider the discrete probability space $\Omega = \{1, ..., n\}$, $\Sigma = \mathcal{P}(\Omega)$, and $\mu = \sum_{k=1}^n \lambda_k \delta_k$. The variance of the function $g : \{1, ..., n\} \to \mathbb{R}$ defined by $g(k) = x_k$ for $k = 1, ..., n$, is

$$\begin{aligned} \operatorname{var}(g) &= E((g - E(g))^2) \\ &= E\left(g^2\right) - (E(g))^2 \\ &= \sum_{n=1}^n \lambda_n x_n^2 - \left(\sum_{n=1}^n \lambda_n x_n\right)^2 \\ &= \frac{1}{2} \sum_{1 \le i,j \le n} \lambda_i \lambda_j (x_i - x_j)^2 = \sum_{1 \le i < j \le n} \lambda_i \lambda_j (x_i - x_j)^2, \end{aligned}$$

and thus the result of Proposition 1.7.7 can be reformulated as

$$\frac{m}{2} \operatorname{var}(g) \le E(f(g)) - f(E(g)) \le \frac{M}{2} \operatorname{var}(g).$$

It is worth noticing that this double estimate works in general. See Exercise 5.

In probability and statistics, an important role is played by the so-called continuous random variables. A random variable X (attached to a probability measure space (Ω, Σ, P)) is called *continuous* if its cumulative distribution function is of the form

$$F_X(x) = P(\{\omega : X(\omega) < x\}) = \int_{-\infty}^x w(t)\, \mathrm{d}t,$$

for a suitable Lebesgue integrable function $w \in L^1(\mathbb{R})$, called the *density* of F_X. In this case, the probability that X takes a value a is 0 and

$$P(\{\omega : a \le X(\omega) \le b\}) = \int_a^b w(t)\, \mathrm{d}t \quad \text{for all } -\infty \le a \le b \le \infty.$$

Moreover, the computation of the expectation and of the variance of X reduces to the computation of certain Lebesgue integrals:

$$E(X) = \int_{-\infty}^{\infty} x w(x) \mathrm{d}x \text{ and } \operatorname{var}(X) = \int_{-\infty}^{\infty} (x - E(X))^2 w(x) \mathrm{d}x.$$

A continuous random variable X is called *normal* if its distribution function is associated to a density of the form

$$w(t, \mu, \sigma) = \frac{1}{\sigma\sqrt{2\pi}} e^{-\frac{(t-\mu)^2}{2\sigma^2}}.$$

In this case, the values of the parameters $\mu \in \mathbb{R}$ and $\sigma > 0$ are precisely the expectation and the standard deviation of X.

1.7.9 Remark (*Upper bounds on the variance*) *As was noticed by D. S. Mitrinović, J. E. Pečarić, and A. M. Fink in* [330], *p.* 296, *if X is a random variable such that $\alpha \leq X \leq \beta$ for two suitable constants α and β, then*

$$\operatorname{var}(X) \leq (\beta - E(X)) (E(X) - \alpha).$$

This remark improves the upper bound previously indicated by T. Popoviciu [405],

$$\operatorname{var}(X) \leq \frac{(\beta - \alpha)^2}{4}.$$

The bound found by Mitrinović, Pečarić, and Fink follows easily from the general properties of expectation:

$$\begin{aligned} 0 \leq E\left((\beta - X)(X - \alpha)\right) &= -\alpha\beta + (\alpha + \beta) E(X) - E\left(X^2\right) \\ &= (\beta - E(X))(E(X) - \alpha) - \operatorname{var}(X). \end{aligned}$$

1.7.10 Remark (*Chebyshev's probabilistic inequality*) *If X is a random variable (associated to a probability measure space (Ω, Σ, μ)), then*

$$\mu\left(\{|X - E(X)| \geq \varepsilon\}\right) \leq \frac{\operatorname{var}(X)}{\varepsilon^2}$$

for all $\varepsilon > 0$.

The *covariance* of two real random variables $X, Y \in L^2(P)$ is defined by

$$\begin{aligned} \operatorname{cov}(X, Y) &= E\left((X - E(X))(Y - E(Y))\right) \\ &= E(XY) - E(X)E(Y). \end{aligned}$$

An instance when the covariance is positive is mentioned in Exercise 8.

Two random variables X and Y whose covariance is zero are called *uncorrelated.* If X and Y are independent, then their covariance is zero. This follows because under independence, $E(XY) = E(X)E(Y)$. However, uncorrelation does not imply in general independence.

The concept of covariance allows us to indicate a new upper estimate of the Jensen gap.

1.7.11 Theorem (The covariance form of Jensen's inequality) *Let (Ω, Σ, μ) be a probability space and let $g\colon \Omega \to \mathbb{R}$ be a μ-integrable function. Suppose also that f is a convex function defined on an interval I that includes the image of g and $\varphi\colon I \to \mathbb{R}$ is a function such that*

(a) $\varphi(x) \in \partial f(x)$ for every $x \in I$; and

(b) $\varphi \circ g$ and $g \cdot (\varphi \circ g)$ are μ-integrable functions.

Then

$$0 \leq E(f \circ g) - f(E(g)) \leq \operatorname{cov}(g, \varphi \circ g).$$

If f is concave, then the last two inequalities work in the reversed direction.

Proof. The first inequality is motivated by Theorem 1.7.3. The second inequality follows by integrating the inequality

$$f(E(g)) \geq f(g(x)) + (E(g) - g(x)) \cdot \varphi(g(x)) \quad \text{for all } x \in \Omega.$$

■

1.7.12 Corollary (S. S. Dragomir and N. M. Ionescu [134]) *If f is a differentiable convex function defined on an open interval I, then*

$$\begin{aligned} 0 &\leq \sum_{k=1}^{n} \lambda_k f(x_k) - f\Big(\sum_{k=1}^{n} \lambda_k x_k\Big) \\ &\leq \sum_{k=1}^{n} \lambda_k x_k f'(x_k) - \Big(\sum_{k=1}^{n} \lambda_k x_k\Big)\Big(\sum_{k=1}^{n} \lambda_k f'(x_k)\Big) \end{aligned}$$

for all $x_1, \ldots, x_n \in I$ and all $\lambda_1, \ldots, \lambda_n \in [0,1]$, with $\sum_{k=1}^{n} \lambda_k = 1$.

The covariance defines a Hermitian product; it differs from a scalar product by the fact that $\operatorname{cov}(X, X) = 0$ implies only that X is constant almost everywhere. Since the Cauchy–Bunyakovsky–Schwarz inequality still works for such products, we have the inequality

$$|\operatorname{cov}(X, Y)| \leq (\operatorname{var}(X))^{1/2} (\operatorname{var}(Y))^{1/2}, \tag{1.22}$$

known as the *covariance form of Cauchy–Bunyakovsky–Schwarz inequality*. This inequality shows that *Pearson's correlation coefficient* of X and Y, that is,

$$\rho_{X,Y} = \frac{\operatorname{cov}(X, Y)}{(\operatorname{var}(X))^{1/2} (\operatorname{var}(Y))^{1/2}}$$

takes values in the interval $[-1, 1]$. A value of 1 (respectively -1) for $\rho_{X,Y}$ implies that the relationship between X and Y is described by a linear equation, for which Y increases (respectively decreases) as X increases. A value of 0 implies that there is no linear correlation between the two random variables.

From the covariance form of Cauchy–Bunyakovsky–Schwarz inequality and Remark 1.7.9 we infer the following classical result:

1.7.13 Theorem (Grüss Inequality [193]) *Suppose that the random variables X and Y are bounded, precisely, $\alpha \leq X \leq \beta$ and $\delta \leq Y \leq \gamma$. Then*

$$|\operatorname{cov}(X, Y)| \leq \frac{1}{4}(\beta - \alpha)(\gamma - \delta).$$

and the constant $1/4$ being sharp.

Exercises

1. (The power means; see Section 1.1, Exercise 10, for the discrete case) Consider a finite measure space (Ω, Σ, μ) and a function $f \in L^1(\mu)$ such that $f \geq 0$. Adopting the convention $\log 0 = -\infty$ and $\mathrm{e}^{-\infty} = 0$, we define the geometric mean of f by the formula

$$M_0(f; \mu) = \exp\left(\frac{1}{\mu(\Omega)} \int_\Omega \log f(x)\, \mathrm{d}\mu\right).$$

According to Jensen's inequality,

$$\frac{1}{\mu(\Omega)} \int_\Omega \log f(x)\, \mathrm{d}\mu \leq \log\left(\frac{1}{\mu(\Omega)} \int_\Omega f(x)\, \mathrm{d}\mu\right),$$

the inequality being strict except for the case when f is a constant function μ-almost everywhere. This fact can be restated as $M_0(f; \mu) \leq M_1(f; \mu)$, and represents the integral form of *AM–GM* inequality.

The *power mean* of order $t \in \mathbb{R} \setminus \{0\}$ is defined for all positive measurable functions f with $f^t \in L^1(\mu)$ via the formula

$$M_t(f; \mu) = \left(\frac{1}{\mu(\Omega)} \int_\Omega f^t\, \mathrm{d}\mu\right)^{1/t}.$$

We also define

$$M_{-\infty}(f; \mu) = \sup\{\alpha \geq 0 \mid \mu(\{x \in \Omega : f(x) < \alpha\}) = 0\}$$
$$M_{\infty}(f; \mu) = \inf\{\alpha \geq 0 \mid \mu(\{x \in \Omega : f(x) > \alpha\}) = 0\}$$

for $f \in L^\infty(\mu)$, $f \geq 0$.

(a) (Jensen's inequality for means) Suppose that $-\infty \leq s \leq t \leq \infty$ and $M_t(f; \mu) < \infty$. Prove that

$$M_s(f; \mu) \leq M_t(f; \mu).$$

(b) Suppose that $f \in L^\infty(\mu)$, $f \geq 0$. Prove that

$$\lim_{t \to -\infty} M_t(f; \mu) = M_{-\infty}(f; \mu) \quad \text{and} \quad \lim_{t \to \infty} M_t(f; \mu) = M_\infty(f; \mu).$$

(c) Suppose that $f \in L^\infty(\mu)$, $f \geq 0$. Prove the convexity of the function $t \mapsto t \log M_t(f;\mu)$ on $\mathbb{R}$.

(d) Notice that $(t^r - 1)/r$ decreases to $\log t$ as $r \downarrow 0$ and apply the dominated convergence theorem of Lebesgue to conclude that

$$\lim_{r \to 0+} M_r(f;\mu) = M_0(f;\mu)$$

for all $f \in L^1(\mu)$, $f \geq 0$.

2. The integral form of the *arithmetic–geometric–harmonic mean inequality* asserts that

$$M_{-1}(f;\mu) \leq M_0(f;\mu) \leq M_1(f;\mu).$$

Infer that $L(a,b) < I(a,b) < A(a,b)$ for all $a, b > 0$, $a \neq b$. Here

$$L(a,b) = \frac{a-b}{\log a - \log b}, I(a,b) = \frac{1}{\mathrm{e}}\left(\frac{b^b}{a^a}\right)^{1/(b-a)}, \text{ and } A(a,b) = \frac{a+b}{2}$$

are, respectively, the logarithmic, the identric, and the arithmetic mean of a and b.

3. Assume that A, B, C are the angles of a triangle (expressed in radians). Prove that

$$0 \leq 3\sqrt{3}/2 - \sum \sin A \leq \sum \left(\frac{\pi}{3} - A\right) \cos A.$$

4. Let f be a convex function defined on $\mathbb{R}$ and X a random variable such that $g(x) = E\left(f\left(x - X\right)\right)$ exists and is finite for all $x \in \mathbb{R}$. Prove that $g(\cdot)$ is a convex function on $\mathbb{R}$.

5. (An estimate of Jensen gap) Let (X, Σ, μ) be a finite measure space and let $g \in L^2(\mu)$. If f is a twice differentiable function given on an interval I that includes the image of g and $\alpha \leq f''/2 \leq \beta$, then

$$\alpha \operatorname{var}(g) \leq M_1(f \circ g) - f(M_1(g)) \leq \beta \operatorname{var}(g).$$

Here $\operatorname{var}(g)$ denotes the *variance* of g.

[*Hint*: Apply Taylor's formula to infer that

$$\begin{aligned}\alpha(g(x) - M_1(g))^2 &\leq f(g(x)) - f(M_1(g)) - f'(M_1(g))(g(x) - M_1(g)) \\ &\leq \beta(g(x) - M_1(g))^2.]\end{aligned}$$

6. Consider the normal random variable X whose density of its cumulative distribution function is

$$w(t,\mu,\sigma) = \frac{1}{\sigma\sqrt{2\pi}} e^{-\frac{(t-\mu)^2}{2\sigma^2}}.$$

Prove that $E(X) = \mu$ and $\operatorname{var}(X) = \sigma^2$.

7. (The Korkine–Andréief identity) Suppose that (Ω, Σ, μ) is a probability measure space and $f, g \in L^2(\mu)$. Prove the identity

$$\int_\Omega f(x)g(x)\mathrm{d}\mu(x) - \left(\int_\Omega f(x)\mathrm{d}\mu(x)\right)\left(\int_\Omega g(x)\mathrm{d}\mu(x)\right) \\ = \frac{1}{2}\int_\Omega\int_\Omega (f(x) - f(y))\,(g(x) - g(y))\,\mathrm{d}\mu(x)\mathrm{d}\mu(y).$$

8. (Chebyshev's algebraic inequality) Suppose that (Ω, Σ, μ) is a probability measure space and $f, g \in L^2(\mu)$ are two functions synchronous in the sense that

$$(g(x) - g(y))(h(x) - h(y)) \geq 0$$

for all $x, y \in \Omega$. Infer from the preceding exercise that $E(g)E(h) \leq E(gh)$ (that is, $\operatorname{cov}(f, g) \geq 0$).

9. (Jensen's inequality at a point; see C. P. Niculescu and H. Stephan [380]) Let I be an interval of $\mathbb{R}$ endowed with a discrete measure $\mu = \sum_{k=1}^n p_k \delta_{x_k}$, whose weights p_k are all nonzero and sum to 1. We assume that the barycenter of μ, $b_\mu = \sum_{k=1}^n p_k x_k$, belongs to $I \setminus \{x_1, x_2, ..., x_n\}$.

   (a) Prove that every function $f : I \to \mathbb{R}$ verifies the identity

$$\sum_{k=1}^n p_k f(x_k) - f(b_\mu) = \sum_{1 \leq k < j \leq n} p_k p_j \left(s(x_k) - s(x_j)\right)(x_k - x_j),$$

   where

$$s(x) = \frac{f(x) - f(b_\mu)}{x - b_\mu} \quad \text{for } x \in I \setminus \{b_\mu\}$$

   is the slope function of the segment joining the points of abscissas x and b_μ.

   (b) Assume that all weights p_k are positive. Prove that if $s(x)$ is increasing, then

$$\sum_{k=1}^n p_k f(x_k) \geq f(b_\mu),$$

   while if $s(x)$ is decreasing, this inequality works in the reversed direction.

   (c) Infer that

$$\max\left\{\sum_{k=1}^n \frac{x_k}{x_k^3 + 8} : x_1, ..., x_n > -2 \text{ and } \frac{1}{n}\sum_{k=1}^n x_k = 1\right\} = \frac{n}{9}.$$

10. (The connection of Jensen's inequality with subdifferentiability) Suppose that μ is a Borel probability measure μ supported by an interval I and

having the barycenter $\mathrm{bar}(\mu)$. Prove that every μ-integrable function $f : I \to \mathbb{R}$ that admits a supporting line at $\mathrm{bar}(\mu)$ verifies the inequality

$$f(\mathrm{bar}(\mu)) \le \int_I f(x)\mathrm{d}\mu(x).$$

This result allows the use of Jensen's inequality in the context of mixed convex functions like xe^x, x^2e^{-x}, $\log^2 x$, $\frac{\log x}{x}$, etc. See C. P. Niculescu and I. Roventa [374].

1.8 Two More Applications of Jensen's Inequality

Jensen's inequality is instrumental in deriving many useful inequalities, including the Rogers–Hölder inequality and Hardy's inequality. In the first case we have to notice that Jensen's inequality (when applied to the function x^p on $(0,\infty)$ and $p \ge 1$) implies that every positive measurable function h (attached to a finite measure space (Ω, Σ, μ) with $\mu(\Omega) = 1$) verifies the inequality

$$\left(\int_\Omega h(x)\mathrm{d}\mu(x)\right)^p \le \int_\Omega h^p(x)\mathrm{d}\mu(x). \tag{1.23}$$

Suppose now that $p, q \in (1,\infty)$ with $1/p + 1/q = 1$, and f, g are two measurable functions such that $f \in L^p(\mu)$ and $g \in L^q(\mu)$. To prove that

$$\int_\Omega |fg|\, \mathrm{d}\mu \le \|f\|_{L^p} \|g\|_{L^q}$$

it suffices to consider the case where f and g are positive and $\int_\Omega g^q(x)\mathrm{d}\mu(x) = 1$. The measure

$$\nu(A) = \int_A g^q(x)\mathrm{d}\mu(x), \quad A \in \Sigma,$$

is a probability measure that acts according to the formula

$$\int_\Omega h\mathrm{d}\nu = \int_\Omega hg^q\mathrm{d}\mu.$$

Taking into account the formula (1.23) we infer that

$$\begin{aligned}\int_\Omega fg\mathrm{d}\mu &= \int_\Omega fg^{1-q}g^q\mathrm{d}\mu = \int_\Omega fg^{1-q}\mathrm{d}\nu \\ &\le \left(\int_\Omega f^p g^{p(1-q)}\mathrm{d}\nu\right)^{1/p} = \left(\int_\Omega f^p\mathrm{d}\mu\right)^{1/p},\end{aligned}$$

and the Rogers–Hölder inequality follows.

We next deal with Hardy's inequality.

1.8.1 Theorem (Hardy's inequality [207]) *Suppose that $f \in L^p(0,\infty)$, $f \geq 0$, where $p \in (1,\infty)$. Then the function*

$$F(x) = \frac{1}{x}\int_0^x f(t)\,\mathrm{d}t, \quad x > 0,$$

also belongs to $L^p(0,\infty)$ and

$$\|F\|_{L^p} \leq \frac{p}{p-1}\,\|f\|_{L^p}; \tag{1.24}$$

equality occurs if and only if $f = 0$ almost everywhere.

Hardy's inequality provides the norm of the averaging operator

$$H : L^p(0,\infty) \to L^p(0,\infty), \quad H(f)(x) = \frac{1}{x}\int_0^x f(t)\,\mathrm{d}t.$$

We have $\|H\| = p/(p-1)$, but the constant $p/(p-1)$, though the best possible, is unattained in inequality (1.24). The optimality can easily be checked by considering the sequence of functions $f_\varepsilon(t) = (t^{-1/p+\varepsilon})\chi_{(0,1)}(t)$, and letting $\varepsilon \to 0+$.

Hardy's inequality can be deduced from the following lemma:

1.8.2 Lemma (L.-E. Persson and N. Samko [396]) *Suppose that $0 < b \leq \infty$ and $-\infty \leq a < c \leq \infty$. If u is a positive convex function on (a,c), then*

$$\int_0^b u\Big(\frac{1}{x}\int_0^x h(t)\,\mathrm{d}t\Big)\frac{\mathrm{d}x}{x} \leq \int_0^b u(h(x))\Big(1 - \frac{x}{b}\Big)\frac{\mathrm{d}x}{x}$$

for all integrable functions $h\colon (0,b) \to (a,c)$. If u is instead concave, then the inequality holds in the reversed direction.

Proof. In fact, by Jensen's inequality,

$$\begin{aligned}
\int_0^b u\left(\frac{1}{x}\int_0^x h(t)\,\mathrm{d}t\right)\frac{\mathrm{d}x}{x} &\leq \int_0^b \left(\frac{1}{x}\int_0^x u(h(t))\,\mathrm{d}t\right)\frac{\mathrm{d}x}{x} \\
&= \int_0^b \frac{1}{x^2}\left(\int_0^b u(h(t))\chi_{[0,x]}(t)\,\mathrm{d}t\right)\mathrm{d}x \\
&= \int_0^b u(h(t))\left(\int_t^b \frac{1}{x^2}\,\mathrm{d}x\right)\mathrm{d}t \\
&= \int_0^b u(h(t))\left(1 - \frac{t}{b}\right)\frac{\mathrm{d}t}{t}
\end{aligned}$$

and the proof is complete. ∎

For $u(x) = |x|^p$, the result of Lemma 1.8.2 can be put in the form

$$\int_0^a \left|\frac{1}{x}\int_0^x f(t)\,\mathrm{d}t\right|^p \mathrm{d}x \leq \left(\frac{p}{p-1}\right)^p \int_0^a |f(x)|^p\left(1 - \left(\frac{x}{a}\right)^{(p-1)/p}\right)\mathrm{d}x, \tag{1.25}$$

where $a = b^{p/(p-1)}$ and $f(x) = h(x^{1-1/p})x^{-1/p}$. This yields an analogue of Hardy's inequality for functions $f \in L^p(0,a)$ (where $0 < a < \infty$), from which Hardy's inequality follows by letting $a \to \infty$.

It is worth noticing that the inequality (1.25) also holds for $p < 0$ if we assume that $f(x) > 0$ for $x \in (0,a)$.

The equality case in Theorem 1.8.1 implies that F and f are proportional, which makes f of the form Cx^r. Since $f \in L^p(0,\infty)$, this is possible only for $C = 0$.

There are known today more than ten different proofs of Theorem 1.8.1. Let us just mention the following one which is built on the well-known fact that $C_c(0,\infty)$ (the space of all continuous functions $f\colon (0,\infty) \to \mathbb{C}$ with compact support) is dense in $L^p(0,\infty)$. This allows us to restrict ourselves to the case where $f \in C_c(0,\infty)$. Then $(xF(x))' = f(x)$, which yields

$$\begin{aligned}\int_0^\infty F^p(t)\,\mathrm{d}t &= -p\int_0^\infty F^{p-1}(t)tF'(t)\,\mathrm{d}t\\ &= -p\int_0^\infty F^{p-1}(t)(f(t)-F(t))\,\mathrm{d}t,\end{aligned}$$

that is, $\int_0^\infty F^p(t)\,\mathrm{d}t = \frac{p}{p-1}\int_0^\infty F^{p-1}(t)f(t)\,\mathrm{d}t$. The proof ends by taking into account the integral form of the Rogers–Hölder inequality.

For a concrete application of Hardy's inequality, let us consider the *Schrödinger operator*

$$H_\lambda = -\frac{\mathrm{d}^2}{\mathrm{d}x^2} - \frac{\lambda}{x^2}$$

on $(0,\infty)$. The natural domain of this operator is the Sobolev space $H_0^1(0,\infty)$, the completion of $C_c(0,\infty)$ with respect to the norm

$$\|f\|_{H^1} = \left(\int_0^\infty \left(|f|^2 + |f'|^2\right)\mathrm{d}x\right)^{1/2}.$$

See Appendix E for a quick introduction to these spaces. The interested reader may found a detailed presentation of the theory of Sobolev spaces in the book of R. A. Adams and J. J. F. Fournier [6].

A natural question is to find out the range of parameter λ for which the operator H_λ is positive as an operator acting on $L^2(0,\infty)$, that is,

$$\langle H_\lambda\varphi,\varphi\rangle \geq 0 \quad \text{for all } \varphi \in C_c(0,\infty).$$

Since the last inequality is equivalent to

$$\int_0^\infty |\varphi'(x)|^2\,\mathrm{d}x \geq \lambda\int_0^\infty \left|\frac{\varphi(x)}{x}\right|^2\mathrm{d}x,$$

the answer is $\lambda \geq 1/4$ and follows from Hardy's inequality for $p = 2$.

Exercises

1. Infer from Theorem 1.8.1 the discrete form of Hardy's inequality:

$$\Big(\sum_{n=1}^{\infty}\Big(\frac{1}{n}\sum_{k=1}^{n}a_k\Big)^p\Big)^{1/p}<\frac{p}{p-1}\Big(\sum_{k=1}^{\infty}a_k^p\Big)^{1/p},$$

for every sequence $(a_n)_n$ of positive numbers (not all zero) and every $p\in(1,\infty)$.

2. (The Pólya–Knopp inequality; see [209], [255]) Prove the following limiting case of Hardy's inequality: for every $f\in L^1(0,\infty)$, $f\geq 0$ and f not identically zero,

$$\int_0^{\infty}\exp\Big(\frac{1}{x}\int_0^x\log f(t)\,\mathrm{d}t\Big)\,\mathrm{d}x<e\int_0^{\infty}f(x)\,\mathrm{d}x.$$

The discrete form of this inequality was previously noticed by T. Carleman [91]:

$$\sum_{n=1}^{\infty}(a_1a_2\cdots a_n)^{1/n}<e\sum_{n=1}^{\infty}a_n$$

for $a_1,a_2,a_3,\ldots\geq 0$, not all zero.

[*Hint*: Apply Lemma 1.8.2 for $h(x)=\log f(x)$.]

3. Formulate and prove the analogue of the Pólya–Knopp inequality for functions defined on bounded intervals of the form $(0,\alpha)$.

4. (L.-E. Persson and N. Samko [396]) Let $0<\ell\leq\infty$ and $p\in\mathbb{R}\backslash\{0\}$.

(a) Prove that the following power weighted form of Hardy's inequality (1.25),

$$\int_0^{\ell}\Big(\frac{1}{x}\int_0^x f(t)\,\mathrm{d}t\Big)^p x^{\alpha}\mathrm{d}x$$
$$\leq\Big(\frac{p}{p-1-\alpha}\Big)^p\int_0^{\ell}f^p(x)\Big(1-\Big(\frac{x}{\ell}\Big)^{(p-\alpha-1)/p}\Big)x^{\alpha}\mathrm{d}x,$$

holds for all positive and measurable functions $f:(0,\ell)\to\mathbb{R}$ and all α such that

$$p\geq 1 \text{ and } \alpha<p-1 \quad\text{or}\quad p<0 \text{ and } \alpha>p-1.$$

(b) Prove that the constant $\left(\frac{p}{p-1-\alpha}\right)^p$ is sharp in both cases.

[*Hint:* Use Lemma 1.8.2 and make the substitution

$$f(x)=h(x^{(p-\alpha-1)/p})x^{-(\alpha+1)/p}.\,]$$

5. Notice that the inequality in Lemma 1.8.2 holds in the reversed direction if u is concave rather than convex. Formulate and prove the variant of Exercise 4 in the case $0<p<1$.

1.9 Applications of Abel's Partial Summation Formula

A powerful device to prove inequalities for convex functions is to take advantage of some approximation results. From this point of view, the class of piecewise linear convex functions appears to be very important. Recall that a function $f\colon [a,b] \to \mathbb{R}$ is called *piecewise linear* if it is continuous and there exists a division $a = x_0 < \cdots < x_n = b$ such that the restriction of f to each partial interval $[x_k, x_{k+1}]$ is an affine function.

1.9.1 Lemma *Every continuous convex function $f : [a,b] \to \mathbb{R}$ is the uniform limit of a sequence of piecewise linear convex functions.*

In a similar manner, every continuous, increasing, and convex function $f : [a,b] \to \mathbb{R}$ is the uniform limit of a sequence of piecewise linear, increasing and convex functions.

Proof. Due to the uniform continuity of f, the sequence of piecewise linear functions f_n obtained by joining the points

$$(a, f(a)),\left(a+\frac{b-a}{n}, f(a+\frac{b-a}{n})\right),\ldots,\left(a+n\cdot\frac{b-a}{n}, f(a+n\cdot\frac{b-a}{n})\right)$$

by linear segments, is uniformly convergent to f. See [107], Theorem 6.8.7, p. 173. The fact that every piecewise linear function inscribed in a convex function is itself a convex function follows from Proposition 1.1.10 (h). ∎

The structure of piecewise linear convex functions is surprisingly simple.

1.9.2 Lemma *Let $f\colon [a,b] \to \mathbb{R}$ be a piecewise linear convex function. Then f is the sum of an affine function and a linear combination, with positive coefficients, of translates of the absolute value function. In other words, f is of the form*

$$f(x) = \alpha x + \beta + \sum_{k=1}^{n} c_k |x - x_k| \tag{1.26}$$

for suitable $\alpha, \beta \in \mathbb{R}$ and suitable positive coefficients $c_1, \ldots, c_n$.

Here the absolute value functions $|x - x_k|$ can be replaced by functions of the form $(x - x_k)^+$ or $(x - x_k)^-$.

The piecewise linear increasing and convex functions can be represented as linear combinations, with positive coefficients, of functions of the form $(x - x_k)^+$.

Proof. Let $a = x_0 < \cdots < x_n = b$ be a division of $[a,b]$ such that the restriction of f to each partial interval $[x_k, x_{k+1}]$ is affine. If $\alpha x + \beta$ is the affine function whose restriction to $[x_0, x_1]$ coincides with $f|_{[x_0,x_1]}$, then it will be a support line for f and $f(x) - (\alpha x + \beta)$ will be an increasing convex function which vanishes on $[x_0, x_1]$. A moment's reflection shows the existence of a constant $c_1 \geq 0$ such that $f(x) - (\alpha x + \beta) = c_1(x - x_1)^+$ on $[x_0, x_2]$. Repeating the argument we arrive at the representation

$$f(x) = \alpha x + \beta + \sum_{k=1}^{n-1} c_k(x - x_k)^+, \tag{1.27}$$

where all coefficients c_k are positive. The proof ends by replacing the translates of the positive part function by translates of the absolute value function. This is made possible by the formula $y^+ = (|y| + y)/2$. ∎

See Exercise 13, Section 1.1 for an alternative argument.

The representation formula (1.26) admits a generalization for all continuous convex functions on intervals. See L. Hörmander [227].

The above technique of approximation is instrumental in deriving many interesting results concerning the convex functions. A first example describes how convex functions relate the arithmetic means of the subfamilies of a given triplet of numbers.

1.9.3 Theorem (Popoviciu's inequality [408]) *If $f: I \to \mathbb{R}$ is a continuous function, then f is convex if and only if*

$$\frac{f(x) + f(y) + f(z)}{3} + f\Big(\frac{x + y + z}{3}\Big) \geq \frac{2}{3}\Big[f\Big(\frac{x + y}{2}\Big) + f\Big(\frac{y + z}{2}\Big) + f\Big(\frac{z + x}{2}\Big)\Big]$$

for all $x, y, z \in I$.

In the variant of strictly convex functions, the above inequality is strict except for $x = y = z$.

Proof. Necessity. According to Lemma 1.9.1 and Lemma 1.9.2, it suffices to consider the case where f is the absolute value function, that is, to show that

$$|x| + |y| + |z| + |x + y + z| \geq |x + y| + |y + z| + |z + x| \tag{1.28}$$

for all $x, y, z \in I$. Taking into account the polynomial identity,

$$x^2 + y^2 + z^2 + (x + y + z)^2 = (x + y)^2 + (y + z)^2 + (z + x)^2,$$

we have

$$\begin{aligned}\big(|x| + |y| + |z| + |x{+}y + z| - |x + y| - |y + z| - |z + x|\big)&\\ \times \big(|x| + |y| + |z| + |x + y + z|\big)&\\ = \big(|x|{+}|y| - |x + y|\big)\big(|z| + |x + y + z| - |x + y|\big)&\\ + \big(|y|{+}|z| - |y + z|\big)\big(|x| + |x + y + z| - |y + z|\big)&\\ + \big(|z| + |x| - |z + x|\big)\big(|y| + |x + y + z| - |z + x|\big) \geq 0,&\end{aligned}$$

and the necessity part is done.

Sufficiency. Popoviciu's inequality (when applied for $y = z$) yields the following substitute for the condition of midpoint convexity:

$$\frac{1}{4}f(x) + \frac{3}{4}f\Big(\frac{x + 2y}{3}\Big) \geq f\Big(\frac{x + y}{2}\Big) \quad \text{for all } x, y \in I. \tag{1.29}$$

The proof ends by following the argument in the proof of Theorem 1.1.8. ∎

Two other proofs of Theorem 1.9.3 may be found in the first edition of this book. See [367], pp. 12 and 33. The above statement of Popoviciu's inequality is only a simplified version of a considerably more general result (also due to T. Popoviciu). See the Comments at the end of this chapter.

Another result whose proof can be considerably simplified by the piecewise linear approximation of convex functions is the following generalization of Jensen's inequality, due to J. F. Steffensen [458]. Unlike Jensen's inequality, it allows the use of negative weights.

1.9.4 Theorem (The Jensen–Steffensen inequality) *Suppose that $x_1, ..., x_n$ is a monotonic family of points in an interval $[a, b]$ and $w_1, ..., w_n$ are real weights such that*

$$\sum_{k=1}^{n} w_k = 1 \quad \text{and} \quad 0 \leq \sum_{k=1}^{m} w_k \leq \sum_{k=1}^{n} w_k \quad \text{for every } m \in \{1, ..., n\}. \tag{1.30}$$

Then every convex function f defined on $[a, b]$ verifies the inequality

$$f\left(\sum_{k=1}^{n} w_k x_k\right) \leq \sum_{k=1}^{n} w_k f(x_k).$$

Our argument for Theorem 1.9.4 combines the piecewise linear approximation of convex functions with a consequence of *Abel's partial summation formula* (also known as *Abel's transformation*).

1.9.5 Theorem (Abel's partial summation formula) *If $(a_k)_{k=1}^n$ and $(b_k)_{k=1}^n$ are two families of complex numbers, then*

$$\sum_{k=1}^{n} a_k b_k = \sum_{k=1}^{n-1} \left[(a_k - a_{k+1}) \left(\sum_{j=1}^{k} b_j \right) \right] + a_n \left(\sum_{j=1}^{n} b_j \right).$$

Abel [1] used his formula to derive a number of important results known today as the Abel criterion of convergence of signed series, the Abel theorem on power series, and the Abel summation method. In what follows we will be interested in another consequence of it:

1.9.6 Corollary (*The Abel–Steffensen inequality* [458]) *If $x_1, x_2, ..., x_n$ and $y_1, y_2, ..., y_n$ are two families of real numbers that verify one of the following two conditions*

(*a*) $x_1 \geq x_2 \geq \cdots \geq x_n \geq 0$ *and* $\sum_{k=1}^{j} y_k \geq 0$ *for all* $j \in \{1, 2, ..., n\}$,

(*b*) $0 \leq x_1 \leq x_2 \leq \cdots \leq x_n$ *and* $\sum_{k=j}^{n} y_k \geq 0$ *for all* $j \in \{1, 2, ..., n\}$, *then*

$$\sum_{k=1}^{n} x_k y_k \geq 0.$$

Therefore, if $x_1, x_2, ..., x_n$ *is a monotonic family and* $y_1, y_2, ..., y_n$ *is a family of real numbers such that*

$$0 \leq \sum_{k=1}^{j} y_k \leq \sum_{k=1}^{n} y_k,$$

for j =1, ... , n, then we have

$$\left(\min_{1 \leq k \leq n} x_k\right) \sum_{k=1}^{n} y_k \leq \sum_{k=1}^{n} x_k y_k \leq \left(\max_{1 \leq k \leq n} x_k\right) \sum_{k=1}^{n} y_k.$$

Proof of Jensen–Steffensen inequality. Taking into account Lemma 1.9.1 and Lemma 1.9.2, we may reduce ourselves to the case of absolute value function. This case can be settled as follows. Assuming the ordering $x_1 \leq \cdots \leq x_n$ (to make a choice), we infer that

$$0 \leq x_1^+ \leq \cdots \leq x_n^+ \text{ and } x_1^- \geq \cdots \geq x_n^- \geq 0.$$

According to Corollary 1.9.6,

$$\sum_{k=1}^{n} w_k x_k^+ \geq 0 \text{ and } \sum_{k=1}^{n} w_k x_k^- \geq 0,$$

which yields

$$\left| \sum_{k=1}^{n} w_k x_k \right| \leq \sum_{k=1}^{n} w_k |x_k|.$$

The proof is done. ■

The integral version of Jensen–Steffensen inequality can be established in the same manner, using integration by parts instead of Abel's partial summation formula.

1.9.7 Theorem (The integral version of Jensen–Steffensen inequality) *Suppose that* $g : [a,b] \to \mathbb{R}$ *is a monotone function and* $w : [a,b] \to \mathbb{R}$ *is an integrable function such that*

$$0 \leq \int_a^x w(t)\mathrm{d}t \leq \int_a^b w(t)\mathrm{d}t = 1 \quad \text{for every } x \in [a,b]. \tag{1.31}$$

Then every convex function f *defined on an interval* I *that includes the range of* g *verifies the inequality*

$$f\left(\int_a^b g(t)w(t)\mathrm{d}t\right) \leq \int_a^b f(g(t))\, w(t)\mathrm{d}t.$$

Another application of Abel's partial summation formula is as follows.

1.9.8 Theorem (The discrete form of Hardy–Littlewood–Pólya rearrangement inequality [209]) *Let $x_1, \ldots, x_n, y_1, \ldots, y_n$ be real numbers. Prove that*

$$\sum_{k=1}^{n} x_k^{\downarrow} y_{n-k+1}^{\downarrow} \leq \sum_{k=1}^{n} x_k y_k \leq \sum_{k=1}^{n} x_k^{\downarrow} y_k^{\downarrow};$$

here $z_1^{\downarrow} \geq \cdots \geq z_n^{\downarrow}$ denotes the decreasing rearrangement of a family $z_1, \ldots, z_n$ of real numbers.

Proof. Notice first that we may assume that $x_1 \geq \cdots \geq x_n \geq 0$. Then, apply Abel's partial summation formula. ■

More on rearrangements will be found in Chapter 5.

Exercises

1. (Hlawka's inequality) The inequality (1.28) that appears in the proof of Popoviciu's inequality can be extended to all inner product spaces H. For this, prove first Hlawka's identity,

$$\|x\|^2 + \|y\|^2 + \|z\|^2 + \|x+y+z\|^2 = \|x+y\|^2 + \|y+z\|^2 + \|z+x\|^2,$$

and infer from it the inequality

$$\|x+y+z\| + \|x\| + \|y\| + \|z\| - \|x+y\| - \|y+z\| - \|z+x\| \geq 0.$$

2. (An illustration of Popoviciu's inequality) Suppose that x_1, x_2, x_3 are positive numbers, not all equal. Prove that:

   (*a*) $27 \prod_{i<j} (x_i + x_j)^2 > 64 x_1 x_2 x_3 (x_1 + x_2 + x_3)^3$;

   (*b*) $x_1^6 + x_2^6 + x_3^6 + 3x_1^2 x_2^2 x_3^2 > 2(x_1^3 x_2^3 + x_2^3 x_3^3 + x_3^3 x_1^3)$.

3. Let $f\colon [0, 2\pi] \to \mathbb{R}$ be a convex function. Prove that

$$a_n = \frac{1}{\pi} \int_0^{2\pi} f(t) \cos nt \, \mathrm{d}t \geq 0 \quad \text{for every } n \geq 1.$$

4. Infer the *AM–GM* inequality from Theorem 1.9.8 (the discrete form of Hardy–Littlewood–Pólya rearrangement inequality).

1.10 The Hermite–Hadamard Inequality

As stated in Proposition 1.1.3, every convex function f on an interval $[a, b]$ can be modified at the endpoints to become convex and continuous. An immediate consequence of this fact is the (Riemann) integrability of f. The arithmetic mean of f can be estimated (both from below and from above) by the *Hermite–Hadamard double inequality*,

$$f\left(\frac{a+b}{2}\right) \leq \frac{1}{b-a}\int_a^b f(x)\,dx \leq \frac{f(a)+f(b)}{2}. \tag{1.32}$$

The left-hand side inequality (denoted (LHH)) is nothing but a special case of Jensen's inequality. See Corollary 1.7.4. A simpler argument (taking into consideration the existence of supporting lines at interior points) consists in choosing a supporting line $y(x) = f\left(\frac{a+b}{2}\right) + \lambda\left(x - \frac{a+b}{2}\right)$ of f at $\frac{a+b}{2}$ and then by integrating the inequality $f(x) \geq y(x)$ over $[a,b]$. This also shows that (LHH) becomes an equality if and only if f is affine on (a,b).

The right-hand side inequality (denoted (RHH)) follows by integrating the inequality (1.2) in Section 1.1 (which says that the graph is under the chord joining the endpoints). As above, the equality case occurs only when f coincides with an affine function on (a,b).

It is worth noticing that each of the two sides of (1.32) actually characterizes convex functions. More precisely, *if I is an interval and $f\colon I \to \mathbb{R}$ is a function continuous on* $\operatorname{int} I$, *whose restriction to each compact subinterval* $[a,b]$ *verifies* (LHH), *then f is convex.* The same works when (LHH) is replaced by (RHH). See Exercises 1 and 2.

Some examples illustrating the usefulness of Hermite–Hadamard inequality are presented below.

1.10.1 Example (Ch. Hermite [213]) *For $f(x) = 1/(1+x)$, $x \geq 0$, the inequality* (1.32) *becomes*

$$x - x^2/(2+x) < \log(1+x) < x - x^2/(2+2x).$$

Particularly,

$$\frac{1}{n+1/2} < \log(n+1) - \log n < \frac{1}{2}\left(\frac{1}{n} + \frac{1}{n+1}\right) \tag{1.33}$$

for all $n \in \mathbb{N}^*$, and this fact is instrumental in deriving Stirling's formula,

$$n! \sim \sqrt{2\pi}\cdot n^{n+1/2}\mathrm{e}^{-n}.$$

1.10.2 Example *For $f = \exp$, the inequality* (1.32) *yields*

$$\mathrm{e}^{(a+b)/2} < \frac{\mathrm{e}^b - \mathrm{e}^a}{b-a} < \frac{\mathrm{e}^a + \mathrm{e}^b}{2} \quad \text{for } a \neq b \text{ in } \mathbb{R},$$

that is,

$$\sqrt{xy} < L(x,y) = \frac{x-y}{\log x - \log y} < \frac{x+y}{2} \quad \text{for } x \neq y \text{ in } (0,\infty), \tag{1.34}$$

which represents the geometric–logarithmic–arithmetic mean inequality. For $f = \log$, we obtain a similar inequality, where the role of the logarithmic mean is taken by the identric mean. See Appendix A, Section A1, for more on these means.

1.10.3 Example *For $f(x) = \sin x$, $x \in [0, \pi]$, we obtain*

$$\frac{\sin a + \sin b}{2} < \frac{\cos a - \cos b}{b - a} < \sin\left(\frac{a+b}{2}\right) \quad \text{for } a \neq b \text{ in } \mathbb{R},$$

and this implies the well-known inequalities $\tan x > x > \sin x$ *(for x in $(0, \pi/2)$).*

The following result yields an estimate of the precision in the Hermite–Hadamard inequality:

1.10.4 Lemma *Let $f\colon [a,b] \to \mathbb{R}$ be a twice differentiable function for which there exist real constants m and M such that*

$$m \leq f'' \leq M.$$

Then

$$m \cdot \frac{(b-a)^2}{24} \leq \frac{1}{b-a}\int_a^b f(x)\,\mathrm{d}x - f\left(\frac{a+b}{2}\right) \leq M \cdot \frac{(b-a)^2}{24},$$

and

$$m \cdot \frac{(b-a)^2}{12} \leq \frac{f(a)+f(b)}{2} - \frac{1}{b-a}\int_a^b f(x)\,\mathrm{d}x \leq M \cdot \frac{(b-a)^2}{12}.$$

In fact, the functions $f - mx^2/2$ and $Mx^2/2 - f$ are convex and thus we can apply to them the Hermite–Hadamard inequality.

For other estimates of the Hermite–Hadamard inequality see the Comments at the end of this chapter.

1.10.5 Remark (An improvement of the Hermite–Hadamard inequality) *Suppose that $f\colon [a,b] \to \mathbb{R}$ is a convex function. By applying the Hermite–Hadamard inequality on each of the intervals $[a, (a+b)/2]$ and $[(a+b)/2, b]$ we get*

$$f\left(\frac{3a+b}{4}\right) \leq \frac{2}{b-a}\int_a^{(a+b)/2} f(x)\,\mathrm{d}x \leq \frac{1}{2}\left(f(a) + f\left(\frac{a+b}{2}\right)\right)$$

and

$$f\left(\frac{a+3b}{4}\right) \leq \frac{2}{b-a}\int_{(a+b)/2}^{b} f(x)\,\mathrm{d}x \leq \frac{1}{2}\left(f\left(\frac{a+b}{2}\right) + f(b)\right).$$

Summing up (side by side), we obtain the following refinement of (1.32) *due to P. C. Hammer* [197]*:*

$$\begin{aligned}
f\left(\frac{a+b}{2}\right) &\leq \frac{1}{2}\left(f\left(\frac{3a+b}{4}\right) + f\left(\frac{a+3b}{4}\right)\right) \\
&\leq \frac{1}{b-a}\int_a^b f(x)\,\mathrm{d}x \leq \frac{1}{2}\left[f\left(\frac{a+b}{2}\right) + \frac{f(a)+f(b)}{2}\right] \\
&\leq \frac{1}{2}(f(a)+f(b)).
\end{aligned}$$

By continuing the division process, the arithmetic mean of f can be approximated as close as desired by convex combinations of values of f at suitable dyadic points of $[a, b]$.

1.10.6 Remark (The case of log-convex functions) *If $f\colon [a, b] \to (0, \infty)$ is a log-convex function, then the Hermite–Hadamard inequality yields*

$$f\Big(\frac{a+b}{2}\Big) \leq \exp\left(\frac{1}{b-a}\int_a^b \log f(x)\,\mathrm{d}x\right) \leq \sqrt{f(a)f(b)}.$$

The middle term is the geometric mean of f, so that it can be bounded above by the arithmetic mean of f. As was noticed by P. M. Gill, C. E. M. Pearce, and J. E. Pečarić [181], *the arithmetic mean of f is bounded from above by the logarithmic mean of $f(a)$ and $f(b)$:*

$$\frac{1}{b-a}\int_a^b f(x)\,\mathrm{d}x = \int_0^1 f((1-t)\,a+tb)\,\mathrm{d}t \leq \int_0^1 f(a)^{1-t} f(b)^t\;\mathrm{d}t = L(f(a), f(b)).$$

The Hermite–Hadamard inequality is an archetype for many interesting results in higher dimensions. Indeed, a careful look at the formula (1.32) shows that by considering the compact interval $K = [a, b]$ endowed with the probability measure $\mu = \frac{1}{b-a}dx$, the integral of any convex function f over K lies between the value of f at the barycenter of μ and the integral of f with respect to the probability measure $\nu = \frac{1}{2}\delta_a + \frac{1}{2}\delta_b$, supported by the boundary $\partial K = \{a, b\}$ of K. This fact can be extended to all Borel probability measures defined on compact convex subsets of locally convex Hausdorff spaces. The details constitute the core of Choquet's theory and will be presented in Chapter 7. At the same time, the Hermite–Hadamard inequality draws attention to an order relation $\prec$ on the Borel probability measures on K. This order relation, known as the *relation of majorization,* is associated to the cone $\mathrm{Conv}(K)$, of continuous convex functions defined on K, via the formula

$$\lambda_1 \prec \lambda_2 \text{ if and only if } \lambda_1(f) \leq \lambda_2(f) \text{ for all } f \in \mathrm{Conv}(K).$$

According to Remark 1.10.5,

$$\begin{aligned}\delta_{(a+b)/2} \prec \frac{1}{2}\left(\delta_{(3a+b)/4} + \delta_{(a+3b)/4}\right) &\prec \frac{1}{b-a}\mathrm{d}x \\ &\prec \frac{1}{2\,(b-a)}\mathrm{d}x + \frac{1}{4}\delta_a + \frac{1}{4}\delta_b \prec \frac{1}{2}\delta_a + \frac{1}{2}\delta_b.\end{aligned}$$

The theory of majorization was initiated by G. H. Hardy, J. E. Littlewood, and G. Pólya [208], [209] (and independently, by J. Karamata [241]). Its main features will be presented in Section 4.1.

Exercises

1. Infer from Theorem 1.1.6 that a (necessary and) sufficient condition for a continuous function f to be convex on an open interval I is that

$$f(x) \leq \frac{1}{2h} \int_{x-h}^{x+h} f(t)\, \mathrm{d}t$$

for all x and h with $[x-h, x+h] \subset I$.

2. Let f be a real-valued continuous function defined on an open interval I. Prove that f is convex if all restrictions $f|_{[a,b]}$ (for $[a,b] \subset I$) verify the right-hand side inequality in (1.32).

3. (An improvement of the left-hand side of formula (1.32)) Let f be a convex function defined on the interval $[a,b]$. Use the existence of supporting lines to show that

$$\frac{1}{2}\left(f\left(\frac{a+b}{2} - c\right) + f\left(\frac{a+b}{2} + c\right)\right) \leq \frac{1}{b-a} \int_a^b f(x)\, \mathrm{d}x$$

for all $c \in [0, (b-a)/4]$, and that $c = (b-a)/4$ is maximal within the class of convex functions on $[a,b]$.

4. (Identities that yield the Hermite–Hadamard inequality). Suppose that $f \in C^2([a,b], \mathbb{R})$. Prove the following two identities related to trapezoidal and midpoint rules of quadrature:

$$\frac{1}{b-a} \int_a^b f(x)\, \mathrm{d}x = \frac{1}{2}\left[f(a) + f(b)\right] - \frac{1}{b-a} \int_a^b \frac{(b-x)(x-a)}{2} f''(x)\, \mathrm{d}x$$

and

$$\frac{1}{b-a} \int_a^b f(x) \mathrm{d}x = f\left(\frac{a+b}{2}\right) + \frac{1}{b-a} \int_a^b \varphi(x) f''(x) \mathrm{d}x,$$

where

$$\varphi(x) = \begin{cases} \frac{(x-a)^2}{2} & \text{if } x \in [a, (a+b)/2] \\ \frac{(b-x)^2}{2} & \text{if } x \in [(a+b)/2, b]. \end{cases}$$

Then, use an approximation argument to cover the general case of the Hermite–Hadamard inequality.

5. (Bullen–Simpson inequality [82]) Prove that every function $f \in C^4([a,b])$ with $f^{(4)}(x) \geq 0$ for all x verifies the double inequality

$$\begin{aligned} 0 \leq \frac{1}{b-a} \int_a^h f(x)\, \mathrm{d}x - \frac{1}{3}\left[2f\left(\frac{3a+b}{4}\right) - f\left(\frac{a+b}{2}\right) + 2f\left(\frac{a+3b}{4}\right)\right] \\ \leq \frac{1}{6}\left[f(a) + 4f\left(\frac{a+b}{2}\right) + f(b)\right] - \frac{1}{b-a} \int_a^b f(x)\, \mathrm{d}x. \end{aligned}$$

[*Hint:* Use Simpson's quadrature formulas

$$\int_0^1 f(x)\mathrm{d}x = \frac{1}{6}\left[f(0) + 4f\left(\frac{1}{2}\right) + f(1)\right] - \frac{1}{2880}f^{(4)}(\xi)$$

and

$$\int_0^1 f(x)\mathrm{d}x = \frac{1}{3}\left[2f\left(\frac{1}{4}\right) - f\left(\frac{1}{2}\right) + 2f\left(\frac{3}{4}\right)\right] + \frac{7}{23040}f^{(4)}(\eta).]$$

6. (The Hermite–Hadamard inequality for arbitrary Borel probability measures; A. M. Fink [164]) Suppose that $f\colon [a,b] \to \mathbb{R}$ is a convex function and μ is Borel probability measure on $[a,b]$. Prove that

$$f(\mathrm{bar}(\mu)) \leq \int_a^b f(x)\,\mathrm{d}\mu(x) \leq \frac{b - \mathrm{bar}(\mu)}{b-a}\cdot f(a) + \frac{\mathrm{bar}(\mu) - a}{b-a}\cdot f(b),$$

where $\mathrm{bar}(\mu) = \int_a^b x\,\mathrm{d}\mu(x)$ represents the barycenter of μ.

[*Hint*: The left-hand side follows from Corollary 1.7.4. For the right-hand side, proceed as in the classical case of Lebesgue measure.]

Remark. An immediate consequence of Exercise 6 is the following result due to L. Fejér: Suppose that $f\colon [a,b] \to \mathbb{R}$ is a convex function and $w : [a,b] \to [0,\infty)$ is an integrable function such that $W = \int_a^b w(x)\mathrm{d}x > 0$ and $w(\frac{a+b}{2} - x) = w(\frac{a+b}{2} + x)$ for all $x \in [0,(a+b)/2]$. Then

$$f\Big(\frac{a+b}{2}\Big) \leq \frac{1}{W}\int_a^b f(x)\,w(x)\mathrm{d}x \leq \frac{f(a)+f(b)}{2}.$$

For more general results see Section 7.5 and the paper by A. Florea and C. P. Niculescu [167].

1.11 Comments

The recognition of convex functions as a class of functions to be studied in its own right generally can be traced back to J. L. W. V. Jensen [231]. However, he was not the first one to deal with convex functions. The discrete form of Jensen's inequality was first proved by O. Hölder [225] in 1889, under the stronger hypothesis that the second derivative is positive. Moreover, O. Stolz [463] proved in 1893 that every midpoint convex continuous function $f\colon [a,b] \to \mathbb{R}$ has left and right derivatives at each point of (a,b). While the usual convex functions are continuous at all interior points (a fact due to J. L. W. V. Jensen [231]), the midpoint convex functions may be discontinuous everywhere. In fact, regard $\mathbb{R}$ as a vector space over $\mathbb{Q}$ and choose (via the axiom of choice) a basis $(b_i)_{i\in I}$ of $\mathbb{R}$ over $\mathbb{Q}$, that is, a maximal linearly independent set. Then every element x of $\mathbb{R}$ has a unique representation $x = \sum_{i\in I} c_i(x)b_i$ with coefficients

$c_i(x)$ in $\mathbb{Q}$ and $c_i(x) = 0$ except for finitely many indices i. The uniqueness of this representation gives rise, for each $i \in I$, to a surjective coordinate projection $\mathrm{pr}_i : x \to c_i(x)$, from $\mathbb{R}$ to $\mathbb{Q}$. As G. Hamel [196] observed in 1905, the functions pr_i are discontinuous everywhere and

$$\mathrm{pr}_i(\alpha x + \beta y) = \alpha\, \mathrm{pr}_i(x) + \beta\, \mathrm{pr}_i(y),$$

for all $x, y \in \mathbb{R}$ and all $\alpha, \beta \in \mathbb{Q}$. H. Blumberg [59] and W. Sierpiński [445] have noted independently that if $f\colon (a, b) \to \mathbb{R}$ is measurable and midpoint convex, then f is also continuous (and thus convex). See [420, pp. 220–221] for related results. The complete understanding of midpoint convexity is due to G. Rodé [423], who proved that a real-valued function is midpoint convex if and only if it is the pointwise supremum of a family of functions of the form $a(x)+c$, where $a(x)$ is additive and c is a real constant.

A nice graphical proof of Jensen's inequality can be found at the blog of M. Reid (post of November 17, 2008).

The classical *AM–GM* inequality can be traced back to C. Maclaurin [292] who obtained it as part of a chain of inequalities concerning the elementary symmetric functions. See Appendix C. The famous *Cours d'Analyse* of A.-L. Cauchy includes a proof of this inequality by forward–backward induction. See [99], Théorème 17, page 457. Later, in 1891, A. Hurwitz made the striking remark that the *AM–GM* inequality follows from an identity. Indeed, he proved that

$$\frac{x_1^{2n} + x_2^{2n} + \cdots + x_n^{2n}}{n} - x_1^2 x_2^2 \cdots x_n^2$$

can be expressed as a sum of squares of real polynomials in the real variables $x_1, x_2, ..., x_n$. For example,

$$\frac{x_1^4 + x_2^4 + x_3^4 + x_4^4}{4} - x_1 x_2 x_3 x_4 = \frac{\left(x_1^2 - x_2^2\right)^2 + (x_3^2 - x_4^2)^2}{4} + \frac{(x_1 x_2 - x_3 x_4)^2}{2}.$$

Can every real multivariate polynomial that takes only positive values over the reals be represented as a sum of squares of real polynomials? The answer is negative but the first concrete example was produced only in 1966 by T. S. Motzkin [340], who noticed that the polynomial $1 + x^4y^2 + x^2y^4 - 3x^2y^2$ (positive, according to the *AM–GM* inequality) can be only represented as a sum of squares of rational functions:

$$\begin{aligned} 1 + x^4y^2 &+ x^2y^4 - 3x^2y^2 \\ &= \left(\frac{x^2y(x^2+y^2-2)}{x^2+y^2}\right)^2 + \left(\frac{xy^2(x^2+y^2-2)}{x^2+y^2}\right)^2 \\ &\quad + \left(\frac{xy(x^2+y^2-2)}{x^2+y^2}\right)^2 + \left(\frac{x^2-y^2}{x^2+y^2}\right)^2. \end{aligned}$$

See, for details, J. M. Steele [456], p. 45.

In 1900, D. Hilbert gave an address in Paris to the second International Congress of Mathematicians, describing a list of problems which he believed to be worth the attention of the world's mathematicians. Hilbert's 17th Problem asked whether every real multivariate polynomial that takes only positive values over the reals can be represented as a sum of squares of real rational functions. This was answered affirmatively in 1927 by E. Artin.

The relationship between inequalities and identities is much more subtle than saying that $a \leq b$ is always the consequence of $a + c = b$ for some $c \geq 0$. Indeed, the problem of finding c in a reasonable class of objects related to a and b and to ascribe to c a meaningful interpretation needs more attention. On the other hand, we use inequalities (even when identities are available) because they extract, in a certain context, the useful part of the identities lying behind them. See, for example, the case of Cauchy–Bunyakovsky–Schwarz inequality.

The restriction to strict monotonicity and continuity in Young's inequality is not necessary if we use the concept of generalized inverse. Suppose that $f : [0, \infty) \longrightarrow [0, \infty)$ is an increasing function such that $f(0) = 0$ and $\lim_{x\to\infty} f(x) = \infty$. The *generalized inverse* of f is the function

$$f^{-1} : [0, \infty) \longrightarrow [0, \infty), \quad f^{-1}(y) = \inf\{x \geq 0 : f(x) \geq y\}.$$

Clearly, f^{-1} is increasing, left continuous, and admits a limit from the right at $y > 0$. Moreover $f^{-1}(f(x)) \leq x$ for all x (with equality if f is strictly increasing). A thorough presentation of the theory of generalized inverses may be found in P. Embrechts and M. Hofert [147]. The paper by F.-C. Mitroi and C. P. Niculescu [332] includes several generalizations of Young's inequality in the context of generalized inverses.

The Rogers–Hölder inequality (known to most mathematicians as the Hölder inequality) was proved in 1888 by L. J. Rogers [424] in a slightly different, but equivalent form. The basic ingredient was his weighted form of the *AM–GM* inequality (as stated in Theorem 1.1.6). One year later, O. Hölder [225] clearly wrote that he, after Rogers, proved the inequality

$$\left(\sum_{k=1}^{n} a_k b_k\right)^t \leq \left(\sum_{k=1}^{n} a_k\right)^{t-1} \left(\sum_{k=1}^{n} a_k b_k^t\right),$$

valid for all $t > 1$, and all $a_k > 0$, $b_k > 0$, $k = 1, \ldots, n$, $n \in \mathbb{N}^*$. His idea was to apply Jensen's inequality to the function $f(x) = x^t$, $x > 0$. However, F. Riesz was the first who stated and used the Rogers–Hölder inequality as we did in Section 1.2. See the paper of L. Maligranda [299] for the complete history.

The inequality of J. Bernoulli, $(1+x)^n \geq 1 + nx$, for all $x \geq -1$ and n in $\mathbb{N}$, appeared in [44]. The generalized form (see Exercise 1, Section 1.2) is due to O. Stolz and J. A. Gmeiner; see L. Maligranda [299]. L. Maligranda also noticed that the classical Bernoulli inequality, the classical *AM–GM* inequality, and the generalized *AM–GM* inequality of Rogers are all equivalent (that is, each one can be used to prove the other ones).

The proof of the integral form of Jensen's inequality can be done using only techniques of measure theory. See L. Ambrosio, G. Da Prato, and A. Mennucci

[15], pp. 55–56. The upper estimate for the discrete form of Jensen's inequality given in Theorem 1.4.4 and the covariance form of Jensen's inequality (see Theorem 1.7.11 above) follow the paper of C. P. Niculescu [354].

A refinement of Jensen's inequality for "more convex" functions was proved by S. Abramovich, G. Jameson, and G. Sinnamon [2]. Call a function $\varphi\colon [0,\infty) \to \mathbb{R}$ *superquadratic* provided that for each $x \geq 0$ there exists a constant $C_x \in \mathbb{R}$ such that $\varphi(y) - \varphi(x) - \varphi(|y-x|) \geq C_x(y-x)$ for all $y \geq 0$. For example, if $\varphi\colon [0,\infty) \to \mathbb{R}$ is continuously differentiable, $\varphi(0) \leq 0$ and either $-\varphi'$ is subadditive or $\varphi'(x)/x$ is increasing, then φ is superquadratic. Particularly, this is the case when $f(x) = x^p$ for $p \geq 2$, or $f(x) = x^2 \log x$. Moreover every superquadratic positive function is convex. Their main result asserts that the inequality

$$\varphi\Big(\int_X f(y)\,\mathrm{d}\mu(y)\Big) \leq \int_X \Big[\varphi(f(x)) - \varphi\Big(\Big|f(x) - \int_X f(y)\,\mathrm{d}\mu(y)\Big|\Big)\Big]\,\mathrm{d}\mu(x)$$

holds for all probability spaces (X, Ω, μ) and all positive μ-measurable functions f if and only if φ is superquadratic.

The prehistory of Hardy's inequality (until G. H. Hardy finally proved his inequality in 1925) is described in details by A. Kufner, L. Maligranda, and L.-E. Persson in [263]. The further development is presented in [264], [265], and [383].

We can arrive at Hardy's inequality via mixed means. For a positive n-tuple $a = (a_1, \ldots, a_n)$, the *mixed arithmetic–geometric inequality* asserts that the arithmetic mean of the numbers

$$a_1, \sqrt{a_1 a_2}, \ldots, \sqrt[n]{a_1 a_2 \cdots a_n}$$

does not exceed the geometric mean of the numbers

$$a_1, \frac{a_1 + a_2}{2}, \ldots, \frac{a_1 + a_2 + \cdots + a_n}{n}$$

(see K. Kedlaya [246]). As noted by B. Mond and J. E. Pečarić [334], the arithmetic and the geometric means can be replaced (in this order) by any pair (M_r, M_s) of power means with $r > s$. For $r = p > 1$ and $s = 1$ this gives us

$$\Big[\frac{1}{n}\sum_{k=1}^{n}\Big(\frac{a_1 + a_2 + \cdots + a_k}{k}\Big)^p\Big]^{1/p} \leq \frac{1}{n}\sum_{k-1}^{n}\Big(\frac{1}{k}\sum_{j=1}^{k} a_j^p\Big)^{1/p}$$

so that $\sum_{k=1}^{n}((a_1 + a_2 + \cdots + a_k)/k)^p$ is less than or equal to

$$n^{1-p}\Big(\sum_{j=1}^{n} a_j^p\Big)\Big[\sum_{k=1}^{n}\Big(\frac{1}{k}\Big)^{1/p}\Big]^p < \Big(\frac{p}{p-1}\Big)^p\Big(\sum_{j=1}^{n} a_j^p\Big),$$

as $\int_0^n x^{-1/p}\,\mathrm{d}x = \frac{p}{p-1}n^{1-1/p}$. The integral case of this result is discussed by A. Čižmešija and J. E. Pečarić [108]. The limit case (known as Carleman's inequality) as well as its ramifications has also received a great deal of attention in

recent years. The reader may consult the papers by J. E. Pečarić and K. B. Stolarsky [390], J. Duncan and C. M. McGregor [137], M. Johansson, L.-E. Persson and
A. Wedestig [234], S. Kaijser, L.-E. Persson and A. Öberg [240], and A. Čižmešija, J. E. Pečarić and L.-E. Persson [109].

Some interesting connections between the weighted inequalities of Hardy type and the spectral problems are mentioned by A. Kufner in his paper [262]. The following variant of Hardy's inequality is a crucial ingredient in the analysis of boundary behavior of self-adjoint, second-order, elliptic differential operators: *If* $f : [a,b] \to \mathbb{C}$ *is a continuously differentiable function with* $f(a) = f(b) = 0$, *then*

$$\int_a^b \frac{|f(x)^2|}{4d(x)^2} \mathrm{d}x \leq \int_a^b |f'(x)|^2 \, \mathrm{d}x$$

where $d(x) = \min\{|x-a|, |x-b|\}$. See E. B. Davies [126], Lemma 1.5.1, p. 25. More applications of Hardy's inequality can be found in the books of A. Kufner, L.-E. Persson and N. Samko [265] and B. Opic and A. Kufner [383].

An account of the history, variations, and generalizations of the Chebyshev inequality can be found in the paper by D. S. Mitrinović and P. M. Vasić [331]. Other complements to Jensen's and Chebyshev's inequalities can be found in the papers by H. Heinig and L. Maligranda [210] and S. M. Malamud [294].

It is worth mentioning here the existence of a nonlinear framework for Jensen's inequality, initiated by B. Jessen [232] in 1931. Let $\mathcal{L}$ be a convex cone of real-valued functions defined on an abstract set S (that is, $\mathcal{L}$ is closed under addition and multiplication by positive scalars). We assume that $\mathcal{L}$ contains the constant function 1. A *normalized isotone sublinear functional* on $\mathcal{L}$ is any functional $A : \mathcal{L} \to \mathbb{R}$ such that

$$\begin{gathered} A(f+g) \leq A(f) + A(g) \text{ for every } f, g \in \mathcal{L} \\ A(\lambda f) = \lambda A(f) \text{ for every } f \in \mathcal{L} \text{ and } \lambda \geq 0 \\ f \leq g \text{ in } \mathcal{L} \text{ implies } A(f) \leq A(g) \\ A(\pm 1) = \pm 1. \end{gathered}$$

Some simple examples are as follows:

(a) $\mathcal{L} = c$, the space of convergent real sequences, and

$$A((a_n)_n) = \frac{1}{2}\left(a_0 + \lim_{n\to\infty} a_n\right);$$

(b) $\mathcal{L} = C([a,b])$ and $A(f) = \frac{1}{b-a}\int_a^b f(t)\mathrm{d}t$ (the case that appears in Hermite–Hadamard inequality);

(c) $\mathcal{L} = C([a,b])$ and $A(f) = \sup_{x\in(a,b]} \frac{1}{x-a}\int_a^x f(t)\mathrm{d}t$;

(d) $\mathcal{L} = \mathcal{BUP}(\mathbb{R})$, the space of Bohr almost periodic functions defined on $\mathbb{R}$ (that is, the closure of the trigonometric polynomials with respect to the sup norm) and $A(f) = \lim_{T\to\infty} \frac{1}{T}\int_0^T f(t)\mathrm{d}t$.

In particular, B. Jessen [232] proved the following generalization of Jensen's inequality: *If* $A : \mathcal{L} \to \mathbb{R}$ *is a normalized isotone sublinear functional and* $\varphi : [a,b] \to \mathbb{R}$ *is a continuous convex function then, for every* $f \in \mathcal{L}$ *with* $\varphi \circ f \in \mathcal{L}$, *we have that* $A(f) \in [a,b]$ *and* $\varphi(A(f)) \leq A(\varphi \circ f)$.

This fact was complemented by S. S. Dragomir, C. E. M. Pearce, and J. E. Pečarić [135], who proved the following generalization of Hermite–Hadamard inequality for isotone sublinear functionals:

$$A(\varphi \circ f) \leq \frac{b\varphi(a) - a\varphi(b)}{b-a} + \frac{|\varphi(b) - \varphi(a)|}{b-a} A(\sigma f),$$

where $\sigma = \operatorname{sgn}(\varphi(b) - \varphi(a))$. See Exercise 7, Section 5.2, for more about the Hermite–Hadamard inequality.

The approximation of convex functions by piecewise linear functions and the simple structure of these approximants was first noticed by Hardy, Littlewood, and Pólya [208], [209], in connection with their inequality of majorization. The proof presented here is due to T. Popoviciu [408]. The fact that a convex function is a superposition of linear functions and the angle functions $(x-a)^+$ is discussed in the book of L. Hörmander [227] (and in the first edition of the present book).

Popoviciu's inequality [408] , as stated in Theorem 1.9.3, was proved by him in a more general form that applies to all continuous convex functions $f : I \to \mathbb{R}$ and all finite families $x_1, \dots, x_n$ of $n \geq 2$ points with equal weights. However, his argument also covers the case of arbitrary weights $\lambda_1, \dots, \lambda_n > 0$, so that the general form of Popoviciu's inequality reads as follows:

$$\sum_{1 \leq i_1 < \cdots < i_k \leq n} (\lambda_{i_1} + \cdots + \lambda_{i_k}) f\Big(\frac{\lambda_{i_1} x_{i_1} + \cdots + \lambda_{i_k} x_{i_k}}{\lambda_{i_1} + \cdots + \lambda_{i_k}}\Big)$$
$$\leq \binom{n-2}{k-2} \Big[\frac{n-k}{k-1} \sum_{i=1}^n \lambda_i f(x_i) + \Big(\sum_{i=1}^n \lambda_i\Big) f\Big(\frac{\lambda_1 x_1 + \cdots + \lambda_n x_n}{\lambda_1 + \cdots + \lambda_n}\Big)\Big].$$

Popoviciu's inequality was extended in various directions by M. Mihai and F.-C. Mitroi in [321] and C. P. Niculescu and I. Rovenţa [374]. An estimate from below of Popoviciu's inequality is available in [369] and [370]. The integral version of this inequality is discussed in [362] and [371]. A several variables analogue of Popoviciu's inequality is discussed by M. Bencze, C. P. Niculescu, and F. Popovici [39].

The proof of Jensen–Steffensen inequality in Section 1.8 is borrowed from C. P. Niculescu and M. M. Stănescu [379].

The dramatic story of the Hermite–Hadamard inequality is told in a short note by D. S. Mitrinović and I. B. Lacković [328]: In a letter sent on November 22, 1881, to *Mathesis* (and published there in 1883), Ch. Hermite [213] noted that every convex function $f : [a,b] \to \mathbb{R}$ satisfies the inequalities

$$f\Big(\frac{a+b}{2}\Big) \leq \frac{1}{b-a} \int_a^b f(x)\,\mathrm{d}x \leq \frac{f(a)+f(b)}{2}$$

and illustrated this with Example 1.9.1 in our text. Ten years later, the left-hand side inequality was rediscovered by J. Hadamard [195]. However the priority of Ch. Hermite was not recorded and his note was not even mentioned in Hermite's *Collected Papers* (published by E. Picard). The precision in the Hermite–Hadamard inequality can be estimated via two classical inequalities that work in the Lipschitz function framework. Suppose that $f\colon [a,b] \to \mathbb{R}$ is a Lipschitz function, with *Lipschitz constant*

$$\operatorname{Lip}(f) = \sup\Big\{\Big|\frac{f(x)-f(y)}{x-y}\Big| \;\Big|\; x \neq y\Big\}.$$

Then the left Hermite–Hadamard inequality can be estimated by the *inequality of Ostrowski*,

$$\left| f(x) - \frac{1}{b-a}\int_a^b f(t)\,\mathrm{d}t \right| \leq M\Big[\frac{1}{4} + \Big(\frac{x-\frac{a+b}{2}}{b-a}\Big)^2\Big](b-a),$$

while the right Hermite–Hadamard inequality can be estimated by the *inequality of Iyengar*,

$$\left| \frac{f(a)+f(b)}{2} - \frac{1}{b-a}\int_a^b f(t)\,\mathrm{d}t \right| \leq \frac{M(b-a)}{4} - \frac{1}{4M(b-a)}\,(f(b)-f(a))^2,$$

where $M = \operatorname{Lip}(f)$. The first inequality is a direct consequence of the triangle inequality. The second one makes the objective of Exercise 5, Section 7.1.

Many classical results related to the Hermite–Hadamard inequality can be found in the monograph of J. E. Pečarić, F. Proschan, and Y. C. Tong [388]. The contributions of Niculescu and his collaborators to this subject are available in [353], [355], [361], [364], [365], and [366].

We end this chapter with a brief discussion on the differentiability properties of convex functions from a generic point of view. Let $\mathcal{P}$ be a property which refers to the elements of a complete metric space X. We say that $\mathcal{P}$ is *generic* (or that most elements of X enjoy $\mathcal{P}$) if those elements not enjoying the property $\mathcal{P}$ form a set of first Baire category, that is, a countable union of nowhere dense sets. The space $C[0,1]$, of all continuous real functions on $[0,1]$, endowed with the usual sup norm, is complete. The same is true for $\operatorname{Conv}([0,1])$, the subset of all continuous convex functions on $[0,1]$. A well-known elegant proof of S. Banach shows that most functions in $C[0,1]$ are nowhere differentiable. The situation in $\operatorname{Conv}([0,1])$ is different. In fact, as noted by V. Klee [252], *most convex functions in* $C[0,1]$ *are differentiable.* The generic aspects of the second differentiability of convex functions are described by T. Zamfirescu [495]: *For most convex functions* $f\colon [0,1] \to \mathbb{R}$,

$$\underline{D}f' = 0 \;\text{ or }\; \overline{D}f' = \infty \quad \text{everywhere.}$$

Moreover, for most convex functions f, *the second derivative* f'' *vanishes wherever it exists, that is, almost everywhere.* Thus, the situation mentioned in Remark 1.5.1 is rather the rule, not the exception.

Chapter 2

Convex Sets in Real Linear Spaces

The natural domain for a convex function is a convex set. In this chapter we review some basic facts, necessary for a deep understanding of the concept of convexity in real linear spaces. For reader's convenience, all results concerning the separation of convex sets in Banach spaces are stated in Section 2.2 with proofs covering only the particular (but important) case of Euclidean spaces. Full details in the general case are to be found in Appendix B.

2.1 Convex Sets

A subset C of a linear space E is said to be *convex* if it contains the *line segment*

$$[\mathbf{x}, \mathbf{y}] = \{(1-\lambda)\mathbf{x} + \lambda\mathbf{y} : \lambda \in [0,1]\},$$

connecting any of its points $\mathbf{x}$ and $\mathbf{y}$. As Figure 2.1 suggests, convexity is a weak form of rotundity (in the sense of M. M. Day [128]). Besides line segments, some other simple examples of convex sets in the Euclidean space $\mathbb{R}^N$ are the lines, the planes, the open discs (plus any part of their boundary), and the N-dimensional rectangles (Cartesian products of N nonempty intervals).

New examples from the old ones can be obtained by considering arbitrary intersections and/or the following two algebraic operations with sets:

$$A + B = \{\mathbf{x} + \mathbf{y} : \mathbf{x} \in A,\ \mathbf{y} \in B\}$$
$$\lambda A = \{\lambda\mathbf{x} : \mathbf{x} \in A\}$$

for $A, B \subset E$ and $\lambda \in \mathbb{R}$. See Figure 2.2. The addition of sets is also known as the *Minkowski addition*. Addition of sets is commutative and associative. One can prove easily that $\lambda A + \mu B$ is a convex set provided that A and B are convex and $\lambda, \mu \geq 0$.

C. P. Niculescu and L.-E. Persson, *Convex Functions and Their Applications*,
CMS Books in Mathematics, https://doi.org/10.1007/978-3-319-78337-6_2

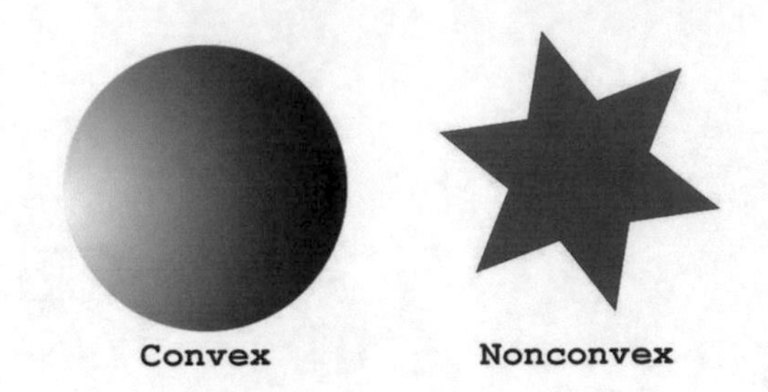

Figure 2.1: Convex and nonconvex planar sets.

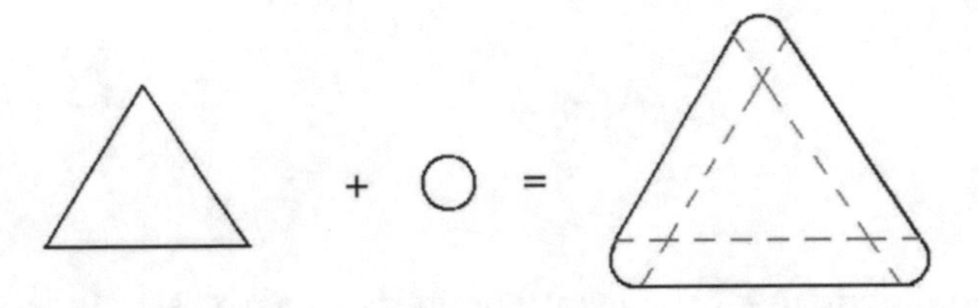

Figure 2.2: Minkowski addition of sets.

A subset A of E is said to be *affine* if it contains the whole line through any two of its points. Algebraically, this means that

$$\mathbf{x}, \mathbf{y} \in A \text{ and } \lambda \in \mathbb{R} \text{ imply } (1-\lambda)\mathbf{x} + \lambda\mathbf{y} \in A.$$

Clearly, any affine subset is also convex (but the converse is not true). It is important to notice that any affine subset A is just the translate of a (unique) linear subspace L (and all translations of a linear space represent affine sets). In fact, for every $\mathbf{a} \in A$, the translate

$$L = A - \mathbf{a}$$

is a linear space and it is clear that $A = L + \mathbf{a}$. For the uniqueness part, notice that if L and M are linear subspaces of E and $a, b \in E$ verify

$$L + \mathbf{a} = M + \mathbf{b},$$

then necessarily $L = M$ and $\mathbf{a} - \mathbf{b} \in L$. This remark allows us to introduce the concept of *dimension* for an affine set (as the dimension of the linear subspace of which it is a translate). Given a finite family $\mathbf{x}_1, \ldots, \mathbf{x}_n$ of points in E, an *affine combination* of them is any point of the form

$$\mathbf{x} = \sum_{k=1}^{n} \lambda_k \mathbf{x}_k$$

where $\lambda_1, \ldots, \lambda_n \in \mathbb{R}$, and $\sum_{k=1}^{n} \lambda_k = 1$. If in addition $\lambda_1, \ldots, \lambda_n \geq 0$, then $\mathbf{x}$ is called a *convex combination* of $\mathbf{x}_1, \ldots, \mathbf{x}_n$.

2.1.1 Lemma *A subset C of E is convex (respectively affine) if and only if it contains every convex (respectively affine) combination of points of C.*

Proof. The sufficiency part is clear, while the necessity part can be proved by mathematical induction. See the remark before Lemma 1.1.11. ■

Given a subset A of E, the intersection $\operatorname{conv}(A)$ of all convex subsets of E containing A is convex and thus it is the smallest set of this nature containing A. We call it the *convex hull* of A. By using Lemma 2.1.1, one can verify easily that $\operatorname{conv}(A)$ consists of all convex combinations of elements of A.

The affine variant of this construction yields the *affine hull* of A, denoted $\operatorname{aff}(A)$. As a consequence we can introduce the concept of dimension for convex sets to be the dimension of their affine hulls.

A nice example of convex hull is offered by the *Gauss–Lucas theorem* on the distribution of the critical points of a polynomial: the roots $(\mu_k)_{k=1}^{n-1}$ of the derivative P' of any complex polynomial P of degree $n \geq 2$ lie in the smallest convex polygon containing the roots $(\lambda_j)_{j=1}^{n}$ of the polynomial P. Indeed, assuming that w is a root of P' and $P(w) \neq 0$, we have

$$0 = \frac{P'(w)}{P(w)} = \sum_{k=1}^{n} \frac{1}{w - \lambda_k} = \sum_{k=1}^{n} \frac{\overline{w} - \overline{\lambda}_k}{|w - \lambda_k|^2},$$

whence

$$w = \sum_{k=1}^{n} \frac{1}{|w - \lambda_k|^2} \lambda_k \Bigg/ \sum_{k=1}^{n} \frac{1}{|w - \lambda_k|^2}.$$

If $(S_i)_{i \in \mathcal{I}}$ is a finite family of subsets of an N-dimensional linear space E, then the convex hull of their Minkowski addition equals the Minkowski addition of their convex hulls:

$$\operatorname{conv}\Big(\sum\nolimits_{i \in \mathcal{I}} S_i\Big) = \sum\nolimits_{i \in \mathcal{I}} \operatorname{conv}(S_i). \tag{2.1}$$

Surprisingly, this simple remark has deep consequences to the geometry of convex sets. See Theorems 2.1.3 and 2.1.8 below. The clue is provided by the following unifying lemma, used by R. M. Anderson in his course on Economic Theory, taught in Spring 2010 at Berkeley.

2.1.2 Lemma *Consider a finite family $(S_i)_{i \in \mathcal{I}}$ of nonempty subsets of $\mathbb{R}^N$. Then every $\mathbf{x} \in \operatorname{conv}\left(\sum_{i \in \mathcal{I}} S_i\right)$ admits a representation of the form*

$$\mathbf{x} = \sum_{i \in \mathcal{I}} \left(\sum_{1 \leq j \leq n_i} \lambda_{ij} \mathbf{x}_{ij} \right),$$

such that

$$\begin{array}{ll} (a) & \sum_{i \in \mathcal{I}} n_i \leq |\mathcal{I}| + N; \\ (b) & \mathbf{x}_{ij} \in S_i \text{ and } \lambda_{ij} > 0 \text{ for all } i, j; \\ (c) & \sum_{j=1}^{n_i} \lambda_{ij} = 1 \text{ for all } i \in \mathcal{I}. \end{array}$$

Proof. According to formula (2.1), every point $\mathbf{x} \in \operatorname{conv}\left(\sum_{i\in\mathcal{I}} S_i\right)$ admits representations of the form $\mathbf{x} = \sum_{i\in\mathcal{I}} \mathbf{x}_i$ with $\mathbf{x}_i \in \operatorname{conv}(S_i)$ for all i. Therefore,

$$\mathbf{x} = \sum_{i\in\mathcal{I}} \left(\sum_{1\leq j\leq n_i} \lambda_{ij}\mathbf{x}_{ij} \right), \tag{2.2}$$

for suitable $\mathbf{x}_{ij} \in S_i$ and $\lambda_{ij} > 0$ with $\sum_{j=1}^{n_i} \lambda_{ij} = 1$. Clearly, one can choose such a representation for which $n = \sum_{i\in\mathcal{I}} n_i$ is minimal. If $n > |\mathcal{I}| + N$, then the vectors $\mathbf{x}_{ij} - \mathbf{x}_{i1}$ for $i \in \mathcal{I}$ and $j \in [2, n_i]$ are linearly dependent in $\mathbb{R}^N$. Then

$$\sum_{i\in\mathcal{I}} \sum_{2\leq j\leq n_i} c_{ij} \left(\mathbf{x}_{ij} - \mathbf{x}_{i1}\right) = 0$$

for some real coefficients, not all zero. Adding to equation (2.2) the last equation multiplied by a real number λ we obtain

$$\mathbf{x} = \sum_{i\in\mathcal{I}} \sum_{1\leq j\leq n_i} \tilde{\lambda}_{ij}\mathbf{x}_{ij}, \tag{2.3}$$

where

$$\begin{aligned} \tilde{\lambda}_{ij} &= \lambda_{ij} + \lambda c_{ij} && \text{if } j \geq 2 \\ \tilde{\lambda}_{i1} &= \lambda_{i1} - \lambda \textstyle\sum_{2\leq k\leq n_i} c_{ik} && \text{if } j = 1. \end{aligned}$$

By a suitable choice of λ one can ensure that $\tilde{\lambda}_{ij} \geq 0$ for all indices i, j and that at least one coefficient $\tilde{\lambda}_{ij}$ is zero. The representation (2.3) eliminates one of the x_{ij}, contrary to the minimality of n. Consequently $n \leq |\mathcal{I}| + N$. ■

2.1.3 Theorem (Carathéodory's theorem) *Suppose that S is a subset of a linear space E and its convex hull* $\operatorname{conv}(S)$ *has dimension m. Then each point of* $\operatorname{conv}(S)$ *is the convex combination of at most $m + 1$ points of S.*

Proof. Clearly, we may assume that $E = \mathbb{R}^m$. Then apply Lemma 2.1.2 for $\mathcal{I} = \{1\}$ and $S_1 = S$. ■

The sets of the form $C = \operatorname{conv}(\{\mathbf{x}_0, \ldots, \mathbf{x}_N\})$ are usually called *polytopes*. If $\mathbf{x}_1 - \mathbf{x}_0, \ldots, \mathbf{x}_N - \mathbf{x}_0$ are linearly independent, then C is called an *N-simplex* (with vertices $\mathbf{x}_0, \ldots, \mathbf{x}_N$). In this case, $\dim C = N$ and every point $\mathbf{x}$ of C has a unique representation $\mathbf{x} = \sum_{k=0}^{N} \lambda_k\mathbf{x}_k$, as a convex combination of vertices; the numbers $\lambda_0, \ldots, \lambda_N$ are called the *barycentric coordinates* of $\mathbf{x}$.

The *standard N-simplex* (or *unit N-simplex*) is the simplex Δ^N whose vertices are the elements of the canonical algebraic basis of $\mathbb{R}^{N+1}$, that is,

$$\Delta^N = \left\{ (\lambda_0, ..., \lambda_N) \in \mathbb{R}^{N+1} : \sum_{k=0}^{N} \lambda_k = 1 \text{ and } \lambda_k \geq 0 \text{ for all } k \right\}.$$

Given an arbitrary N-simplex C with vertices $(\mathbf{x}_0, \ldots, \mathbf{x}_N)$, the map

$$\omega : \Delta^N \to C, \quad \omega\left(\lambda_0, ..., \lambda_N\right) = \sum_{k=0}^{N} \lambda_k\mathbf{x}_k$$

is affine and bijective.

Notice that any polytope $\operatorname{conv}(\{\mathbf{x}_0, \ldots, \mathbf{x}_N\})$ is a union of simplices whose vertices belong to $\{\mathbf{x}_0, \ldots, \mathbf{x}_N\}$.

2.1.4 Remark (Lagrange's barycentric identity) *Consider a finite system $\mathcal{S}$ of mass points $(\mathbf{x}_k, m_k)$ in $\mathbb{R}^{N+1}$, for $k = 1, ..., n$; $\mathbf{x}_k$ indicates position and m_k the mass. In mechanics and physics, one defines the* barycenter (or center of mass) *of the system by*

$$\operatorname{bar}(\mathcal{S}) = \frac{\sum_{k=1}^{n} m_k \mathbf{x}_k}{\sum_{k=1}^{n} m_k}.$$

The mass point $(\operatorname{bar}(\mathcal{S}), \sum_{k=1}^{n} m_k)$ *represents the* resultant *of the system* $\mathcal{S}$. *Notice that* $\operatorname{bar}(\mathcal{S}) \in \operatorname{conv}(\{\mathbf{x}_1, ..., \mathbf{x}_n\})$. *A practical way to determine the barycenter was found by J. L. Lagrange [267], who proved the following identity:* For every family of points $\mathbf{x}, \mathbf{x}_1, ..., \mathbf{x}_n$ in $\mathbb{R}^N$ and every family of real weights $m_1, ..., m_n$ with $M = \sum_{k=1}^{n} m_k > 0$, we have

$$\sum_{k=1}^{n} m_k \|\mathbf{x} - \mathbf{x}_k\|^2 = M \left\| \mathbf{x} - \frac{1}{M} \sum_{k=1}^{n} m_k \mathbf{x}_k \right\|^2 + \frac{1}{M} \cdot \sum_{1 \le i < j \le n} m_i m_j \|\mathbf{x}_i - \mathbf{x}_j\|^2. \tag{2.4}$$

For the proof, use the formula $\|\mathbf{z}\|^2 = \langle \mathbf{z}, \mathbf{z} \rangle$. *The formula* (2.4) *yields the following variational definition of barycenter:* $\operatorname{bar}(\mathcal{S})$ *is the unique point that minimizes the function* $\mathbf{x} \to \sum_{k=1}^{n} m_k \|\mathbf{x} - \mathbf{x}_k\|^2$, *that is,*

$$\operatorname{bar}(\mathcal{S}) = \arg\min_{\mathbf{x} \in \mathbb{R}^N} \sum_{k=1}^{n} m_k \|\mathbf{x} - \mathbf{x}_k\|^2.$$

In Section 3.9, *we will present the concept of barycenter associated to an arbitrary Borel probability measure.*

Another important class of convex sets are the convex cones. A *convex cone* in the real linear space E is a subset C with the following two properties:

$$C + C \subset C$$
$$\lambda C \subset C \quad \text{for all } \lambda \geq 0.$$

Examples of convex cones:

$\mathbb{R}_+^N = \{(x_1, \ldots, x_N) \in \mathbb{R}^N : x_1, \ldots, x_N \geq 0\}$, the *positive orthant*;

$\mathbb{R}_{\geq}^N = \{(x_1, \ldots, x_N) \in \mathbb{R}^N : x_1 \geq \cdots \geq x_N\}$, the *monotone cone*;

$\{(x, y, z) \in \mathbb{R}^3 : z \geq 0\}$, the *closed upper half-space*;

$\{(x, y, z, t) \in \mathbb{R}^3 \times [0, \infty) : x^2 + y^2 + z^2 \leq c^2 t^2\}$, the *future light cone* (c is the speed of light);

$\mathrm{Sym}^+(N, \mathbb{R})$, the set of all *positive* matrices A of $\mathrm{M}_N(\mathbb{R})$, that is, of real symmetric matrices such that

$$\langle A\mathbf{x}, \mathbf{x}\rangle \geq 0 \quad \text{for all } \mathbf{x} \in \mathbb{R}^N.$$

An important source of convex cones is provided by the theory of ordered linear spaces, briefly summarized in Section 2.4.

So far we have not used any topology; only the linear properties of the space E have played a role. Suppose now that E is a linear normed space. The following two lemmas relate convexity and topology:

2.1.5 Lemma *If U is a convex set in a linear normed space, then its interior $\operatorname{int} U$ and its closure $\overline{U}$ are convex as well.*

Proof. For example, if $\mathbf{x}$, $\mathbf{y} \in \operatorname{int} U$ and $\lambda \in (0,1)$, then

$$\lambda\mathbf{x} + (1-\lambda)\mathbf{y} + \mathbf{u} = \lambda(\mathbf{x}+\mathbf{u}) + (1-\lambda)(\mathbf{y}+\mathbf{u}) \in U$$

for all $\mathbf{u}$ in a suitable ball $B_\varepsilon(0)$. This shows that $\operatorname{int} U$ is a convex set. Now let $\mathbf{x}, \mathbf{y} \in \overline{U}$. Then there exist sequences $(\mathbf{x}_k)_k$ and $(\mathbf{y}_k)_k$ in U, converging to $\mathbf{x}$ and $\mathbf{y}$ respectively. This yields $\lambda\mathbf{x} + (1-\lambda)\mathbf{y} = \lim_{k\to\infty}[\lambda\mathbf{x}_k + (1-\lambda)\mathbf{y}_k] \in \overline{U}$ for all $\lambda \in [0,1]$, that is, $\overline{U}$ is convex as well. ∎

Notice that affine sets in $\mathbb{R}^N$ are closed because finite-dimensional subspaces are always closed. Also, all cones mentioned above are closed. The case of $\mathrm{Sym}^+(N, \mathbb{R})$ needs an explanation concerning the topology on the matrix space $\mathrm{M}_N(\mathbb{R})$. Since all norms on a finite- dimensional space are equivalent (see, e.g., R. Bhatia [51], p. 16), they produce the same topology. For convenience, we will use the Hilbertian norm inherited from $\mathbb{R}^{N\times N}$,

$$\|A\|_{HS} = \left(\sum_{i,j=1}^{N} a_{ij}^2\right)^{1/2} \qquad \text{for } A = (a_{ij})_{i,j=1}^N, \tag{2.5}$$

usually known as the *Hilbert–Schmidt norm.* This norm is associated to the inner product

$$\langle A, B\rangle_{HS} = \operatorname{trace}(B^*A),$$

where B^* denotes the transpose of B.

In Section 2.5 we will present the *operator norm* on $\mathrm{M}_N(\mathbb{R})$, the standard norm for linear operators from $\mathbb{R}^N$ into itself.

The interior of $\mathbb{R}^N_+$ is the set

$$\mathbb{R}^N_{++} = \{(x_1, \ldots, x_N) \in \mathbb{R}^N : x_1, \ldots, x_N > 0\},$$

while the interior of $\mathrm{Sym}^+(N, \mathbb{R})$ is

$$\mathrm{Sym}^{++}(N, \mathbb{R}),$$

the set of all *strictly positive* (or *positive definite*) matrices; these are the symmetric matrices $A \in \mathrm{M}_N(\mathbb{R})$ such that

$$\langle A\mathbf{x}, \mathbf{x}\rangle > 0 \quad \text{for all } \mathbf{x} \in \mathbb{R}^N,\ \mathbf{x} \neq 0.$$

Characteristic for a strictly positive matrix A is the existence of a constant $C > 0$ such that

$$\langle A\mathbf{x}, \mathbf{x}\rangle \geq C\,\|\mathbf{x}\|^2 \quad \text{for all } \mathbf{x} \in \mathbb{R}^n;$$

to check this, put $C = \max\left\{\langle A\mathbf{x}, \mathbf{x}\rangle : \mathbf{x} \in \mathbb{R}^N, \|\mathbf{x}\| = 1\right\}$ and take into account the Weierstrass extreme value theorem.

As is well known from linear algebra, every symmetric matrix has only real eigenvalues and is diagonalizable with respect to an orthonormal basis consisting of eigenvectors. Therefore a symmetric matrix A belongs to $\mathrm{Sym}^+(N,\mathbb{R})$ (respectively to $\mathrm{Sym}^{++}(N,\mathbb{R})$) if and only if all its eigenvalues are positive (respectively strictly positive). See R. A. Horn and C. R. Johnson [223], Theorem 4.1.5, p. 171. We will come back to these facts in Section 2.5.

If $A \in \mathrm{Sym}^{++}(N,\mathbb{R})$, then the formula

$$\|\mathbf{x}\|_A = (\langle A\mathbf{x}, \mathbf{x}\rangle)^{1/2}$$

defines a norm on $\mathbb{R}^N$, equivalent to the Euclidean norm. The associated closed balls,

$$Q = \{\mathbf{x} : \langle A(\mathbf{x} - \mathbf{v}), \mathbf{x} - \mathbf{v}\rangle \leq 1\},$$

represent *ellipsoids* centered at $\mathbf{v}$.

2.1.6 Lemma *If U is an open set in a normed linear space E, then its convex hull is open. If E is finite dimensional and K is a compact subset, then its convex hull is compact.*

Proof. For the first assertion, let $\mathbf{x} = \sum_{k=0}^{m} \lambda_k \mathbf{x}_k$ be a convex combination of elements of the open set U. Then

$$\mathbf{x} + \mathbf{u} = \sum_{k=0}^{m} \lambda_k(\mathbf{x}_k + \mathbf{u}) \quad \text{for all } \mathbf{u} \in E$$

and since U is open it follows that $\mathbf{x}_k + \mathbf{u} \in U$ for all k, provided that $\|\mathbf{u}\|$ is small enough. Consequently, $\mathbf{x} + \mathbf{u} \in \mathrm{conv}(U)$ for $\mathbf{u}$ in a ball $B_\varepsilon(\mathbf{0})$. We pass now to the second assertion. Clearly, we may assume that $E = \mathbb{R}^N$. Then, consider the map defined by

$$f(\lambda_0, \ldots, \lambda_N, \mathbf{x}_0, \ldots, \mathbf{x}_N) = \sum_{k=0}^{N} \lambda_k \mathbf{x}_k,$$

where $\lambda_0, \ldots, \lambda_N \in [0,1]$, $\sum_{k=0}^{N} \lambda_k = 1$, and $\mathbf{x}_0, \ldots, \mathbf{x}_N \in K$. Since f is continuous and its domain of definition is a compact space, so is the range of f.

According to Carathéodory's theorem, the range of f is precisely $\operatorname{conv}(K)$, and this ends the proof. ■

The union of a line and a point not on it is a closed set whose convex hull fails to be closed.

Notice also that the convex hull of a compact subset of an infinite- dimensional Banach space needs not to be compact. For example, consider the case of the Hilbert space ℓ^2 (of all square summable sequences of real numbers) and the compact set K consisting of elements $\mathbf{v}_n = \frac{1}{2^n}(\delta_{kn})_k$ for $n = 1, 2, 3, ...$ and their limit, the sequence having all terms equal to zero. Then

$$\mathbf{v} = \sum_{n=1}^{\infty} \frac{1}{2^n} \mathbf{v}_n \in \overline{\operatorname{conv}(K)} \setminus \operatorname{conv}(K).$$

2.1.7 Lemma (Accessibility lemma) *Suppose that C is a convex set in a normed linear space E. If $\mathbf{x} \in \operatorname{int} C$ and $\mathbf{y} \in \overline{C}$, then the linear segment $[\mathbf{x}, \mathbf{y})$ is contained in* $\operatorname{int} C$.

Proof. Let $\varepsilon > 0$ such that $B_\varepsilon(\mathbf{x}) \subset C$. We have to show that $\lambda\mathbf{x} + (1 - \lambda)\mathbf{y} \in \operatorname{int} C$ for all $\lambda \in (0, 1)$. Since $\mathbf{y} \in \overline{C}$, there is a point $\mathbf{z} \in C$ such that $\|\mathbf{y} - \mathbf{z}\| < \varepsilon\lambda/(2 - 2\lambda)$. Suppose that $\mathbf{w} \in E$ and $\|\mathbf{w}\| < \varepsilon/2$. We have

$$\begin{aligned}\lambda\mathbf{x} + (1 - \lambda)\mathbf{y} + \lambda\mathbf{w} &= \lambda\mathbf{x} + (1 - \lambda)(\mathbf{y} - \mathbf{z}) + \lambda\mathbf{w} + (1 - \lambda)\mathbf{z} \\ &= \lambda\left[\mathbf{x} + \mathbf{w} + (1 - \lambda)\frac{\mathbf{y} - \mathbf{z}}{\lambda}\right] + (1 - \lambda)\mathbf{z}.\end{aligned}$$

The point $\mathbf{w} + (1 - \lambda)(\mathbf{y} - \mathbf{z})/\lambda$ belongs to $B_\varepsilon(\mathbf{0})$ since

$$\begin{aligned}\left\|\mathbf{w} + (1 - \lambda)\frac{\mathbf{y} - \mathbf{z}}{\lambda}\right\| &\leq \|\mathbf{w}\| + (1 - \lambda)\|\mathbf{y} - \mathbf{z}\|/\lambda \\ &< \varepsilon/2 + \varepsilon/2 = \varepsilon.\end{aligned}$$

Hence $\mathbf{x}+\mathbf{w}+(1 - \lambda)(\mathbf{y} - \mathbf{z})/\lambda \in B_\varepsilon(\mathbf{x}) \subset C$. As a consequence, $\lambda\mathbf{x}+(1-\lambda)\mathbf{y}+\lambda\mathbf{w} \in C$, that is, $\lambda\mathbf{x}+(1-\lambda)\mathbf{y}+\lambda B_{\varepsilon/2}(\mathbf{0}) \subset C$. Therefore, $\lambda\mathbf{x}+(1-\lambda)\mathbf{y} \in \operatorname{int} C$. ■

See Exercises 1, 6, and 12 for applications.

We next discuss an important feature of Minkowski addition of nonconvex sets in $\mathbb{R}^N$: approximate convexity of vector sums $\sum_{i\in\mathcal{I}} S_i$ of a large family of sets.

2.1.8 Theorem (Shapley–Folkman theorem; see [455]). *Consider a finite family $(S_i)_{i\in\mathcal{I}}$ of nonempty subsets of $\mathbb{R}^N$ with $|\mathcal{I}| > N$. If $\mathbf{x} \in \operatorname{conv}\left(\sum_{i\in\mathcal{I}} S_i\right)$, then there exists a subset J of $\mathcal{I}$, of cardinality at most N, such that*

$$\mathbf{x} \in \sum_{i\notin J} S_i + \operatorname{conv}\left(\sum_{i\in J} S_i\right).$$

Proof. Let $\mathbf{x} = \sum_{i\in\mathcal{I}}\left(\sum_{1\leq j\leq n_i} \lambda_{ij}\mathbf{x}_{ij}\right)$ be the decomposition given by Lemma 2.1.2. Because $\sum_{i\in\mathcal{I}} n_i \leq |\mathcal{I}| + N$, we have $n_i = 1$ except for at most N values of i. If $n_i = 1$, then $\lambda_{i1} = 1$ and $\sum_{1\leq j\leq n_i} \lambda_{ij}\mathbf{x}_{ij} = \mathbf{x}_{i1} \in S_i$, which yields the conclusion of the theorem. ■

The Shapley–Folkman theorem can be connected with the following measure for the *deviation* of a nonempty compact subset S of $\mathbb{R}^N$ from its convex hull:

$$\rho(S) = \sup_{\mathbf{x}\in \operatorname{conv}(S)} \inf_{\mathbf{y}\in S} \|\mathbf{x}-\mathbf{y}\|.$$

In terms of the Pompeiu–Hausdorff distance between sets, $\rho(S) = d_H(S, \operatorname{conv}(S))$. Recall that for any two nonempty subsets A and B of a metric space $M = (M, d)$, their *Pompeiu–Hausdorff distance* is defined by

$$d_H(A,B) = \max\{\sup_{x\in A}\inf_{y\in B} d(x,y), \sup_{y\in B}\inf_{x\in A} d(x,y)\}.$$

A compact set S is convex if and only if $\rho(S) = 0$. It is immediate that

$$\operatorname{diam}(S) = \operatorname{diam}\operatorname{conv}(S) \text{ and } \rho(S) \leq \operatorname{diam}(S).$$

2.1.9 Corollary *Let $(S_i)_{i\in\mathcal{I}}$ be a finite family of compact subsets of $\mathbb{R}^N$ and $L > 0$ a real number such that $\rho(S_i) \leq L$ for all i. Then for any $\mathbf{x}$ in* conv $\left(\sum_{i\in\mathcal{I}} S_i\right)$ *there is $\mathbf{y} \in \sum_{i\in\mathcal{I}} S_i$ for which $\|\mathbf{x}-\mathbf{y}\| \leq LN$.*

Proof. Given $\mathbf{x} \in \operatorname{conv}\left(\sum_{i\in\mathcal{I}} S_i\right)$, we infer from the Shapley–Folkman theorem the existence of a subset J of $\mathcal{I}$, of cardinality at most N, such that $\mathbf{x} = \sum_{i\in\mathcal{I}} \mathbf{x}_i$, where $\mathbf{x}_i \in S_i$ for $i \notin J$ and $\mathbf{x}_i \in \operatorname{conv}(S_i)$ for $i \in J$. According to the definition of the function ρ, for each $i \in J$ there is $\mathbf{y}_i \in S_i$ such that $\|\mathbf{x}_i - \mathbf{y}_i\| \leq \rho(S_i)$. Put $\mathbf{y} = \sum_{i\in J} \mathbf{y}_i + \sum_{i\notin J} \mathbf{x}_i$. Then

$$\|\mathbf{x}-\mathbf{y}\| = \left\|\sum_{i\in J} (\mathbf{x}_i - \mathbf{y}_i)\right\| \leq \sum_{i\in J} \|\mathbf{x}_i - \mathbf{y}_i\| \leq \sum_{i\in J} \rho(S_i) \leq NL$$

and the proof is done. ■

The estimate of Corollary 2.1.9 is independent of the size of $\mathcal{I}$ for $|\mathcal{I}| > N$. As a consequence, if $(S_n)_n$ is a sequence of compact subsets of $\mathbb{R}^N$ with $\sup_n \rho(S_n) < \infty$, then

$$d_H\left(\frac{1}{n}(S_1 + \cdots + S_n), \frac{1}{n}\operatorname{conv}(S_1 + \cdots + S_n)\right) \to 0$$

as $n \to \infty$. According to R. Schneider [436], p. 142, the averaging of compact sets in the Minkowski sense is "asymptotically convexifying".

2.1.10 Remark (Relativization with respect to affine hull) *While working with a convex subset C of $\mathbb{R}^N$, the natural space containing it is often* aff(C), *not $\mathbb{R}^N$, which may be far too large. For example, if $\dim C = k < n$, then C has empty interior. We can talk more meaningfully about the topological notions of interior and boundary by using the notions of relative interior and relative* *boundary. If C is a convex subset of $\mathbb{R}^N$, the relative interior of C, denoted* $\mathrm{ri}(C)$, *is the interior of C relative to* $\mathrm{aff}(C)$. *That is, $a \in \mathrm{ri}(C)$ if and only if there is an $\varepsilon > 0$ such that $B_\varepsilon(a) \cap \mathrm{aff}(C) \subset C$. Every nonempty convex subset of $\mathbb{R}^N$ has a nonempty relative interior. The relative boundary of A, denoted* $\mathrm{rbd}(A)$, *is defined as* $\mathrm{rbd}(A) = \bar{A} \setminus \mathrm{ri}(A)$. *Many results concerning convex sets C in a finite- dimensional real space can be simplified by replacing E to a subspace F with* $\dim F = \dim C$ *(a fact that transforms* $\mathrm{ri}(C)$ *into* $\mathrm{int}\, C$*). Indeed, for every affine subset A of $\mathbb{R}^N$ with $\dim A = n$, there is a one-to-one affine transform of $\mathbb{R}^N$ into itself that maps A onto the subspace*

$$L = \left\{ \mathbf{x} \in \mathbb{R}^N : \mathbf{x} = (x_1, ..., x_N), x_{n+1} = \cdots = x_N = 0 \right\}.$$

For details, see R. T. Rockafellar [421], *Corollary* 1.6.1, *p.* 8. *More information on the relativization with respect to the affine hull is made available by Exercises* 4–7 *and Remark* 2.3.7 *below.*

2.1.11 Remark (Measurability) *A convex set in $\mathbb{R}^N$ is not necessarily a Borel set. Think of the open unit ball $B_1(\mathbf{0})$ to which it is added a non-Borel subset from its boundary. However, every convex set C in $\mathbb{R}^N$ is Lebesgue measurable. Indeed, since $C = \cup_n (C \cap B_n(0))$, we may reduce ourselves to the case where C is a convex and bounded set. If* $\mathrm{int}\, C = \emptyset$, *then C cannot contain any N-simplex and in this case C must lie in a subspace of dimension at most $N-1$, hence C is a Lebesgue null set (in particular, it is Lebesgue measurable). The argument in the case where* $\mathrm{int}\, C \neq \emptyset$ *follows from Exercise* 12.

Exercises

1. Suppose that C is a convex set in $\mathbb{R}^N$. Infer from the accessibility lemma (Lemma 2.1.7) that:

   (a) $\mathrm{int}\, C = \mathrm{int}\, \overline{C}$;

   (b) $\overline{C} = \overline{\mathrm{int}\, C}$ if the interior of C is nonempty.

2. Consider the disk $D = \{(x, y, z) : x^2 + y^2 \leq 1, z = 0\}$. Describe the relative interior and the relative boundary of D. Observe that in the standard topology D has empty interior and $\partial D = D$.

3. Prove that:

   (a) the convex hull of a closed set might be not closed;

   (b) if S is a bounded set, then $\mathrm{diam}\,(S) = \mathrm{diam}\,(\mathrm{conv}\,(S))$ and $\mathrm{conv}\,(\overline{S}) = \overline{\mathrm{conv}\,(S)}$.

4. Let $S = \{\mathbf{x}_0, \ldots, \mathbf{x}_m\}$ be a finite subset of $\mathbb{R}^N$. Prove that

$$\mathrm{ri}(\mathrm{conv}\, S) = \Big\{ \sum_{k=0}^{m} \lambda_k \mathbf{x}_k : \lambda_k \in (0,1),\ \sum_{k=0}^{m} \lambda_k = 1 \Big\}.$$

5. Suppose that A and B are convex subsets of $\mathbb{R}^N$, with $A \subset B$ and $\operatorname{aff}(A) = \operatorname{aff}(B)$. Prove that $\operatorname{ri}(A) \subset \operatorname{ri}(B)$.

6. (Extension of Accessibility Lemma 2.1.7) Suppose that A is a nonempty convex subset of $\mathbb{R}^N$, $\mathbf{x} \in \operatorname{ri}(A)$ and $\mathbf{y} \in \bar{A}$. Prove that

$$[\mathbf{x}, \mathbf{y}) = \{(1-\lambda)\mathbf{x} + \lambda\mathbf{y} : \lambda \in [0,1)\} \subset \operatorname{ri}(A).$$

Infer that $\operatorname{ri}(A)$ and $\operatorname{int}(A)$ are both convex sets and every point of $\bar{A}$ can be approximated by points in $\operatorname{ri}(A)$. In particular, $\operatorname{ri}(A)$ is nonempty.

7. Suppose that C is a nonempty convex subset of $\mathbb{R}^N$. Prove that:

(a) $\operatorname{ri}(C)$ and $\overline{C}$ have the same affine hull (and hence the same dimension);

(b) $\operatorname{ri}(C) = \operatorname{ri}(\overline{C})$ and $\overline{C} = \overline{\operatorname{ri} C}$.

8. Prove that:

(a) the map $\mathbf{x} \to \mathbf{x}/(1 - \|\mathbf{x}\|^2)^{1/2}$ provides a homeomorphism between the open unit ball of the Euclidean space $\mathbb{R}^N$ and $\mathbb{R}^N$;

(b) the map

$$(t_0, t_1, ..., t_N) \to \left(\log \frac{t_1}{t_0}, \log \frac{t_2}{t_0}, ..., \log \frac{t_N}{t_0}\right)$$

provides a homeomorphism between the interior of the standard N-*simplex* Δ^N and $\mathbb{R}^N$.

Remark. One can prove that all (nonempty) open convex subsets of $\mathbb{R}^N$ are homeomorphic to each other. For an argument, adapt the proof of Corollary 3.4.6.

9. A subset $S = \{\mathbf{x}_0, \ldots, \mathbf{x}_n\}$ of $\mathbb{R}^N$ is said to be *affinely independent* if the family $\{\mathbf{x}_1 - \mathbf{x}_0, \ldots, \mathbf{x}_n - \mathbf{x}_0\}$ is linearly independent. Prove that this means

$$\sum_{k=0}^{n} \lambda_k \mathbf{x}_k = 0 \text{ and } \sum_{k=0}^{n} \lambda_k = 0 \text{ imply } \lambda_k = 0 \text{ for all } k \in \{0, \ldots, n\}.$$

Infer that an affinely independent set in $\mathbb{R}^N$ can have at most $N+1$ points.

10. (Radon's theorem) Prove that each finite set of at least $N+2$ points in $\mathbb{R}^N$ can be expressed as the union of two disjoint sets whose convex hulls have a common point.

[*Hint*: If $(\mathbf{x}_i)_{i=1}^n$ is a family of $n \geq N+2$ points in $\mathbb{R}^N$, then these points are affinely dependent, which yields real scalars λ_i, not all null, such that

$$\sum_{i=1}^{n} \lambda_i \mathbf{x}_i = 0 \quad \text{and} \quad \sum_{i=1}^{n} \lambda_i = 0.$$

We may assume (renumbering the indices) that $\lambda_1, ..., \lambda_m > 0$ and $\lambda_{m+1}, ...,$ $\lambda_n \leq 0$. Then $\lambda = \sum_{i=1}^m \lambda_i > 0$ and

$$\mathbf{x} = \sum_{i=1}^{m} (\lambda_i/\lambda)\mathbf{x}_i = \sum_{i=m+1}^{n} (-\lambda_i/\lambda)\mathbf{x}_i$$

belongs to $\operatorname{conv}(\{\mathbf{x}_1, ..., \mathbf{x}_m\}) \cap \operatorname{conv}(\{\mathbf{x}_{m+1}, ..., \mathbf{x}_n\})$.]

11. (Helly's theorem) Let C_k be convex subsets of $\mathbb{R}^N$ for $1 \leq k \leq m$, where $m \geq N+1$. Suppose that every subcollection of at most $N+1$ sets has a nonempty intersection. Prove that the whole collection has a nonempty intersection.

    [*Hint:* Use induction on $m \geq N+1$. For $m = N+1$ things are clear. Suppose that $m \geq N+2$ and that the statement is true for families of $m-1$ sets. Choose $\mathbf{x}_i$ in the intersection of $C_1 \cap \cdots \cap C_{i-1} \cap \widehat{C}_i \cap C_{i+1} \cap \cdots \cap C_m$, where $\widehat{C}_i$ means that the set C_i is omitted. Then apply Radon's theorem to the family of $m \geq N+2$ points $\mathbf{x}_1, ..., \mathbf{x}_m$.]

    Remark. Helly's theorem also works for infinite families of subsets of $\mathbb{R}^N$ provided that they are convex and compact. An interesting reformulation of Helly's theorem in terms of convex inequalities is as follows: *Suppose there is given a system of inequalities*

    $$f_1(\mathbf{x}) < 0, ..., f_k(\mathbf{x}) < 0, \ f_{k+1}(\mathbf{x}) \leq 0, ..., f_m(\mathbf{x}) \leq 0,$$

    where the f_k are convex functions on $\mathbb{R}^N$, and the inequalities may be all strict or all weak. If every subsystem of $N+1$ or fewer inequalities has a solution in a given convex set C, then the entire system has a solution in C.

12. (Measurability of convex sets in $\mathbb{R}^N$) The aim of this exercise is to prove that every bounded convex subset $C \subset \mathbb{R}^N$ having interior points is Lebesgue measurable. Since $C = (C \cap \partial C) \cup \operatorname{int} C$, it suffices to prove that ∂C is a Lebesgue null set. Using the translation invariance of Lebesgue measure, we may assume that the origin 0 is interior to C.

    (a) Infer from the accessibility lemma (Lemma 2.1.7) that $\partial C \subset \frac{1}{1-\varepsilon} \operatorname{int} C$, for every $\varepsilon \in (0,1)$. Thus $\partial C \subset \left(\frac{1}{1-\varepsilon} \operatorname{int} C\right) \setminus \operatorname{int} C$.

    (b) Prove that $\operatorname{int} C \subset \frac{1}{1-\varepsilon} \operatorname{int} C$ for every $\varepsilon \in (0,1)$.

    (c) Prove that the Lebesgue measure of $\left(\frac{1}{1-\varepsilon} \operatorname{int} C\right) \setminus \operatorname{int} C$ is at most $\left(\frac{1}{1-\varepsilon}\right)^N \mathcal{L}(\operatorname{int} C) - \mathcal{L}(\operatorname{int} C)$ for every $\varepsilon \in (0,1)$.

    (d) Conclude that ∂C is a Lebesgue null set.

13. A practical method for establishing convexity of a set C is to show that C is obtained from simple convex sets by operations that preserve convexity.

Prove that images and inverse images of convex sets under affine functions of the form

$$f : \mathbb{R}^M \to \mathbb{R}^N, \quad f(\mathbf{x}) = A\mathbf{x} + \mathbf{b}$$

(where A is an $N \times M$-dimensional matrix and $\mathbf{b} \in \mathbb{R}^N$) are convex sets. Applications: by scaling, translating, or projecting convex sets one obtains convex sets.

14. Extend the preceding exercise to the case of linear-fractional functions of the form

$$f : \{\mathbf{x} \in \mathbb{R}^M : \langle \mathbf{c}, \mathbf{x} \rangle + d > 0\} \to \mathbb{R}^N, \quad f(\mathbf{x}) = \frac{A\mathbf{x} + \mathbf{b}}{\langle \mathbf{c}, \mathbf{x} \rangle + d},$$

where A is an $N \times M$-dimensional matrix, $\mathbf{b} \in \mathbb{R}^N$, $\mathbf{c} \in \mathbb{R}^M$, and $d \in \mathbb{R}$.

2.2 The Orthogonal Projection

In any normed linear space E we can speak about the *distance* from a point $\mathbf{x} \in E$ to a subset $A \subset E$. This is defined by the formula

$$d(\mathbf{x}, A) = \inf\{\|\mathbf{x} - \mathbf{a}\| : \mathbf{a} \in A\},$$

which represents a numerical indicator of how well $\mathbf{x}$ can be approximated by the elements of A. When $E = \mathbb{R}^3$ and A is the xy plane, the Pythagorean theorem shows that $d(\mathbf{x}, A)$ is precisely the distance between $\mathbf{x}$ and its orthogonal projection on that plane. This remark extends to the context of Hilbert spaces.

2.2.1 Theorem *Let C be a nonempty closed convex subset of a Hilbert space H (particularly, of the Euclidean space $\mathbb{R}^N$). Then for every $\mathbf{x} \in H$ there is a unique point $P_C(\mathbf{x})$ of C such that*

$$d(\mathbf{x}, C) = \|\mathbf{x} - P_C(\mathbf{x})\|. \tag{2.6}$$

The point $P_C(\mathbf{x})$ is also characterized by the following variational inequality:

$$\langle \mathbf{x} - P_C(\mathbf{x}), \mathbf{y} - P_C(\mathbf{x}) \rangle \leq 0 \quad \text{for all } \mathbf{y} \in C. \tag{2.7}$$

We call $P_C(\mathbf{x})$ the *orthogonal projection* of $\mathbf{x}$ onto C (or the *nearest point* of C to $\mathbf{x}$). Accordingly, we call the map $P_C : \mathbf{x} \to P_C(\mathbf{x})$, from H into itself, the *orthogonal projection* associated to C. Clearly,

$$P_C(\mathbf{x}) \in C \quad \text{for every } \mathbf{x} \in H$$

and

$$P_C(\mathbf{x}) = \mathbf{x} \quad \text{if, and only if, } \mathbf{x} \in C.$$

In particular,

$$P_C^2 = P_C.$$

The orthogonal projection P_C is a *monotone* map in the sense that

$$\langle P_C(\mathbf{x}) - P_C(\mathbf{y}), \mathbf{x} - \mathbf{y}\rangle \geq 0 \quad \text{for all } \mathbf{x}, \mathbf{y} \in H. \tag{2.8}$$

This inequality follows by adding the inequalities $\|\mathbf{x} - P_C(\mathbf{x})\|^2 \leq \|\mathbf{x} - P_C(\mathbf{y})\|^2$ and $\|\mathbf{y} - P_C(\mathbf{y})\|^2 \leq \|\mathbf{y} - P_C(\mathbf{x})\|^2$ (and replacing the norms by the corresponding inner products).

The terminology is motivated by the particular case of closed linear subspaces M of H. In their case, the variational inequality (2.7) shows that $\langle \mathbf{x} - P_M(\mathbf{x}), \mathbf{v}\rangle = 0$ for all $\mathbf{v} \in M$, that is, $\mathbf{x} - P_M(\mathbf{x})$ is orthogonal on M. Denoting by

$$M^{\perp} = \{\mathbf{y} \in H : \langle \mathbf{x}, \mathbf{y}\rangle = 0 \text{ for all } \mathbf{x} \in M\}$$

the *orthogonal complement* of M, one can easily verify that P_M is a linear self-adjoint projection and $\mathrm{I} - P_M$ equals the orthogonal projection $P_{M^{\perp}}$. This fact is basic for the entire theory of orthogonal decompositions.

Proof of Theorem 2.2.1. The existence of $P_C(\mathbf{x})$ follows from the definition of the distance from a point to a set and the special geometry of the ambient space. In fact, any sequence $(\mathbf{y}_n)_n$ in C such that $\|\mathbf{x} - \mathbf{y}_n\| \to \alpha = d(\mathbf{x}, C)$ is a Cauchy sequence. This is a consequence of the following identity,

$$\|\mathbf{y}_m - \mathbf{y}_n\|^2 + 4\left\|\mathbf{x} - \frac{\mathbf{y}_m + \mathbf{y}_n}{2}\right\|^2 = 2(\|\mathbf{x} - y_m\|^2 + \|\mathbf{x} - y_n\|^2)$$

(motivated by the parallelogram law), and the definition of α as an infimum; notice that $\|\mathbf{x} - \frac{\mathbf{y}_m + \mathbf{y}_n}{2}\| \geq \alpha$, which forces $\limsup_{m,n\to\infty} \|\mathbf{y}_m - \mathbf{y}_n\|^2 = 0$. Since H is complete, there must exist a point $\mathbf{y} \in C$ at which $(\mathbf{y}_n)_n$ converges. Then necessarily $d(\mathbf{x}, \mathbf{y}) = d(\mathbf{x}, C)$. The uniqueness of $\mathbf{y}$ with this property follows again from the parallelogram law. If $\mathbf{y}'$ is another point of C such that $d(\mathbf{x}, \mathbf{y}') = d(\mathbf{x}, C)$, then

$$\|\mathbf{y} - \mathbf{y}'\|^2 + 4\left\|\mathbf{x} - \frac{\mathbf{y} + \mathbf{y}'}{2}\right\|^2 = 2(\|\mathbf{x} - \mathbf{y}\|^2 + \|\mathbf{x} - \mathbf{y}'\|^2),$$

which gives us $\|\mathbf{y} - \mathbf{y}'\|^2 \leq 0$, a contradiction since it was assumed that the points $\mathbf{y}$ and $\mathbf{y}'$ are distinct. The implication (2.6) $\Rightarrow$ (2.7) follows from the fact that

$$\begin{aligned}\|\mathbf{x} - P_C(\mathbf{x})\|^2 &\leq \|\mathbf{x} - [(1-t)\, P_C(\mathbf{x}) + t\mathbf{y}]\,\|^2 = \|\mathbf{x} - P_C(\mathbf{x}) + t\,(P_C(\mathbf{x}) - \mathbf{y})\|^2 \\ &= \|\mathbf{x} - P_C(\mathbf{x})\|^2 + 2t\langle \mathbf{x} - P_C(\mathbf{x}), P_C(\mathbf{x}) - \mathbf{y}\rangle + t^2\,\|P_C(\mathbf{x}) - \mathbf{y}\|^2\end{aligned}$$

for all $t \in [0, 1]$ and $\mathbf{y} \in C$. Indeed this condition can be restated as

$$2\langle \mathbf{x} - P_C(\mathbf{x}), P_C(\mathbf{x}) - \mathbf{y}\rangle + t\,\|P_C(\mathbf{x}) - \mathbf{y}\|^2 \geq 0$$

for all $t \in [0, 1]$ and $\mathbf{y} \in C$ and we have to make $t \to 0$. For the implication (2.7) $\Rightarrow$ (2.6), notice that

$$0 \geq \langle \mathbf{x} - P_C(\mathbf{x}), \mathbf{y} - P_C(\mathbf{x})\rangle = \|\mathbf{x} - P_C(\mathbf{x})\|^2 - \langle \mathbf{x} - P_C(\mathbf{x}), \mathbf{x} - \mathbf{y}\rangle,$$

which yields (via the Cauchy–Bunyakovsky–Schwarz inequality),

$$\|\mathbf{x} - P_C(\mathbf{x})\| \cdot \|\mathbf{x} - \mathbf{y}\| \geq \langle \mathbf{x} - P_C(\mathbf{x}), \mathbf{x} - \mathbf{y} \rangle \geq \|\mathbf{x} - P_C(\mathbf{x})\|^2.$$

Thus $\|\mathbf{x} - \mathbf{y}\| \geq \|\mathbf{x} - P_C(\mathbf{x})\|$ for all $\mathbf{y} \in C$ and the proof is complete. ■

A simple geometric characterization of projection on a convex set (as well as an algorithm for finding the projection of a point on a simplex) can be found in the paper of A. Causa and F. Raciti [100].

It is important to reformulate Theorem 2.2.1 in the framework of approximation theory. Suppose that C is a nonempty closed subset in a real linear normed space E. The *set of best approximation* to $\mathbf{x} \in E$ from C is defined by

$$\mathcal{P}_C(\mathbf{x}) = \{\mathbf{z} \in C : d(\mathbf{x}, C) = \|\mathbf{x} - \mathbf{z}\|\};$$

We say that C is a *Chebyshev set* if $\mathcal{P}_C(\mathbf{x})$ is a singleton for all $\mathbf{x} \in E$, and a *proximinal set* if all the sets $\mathcal{P}_C(\mathbf{x})$ are nonempty. Theorem 2.2.1 asserts that all nonempty closed convex sets in a Hilbert space are Chebyshev sets. Is the converse true? One proves easily that every Chebyshev subset of a Banach space is necessarily (norm) closed. See Exercise 5. Convexity of such sets can be established only under additional assumptions.

2.2.2 Theorem (L. N. H. Bunt and T. S. Motzkin) *Every Chebyshev subset of* $\mathbb{R}^N$ *is convex.*

See R. Webster [483, pp. 362–365] for a proof based on Brouwer's fixed point theorem. Proofs based on the differentiability properties of the distance function $d_C\colon \mathbf{x} \to d(\mathbf{x}, C)$, are available in the paper by J.-B. Hiriart-Urruty [215], and in the monograph by L. Hörmander [227, pp. 62–63]. They are sketched in Exercises 5 and 6, Section 6.2.

V. Klee [253] proved in 1961 that all weakly closed Chebyshev subsets of a real Hilbert space are convex, but it is still unknown if the same works for all Chebyshev subsets. See [69], [215], and [480] for more information on this problem.

Hilbert spaces are not the only Banach spaces with the property that all nonempty closed convex subsets are Chebyshev sets. This is also true for all spaces $L^p(\mu)$, with $1 < p < \infty$ and, more generally, for all uniformly convex Banach spaces.

2.2.3 Definition *A Banach space is called* uniformly convex *if for every* $\varepsilon \in (0, 2]$, *there exists* $\delta > 0$ *such that*

$$\|\mathbf{x}\| = \|\mathbf{y}\| = 1 \text{ and } \|\mathbf{x} - \mathbf{y}\| \geq \varepsilon \text{ imply } \left\| \frac{\mathbf{x} + \mathbf{y}}{2} \right\| \leq 1 - \delta.$$

The condition $\|\mathbf{x}\| = \|\mathbf{y}\| = 1$ in the definition of uniformly convex spaces can be replaced by $\|\mathbf{x}\| \leq 1$ and $\|\mathbf{y}\| \leq 1$. See B. Beauzamy [36], Proposition 1, p. 190.

The uniform convexity of the spaces $L^p(\mu)$, with $1 < p < \infty$, follows from *Clarkson's inequalities* and the elementary fact that $(1-x)^{1/p} < 1 - x/p$ if $x \in (0,1)$ and $p > 1$. According to them, every pair of functions $u, v \in L^p(\mu)$ verifies

$$\left(\frac{\|u\|_{L^p}^p + \|v\|_{L^p}^p}{2}\right)^{1/p} \geq \left(\left\|\frac{u+v}{2}\right\|_{L^p}^q + \left\|\frac{u-v}{2}\right\|_{L^p}^q\right)^{1/q} \quad \text{for } p \in (1,2] \tag{2.9}$$

and

$$\frac{\|u\|_{L^p}^p + \|v\|_{L^p}^p}{2} \geq \left\|\frac{u+v}{2}\right\|_{L^p}^p + \left\|\frac{u-v}{2}\right\|_{L^p}^p \quad \text{for } p \in [2,\infty); \tag{2.10}$$

here $1/p + 1/q = 1$. For details, see Lemma A.2.4, Appendix A. Another simple proof is available in N. L. Carothers [94], pp. 117–119. Notice that only the case $1 < p \leq 2$ needs some effort because the case $2 \leq p < \infty$ follows from a combination between the triangle inequality and the convexity of the function x^p:

$$\begin{aligned}
\left(|a+b|^p + |a-b|^p\right)^{1/p} &\leq \left(|a+b|^2 + |a-b|^2\right)^{1/2} \\
&= 2^{1/2}\left(|a|^2 + |b|^2\right)^{1/2} \leq 2^{1/2}2^{1/2-1/p}\left(|a|^p + |b|^p\right)^{1/p} \\
&\leq 2^{1-1/p}\left(|a|^p + |b|^p\right)^{1/p}.
\end{aligned}$$

Clarkson's inequalities provide a substitute of the parallelogram law in the spaces $L^p(\mu)$, with $1 < p < \infty$, $p \neq 2$.

A couple of finer inequalities in L^p-spaces, the Hanner inequalities, make the object of Exercise 7, Section 3.5.

Every uniformly convex space is *strictly convex* in the sense that

$$\|\mathbf{x}\| = \|\mathbf{y}\| = 1 \text{ and } \left\|\frac{\mathbf{x}+\mathbf{y}}{2}\right\| = 1 \text{ imply } \mathbf{x} = \mathbf{y}.$$

See Exercise 8. The converse is true in the context of finite- dimensional Banach spaces and makes the object of Exercise 9. A condition equivalent to strict convexity is given in Exercise 7.

2.2.4 Theorem *Suppose that E is a uniformly convex Banach space and C is a nonempty closed convex subset of it. Then, for every $\mathbf{x} \in E$ there is a unique point $P_C(\mathbf{x})$ of C such that*

$$d(\mathbf{x}, C) = \|\mathbf{x} - P_C(\mathbf{x})\|.$$

In this context, the map $P_C : \mathbf{x} \to P_C(\mathbf{x})$ is called the *metric projection* of C.

Proof. The nontrivial case is where $\mathbf{x} \notin C$. Then $d = d(\mathbf{x}, C) > 0$ and there exists a sequence $(\mathbf{x}_n)_n$ of elements of C such that $\|\mathbf{x} - \mathbf{x}_n\| \to d$. Since C is convex,

$$d \leq \left\|\mathbf{x} - \frac{\mathbf{x}_m + \mathbf{x}_n}{2}\right\| \leq \frac{\|\mathbf{x} - \mathbf{x}_m\| + \|\mathbf{x} - \mathbf{x}_n\|}{2},$$

which yields $\left\| \frac{(\mathbf{x}-\mathbf{x}_m)+(\mathbf{x}-\mathbf{x}_n)}{2} \right\| \to d$ as $m, n \to \infty$. Then

$$\left\| \frac{\frac{\mathbf{x}-\mathbf{x}_m}{\|\mathbf{x}-\mathbf{x}_m\|} + \frac{\mathbf{x}-\mathbf{x}_n}{\|\mathbf{x}-\mathbf{x}_n\|}}{2} \right\| \to 1,$$

and taking into account the uniform convexity of E we infer that

$$\left\| \frac{\mathbf{x}-\mathbf{x}_m}{\|\mathbf{x}-\mathbf{x}_m\|} - \frac{\mathbf{x}-\mathbf{x}_n}{\|\mathbf{x}-\mathbf{x}_n\|} \right\| \to 0$$

as $m, n \to \infty$. Since

$$\begin{aligned} \|\mathbf{x}_m - \mathbf{x}_n\| &= \|(\mathbf{x}-\mathbf{x}_m) - (\mathbf{x}-\mathbf{x}_n)\| \\ &= \left\| \|\mathbf{x}-\mathbf{x}_m\| \frac{\mathbf{x}-\mathbf{x}_m}{\|\mathbf{x}-\mathbf{x}_m\|} - \|\mathbf{x}-\mathbf{x}_n\| \frac{\mathbf{x}-\mathbf{x}_n}{\|\mathbf{x}-\mathbf{x}_n\|} \right\| \\ &\le \|\mathbf{x}-\mathbf{x}_m\| \left\| \frac{\mathbf{x}-\mathbf{x}_m}{\|\mathbf{x}-\mathbf{x}_m\|} - \frac{\mathbf{x}-\mathbf{x}_n}{\|\mathbf{x}-\mathbf{x}_n\|} \right\| + \left| \|\mathbf{x}-\mathbf{x}_m\| - \|\mathbf{x}-\mathbf{x}_n\| \right|, \end{aligned}$$

we conclude that $(\mathbf{x}_n)_n$ is a Cauchy sequence in E; its limit, say $\mathbf{z}$, verifies the condition $d = \|\mathbf{x} - \mathbf{z}\|$. The uniqueness of the point $\mathbf{z}$ of best approximation follows from the fact that E is strictly convex. ∎

One can prove that all nonempty closed convex subsets of a reflexive Banach space are proximinal. See Corollary B.1.9, Appendix B. This fact extends Theorem 2.2.4 since every uniformly convex space is reflexive according to the Milman–Pettis theorem in Banach space theory. For a nice short argument of the Milman–Pettis theorem see W. B. Johnson and J. Lindenstrauss [235], p. 31.

Exercises

1. Prove that:

   (a) the orthogonal projection of a vector $\mathbf{x} \in \mathbb{R}^N$ onto the closed unit ball of $\mathbb{R}^N$ is given by $P(\mathbf{x}) = \mathbf{x}$ if $\|\mathbf{x}\| \le 1$ and $P(\mathbf{x}) = \mathbf{x}/\|\mathbf{x}\|$ if $\|\mathbf{x}\| > 1$;

   (b) the orthogonal projection of a vector $\mathbf{x} \in \mathbb{R}^N$ onto the affine set $C = \{\mathbf{x} : \langle \mathbf{x}, \mathbf{a} \rangle = b\}$ (with $\mathbf{a} \in \mathbb{R}^N \backslash \{0\}$ and $b \in \mathbb{R}$) is given by the formula $P_C(\mathbf{x}) = \mathbf{x} + (b - \langle \mathbf{x}, \mathbf{a} \rangle)\, \mathbf{a} / \|\mathbf{a}\|^2$;

   (c) the orthogonal projection of a vector $\mathbf{x} \in \mathbb{R}^N$ onto the positive orthant $\mathbb{R}^N_+$ is found by replacing each strictly negative component of $\mathbf{x}$ with 0.

   Remark. In connection with Exercise 1 (a) it is worth mentioning the following inequality due to J. L. Massera and J. J. Schäffer: if $\mathbf{x}$ and $\mathbf{y}$ are nonzero elements in a normed linear space, then

   $$\left\| \frac{\mathbf{x}}{\|\mathbf{x}\|} - \frac{\mathbf{y}}{\|\mathbf{y}\|} \right\| \le \frac{2}{\max\{\|\mathbf{x}\|, \|\mathbf{y}\|\}} \|\mathbf{x} - \mathbf{y}\|.$$

As was noticed by C. F. Dunkl and K. S. Williams, in the context of inner product space, the factor $2/\max\{\|\mathbf{x}\|, \|\mathbf{y}\|\}$ can be replaced by $2/(\|\mathbf{x}\| + \|\mathbf{y}\|)$. See J.-B. Hiriart-Urruty [216] for an elegant argument.

2. Find an explicit formula for the orthogonal projection $P_{\mathcal{E}}$ when $\mathcal{E}$ is the closed convex set in $\mathbb{R}^2$ bounded by the ellipse $x^2/a^2 + y^2/b^2 = 1$. Prove that for every point $A \notin \mathcal{E}$, the point $P_{\mathcal{E}}(A)$ is precisely the point X of the ellipse where the tangent to the ellipse is perpendicular to the line determined by A and X.

3. Let $\ell^\infty(2, \mathbb{R})$ be the space $\mathbb{R}^2$ endowed with the sup norm,
$$\|(x_1, x_2)\| = \sup\{|x_1|, |x_2|\},$$
and let C be the set of all vectors (x_1, x_2) such that $x_2 \geq x_1 \geq 0$. Prove that C is a nonconvex Chebyshev set.

4. Consider in $\mathbb{R}^2$ the nonconvex set $C = \{(x_1, x_2) : x_1^2 + x_2^2 \geq 1\}$. Prove that all points of $\mathbb{R}^2$, except the origin, admit a unique closest point in C.

5. (The nonexistence of metric projections in arbitrary Banach spaces) Consider the Banach space $C([0,1])$, of all continuous functions $f : [0,1] \to \mathbb{R}$, endowed with the sup norm. Prove that
$$C = \left\{ f \in C([0,1]) : \int_0^1 f(t)dt = 1, f(0) = 0, f(1) = 1 \right\}$$
is a closed convex cone in $C([0,1])$, and $d(0, C) = \inf\{\|f\| : f \in C\}$ is not attained.

6. Prove that every Chebyshev subset of a Banach space is necessarily closed.

7. Prove that a Banach space E is strictly convex if and only if whenever $\mathbf{x}$ and $\mathbf{y}$ are nonzero vectors in E, the equality $\|\mathbf{x} + \mathbf{y}\| = \|\mathbf{x}\| + \|\mathbf{y}\|$ implies $\mathbf{x} = c\mathbf{y}$ for some strictly positive number c.

   [*Hint:* Suppose $\|\mathbf{x} + \mathbf{y}\| = \|\mathbf{x}\| + \|\mathbf{y}\|$, $0 < \|\mathbf{x}\| \leq 1$ and $\|\mathbf{y}\| = 1$. Then the functions $\varphi_1(t) = \|t\mathbf{x} + \mathbf{y}\|$ and $\varphi_2(t) = t\|\mathbf{x}\| + \|\mathbf{y}\|$ verify $\varphi_1(0) = \varphi_2(0)$ and $\varphi_1(1) = \varphi_2(1)$. Since φ_1 is convex and φ_2 is affine, they must be equal for all $t \geq 0$. In particular, $\left\| \frac{\mathbf{x}}{\|\mathbf{x}\|} + \mathbf{y} \right\| = 2$.]

8. Prove that every uniformly convex Banach space is also strictly convex.

9. Let E be a finite-dimensional strictly convex space. Prove that E is uniformly convex.
   [*Hint*: Use the fact that every bounded sequence of elements of E has a convergent subsequence.]

2.3 Hyperplanes and Separation Theorems in Euclidean Spaces

The notion of a hyperplane represents a natural generalization of the notion of a line in $\mathbb{R}^2$ or a plane in $\mathbb{R}^3$. Hyperplanes are useful to split the whole space into two pieces (called half-spaces). A *hyperplane* in a normed linear space E is any set of constancy of a nonzero linear functional. In other words, a hyperplane is a set of the form

$$H = \{\mathbf{x} \in E : h(\mathbf{x}) = \alpha\}, \tag{2.11}$$

where $h\colon E \to \mathbb{R}$ is a suitable nonzero linear functional and α is a suitable scalar. In this case the sets

$$\{\mathbf{x} \in E : h(\mathbf{x}) \leq \alpha\} \quad \text{and} \quad \{\mathbf{x} \in E : h(\mathbf{x}) \geq \alpha\}$$

are called the *half-spaces* determined by H. We say that H *separates* two sets U and V if they lie in opposite half-spaces (and *strictly separates* U and V if one set is contained in $\{\mathbf{x} \in E : h(\mathbf{x}) < \alpha\}$ and the other in $\{\mathbf{x} \in E : h(\mathbf{x}) \geq \alpha\}$). When the functional h which appears in the representation formula (2.11) is continuous (that is, when h belongs to the dual space E^*) we say that the corresponding hyperplane H is *closed.*

In the context of $\mathbb{R}^N$, all linear functionals are continuous and thus all hyperplanes are closed. In fact, any linear functional $h\colon \mathbb{R}^N \to \mathbb{R}$ has the form $h(\mathbf{x}) = \langle \mathbf{x}, \mathbf{z}\rangle$, for some $\mathbf{z} \in \mathbb{R}^N$, uniquely determined by h. This follows directly from the linearity of h and the representation of $\mathbb{R}^N$ with respect to the canonical basis:

$$\begin{aligned} h(\mathbf{x}) = h\Big(\sum_{k=1}^{N} \mathbf{x}_k e_k\Big) &= \sum_{k=1}^{N} \mathbf{x}_k h(e_k) \\ &= \langle \mathbf{x}, \mathbf{z}\rangle, \end{aligned}$$

where $\mathbf{z} = \sum_{k=1}^{N} h(e_k) e_k$ is the *gradient* of h. Therefore, every hyperplane H in $\mathbb{R}^N$ is closed and has the form

$$H = \{\mathbf{x} : \langle \mathbf{x}, \mathbf{z}\rangle = \alpha\}.$$

Some authors define the hyperplanes as the maximal proper affine subsets H of E. Here *proper* means different from E. One can prove that the hyperplanes are precisely the translates of codimension-1 linear subspaces, and this explains the equivalence of the two definitions. The following results on the separation of convex sets by closed hyperplanes are part of a much more general theory that will be presented in Appendix B:

2.3.1 Theorem (Separation theorem) *Let U and V be two convex sets in a normed linear space E, with $\operatorname{int} U \neq \emptyset$ and $V \cap \operatorname{int} U = \emptyset$. Then there exists a closed hyperplane that separates U and V.*

2.3.2 Theorem (Strong separation theorem) *Let K and C be two disjoint nonempty convex sets in a normed linear space E with K compact and C closed. Then there exists a closed hyperplane that separates strictly K and C.*

The special case of Theorem 2.3.2 when K is a singleton is known as the *basic separation theorem*. See Exercises 3 and 4 for immediate applications.

The proofs of Theorems 2.3.1 and 2.3.2 in the finite-dimensional case are sketched in Exercises 1 and 2.

Notice that not every pair of disjoint closed convex sets can be strictly separated. An example in $\mathbb{R}^2$ is offered by the sets $K = \{(x, 0) : x \in \mathbb{R}\}$ and $C = \{(x, y) : x \in \mathbb{R}, y \geq e^x\}$. This shows that it is essential that at least one of the sets K and C is compact.

Next we introduce the notion of a *supporting hyperplane* to a convex set A in a normed linear space E.

2.3.3 Definition *We say that the hyperplane H* supports *A at a point $\mathbf{a}$ in A if $\mathbf{a} \in H$ and A is contained in one of the half-spaces determined by H.*

Theorem 2.3.1 assures the existence of a supporting hyperplane to any convex set A at a boundary point, provided that A has nonempty interior. When $E = \mathbb{R}^N$, the existence of a supporting hyperplane of U at a boundary point $\mathbf{a}$ will mean the existence of a vector $z \in \mathbb{R}^N$ and of a real number α such that

$$\langle \mathbf{a}, \mathbf{z} \rangle = \alpha \quad \text{and} \quad \langle \mathbf{x}, \mathbf{z} \rangle \leq \alpha \quad \text{for all } \mathbf{x} \in U.$$

A direct argument for the existence of a supporting hyperplane in the finite-dimensional case is given in Exercise 5. We end this section with a discussion on the geometry of convex sets in finite-dimensional spaces.

2.3.4 Definition *Let U be a convex subset of a linear space E. A point $\mathbf{z}$ in U is an* extreme point *if it is not an interior point of any linear segment in U, that is, if there do not exist distinct points $\mathbf{x}, \mathbf{y} \in U$ and numbers $\lambda \in (0, 1)$ such that*

$$\mathbf{z} = (1 - \lambda)\mathbf{x} + \lambda \mathbf{y}.$$

The extreme points of a triangle are its vertices. More generally, every polytope $A = \operatorname{conv}\{a_0, \ldots, a_m\}$ has finitely many extreme points, and they are among the points $a_0, \ldots, a_m$. All boundary points of a disc $\overline{D}_R(0) = \{(x, y) : x^2 + y^2 \leq R^2\}$ are extreme points; this is an expression of the rotundity of discs. The closed upper half-plane $\mathbf{y} \geq 0$ in $\mathbb{R}^2$ has no extreme point. The extreme points are the landmarks of compact convex sets in $\mathbb{R}^N$:

2.3.5 Theorem (H. Minkowski) *Every nonempty convex and compact subset K of $\mathbb{R}^N$ is the convex hull of its extreme points.*

Proof. We use induction on the dimension m of K. If $m = 0$ or $m = 1$, that is, when K is a point or a closed segment, the above statement is obvious. Assume the theorem is true for all compact convex sets of dimension at most $m \leq N - 1$.

Consider now a compact convex set K whose dimension is $m+1$ and embed it into a linear subspace E of dimension $m+1$. If $\mathbf{z}$ is a boundary point of K, then we can choose a supporting hyperplane $H \subset E$ for K through $\mathbf{z}$. The set $K \cap H$ is compact and convex and its dimension is less than or equal to m. By the induction hypothesis, $\mathbf{z}$ is a convex combination of extreme points of $K \cap H$. Or, any extreme point e of $K \cap H$ is also an extreme point of K. In fact, letting $H = \{t \in E : \varphi(t) = \alpha\}$, we may assume that K is included in the half-space $\varphi(t) \leq \alpha$. If $e = (1-\lambda)\mathbf{x} + \lambda\mathbf{y}$ with $\mathbf{x} \neq \mathbf{y}$ in K and $\lambda \in (0,1)$, then necessarily $\varphi(\mathbf{x}) = \varphi(\mathbf{y}) = \alpha$, that is, $\mathbf{x}$ and $\mathbf{y}$ should be in $K \cap H$, in contradiction with the choice of e. If $\mathbf{z}$ is an interior point of K, then each line through $\mathbf{z}$ intersects K in a segment whose endpoints belong to the boundary of K. Consequently, $\mathbf{z}$ is a convex combination of boundary points that in turn are convex combinations of extreme points. This ends the proof. ■

The result of Theorem 2.3.5 can be made more precise: *every point in a compact convex subset K of $\mathbb{R}^N$ is the convex combination of at most $N+1$ extreme points.* See Theorem 2.1.3.

The Krein–Milman theorem provides an extension of Minkowski's theorem to the framework of infinite-dimensional spaces. See Appendix B, Section B.3, for details.

Matrix theory related to probability and combinatorics emphasizes the importance of the *doubly stochastic* matrices. These are the square matrices $P = (p_{ij})_{i,j} \in \mathrm{M}_N(\mathbb{R})$ of positive real numbers, each of whose rows and columns sum to 1, that is,

$$\sum_{j=1}^{N} p_{ij} = 1 \text{ for } i = 1, ..., N \text{ and } \sum_{i=1}^{N} p_{ij} = 1 \text{ for } j = 1, ..., N.$$

The $N \times N$-dimensional doubly stochastic matrices constitute a compact convex set $\mathcal{B}_N$, known as the *Birkhoff polytope.* Since the conditions above form a set of $2N-1$ independent linear constraints, the dimension of this polytope is $(N-1)^2 = N^2 - (2N-1)$. Among the elements of $\mathcal{B}_N$ we distinguish the *permutation* matrices, that is, the $N!$ matrices obtained by permuting the rows of the identity matrix. Any permutation matrix $P = (p_{ij})_{i,j=1}^{N} \in \mathrm{M}_N(\mathbb{R})$ is associated to a permutation π of the set $\{1, ..., N\}$ via the formula

$$p_{ij} = \delta_{i\pi(j)}.$$

2.3.6 Theorem (G. Birkhoff [55]) *The Birkhoff polytope $\mathcal{B}_N$ is the convex hull of permutation matrices.*

Proof. According to the Krein–Milman theorem, the proof of this result can be reduced to the fact that the permutation matrices are precisely the extreme points of $\mathcal{B}_N$. The fact that every permutation matrix is an extreme point of $\mathcal{B}_N$ is simple. See Exercise 9. To end the proof it remains to prove that every extreme point of $\mathcal{B}_N$ is a permutation matrix, that is, a matrix having exactly one entry $+1$ in each row and in each column and all other entries are zero. The following ingenious argument is due to C. Villani [477], p. 15.

Assume that $P = (p_{ij})_{i,j}$ is an element of $\mathcal{B}_N$ having an entry $p_{i_0 j_0}$ that lies in $(0,1)$. Then choose distinct entries $p_{i_0 j_1}, p_{i_1 j_1}, p_{i_1 j_2}, p_{i_2 j_2}, \ldots$, that also lie in $(0,1)$, and continue till $j_k = j_0$ or $i_k = i_0$. In the first case, we have an even number of entries and we can perturb these entries (and only them) to produce two new doubly stochastic matrices whose arithmetic mean is P. In the second case, leave $p_{i_0 j_0}$ out of the list and repeat the previous step with the others entries. In both cases we conclude that P is not an extreme point. ■

The doubly stochastic matrices represent an important tool in the study of majorization. This will be presented in Chapter 4.

2.3.7 Remark (Relativization with respect to the affine hull and separation theorems) *One of the useful consequences of relative interior is that it allows us to relax interiority conditions in certain Euclidean space results. An easy application of the separation theorem yields the following fact: If C is finite-dimensional convex set in a finite-dimensional real Banach space E and $\mathbf{x} \in \operatorname{aff}(C) \setminus \operatorname{ri}(C)$, then there exists $x^* \in E^*$ such that $\sup x^*(C) \leq x^*(\mathbf{x})$ and x^* is nonconstant on C. Also, the separation theorem 2.3.1 admits the following "relative" variant: If C_1 and C_2 are convex subsets of $\mathbb{R}^N$ and $\operatorname{ri} C_1 \cap C_2 = \emptyset$, then there exists a nonzero vector $\mathbf{z} \in \mathbb{R}^N$ such that $\langle \mathbf{z}, \mathbf{x} \rangle \leq \langle \mathbf{z}, \mathbf{y} \rangle$ for all $\mathbf{x} \in C_1$ and all $\mathbf{y} \in C_2$.*

Exercises

1. Complete the following sketch of the proof of Theorem 2.3.2 in the case when $E = \mathbb{R}^N$:

   (a) First prove that the distance

   $$d = \inf\{\|\mathbf{x} - \mathbf{y}\| : \mathbf{x} \in K, \mathbf{y} \in C\}$$

   is attained for a pair $\mathbf{x}_0 \in K$ and $\mathbf{y}_0 \in C$, that is, $d = \|\mathbf{x}_0 - \mathbf{y}_0\|$.

   (b) Then notice that the hyperplane through $\mathbf{x}_0$, orthogonal to the linear segment $[\mathbf{x}_0, \mathbf{y}_0]$, determined by $\mathbf{x}_0$ and $\mathbf{y}_0$, has the equation $\langle \mathbf{y}_0 - \mathbf{x}_0, \mathbf{z} - \mathbf{x}_0 \rangle = 0$.

   (c) Fix arbitrarily a point $\mathbf{x} \in K$ and prove that $\langle \mathbf{y}_0 - \mathbf{x}_0, \mathbf{z} - \mathbf{x}_0 \rangle \leq 0$ for every point $\mathbf{z} \in [\mathbf{x}_0, \mathbf{x}]$ (and thus for every $\mathbf{z} \in K$). Conclude that every hyperplane through any point inside the segment $[\mathbf{x}_0, \mathbf{y}_0]$, orthogonal to this segment, separates strictly K and C.

2 Infer the finite-dimensional case of Theorem 2.3.1 from Theorem 2.3.2.

   [*Hint*: It suffices to assume that both sets U and V are closed. Then choose a point $\mathbf{x}_0 \in \operatorname{int} U$ and apply the preceding exercise to V and to the compact sets

   $$K_n = \{\mathbf{x}_0 + (1 - 1/n)(\mathbf{x} - \mathbf{x}_0) : \mathbf{x} \in U\} \cap \overline{B}_n(0)$$

for $n \in \mathbb{N}^*$. This gives us a sequence of unit vectors $\mathbf{u}_n$ and numbers α_n such that $\langle \mathbf{u}_n, \mathbf{x} \rangle \leq \alpha_n$ for $\mathbf{x} \in K_n$ and $\langle \mathbf{u}_n, \mathbf{y} \rangle \geq \alpha_n$ for $\mathbf{y} \in V$. Since $(\mathbf{u}_n)_n$ and $(\alpha_n)_n$ are bounded, they admit converging subsequences, say to $\mathbf{u}$ and α respectively. Conclude that $H = \{\mathbf{z} : \langle \mathbf{u}, \mathbf{z} \rangle = \alpha\}$ is the desired separation hyperplane.]

3. (Farkas' alternative theorem) Suppose that A is an $n \times m$-dimensional real matrix and $\mathbf{b} \in \mathbb{R}^n$. Prove that precisely one of the following two alternatives is true:

   (F1) the linear system $A\mathbf{x} = \mathbf{b}$ admits a solution $\mathbf{x} \in \mathbb{R}^m_+$;

   (F2) there is $\mathbf{y} \in \mathbb{R}^n$ such that $A^*\mathbf{y} \geq \mathbf{0}$ and $\langle \mathbf{y}, \mathbf{b} \rangle < 0$.
   As an application, infer that the following system

$$\begin{pmatrix} 4 & 1 & -5 \\ 1 & 0 & 2 \end{pmatrix} \begin{pmatrix} x_1 \\ x_2 \\ x_3 \end{pmatrix} = \begin{pmatrix} 1 \\ 1 \end{pmatrix}$$

   has a positive solution.

   [*Hint:* Notice first that (F1) and (F2) can't be true simultaneously. Thus, Farkas' result is equivalent to the following statement: there is a vector $\mathbf{y}$ such that $A^*\mathbf{y} \geq 0$ and $\langle \mathbf{y}, \mathbf{b} \rangle < 0$ if and only if b is not in the closed convex set $C = \left\{A\mathbf{x} : \mathbf{x} \in \mathbb{R}^m_+\right\}$. Then apply the basic separation theorem.]

4. (Gordan's alternative theorem) Under the assumptions of Exercise 3, prove that exactly one of the following two alternatives is true:

   (G1) there is $\mathbf{x} \in \mathbb{R}^m$ such that $A\mathbf{x} < 0$;

   (G2) there is $\mathbf{y} \in \mathbb{R}^n_+$, $\mathbf{y} \neq 0$, such that $A^*\mathbf{y} = 0$.

   [*Hint:* Use Farkas' alternative theorem.]

5. (The spectrum of Markov matrices) A *Markov matrix* is a matrix $A \in \mathrm{M}_N(\mathbb{R})$ whose elements a_{ij} are positive and the sum of the elements in each column is one, that is,

$$a_{ij} \geq 0 \text{ and } \sum_{i=1}^{N} a_{ij} = 1 \text{ for all } j = 1, 2, ..., N.$$

   (a) Prove that all eigenvalues λ of a Markov matrix verify $|\lambda| \leq 1$.

   (b) Prove that if $A \in \mathrm{M}_N(\mathbb{R})$ is a Markov matrix, then 1 is an eigenvalue to which it corresponds an eigenvector $v \in \mathbb{R}^N_+$.

   [*Hint:* (b) We have to prove the existence of a vector $\mathbf{x} \geq 0$ such that $(A - \mathrm{I})\mathbf{x} = 0$ and $\langle \mathbf{x}, \mathbf{e} \rangle = 1$, where $\mathbf{e} = (1, 1, ..., 1)$. Apply Farkas' alternative for the $(N+1, N)$-dimensional matrix

$$\tilde{A} = \begin{pmatrix} A - \mathrm{I} \\ \mathbf{e} \end{pmatrix}$$

and the vector $\tilde{\mathbf{b}} = (\mathbf{0}, 1) \in \mathbb{R}^{N+1}$.]

6. (The support theorem) Assume that $E = \mathbb{R}^N$ and $\mathbf{a}$ is a point in the relative boundary of the convex subset A of E. Prove that there exists a supporting hyperplane H to A at $\mathbf{a}$ which differs from $\operatorname{aff}(A)$.

   [*Hint:* We may assume that A is closed, by replacing A with $\overline{A}$. Choose a point $\mathbf{x}_0 \in S_1(\mathbf{a}) = \{\mathbf{x} : \|\mathbf{x} - \mathbf{a}\| = 1\}$ such that

   $$d(\mathbf{x}_0, A) = \sup\{d(\mathbf{x}, A) : \mathbf{x} \in S_1(\mathbf{a})\},$$

   that is, $\mathbf{x}_0$ is the farthest point from A. Notice that $\mathbf{a}$ is the point of A closest to $\mathbf{x}_0$ and conclude that the hyperplane $H = \{\mathbf{z} : \langle \mathbf{x}_0 - \mathbf{a}, \mathbf{z} - \mathbf{a} \rangle = 0\}$ supports U at $\mathbf{a}$.]

7. Prove that every closed convex set in $\mathbb{R}^N$ is the intersection of closed half-spaces which contain it.

8. A set in $\mathbb{R}^N$ is a *polyhedron* if it is a finite intersection of closed half-spaces. Prove that:

   (a) Every compact polyhedron is a polytope (the converse is also true);

   (b) Every polytope has finitely many extreme points;

   (c) $\operatorname{Sym}^+(2, \mathbb{R})$ is a closed convex cone (with interior $\operatorname{Sym}^{++}(2, \mathbb{R})$) but not a polyhedron.

9. Prove that every permutation matrix is an extreme point of the Birkhoff polytope $\mathcal{B}_N$.

   [*Hint:* A permutation matrix has exactly one entry $+1$ in each row and in each column and all other entries are zero.]

10. Let C be a nonempty subset of a real Hilbert space H. The *polar* set of C is the set

    $$C^\circ = \{\mathbf{y} \in H : \langle \mathbf{x}, \mathbf{y} \rangle \leq 1 \text{ for every } \mathbf{x} \in C\}.$$

    (a) Prove that C° is a closed convex set containing $\mathbf{0}$ and $C \subset D$ implies $D^\circ \subset C^\circ$.

    (b) (The bipolar theorem) Infer from the basic separation theorem 2.3.2 that the bipolar $C^{\circ\circ} = (C^\circ)^\circ$ of C verifies the formula

    $$C^{\circ\circ} = \overline{\operatorname{conv}}(C \cup \{\mathbf{0}\}).$$

    This implies that the polarity is an involution of the set of all closed and convex subsets of H that contain the origin.

    (c) Infer that the polar of the closed unit ball B of H is B itself and the polar of a linear subspace M is the orthogonal space $M^\perp$.

11. (The dual cone) Suppose that C is a cone in a Hilbert space H. The *dual cone* of C is the cone

$$C^* = \{\mathbf{y} \in H : \langle \mathbf{x}, \mathbf{y} \rangle \geq 0 \text{ for all } \mathbf{x} \in C\}.$$

Prove that:

(a) The dual cone is equal to the negative of the polar cone;

(b) $\mathbb{R}_+^N$ and $\mathrm{Sym}^+(N, \mathbb{R})$ are self-dual cones, that is, equal to their duals;

(c) The dual of the monotone cone $\mathbb{R}_{\geq}^N = \{(x_1, \dots, x_N) \in \mathbb{R}^N : x_1 \geq \dots \geq x_N\}$ is the cone

$$\left(\mathbb{R}_{\geq}^N\right)^* = \left\{\mathbf{y} \in \mathbb{R}^N : \sum_{k=1}^{m} y_k \geq 0 \text{ for } m = 1, 2, ..., N-1, \text{ and } \sum_{k=1}^{N} y_k = 0\right\}.$$

[*Hint:* (b) Use Abel's partial summation formula.]

12. Prove that the *norm cone* $C = \{(\mathbf{x}, t) \in \mathbb{R}^N \times [0, \infty) : \|\mathbf{x}\| \leq t\}$ is self-dual. What is happening if the natural norm on $\mathbb{R}^N$ (that is, the Euclidean norm) is replaced with an equivalent norm, for example, with $\|\mathbf{x}\|_p = \left(\sum_{k=1}^N |x_k|^p\right)^{1/p}$, for $p \in [1, \infty)$?

13. Suppose that C is a closed cone in a Hilbert space H. Prove that $P_{C^*}(\mathbf{x}) = \mathbf{x} - P_C(\mathbf{x})$ for all $\mathbf{x} \in H$ (which implies that P_C and P_{C^*} determine each other).

14. Prove that every nonempty compact and convex set K in a Hilbert space contains extreme points.

[*Hint.* Consider the case of the points $\mathbf{z} \in K$ such that $\|\mathbf{z}\|^2 = \sup\{\|\mathbf{x}\|^2 : \mathbf{x} \in K\}$.]

2.4 Ordered Linear Spaces

An *ordered linear space* is any real linear space E endowed with a (partial) order relation $\leq$ such that for all $\mathbf{x}, \mathbf{y}, \mathbf{z} \in E$ and all numbers $\lambda \geq 0$ we have

$$\mathbf{x} \leq \mathbf{y} \text{ implies } \mathbf{x} + \mathbf{z} \leq \mathbf{y} + \mathbf{z}$$
$$\mathbf{x} \leq \mathbf{y} \text{ implies } \lambda\mathbf{x} \leq \lambda\mathbf{y}.$$

We write $\mathbf{x} < \mathbf{y}$ for $\mathbf{x} \leq \mathbf{y}$ and $\mathbf{x} \neq \mathbf{y}$, $\mathbf{y} \geq \mathbf{x}$ (respectively $\mathbf{y} > \mathbf{x}$) means $\mathbf{x} \leq \mathbf{y}$ ($\mathbf{x} < \mathbf{y}$). The set

$$E_+ = \{\mathbf{x} \in E : \mathbf{x} \geq 0\},$$

of positive elements, is called the *positive cone*. Necessarily, the positive cone is proper, that is, $C \cap (-C) = \{0\}$. We will always assume that the positive cone is also *generating* in the sense that $E = C - C$; this fact assures that every element of E is the difference of two positive elements.

Conversely, every proper and generating convex cone C in a linear space E makes it an ordered linear space with respect to the ordering

$$\mathbf{x} \leq \mathbf{y} \quad \text{if and only if } \mathbf{y} - \mathbf{x} \in C.$$

The natural order relation on the N-dimensional space $\mathbb{R}^N$ is that defined coordinatewise by the formula

$$\mathbf{x} \leq \mathbf{y} \text{ in } \mathbb{R}^N \text{ if and only if } x_k \leq y_k \text{ for all } k \in \{1, ..., N\},$$

where $\mathbf{x} = (x_1, ..., x_N)$ and $\mathbf{y} = (y_1, ..., y_N)$. In this case the positive cone is precisely the positive orthant. Every pair of elements $\mathbf{x}$ and $\mathbf{y}$ admits an infimum and a supremum given by

$$\inf \{\mathbf{x}, \mathbf{y}\} = (\min \{x_1, y_1\}, ..., \min \{x_N, y_N\})$$

and

$$\sup \{\mathbf{x}, \mathbf{y}\} = (\max \{x_1, y_1\}, ..., \max \{x_N, y_N\});$$

$\inf \{\mathbf{x}, \mathbf{y}\}$ and $\sup \{\mathbf{x}, \mathbf{y}\}$ may differ from both $\mathbf{x}$ and $\mathbf{y}$ when $N \geq 2$. For example, $\mathbf{x} = (2, 1)$ and $\mathbf{y} = (1, 2)$ are not comparable to each other in $\mathbb{R}^2$, a phenomenon that marks a serious difference between $\mathbb{R}$ and the other N-dimensional spaces $\mathbb{R}^N$. Also, in higher dimensions, balls and order intervals are distinct. The open and respectively the closed interval of endpoints $\mathbf{x} \leq \mathbf{y}$ in $\mathbb{R}^N$ are defined by the formulas $(\mathbf{x}, \mathbf{y}) = \{\mathbf{z} \in \mathbb{R}^N : \mathbf{x} < \mathbf{z} < \mathbf{y}\}$ and $[\mathbf{x}, \mathbf{y}] = \{\mathbf{z} \in \mathbb{R}^N : \mathbf{x} \leq \mathbf{z} \leq \mathbf{y}\}$. Both are convex sets.

Ordered linear spaces such that every pair of elements admits an infimum and a supremum are called *linear lattices*. Every element $\mathbf{x}$ in a linear lattice admits positive part $\mathbf{x}^+ = \sup \{\mathbf{x}, 0\}$, negative part $\mathbf{x}^- = \sup \{-\mathbf{x}, 0\}$, and absolute value $|\mathbf{x}| = \sup \{-\mathbf{x}, \mathbf{x}\}$.

$\mathbb{R}^N$ and most of the usual function spaces (such as $C(K)$ and $L^p(\mu)$ for $1 \leq p \leq \infty$) are not only linear lattices but also *Banach lattices*, because they verify the following condition of compatibility between order and norm:

$$|\mathbf{x}| \leq |\mathbf{y}| \text{ implies } \|\mathbf{x}\| \leq \|\mathbf{y}\|.$$

The linear space $\mathrm{M}_N(\mathbb{R})$ endowed with the Hilbert–Schmidt norm and the order inherited from $\mathbb{R}^{N \times N}$ is also a Banach lattice; the positive cone is constituted by all matrices with positive coefficients.

If E is a Banach lattice, then its dual E^* is also a Banach lattice with respect to the cone

$$E^*_+ = \{x^* \in E^* : x^*(\mathbf{x}) \geq 0 \quad \text{for every } \mathbf{x} \geq \mathbf{0}\}.$$

See Exercise 2.

The theory of Banach lattices can be covered from many sources. We recommend the monograph of H. H. Schaefer [434].

2.4.1 Remark *It is important to notice that all numerical inequalities of the form*

$$f(x_1, \ldots, x_n) \geq 0 \quad \text{for } x_1, \ldots, x_n \geq 0, \tag{2.12}$$

where f is a continuous and positively homogeneous function of degree 1 *(that is, $f(\lambda x_1, \ldots, \lambda x_n) = \lambda f(x_1, \ldots, x_n)$ for $\lambda \geq 0$), extend to the context of Banach lattices, via a functional calculus invented by A. J. Yudin and J. L. Krivine. This allows us to replace the real variables of f by positive elements of a Banach lattice. See* [288, pp. 40–43]. *Particularly, this is the case of the AM–GM inequality, Rogers–Hölder inequality and Minkowski inequality. Also, all numerical inequalities of the form* (2.12)*, attached to continuous functions, extend (via the functional calculus with self-adjoint elements) to the context of $C^\star$-algebras. In fact, the n-tuples of real numbers can be replaced by n-tuples of mutually commuting positive elements of a $C^\star$-algebra. See* [115].

Exercises

1. Suppose that E is a Banach lattice and $x^* : E \to \mathbb{R}$ is a positive linear functional, that is, a linear functional such that

$$\mathbf{x} \in E_+ \text{ implies } x^*(\mathbf{x}) \geq 0.$$

   Prove that x^* is necessarily continuous and

$$\|x^*\| = \sup \{x^*(\mathbf{x}) : \mathbf{x} \in E_+, \|\mathbf{x}\| \leq 1\}.$$

2. Suppose that E is a Banach lattice. Prove that E^* is also Banach lattice, where the absolute value of a functional x^* verifies the formula

$$|x^*|(\mathbf{x}) = \sup \{x^*(\mathbf{y}) : \mathbf{y} \in E, |\mathbf{y}| \leq \mathbf{x}\} \qquad \text{for all } \mathbf{x} \geq 0.$$

   Moreover, every family $(x^*_\alpha)_\alpha$ of elements of E^* which is bounded above by a functional $z^* \in E^*$ admits a least upper bound.

   [*Hint.* Without loss of generality we may assume that the family $\mathcal{X} = (x^*_\alpha)_\alpha$ is upward directed (that is, every pair of elements is bounded above by an element of the family $\mathcal{X}$). For this, add to $\mathcal{X}$ the set of suprema of all finite subfamilies. Then $(\sup_\alpha x^*_\alpha)(\mathbf{x}) = \sup_\alpha x^*_\alpha(\mathbf{x})$ for all $\mathbf{x} \geq 0$.]

2.5 $\mathrm{Sym}(N, \mathbb{R})$ as a Regularly Ordered Banach Space

An important example of ordered Banach space which is not a Banach lattice is provided by the space $\mathrm{Sym}(N, \mathbb{R})$, of all symmetric matrices $A \in \mathrm{M}_N(\mathbb{R})$, when endowed with the positive cone $\mathrm{Sym}^+(N, \mathbb{R})$ and the *operator norm*

$$\|A\| = \sup_{\|\mathbf{x}\| \leq 1} \|A\mathbf{x}\| = \sup_{\|\mathbf{x}\| = 1} \|A\mathbf{x}\|.$$

The computation of the operator norm in the case of a symmetric matrix A can be done more conveniently via the formulas

$$\|A\| = \sup_{\|\mathbf{x}\|\leq 1} \langle A\mathbf{x}, \mathbf{x}\rangle = \sup_{\|\mathbf{x}\|=1} \langle A\mathbf{x}, \mathbf{x}\rangle; \tag{2.13}$$

see R. Bhatia [51], p. 120. An immediate consequence is the formula

$$\pm A \leq \|A\|\,\mathrm{I} \quad \text{for every } A \in \mathrm{Sym}(N, \mathbb{R}), \tag{2.14}$$

where I is the identity matrix of $\mathbb{R}^N$. Therefore

$$\|A\| = \inf\{\|B\| : -B \leq A \leq B\} \quad \text{for every } A \in \mathrm{Sym}(N, \mathbb{R}). \tag{2.15}$$

A remarkable fact about $\mathrm{Sym}(N, \mathbb{R})$ is the possibility to define functions of a symmetric matrix in a natural way. Indeed, every matrix $A \in \mathrm{Sym}(N, \mathbb{R})$ has N real eigenvalues $\lambda_1(A), ..., \lambda_N(A)$, counted by multiplicity, and the space $\mathbb{R}^N$ admits an orthonormal basis consisting of eigenvectors $\mathbf{v}_1, ..., \mathbf{v}_N$ such that $A\mathbf{v}_k = \lambda_k(A)\mathbf{v}_k$ for all k. This gives rise to the following representation of A :

$$A\mathbf{x} = \sum_{k=1}^{N} \lambda_k(A)\langle \mathbf{x}, \mathbf{v}_k\rangle \mathbf{v}_k \quad \text{for every } \mathbf{x} \in \mathbb{R}^N.$$

The matrix A can be represented *uniquely* as

$$A = \sum_{i=1}^{r} \lambda_i(A) P_i, \tag{2.16}$$

where $\lambda_1(A), ..., \lambda_r(A)$ are the distinct eigenvalues of A and the P_i 's are the orthogonal projections onto the corresponding eigenspaces

$$E_i = \left\{\mathbf{x} \in \mathbb{R}^N : A\mathbf{x} = \lambda_i(A)\mathbf{x}\right\}.$$

One can easily show that $P_iP_j = 0$ for $i \neq j$ and $\sum_{i=1}^{r} P_i = \mathrm{I}$.

The representation (2.16) is known as the *spectral decomposition* of A. It allows us to associate to *every* real function f defined on the spectrum $\sigma(A)$ of A the self-adjoint operator

$$f(A) = \sum_{i=1}^{r} f(\lambda_i(A)) P_i.$$

The main features of this construction are listed as follows:

$$\begin{cases} (f+g)(A) = f(A) + g(A) \\ (fg)(A) = f(A)g(A) = g(A)f(A) \\ f \geq 0 \text{ implies } f(A) \geq 0 \\ \|f(A)\| = \sup\{|f(\lambda)| : \lambda \in \sigma(A)\}. \end{cases} \tag{2.17}$$

As a consequence, one can define the positive part A^+, the negative part A^-, and modulus $|A|$ for every symmetric matrix A. Clearly, $A = A^+ - A^-$ and $|A| = A^+ + A^-$, but in general it is not true that $|A + B| \leq |A| + |B|$. See Exercise 1 for an example.

2.5.1 Remark *The* diagonal map

$$\mathrm{Diag} : \mathbb{R}^N \to \mathrm{Sym}(N, \mathbb{R})$$

associates to every vector $\mathbf{x}$ *the diagonal matrix* $\mathrm{Diag}(\mathbf{x})$, *with diagonal entries* $x_1 ..., x_N$, *the coordinates of* $\mathbf{x}$. *Using it, every matrix* $A \in \mathrm{Sym}(n, \mathbb{R})$ *can be factorized as*

$$A = UDU^{-1},$$

where

$$D = \begin{pmatrix} \lambda_1(A) & & 0 \\ & \ddots & \\ 0 & & \lambda_N(A) \end{pmatrix},$$

and $U \in \mathrm{M}_N(\mathbb{R})$ *is an orthogonal matrix, that is, a real square matrix with orthonormal columns. With respect to this decomposition, the functions of* A *can be represented as*

$$f(A) = U \begin{pmatrix} f(\lambda_1(A)) & & 0 \\ & \ddots & \\ 0 & & f(\lambda_N(A)) \end{pmatrix} U^{-1}.$$

Given an arbitrary matrix $A \in \mathrm{M}_N(\mathbb{R})$, the matrix A^*A is always positive and its square root,

$$|A| = (A^*A)^{1/2}, \tag{2.18}$$

is called the *modulus* of A. For a symmetric matrix, this definition of modulus agrees with that introduced before in this section. The reason for taking A^*A rather than AA^* in (2.18) is the formula

$$\| |A| \mathbf{x} \| = \| A\mathbf{x} \| \quad \text{for all } \mathbf{x},$$

a fact that leads to the polar decomposition of a matrix.

The eigenvalues $s_1(A), s_2(A), ..., s_N(A)$ of $|A|$ counted by multiplicity are called the *singular values* of A.

2.5.2 Theorem (Polar decomposition of a matrix) *Every matrix* $A \in \mathrm{M}_N(\mathbb{R})$ *admits a decomposition of the form* $A = U|A|$, *where* U *is a suitable unitary matrix. If* $A > 0$, *then* U *is uniquely determined by* A.

Proof. Notice first that $\|A\mathbf{x}\|^2 = \langle A\mathbf{x}, A\mathbf{x} \rangle = \langle A^T A\mathbf{x}, \mathbf{x} \rangle = \| |A| \mathbf{x} \|^2$, so $\ker A = \ker |A|$ and A and $|A|$ have the same rank n. Since $|A|$ is symmetric, there exists an orthonormal basis $\varphi_1, ..., \varphi_N$ of $\mathbb{R}^N$ consisting of eigenvectors of $|A|$ and $|A| \varphi_k = s_k(A) \varphi_k$ for all k. The vectors $\frac{1}{s_1(A)} A\varphi_1, ... \frac{1}{s_n(A)} A\varphi_n$ form an orthonormal family since

$$\langle \frac{1}{s_i(A)} A\varphi_i, \frac{1}{s_j(A)} A\varphi_j \rangle = \frac{1}{s_i(A)\, s_j(A)} \langle A\varphi_i, A\varphi_j \rangle = \frac{s_i^2(A)}{s_i(A)\, s_j(A)} \langle \varphi_i, \varphi_j \rangle = \delta_{ij}$$

for all $i, j \in \{1, ..., n\}$. If $n < N$ we complete this family with vectors $\psi_{n+1}, ..., \psi_N$ in $\ker A = \ker |A|$ to obtain an orthonormal basis of $\mathbb{R}^N$. The proof ends by choosing U as the matrix that acts on this new basis as follows:

$$U(\varphi_k) = \frac{1}{s_k(A)} A\varphi_k \text{ for } 1 \leq k \leq n \text{ and } U\varphi_k = \psi_k \text{ for } n+1 \leq k \leq N.$$

■

Polar decomposition easily yields the *singular value decomposition:* for every matrix $A \in \mathrm{M}_N(\mathbb{R})$ there exist two orthogonal matrices U and V such that

$$A = UDV^*,$$

where D is the diagonal matrix whose diagonal entries are the singular values of A. Equivalently, there exist two orthonormal families $(\mathbf{u}_k)_k$ and $(\mathbf{v}_k)_k$ such that

$$A = \sum_{k=1}^{N} s_k(A)\langle \cdot, \mathbf{u}_k \rangle \mathbf{v}_k.$$

According to E. B. Davies [125], a real Banach space E endowed with a closed and generating cone E_+ such that

$$\|\mathbf{x}\| = \inf \{\|\mathbf{y}\| : -\mathbf{y} \leq \mathbf{x} \leq \mathbf{y}\} \quad \text{for all } \mathbf{x} \in E,$$

is called a *regularly ordered Banach space*. Examples are Banach lattices and some other spaces such as $\mathrm{Sym}(n, \mathbb{R})$.

2.5.3 Theorem *If E is a* regularly ordered Banach space, then

$$\mathbf{x} \leq \mathbf{y} \textit{ in } E \textit{ is equivalent to } x^*(\mathbf{x}) \leq x^*(\mathbf{y}) \textit{ for all } x^* \in E_+^*.$$

Proof. The necessity part is trivial. For the sufficiency part assume that $\mathbf{y} - \mathbf{x}$ is not in E_+ and apply the basic separation theorem 2.3.2 to arrive at a contradiction. ■

For a not necessarily linear map $A : E \to F$, acting on regularly ordered Banach spaces, one can introduce the concepts of positivity and order monotonicity as follows: A is called *positive* if it maps positive elements into positive elements and *order-preserving* (or monotonically increasing) if

$$\mathbf{x} \leq \mathbf{y} \text{ in } E \text{ implies } A(\mathbf{x}) \leq A(\mathbf{y});$$

the concept of *order-reversing* (or monotonically decreasing) map can be introduced in a similar way, reversing the inequality between $A(\mathbf{x})$ and $A(\mathbf{y})$. These concepts are important in nonlinear analysis. See Exercise 8.

When A is linear, positivity and monotone increasing are equivalent. It is worth noticing that in the case of symmetric matrices, positivity in this sense and positivity in the sense of membership to the cone $\mathrm{Sym}^+(N, \mathbb{R})$ are distinct facts. See Exercises 4 and 5.

Exercises

1. The modulus of a matrix is not a genuine modulus because the inequality $|A+B| \leq |A| + |B|$ may fail. Here is an example in $\mathrm{M}_2(\mathbb{R})$, due to E. Nelson. Let

$$A = \begin{pmatrix} 1 & 0 \\ 0 & -1 \end{pmatrix} \quad \text{and} \quad B = \begin{pmatrix} 0 & 1 \\ 1 & 0 \end{pmatrix}.$$

Prove that $|(A+\mathrm{I}) + (B-\mathrm{I})| \not\leq |A+\mathrm{I}| + |B-\mathrm{I}|$.

2. (The trace functional) Suppose $A \in \mathrm{M}_N(\mathbb{R})$, $A = (a_{ij})_{i,j}$. By definition, the trace of A is the sum $\operatorname{trace}(A) = \sum_{k=1}^N a_{kk}$, of its diagonal elements. From the expression of the characteristic polynomial of A we also know that that $\operatorname{trace}(A)$ equals the sum of all eigenvalues of A, counted with their multiplicity.

   (a) Prove that $\operatorname{trace}\colon \mathrm{M}_N(\mathbb{R}) \to \mathbb{R}$ is a linear map such that $\operatorname{trace}(A) \geq 0$ whenever $A \in \mathrm{Sym}^+(N,\mathbb{R})$ and $\operatorname{trace}(AB) = \operatorname{trace}(BA)$ for all $A, B \in \mathrm{M}_N(\mathbb{R})$. The inequality $\frac{1}{N}\operatorname{trace}(A) \geq (\det(A))^{1/N}$ for $A \in \mathrm{Sym}^+(N,\mathbb{R})$ is a matrix variant of the *AM–GM* inequality.

   (b) Prove that for every orthonormal basis $(\mathbf{u}_k)_k$ of $\mathbb{R}^N$ we have

$$\operatorname{trace}(A) = \sum_{k=1}^{N} \langle A\mathbf{u}_k, \mathbf{u}_k \rangle.$$

3. The Hilbert–Schmidt norm $\|\cdot\|_{HS}$ on $\mathrm{M}_N(\mathbb{R})$ was introduced in Section 2.1 via formula (2.5). This norm is associated to the inner product $\langle A, B \rangle_{HS} = \operatorname{trace}(B^* A)$, where B^* denotes the transpose of B.

   (a) Using spectral decompostion, show that $\|A\|_{HS} = \left(\sum_{k=1}^N \sigma_k^2\right)^{1/2}$, where the σ_ks are the singular values of A, counted by multiplicity.

   (b) Prove that every linear functional f on $\mathrm{M}_N(\mathbb{R})$ can be represented uniquely as $f(X) = \operatorname{trace}(A^* X)$, for some $A \in \mathrm{M}_N(\mathbb{R})$.

4. (Spectral characterization of the membership to the cone $\mathrm{Sym}^+(N,\mathbb{R})$) Prove that:

   (a) A symmetric matrix A belongs to $\mathrm{Sym}^+(N,\mathbb{R})$ if and only if its eigenvalues are positive;

   (b) A symmetric matrix A belongs to $\mathrm{Sym}^{++}(N,\mathbb{R})$ if and only if its eigenvalues are strictly positive

   Remark. A practical criterion to decide whether a symmetric matrix is (strictly) positive is due to Sylvester. See Theorem 3.8.7. This criterion is especially useful when employing the Second Derivative Test for local extrema of functions of several variables.

5. (a) Prove that a matrix $A \in \mathrm{M}_N(\mathbb{R})$ maps $\mathbb{R}^N_+$ into $\mathbb{R}^N_+$ if and only if all its coefficients are positive. These matrices constitute a proper generating and convex cone.
(b) Prove that the order induced by aforementioned cone makes $\mathrm{M}_N(\mathbb{R})$ a Banach lattice when endowed with the Hilbert–Schmidt norm. Describe A^+, A^-, and $|A|$ for $A \in M_N(\mathbb{R})$.
(c) Consider the case of the matrix $\begin{pmatrix} 1 & 2 \\ 1 & 1 \end{pmatrix}$, to see that the spectrum of a matrix with strictly positive coefficients may contain negative eigenvalues.

6. (A matrix extension of the Cauchy–Bunyakovsky–Schwarz inequality) Suppose that $A, B \in \mathrm{Sym}(N, \mathbb{R})$ are two matrices.
(a) Prove that $|\langle A\mathbf{x}, B\mathbf{y}\rangle|^2 \leq \langle A^2\mathbf{x}, \mathbf{x}\rangle \langle B^2\mathbf{y}, \mathbf{y}\rangle$ for all $\mathbf{x}, \mathbf{y} \in \mathbb{R}^N$.
(b) Infer that $|\langle A\mathbf{x}, \mathbf{y}\rangle|^2 \leq \langle A\mathbf{x}, \mathbf{x}\rangle \langle A\mathbf{y}, \mathbf{y}\rangle$ in the case where A is positive.
(c) Suppose that A is strictly positive and $\mathbf{y} \in \mathbb{R}^N$. Prove that

$$\max_{\mathbf{x}\neq 0} \frac{|\langle \mathbf{x}, \mathbf{y}\rangle|^2}{\langle A\mathbf{x}, \mathbf{x}\rangle} = \langle A^{-1}\mathbf{y}, \mathbf{y}\rangle.$$

The maximum is attained for $\mathbf{x} = cA^{-1}\mathbf{y}$, with $c \neq 0$ arbitrary. This solves a filtering problem in signal processing. See Ch. L. Byrne [87], pp. 10–11.

7. The *numerical range* of a symmetric matrix $A \in \mathrm{M}_N(\mathbb{R})$ is the set

$$W(A) = \{\langle A\mathbf{x}, \mathbf{x}\rangle : \mathbf{x} \in \mathbb{R}^N, \|\mathbf{x}\| = 1\}.$$

(a) Infer from the spectral decomposition of symmetric matrices that the spectral bounds

$$\lambda_{\min}(A) = \inf\{\lambda : \lambda \in \sigma(A)\} \text{ and } \lambda_{\max}(A) = \max\{\lambda : \lambda \in \sigma(A)\}$$

verify $\lambda_{\min}(A) = \inf_{\|\mathbf{x}\|=1}\langle A\mathbf{x}, \mathbf{x}\rangle$ and $\lambda_{\max}(A) = \max_{\|\mathbf{x}\|=1}\langle A\mathbf{x}, \mathbf{x}\rangle$.
(b) Prove that $W(A) = W(UAU^{-1})$ for every orthogonal matrix U and conclude that $W(A)$ is an interval.

Remark. Part (b) is a particular case of the Toeplitz–Hausdorff theorem on the convexity of numerical range: The numerical range, $W(T) = \{\langle T\mathbf{x}, \mathbf{x}\rangle : \mathbf{x} \in H, \|\mathbf{x}\| = 1\}$, of an arbitrary (possibly unbounded and not densely defined) linear operator T in a Hilbert space H (real or complex) is convex. See K. Gustafson [194] for a very short proof.

8. (A nonlinear Perron type theorem) Suppose that $A : \mathbb{R}^N \to \mathbb{R}^N$ is a continuous map such that the following three conditions are fulfilled:
(a) $A(\mathbf{x} + \mathbf{y}) \geq A(\mathbf{x}) + A(\mathbf{y})$ for all $\mathbf{x}, \mathbf{y} \in \mathbb{R}^N$;
(b) $A(\lambda\mathbf{x}) = \lambda A(\mathbf{x})$ for all $\mathbf{x} \in \mathbb{R}^N$ and $\lambda \in \mathbb{R}_+$;

(c) if $\mathbf{x} \in \mathbb{R}_+^N$, $\mathbf{x} \neq \mathbf{0}$, then $A(\mathbf{x}) > \mathbf{0}$.

Prove that there is a strictly positive number λ and a vector $\mathbf{v} > \mathbf{0}$ such that $A\mathbf{v} = \lambda \mathbf{v}$.

[*Hint:* The set

$$S = \{\mu : \mu \in \mathbb{R}_+ \text{ and } A\mathbf{x} \geq \mu \mathbf{x} \text{ for some } \mathbf{x} \in \mathbb{R}_+^N, \ \mathbf{x} \neq \mathbf{0}\}$$

is nonempty and compact and one can choose λ as the maximum of S.]

Remark. The classical Perron theorem asserts that if a matrix has strictly positive entries, then its largest eigenvalue is strictly positive, occurs with multiplicity 1 and the corresponding eigenvector can be chosen to have strictly positive components. See R. A. Horn and C. R. Johnson [223], Theorem 8.2.2, p. 496.

2.6 Comments

The early history of the theorems of Carathéodory, Radon, and Helly, and many generalizations, ramifications, and analogues of these theorems forming an essential part of combinatorial convexity can be found in the survey article of L. Danzer, B. Grünbaum and V. Klee [122].

The Shapley–Folkman theorem (due to L. S. Shapley and J. H. Folkman), was first published in R. Starr [455]. The unifying Lemma 2.1.2 appears in the notes prepared by R. M. Anderson for his students in economics. They are available online at

http://eml.berkeley.edu/~anderson/Econ201B/NonconvexHandout.pdf

Helly's theorem has many amazing consequences. Here is an example: Given s ($s \geq N+1$) points in the N-dimensional space such that every $N+1$ of them are contained in a ball of radius 1, then the whole family of points is contained in a ball of radius 1. This fact easily yields Jung's theorem: *If S is a finite set of points in the N-dimensional space, with all pairwise distances between them not exceeding* 1, *then S is contained in a ball of radius* $\sqrt{N/(2N+2)}$.

The Pompeiu–Hausdorff metric between two subsets of a metric space,

$$d_H(K_1, K_2) = \max\{\sup_{x \in K_1} \inf_{y \in K_2} d(x,y), \sup_{y \in K_2} \inf_{x \in K_1} d(x,y)\},$$

was introduced by D. Pompeiu in his PhD thesis defended at Sorbonne in 1905 (and published in the same year in Annales de la Faculté de Sciences de Toulouse) and by F. Hausdorff in his book *Grundzüge der Mengenlehre*, first published in 1914. This metric plays an important role in image processing, face detection, convex geometry etc. For example, one can show that the set $\mathcal{C}^N$, of all nonempty compact convex subsets of $\mathbb{R}^N$, can be made into a complete metric space via the Pompeiu–Hausdorff metric. For this, use the fact that

$$d_H(K_1, K_2) = \inf\{\delta > 0 : K_1 \subset K_2 + \delta B \text{ and } K_2 \subset K_1 + \delta B\},$$

where B denotes the open unit ball of $\mathbb{R}^N$. The main feature of the space $\mathcal{C}^N$ is described by the following important result due to W. Blaschke:

2.6.1 Theorem (The Blaschke selection theorem) *Every bounded sequence in $\mathcal{C}^N$ contains a convergent subsequence.*

See R. Schneider [436], pp. 62–63, for details. An important application of the Blaschke selection theorem is the existence of a solution to the isoperimetric problem.

Interestingly, the Blaschke selection theorem can be deduced from the well known criterion of compactness due to Arzelà and Ascoli. Indeed, as it will be shown in Section 3.4, each nonempty compact convex subset K of the closed unit ball $\bar{B}$ of $\mathbb{R}^N$ can be associated uniquely to its support function,

$$f_K : \bar{B} \to \mathbb{R}, \quad f_K(x) = \max_{k \in K} \langle x, k \rangle,$$

and this association provides an isometry between $\{K \in \mathcal{C}^N : K \subset \bar{B}\}$ and a subset $\mathcal{K}$ of $C(\bar{B})$. Clearly, $\mathcal{K}$ verifies the two requirements of the Arzelà–Ascoli criterion: boundedness and uniform equicontinuity.

Theorem 2.2.2 (about convexity of Chebyshev sets in $\mathbb{R}^N$) was proved independently by L. N. H. Bunt (as part of his Ph.D. thesis in 1934) and by T. S. Motzkin (1935). The present status of Klee's problem is discussed by J.-B. Hiriart-Urruty [215] and J. M. Borwein and J. Vanderwerff [73].

A substitute of orthogonal projection in the case of arbitrary Banach spaces is the following result due to F. Riesz [419], concerning the existence of *almost orthogonal elements*:

2.6.2 Theorem *Suppose that F is a proper closed subspace of the Banach space E. Then for every $\varepsilon > 0$ there exists $\mathbf{x}_\varepsilon \in E$ such that $\|\mathbf{x}_\varepsilon\| = 1$ and $d(\mathbf{x}_\varepsilon, F) \geq 1 - \varepsilon$.*

2.6.3 Corollary *A normed linear space whose closed unit ball is compact is necessarily finite dimensional.*

See R. Bhatia [51], pp. 17–18, for details.

All results in Section 2.2 on hyperplanes and separation theorems in $\mathbb{R}^N$ are due to H. Minkowski [325]. Their extension to the general context of linear topological spaces is presented in Appendix B.

The critical role played by finite dimensionality in a number of important results on convex functions is discussed by J. M. Borwein and A. S. Lewis in [72, Chapter 9]. See also J. M. Borwein and J. Vanderwerff [73].

The comment in Remark 2.1.11 concerning the Lebesgue measurability of convex sets is generalized by the following result due to R. Lang [269]: every convex set A in $\mathbb{R}^N$ is measurable with respect to any σ-finite product measure $\mu = \otimes_{k=1}^N \mu_k$ defined on the Borel σ-algebra of $\mathbb{R}^N$.

Farkas' alternative theorem is due to J. Farkas [157] who published it in a paper laying the foundation of linear and nonlinear programming. It yields not only Gordan's theorem of alternative but also many others. See Section 4.9 of the book of C. L. Byrne [87].

The infinite-dimensional analogue of the space $\mathrm{Sym}(n, \mathbb{R})$ is represented by the space $\mathcal{A}(H)$, of all self-adjoint bounded operators acting on a Hilbert space H. If $A \in \mathcal{A}(H)$, then clearly $\langle A\mathbf{x}, \mathbf{x}\rangle \in \mathbb{R}$ for all $\mathbf{x}$. Since the key formula

$$\|A\| = \sup_{\|\mathbf{x}\| \leq 1} \langle A\mathbf{x}, \mathbf{x}\rangle = \sup_{\|\mathbf{x}\|=1} \langle A\mathbf{x}, \mathbf{x}\rangle \tag{2.19}$$

and its consequences (2.14) and (2.15) remain valid in this extended context, it follows that $\mathcal{A}(H)$ constitutes a regularly ordered Banach space when endowed with the operator norm and the positive cone

$$\mathcal{A}(H)_+ = \{A \in \mathcal{A}(H) : \langle A\mathbf{x}, \mathbf{x}\rangle \geq 0 \quad \text{for all } \mathbf{x} \in H\}.$$

The interior of $\mathcal{A}(H)_+$ is the set of all strictly positive operators,

$$\mathcal{A}(H)_{++} = \{A \in \mathcal{A}(H) : \langle A\mathbf{x}, \mathbf{x}\rangle > 0 \quad \text{for all } \mathbf{x} \in H \setminus \{0\}\}.$$

As in the finite-dimensional case, the spectrum $\sigma(A)$ of a self-adjoint bounded operator A is part of the real line and A is positive if and only if $\sigma(A) \subset [0, \infty)$.

2.6.4 Theorem *Suppose that $A \in \mathcal{A}(H)$. Then its spectrum $\sigma(A)$ is included in the interval $[\omega_A, \Omega_A]$, where $\omega_A = \inf_{\|\mathbf{x}\|=1} \langle A\mathbf{x}, \mathbf{x}\rangle$ and $\Omega_A = \sup_{\|\mathbf{x}\|=1} \langle A\mathbf{x}, \mathbf{x}\rangle$. Moreover, ω_A and Ω_A belongs to $\sigma(A)$ and*

$$\|A\| = \sup_{\lambda \in \sigma(A)} |\lambda|. \tag{2.20}$$

Proof. Let λ be a complex number such that $d = d(\lambda, [\omega_A, \Omega_A]) > 0$. By the Cauchy–Bunyakovsky–Schwarz inequality, for every vector $\mathbf{x} \in H$, $\|\mathbf{x}\| = 1$, we have

$$\|(A - \lambda \mathrm{I})\, \mathbf{x}\| \geq |\langle (A - \lambda \mathrm{I})\, \mathbf{x}, \mathbf{x}\rangle| = |\langle A\mathbf{x}, \mathbf{x}\rangle - \lambda| \geq d, \tag{2.21}$$

whence $\|(A - \lambda \mathrm{I})\, \mathbf{x}\| \geq d\, \|\mathbf{x}\|$ for all $\mathbf{x} \in H$. An immediate consequence is the fact that $A - \lambda \mathrm{I}$ is injective and has a closed range. The range of $A - \lambda \mathrm{I}$ is also dense in H since every vector $\mathbf{y}$ orthogonal on $(A - \lambda \mathrm{I})\, (H)$ is null. Indeed, if for each $\mathbf{x} \in H$ we have $0 = \langle (A - \lambda \mathrm{I})\, \mathbf{x}, \mathbf{y}\rangle = \langle \mathbf{x}, \left(A - \bar{\lambda} \mathrm{I}\right) \mathbf{y}\rangle$, which yields $A\mathbf{y} = \bar{\lambda}\mathbf{y}$. This implies $\mathbf{y} = 0$, since otherwise $\mathbf{y}/\, \|\mathbf{y}\|$ will contradict the inequalities (2.21). Therefore $A - \lambda \mathrm{I}$ is a continuous and bijective operator, so by the open mapping theorem in functional analysis (see R. Bhatia [51], p. 42) it follows that the inverse of $A - \lambda \mathrm{I}$ is also continuous (that is, $\lambda \notin \sigma(A)$). This ends the proof of the inclusion $\sigma(A) \subset [\omega_A, \Omega_A]$.

The fact that ω_A and Ω_A belong to $\sigma(A)$ is immediate by noticing that $\lambda \in \sigma(A)$ if and only if there exists a sequence of elements $\mathbf{x}_n$ with $\|\mathbf{x}_n\| = 1$ such that $\|A\mathbf{x}_n - \lambda \mathbf{x}_n\| \to 0$. Now, the spectral characterization of $\|A\|$ (given by formula (2.20)) follows easily from formula (2.19). ∎

Given an operator $A \in \mathcal{A}(H)$, one can construct a map π (called continuous functional calculus) that transforms every real polynomial

$$P(t) = t^n + c_1 t^{n-1} + \cdots + c_n$$

into the self-adjoint operator $P(A) = A^n + c_1 A^{n-1} + \cdots + c_n \mathrm{I}$ and, in general, every continuous real function $f \in C(\sigma(A))$ into an operator $f(A) \in \mathcal{A}(H)$ such that the formulas (2.17) hold true. See S. Strătilă and L. Zsidó [465], Chapter 2, for details and a thorough presentation of the theory of self-adjoint bounded operators.

The spectral decomposition of symmetric matrices admits an infinite- dimensional analogue in the case of compact self-adjoint operators. See R. Bhatia [51], p. 183, or M. Willem [489], p. 68.

Chapter 3

Convex Functions on a Normed Linear Space

Convex functions and their relatives are ubiquitous in a large variety of applications such as optimization theory, mass transportation, mathematical economics, and geometric inequalities related to isoperimetric problems. This chapter is devoted to a succinct presentation of their theory in the context of *real* normed linear spaces, but most of the illustrations will refer to the Euclidean space $\mathbb{R}^N$, the matrix space $\mathrm{M}_N(\mathbb{R})$ of all $N \times N$-dimensional real matrices (endowed with the Hilbert–Schmidt norm or with the operator norm), and the Lebesgue spaces $L^p\left(\mathbb{R}^N\right)$ with $p \in [1, \infty]$.

3.1 Generalities and Basic Examples

The notion of a convex function introduced in Chapter 1 can be extended verbatim to the framework of real-valued functions defined on arbitrary convex sets.

In what follows U will be a convex set in a real linear space E.

3.1.1 Definition *A function $f\colon U \to \mathbb{R}$ is said to be convex if*

$$f((1-\lambda)\mathbf{x}+\lambda\mathbf{y}) \leq (1-\lambda)f(\mathbf{x})+\lambda f(\mathbf{y}) \tag{3.1}$$

for all $\mathbf{x}, \mathbf{y} \in U$ and $\lambda \in [0,1]$.

The function f is called strictly convex if the above inequality is strict whenever $\mathbf{x} \neq \mathbf{y}$ and $\lambda \in (0,1)$, and it is called strongly convex with parameter $C > 0$ if

$$f((1-\lambda)\mathbf{x}+\lambda\mathbf{y}) \leq (1-\lambda)f(\mathbf{x})+\lambda f(\mathbf{y}) - \frac{C}{2}\lambda(1-\lambda)\left\|\mathbf{x}-\mathbf{y}\right\|^2$$

for all $\mathbf{x}, \mathbf{y} \in U$ and $\lambda \in [0,1]$.

C. P. Niculescu and L.-E. Persson, *Convex Functions and Their Applications*, CMS Books in Mathematics, https://doi.org/10.1007/978-3-319-78337-6_3

The related notions such as *concave function*, *strictly concave function*, *strongly concave function*, *affine function*, *quasiconvex/quasiconcave function*, *log-convex/log-concave function*, etc., can be introduced as in the one real variable case. To each of them one can ascribe a basic inequality like (3.1), that can be extended to the case of arbitrary finite convex combinations. We shall refer to these extensions as *discrete Jensen-type inequalities.*

Clearly, strong convexity (concavity) implies strict convexity (concavity), that in turn implies convexity (concavity). On the other hand, log-convexity implies convexity, and this one implies quasiconvexity. Notice that every concave function is also quasiconcave and every strictly positive concave function is log-concave.

In the context of several variables, the property of convexity in the sense of Definition 3.1.1 is sometimes called *joint convexity* in order to distinguish it from the weaker property of *separate convexity* (that is, convexity with respect to each variable when the others are kept fixed). A simple example of a nonconvex function that is separately convex is provided by the polynomial xy. Separate convexity has a number of interesting features, for example, the supremum of such a function over a compact convex set equals the supremum over the boundary. See Exercise 7. A special case when separate convexity coincides with convexity makes the object of Exercise 2, Section 3.4.

As in the case of one real variable, the basic criterion of convexity is the second differentiability test. It will be discussed in Section 3.8, but some basic examples can be derived directly using the definition:

- $f(x_1,\ldots,x_N) = \sum_{k=1}^{N}\varphi(x_k)$ is (strictly) convex and continuous on $\mathbb{R}^N$, whenever φ is a (strictly) convex function on $\mathbb{R}$.

- The function $f(\mathbf{x}) = \sum_{k=1}^{n} c_k e^{\langle a_k,\mathbf{x}\rangle}$ is continuous and convex on $\mathbb{R}^N$, whenever $a_1, ..., a_n \in \mathbb{R}^N$ and $c_1, ..., c_n \geq 0$;

- The norm function $f(\mathbf{x}) = \|\mathbf{x}\|$ in any normed linear space E is continuous and convex. The same works for the *distance function* $d_U(\mathbf{x}) = d(\mathbf{x}, U)$, from a nonempty convex set U in E.

As concerns the *distance function*, $d : E \times E \to \mathbb{R}$, $d(\mathbf{x}, \mathbf{y}) = \|\mathbf{x} - \mathbf{y}\|$, this is continuous and jointly convex.

- The squared norm $f(\mathbf{x}) = \|\mathbf{x}\|^2$ in the Euclidean space $\mathbb{R}^N$ is strongly convex (with constant $C = 1$). Renorming $\mathbb{R}^N$ with an equivalent norm, the squared function remains convex, but not necessarily strongly convex or strictly convex. See the case of sup norm.

It is worth noticing that the property of convexity is equivalent with convexity on each line segment included in the domain of definition:

3.1.2 Proposition *A function* $f\colon U \to \mathbb{R}$ *is convex* (*respectively strictly convex, strongly convex, log-convex, concave etc.*) *if and only if for every pair of* *distinct points* $\mathbf{x}$ *and* $\mathbf{y}$ *in* U *the function*

$$\varphi\colon [0,1] \to \mathbb{R}, \quad \varphi(t) = f((1-t)\mathbf{x} + t\mathbf{y})$$

is convex (respectively strictly convex, strongly convex, log-convex, concave, etc.)

Letting $v = \mathbf{y} - \mathbf{x}$, one can reformulate Proposition 3.1.2 by asking the respective condition of convexity for the function

$$\psi(t) = f(\mathbf{x} + tv), \quad t \in \operatorname{dom}\psi = \{t \in \mathbb{R} : \mathbf{x} + tv \in U\},$$

whenever $\mathbf{x} \in U$ and $v \in E$ (the ambient linear space).

3.1.3 Corollary (The extension of Jensen's criterion of convexity) *A continuous function* $f\colon U \to \mathbb{R}$ *is convex if and only if it is midpoint convex, that is,*

$$f\left(\frac{\mathbf{x}+\mathbf{y}}{2}\right) \leq \frac{f(\mathbf{x}) + f(\mathbf{y})}{2} \tag{3.2}$$

for all $\mathbf{x}, \mathbf{y} \in U$.

Notice that continuity at the interior points of U suffices as well. See Theorem 1.1.8.

3.1.4 Example *The log-sum-exp function in* $\mathbb{R}^N$ *is a function of the form*

$$f(\mathbf{x}) = \log\left(\sum_{k=1}^{n} c_k e^{\langle a_k, \mathbf{x}\rangle}\right), \quad \mathbf{x} \in \mathbb{R}^N,$$

where the coefficients c_k *are strictly positive and the vectors* a_k *belong to* $\mathbb{R}^N$. *This function is of class* C^∞ *and its convexity follows easily from Corollary 3.1.3 and Cauchy–Bunyakovsky–Schwarz inequality.*

3.1.5 Example (*Minkowski's inequality of order* $p = 0$) *According to Corollary 3.1.3, the proof of the property of concavity of the geometric mean* $\left(\prod_{k=1}^n x_k\right)^{1/n}$ *on* $\mathbb{R}^n_+$ *reduces to the elementary fact that*

$$\left(\prod\nolimits_{k=1}^{n} (x_k + y_k)\right)^{1/n} \geq \left(\prod\nolimits_{k=1}^{n} x_k\right)^{1/n} + \left(\prod\nolimits_{k=1}^{n} y_k\right)^{1/n} \tag{3.3}$$

for all $x_1, \ldots, x_n, y_1, \ldots, y_n \geq 0$. *By homogeneity one can assume that the left-hand side equals 1, and the conclusion follows from the AM–GM inequality. Alternatively, one can use the following variational formula:*

$$\left(\prod_{k=1}^{n} x_k\right)^{1/n} = \inf\left\{\frac{\alpha_1 x_1 + \cdots + \alpha_n x_n}{n} \;\middle|\; \alpha_1, \ldots, \alpha_n \geq 0, \prod_{k=1}^{n} \alpha_k = 1\right\}.$$

The inequality (3.3) *has a matrix companion, Minkowski's determinant inequality, which asserts that the continuous function*

$$f\colon \operatorname{Sym}^{+}(n,\mathbb{R}) \to \mathbb{R}, \quad f(A) = (\det A)^{1/n}$$

is concave. Using an approximation argument (see Exercise 10 (d)) *we may restrict to the case of strictly positive matrices. Then the proof is immediate by taking into account that for every pair* A, B *of such matrices there exists a nonsingular matrix* T *such that* T^*AT *and* T^*BT *are both of diagonal form. See Lemma* 5.3.1*. The same argument together with the generalized AM–GM inequality yields*

$$\det\left((1-\lambda)\,A + \lambda B\right) \geq (\det A)^{1-\lambda}\,(\det B)^{\lambda}$$

for all $A, B \in \operatorname{Sym}^{+}(n,\mathbb{R})$ *and* $\lambda \in [0,1]$*. In turn, this implies that the log-determinant function,*

$$f\colon \operatorname{Sym}^{++}(n,\mathbb{R}) \to \mathbb{R}, \quad f(A) = \log(\det A)$$

is concave. This feature of the log-determinant function can be also proved via Proposition 3.1.2*. Indeed, for every* $A \in \operatorname{Sym}^{++}(n,\mathbb{R})$, $B \in \operatorname{Sym}(n,\mathbb{R})$ *and* $t \in \mathbb{R}$ *with* $A + tB \in \operatorname{Sym}^{++}(n,\mathbb{R})$, *we have*

$$\begin{aligned}\log\det(A+tB) &= \log\det(A) + \log\det\left[I + tA^{-1/2}BA^{-1/2}\right]\\ &= \log\det(A) + \sum\nolimits_{k=1}^{n}\log\left(1+t\lambda_k\right),\end{aligned}$$

where $\lambda_1, \ldots, \lambda_n$ *are the eigenvalues of* $A^{-1/2}BA^{-1/2}$*. Consequently, the function* $\log\left(\det(A+tB)\right)$ *is the sum of a finite family of concave functions.*

The case of the log-determinant function is an archetype for the so-called concave spectral functions. See Section 5.1*.*

3.1.6 Remark (Operations preserving convexity of functions) *All assertions of Proposition* 1.1.10 *remain valid in the context of several variables with only one exception, statement* (c), *where the connection between synchronicity and monotonicity in the same way should be eliminated. Some other techniques to construct convex functions from old ones are presented in Exercises 3, 5, and 11.*

In Section 1.1 we proved that every convex function defined on an open interval is continuous. This fact can be easily extended to the context of several variables, due to the following lemma.

3.1.7 Lemma *Every convex function* f *defined on an open convex set* U *in* $\mathbb{R}^N$ *is locally bounded (that is, each point* $a \in U$ *admits a neighborhood on which* f *is bounded).*

Proof. For $\mathbf{a} \in U$ arbitrarily fixed, choose a cube K in U, centered at $\mathbf{a}$, with vertices $\mathbf{v}_1, \ldots, \mathbf{v}_{2^N}$. Clearly, K is a neighborhood of $\mathbf{a}$. Every $\mathbf{x} \in K$ is a convex combination of vertices and thus

$$f(\mathbf{x}) = f\Big(\sum_{k=1}^{2^N} \lambda_k \mathbf{v}_k\Big) \leq M = \sup_{1 \leq k \leq 2^N} f(\mathbf{v}_k),$$

which shows that f is bounded above on K. By the symmetry of K, for every $\mathbf{x} \in K$ there is a $\mathbf{y} \in K$ such that $\mathbf{a} = (\mathbf{x}+\mathbf{y})/2$. Then $f(\mathbf{a}) \leq (f(\mathbf{x})+f(\mathbf{y}))/2$, whence $f(\mathbf{x}) \geq 2f(\mathbf{a}) - f(\mathbf{y}) \geq 2f(\mathbf{a}) - M$, and the proof is complete. ∎

3.1.8 Proposition *If f is a convex function on an open convex set U in $\mathbb{R}^N$, then f is locally Lipschitz. In particular, f is continuous on U.*

According to a theorem due to Rademacher (see Appendix D), one can infer from Proposition 3.1.8 that every convex function on an open convex set U in $\mathbb{R}^N$ is almost everywhere differentiable. A direct proof will be given in Section 3.7 (see Theorem 3.7.3).

Proof. According to the preceding lemma, given $\mathbf{a} \in U$, we may find a ball $B_{2r}(\mathbf{a}) \subset U$ on which f is bounded above, say by M. For $\mathbf{x} \neq \mathbf{y}$ in $B_r(\mathbf{a})$, put $\mathbf{z} = \mathbf{y} + (r/\alpha)(\mathbf{y} - \mathbf{x})$, where $\alpha = \|\mathbf{y} - \mathbf{x}\|$. Clearly, $\mathbf{z} \in B_{2r}(\mathbf{a})$. As

$$\mathbf{y} = \frac{r}{r+\alpha}\mathbf{x} + \frac{\alpha}{r+\alpha}\mathbf{z},$$

from the convexity of f we infer that

$$f(\mathbf{y}) \leq \frac{r}{r+\alpha}f(\mathbf{x}) + \frac{\alpha}{r+\alpha}f(\mathbf{z}).$$

Then

$$\begin{aligned} f(\mathbf{y}) - f(\mathbf{x}) &\leq \frac{\alpha}{r+\alpha}[f(\mathbf{z}) - f(\mathbf{x})] \\ &\leq \frac{\alpha}{r}[f(\mathbf{z}) - f(\mathbf{x})] \leq \frac{2M}{r}\|\mathbf{y} - \mathbf{x}\| \end{aligned}$$

and the proof ends by interchanging the roles of $\mathbf{x}$ and $\mathbf{y}$. ∎

3.1.9 Corollary *Let f be a convex function defined on a convex set A in $\mathbb{R}^N$. Then f is Lipschitz on each compact convex subset of* ri(A)*. In particular, f is continuous on* ri(A).

Proof. Clearly, we may assume that $\operatorname{aff}(A) = \mathbb{R}^N$. In this case, $\operatorname{ri}(A) = \operatorname{int}(A)$ and Proposition 3.1.8 applies. ∎

3.1.10 Corollary *Let the convex functions $f_n : \mathbb{R}^N \to \mathbb{R}$ converge pointwise for $n \to \infty$ to $f : \mathbb{R}^N \to \mathbb{R}$. Then f is convex and, for each compact set K, the convergence of f_n to f is uniform on K.*

Proof. The convexity of f is clear. As concerns the second assertion, choose $R > 0$ such that $K \subset B_R(\mathbf{0})$. The function $g(\mathbf{x}) = \sup_n f_n(\mathbf{x})$ is real valued and convex, so by Proposition 3.1.8 convex, it must be continuous on the compact ball $\overline{B}_{2R}(\mathbf{0})$. Hence g is bounded above on this ball and an inspection of the argument of Proposition 3.1.8 shows that g is Lipschitz on $B_R(\mathbf{0})$. As a consequence, there is some $L > 0$ (independent of n) such that $\|f_n(\mathbf{x}) - f_n(\mathbf{y})\| \leq L\|\mathbf{x} - \mathbf{y}\|$ for all $\mathbf{x}, \mathbf{y} \in B_R(\mathbf{0})$. By passing to the limit as $n \to \infty$ we infer that f also verifies this inequality. To end the proof, consider a compact set K in $\mathbb{R}^N$ and a strictly positive number ε arbitrarily fixed. Since K is compact, it admits a finite covering with balls $B_\varepsilon(\mathbf{x}_1), ..., B_\varepsilon(\mathbf{x}_m)$. Then each $\mathbf{x} \in K$ belongs to a ball $B_\varepsilon(\mathbf{x}_{k(\varepsilon)})$ and this fact yields

$$\begin{aligned}
|f_n(\mathbf{x}) - f(\mathbf{x})| &\leq \left|f_n(\mathbf{x}) - f_n(\mathbf{x}_{k(\varepsilon)})\right| + \left|f_n(\mathbf{x}_{k(\varepsilon)}) - f(\mathbf{x}_{k(\varepsilon)})\right| + \left|f(\mathbf{x}_{k(\varepsilon)}) - f(\mathbf{x})\right| \\
&\leq L\left\|\mathbf{x} - \mathbf{x}_{k(\varepsilon)}\right\| + \left|f_n(\mathbf{x}_{k(\varepsilon)}) - f(\mathbf{x}_{k(\varepsilon)})\right| + L\left\|\mathbf{x}_{k(\varepsilon)} - \mathbf{x}\right\| \\
&\leq 2L\varepsilon + \max_{1\leq k\leq m} |f_n(\mathbf{x}_k) - f(\mathbf{x}_k)|.
\end{aligned}$$

Due to the pointwise convergence of the functions f_n to f, there is a rank N_ε such that $\max_{1\leq k\leq m} |f_n(\mathbf{x}_k) - f(\mathbf{x}_k)| < \varepsilon$ for all $n \geq N_\varepsilon$. Therefore

$$|f_n(\mathbf{x}) - f(\mathbf{x})| \leq (2L + 1)\varepsilon$$

for all $\mathbf{x} \in K$ and $n \geq N_\varepsilon$. The proof is complete. ■

Another instance of uniform convergence is made available in Exercise 12.

The infinite-dimensional analogue of Proposition 3.1.8 is as follows:

3.1.11 Proposition (Local Lipschitz continuity via convexity) *Let f be a convex function on an open convex set U in a normed linear space. If f is bounded above in a neighborhood of one point of U, then f is locally Lipschitz on U. In particular, f is a continuous function.*

The proof is similar to that of Proposition 3.1.8, with the difference that the role of Lemma 3.1.7 is taken by the following lemma:

3.1.12 Lemma *Let f be a convex function on an open convex set U in a normed linear space. If f is bounded above in a neighborhood of one point of U, then f is locally bounded on U.*

Proof. Suppose that f is bounded above by M on a ball $B_r(\mathbf{a})$. Let $\mathbf{x} \in U$ and choose $\rho > 1$ such that $\mathbf{z} = \mathbf{a} + \rho(\mathbf{x} - \mathbf{a}) \in U$. If $\lambda = 1/\rho$, then

$$V = \{\mathbf{v} : \mathbf{v} = (1 - \lambda)\mathbf{y} + \lambda\mathbf{z},\ \mathbf{y} \in B_r(\mathbf{a})\}$$

equals the ball with center $\mathbf{x} = (1 - \lambda)\mathbf{a} + \lambda\mathbf{z}$ and radius $(1 - \lambda)r$. Moreover, for $\mathbf{v} \in V$ we have

$$f(\mathbf{v}) \leq (1 - \lambda)f(\mathbf{y}) + \lambda f(\mathbf{z}) \leq (1 - \lambda)M + \lambda f(\mathbf{z}).$$

To show that f is bounded below in the same neighborhood V, choose arbitrarily $\mathbf{v} \in V$ and notice that $2\mathbf{x} - \mathbf{v} \in V$. Consequently, $f(\mathbf{x}) \leq f(\mathbf{v})/2 + f(2\mathbf{x} - \mathbf{v})/2$, which yields $f(\mathbf{v}) \geq 2f(\mathbf{x}) - f(2\mathbf{x} - \mathbf{v}) \geq 2f(\mathbf{x}) - M$. The proof is complete. ■

3.1.13 Remark *On every infinite-dimensional normed linear space E there are discontinuous linear functions (and a fortiori discontinuous convex functions). Indeed, in this case E contains an infinite linearly independent set $\{\mathbf{x}_n : n \in \mathrm{N}\}$ of elements $\mathbf{x}_n$. One can assume that $\|\mathbf{x}_n\| = 1$ for all indices n (since normalizing the vectors does not affect the linear independence). According to the Axiom of Choice, this family can be completed to a normalized algebraic basis $\left\{\mathbf{x}_n : n \in \widetilde{\mathrm{N}}\right\}$ of E (that is, to a maximal family of linearly independent norm-1 elements). See [214], Theorem 3.19, p. 18. The functional $L : \left\{\mathbf{x}_n : n \in \widetilde{\mathrm{N}}\right\} \to \mathbb{R}$ defined by*

$$L(\mathbf{x}_n) = \begin{cases} n & \text{if } n \in \mathrm{N} \\ 0 & \text{if } n \in \widetilde{\mathrm{N}} \backslash \mathrm{N} \end{cases}$$

admits a unique extension by linearity to $E = \operatorname{span}\left\{\mathbf{x}_n : n \in \widetilde{\mathrm{N}}\right\}$ and this extension is discontinuous because it is not bounded on the unit ball.

Exercises

1. Prove that the general form of an affine function $f\colon \mathbb{R}^N \to \mathbb{R}$ is $f(\mathbf{x}) = \langle \mathbf{x}, \mathbf{u}\rangle + a$, where $\mathbf{u} \in \mathbb{R}^N$ and $a \in \mathbb{R}$. This fact can be extended to all affine functions $g : \mathbb{R}^N \to \mathbb{R}^M$. Precisely, $g((1-\lambda)\mathbf{x} + \lambda \mathbf{y}) = (1-\lambda) g(\mathbf{x}) + \lambda g(\mathbf{y})$ for all $\mathbf{x}, \mathbf{y} \in \mathbb{R}^N$ and $\lambda \in [0,1]$ if and only if $g(\mathbf{x}) = A\mathbf{x} + \mathbf{b}$ for some $M \times N$-dimensional real matrix A and some $\mathbf{b} \in \mathbb{R}^M$.

2. The quadratic function associated to a symmetric matrix $A \in \operatorname{Sym}(N, \mathbb{R})$, $A \neq 0$, is $f(\mathbf{x}) = \langle A\mathbf{x}, \mathbf{x}\rangle$ for $\mathbf{x} \in \mathbb{R}^N$. Prove that f is convex if and only if $A \in \operatorname{Sym}^+(N, \mathbb{R})$.

   Extend this result, by considering a real Banach space E and a continuous bilinear map $B : E \times E \to \mathbb{R}$. Then $B(\mathbf{x}, \mathbf{x})$ is a convex function on E if and only if $B(\mathbf{x}, \mathbf{x}) \geq 0$ for all $\mathbf{x}$.

   [*Hint:* Use Corollary 3.1.3.]

3. (Convex monotone superposition) Let $f(\mathbf{x}) = (f_1(\mathbf{x}), ..., f_M(\mathbf{x}))$ be a vector-valued function on $\mathbb{R}^N$ with convex components f_k, and assume that F is a convex function on $\mathbb{R}^M$ which is monotone increasing (that is, $\mathbf{u} \leq \mathbf{v}$ in $\mathbb{R}^M$ implies $F(\mathbf{u}) \leq F(\mathbf{v})$). Prove that their superposition $\varphi(\mathbf{x}) = F(f(\mathbf{x}))$ is convex on $\mathbb{R}^N$. This result easily yields the convexity of the log-sum-exp function and of the powers $\|\cdot\|^\alpha$ of any norm for $\alpha \geq 1$.

4. Let f be a real-valued function defined on a convex set A in $\mathbb{R}^N$. Prove that f is convex if and only if

$$f((1-\lambda)\mathbf{x} + \lambda \mathbf{y}) < (1-\lambda)\alpha + \lambda\beta$$

   whenever $0 < \lambda < 1$, $f(\mathbf{x}) < \alpha$ and $f(\mathbf{y}) < \beta$.

5. (Partial minimization) Suppose that $f : \mathbb{R}^M \times \mathbb{R}^N \to \mathbb{R}$, $f = f(\mathbf{x}, \mathbf{y})$ is jointly convex and $\inf_{\mathbf{y}\in\mathbb{R}^N} f(\mathbf{x}, \mathbf{y}) > -\infty$ for all $\mathbf{x} \in \mathbb{R}^M$. Infer from the preceding exercise that the function $g(\mathbf{x}) = \inf_{\mathbf{y}\in\mathbb{R}^N} f(\mathbf{x}, \mathbf{y})$ is convex on $\mathbb{R}^M$. Notice that the quasiconvex variant of this assertion also works.

6. Prove that the functions $f(x_1, \ldots, x_N) = \max\{x_1, \ldots, x_N\}$ and
$$g(x_1, \ldots, x_N) = \max\{x_1, \ldots, x_N\} - \min\{x_1, \ldots, x_N\}$$
are convex on $\mathbb{R}^N$.

7. (Separately convex functions). A function of several variables is called *separately convex* if it is convex with respect to each variable (when the others are kept fixed). Prove that the supremum of any such function over a compact convex set equals the supremum over the boundary. Apply this result in the case of the function
$$f(\mathbf{x}, \mathbf{y}) = \sum_{1\leq i\leq j\leq N} (x_i y_j - x_j y_i)^2,$$
defined for pair of vectors $\mathbf{x}, \mathbf{y}$ with $x_1, ..., x_N \in [m_1, M_1]$ and $y_1, ..., y_N \in [m_2, M_2]$.

8. Find the maximum of the function
$$f(a, b, c) = \left[3\left(a^5 + b^7 \sin\frac{\pi a}{2} + c\right) - 2(bc + ca + ab)\right]$$
for $a, b, c \in [0, 1]$.
[*Hint*: Notice that $f(a, b, c) \leq \sup[3(a + b + c) - 2(bc + ca + ab)] = 4$.]

9. It was shown in Example 3.1.5 that
$$(\det(A + B))^{1/n} \geq (\det A)^{1/n} + (\det B)^{1/n}$$
for all $A, B \in \mathrm{Sym}^+(n, \mathbb{R})$. Prove that when $A, B \in \mathrm{Sym}^{++}(n, \mathbb{R})$, equality holds if and only if there is a positive constant c so that $B = cA$. In particular, if equality holds and $\det A = \det B$, then $A = B$.

10. (An alternative approach of the log-concavity of the function det)
(a) Prove that every matrix $A \in \mathrm{Sym}^{++}(n, \mathbb{R})$ verifies the formula
$$\int_{\mathbb{R}^n} \mathrm{e}^{-\langle A\mathbf{x},\mathbf{x}\rangle}\, d\mathbf{x} = \pi^{n/2}/\sqrt{\det A};$$
there is no loss of generality assuming that A is diagonal.
(b) Infer from the Rogers–Hölder inequality that
$$\int_{\mathbb{R}^n} \mathrm{e}^{-\langle[\alpha A+(1-\alpha)B]\mathbf{x},\mathbf{x}\rangle}\, d\mathbf{x} \leq \left(\int_{\mathbb{R}^n} \mathrm{e}^{-\langle A\mathbf{x},\mathbf{x}\rangle}\, d\mathbf{x}\right)^{\alpha} \left(\int_{\mathbb{R}^n} \mathrm{e}^{-\langle B\mathbf{x},\mathbf{x}\rangle}\, d\mathbf{x}\right)^{1-\alpha},$$

for every $A, B \in \mathrm{Sym}^{++}(n, \mathbb{R})$ and every $\alpha \in (0, 1)$.

(c) Conclude that $\det(\alpha A + (1-\alpha)B) \geq (\det A)^{\alpha}(\det B)^{1-\alpha}$.

(d) Extend this conclusion for $A, B \in \mathrm{Sym}^{+}(n, \mathbb{R})$, using perturbations of the form $A + \varepsilon \mathrm{I}$ and $B + \varepsilon \mathrm{I}$ with $\varepsilon > 0$ arbitrarily small.

11. (a) Prove that the limit of a pointwise converging sequence of (quasi)convex functions defined on the same convex set U is a (quasi)convex function.

    (b) Let $(f_\alpha)_\alpha$ be a family of (quasi)convex functions defined on the same convex set U, such that $f(\mathbf{x}) = \sup_\alpha f_\alpha(\mathbf{x}) < \infty$ for all $\mathbf{x} \in U$. Prove that f is (quasi)convex.

    (c) Suppose that $(f_\alpha)_\alpha$ is a family of (quasi)concave functions defined on the same convex set U, such that $f(\mathbf{x}) = \inf_\alpha f_\alpha(\mathbf{x}) > -\infty$ for all $\mathbf{x} \in U$. Prove that f is (quasi)concave.

12. Prove that if $g : \mathbb{R}^N \to \mathbb{R}$ is quasiconvex and $h : \mathbb{R} \to \mathbb{R}$ is increasing, then $f = h \circ g$ is quasiconvex.

13. Let $f : \mathbb{R}^N \to \mathbb{R}$ be a convex function. Prove that there is a constant $C > 0$ depending only on N, such that

$$\sup_{\mathbf{x} \in \overline{B}_{r/2}(\mathbf{a})} |f(\mathbf{x})| \leq \frac{C}{\mathrm{Vol}(B_r(\mathbf{a}))} \int_{B_r(\mathbf{a})} |f(y)| \, \mathrm{d}y,$$

    for every $r > 0$ and $\mathbf{a} \in \mathbb{R}^N$.

    [*Hint:* We have $f(\mathbf{x}) \leq (f(\mathbf{x}+\mathbf{y}) + f(\mathbf{x}-\mathbf{y}))/2$ for all $\mathbf{x} \in K = \overline{B}_{r/2}(\mathbf{a})$ and all $\mathbf{y}$ with $\|\mathbf{y}\| < r/2$; integrating this inequality over $B_{r/2}(\mathbf{0})$, infer that $f(\mathbf{x}) \leq A = \left(\mathrm{Vol}\, B_{r/2}(\mathbf{0})\right)^{-1} \int_{B_r(\mathbf{a})} |f(\mathbf{y})| \, \mathrm{d}\mathbf{y}$. A lower bound for $f(\mathbf{x})$ results by integrating the inequality $f(\mathbf{x}) \geq 2f(\mathbf{x}+\mathbf{y}) - f(\mathbf{x}+2\mathbf{y}) \geq f(\mathbf{x}+\mathbf{y}) - A$ for $\mathbf{y} \in B_{r/2}(\mathbf{0})$.]

14. (Convexity via substitution) Consider the function

$$f(x_1, \ldots, x_N) = \sum_{k=1}^{m} a_k x_1^{r_{1k}} \cdots x_N^{r_{Nk}}, \quad (x_1, \ldots, x_N) \in \mathbb{R}^N_{++},$$

    where $a_k > 0$ and $r_{ij} \in \mathbb{R}$. Prove that $g(y_1, \ldots, y_N) = \log f(\mathrm{e}^{y_1}, \ldots, \mathrm{e}^{y_N})$ is convex on $\mathbb{R}^N$.

3.2 Convex Functions and Convex Sets

The concepts of convex function and convex set can be related to each other in many ways. Of special interest is the characterization of convex functions in terms of their epigraphs.

By definition, the *epigraph* of a function $f \colon U \to \mathbb{R}$ is the set

$$\mathrm{epi}(f) = \{(\mathbf{x}, \alpha) : \mathbf{x} \in U, \ \alpha \in \mathbb{R} \text{ and } f(\mathbf{x}) \leq \alpha\}.$$

3.2.1 Theorem *A real-valued function f defined on a convex subset U (of a linear space E) is convex if and only if* $\operatorname{epi}(f)$ *is convex as a subset of $E \times \mathbb{R}$.*

The proof is straightforward and we omit the details.

For concave functions $f : U \to \mathbb{R}$, it is the hypograph

$$\operatorname{hyp}(f) = \{(\mathbf{x}, \alpha) : \mathbf{x} \in U,\ \alpha \in \mathbb{R} \text{ and } \alpha \leq f(\mathbf{x})\}$$

rather than the epigraph that is a convex subset of $E \times \mathbb{R}$.

As we already noticed, the discrete case of Jensen's inequality (as well as its argument) extends verbatim to the case of convex functions of a vector variable. Theorem 3.2.1 offers an alternative simple argument.

3.2.2 Proposition (The discrete case of Jensen's inequality) *If $f : U \to \mathbb{R}$ is a convex function and $\sum_{k=1}^{n} \lambda_k \mathbf{x}_k$ is a convex combination of points of U, then*

$$f\left(\sum_{k=1}^{n} \lambda_k \mathbf{x}_k\right) \leq \sum_{k=1}^{n} \lambda_k f(\mathbf{x}_k).$$

For concave functions, the inequality works in the reverse way.

Proof. Since f is convex, its epigraph is a convex set, so considering n points $(\mathbf{x}_1, f(\mathbf{x}_1)), ..., (\mathbf{x}_n, f(\mathbf{x}_n))$ in $\operatorname{epi}(f)$, any convex combination of them,

$$\left(\sum_{k=1}^{n} \lambda_k \mathbf{x}_k, \sum_{k=1}^{n} \lambda_k f(\mathbf{x}_k)\right),$$

also belongs to $\operatorname{epi}(f)$. By the definition of the epigraph, this is equivalent to Jensen's inequality. ■

3.2.3 Corollary *If $f : U \to \mathbb{R}$ is a convex function and $\mathbf{x}_1, ..., \mathbf{x}_n$ are points in U, then*

$$f(\mathbf{x}) \leq \max\{f(\mathbf{x}_1), ..., f(\mathbf{x}_n)\} \quad \textit{for every } \mathbf{x} \in \operatorname{conv}(\{\mathbf{x}_1, ..., \mathbf{x}_n\}).$$

As a consequence, the function f attains its supremum on $\operatorname{conv}(\{\mathbf{x}_1, ..., \mathbf{x}_n\})$ *at one of the points $\mathbf{x}_1, ..., \mathbf{x}_n$.*

The property stated in Corollary 3.2.3 actually characterizes the *quasiconvex functions*, that is, the functions $f : U \to \mathbb{R}$ such that

$$f((1-\lambda)\mathbf{x} + \lambda\mathbf{y}) \leq \max\{f(\mathbf{x}), f(\mathbf{y})\} \tag{3.4}$$

for all $\mathbf{x}, \mathbf{y} \in U$ and all $\lambda \in [0, 1]$. This corollary represents the analogue of Jensen's discrete inequality in the context of quasiconvex functions.

Given a function $f : U \to \mathbb{R}$ and a real number α, the *level set* of height α of f is defined as

$$\operatorname{level}_\alpha(f) = \{\mathbf{x} \in U : f(\mathbf{x}) \leq \alpha\}.$$

3.2.4 Lemma *The quasiconvex functions are precisely the functions whose level sets are convex.*

Reversing the inequality sign in (3.4) and replacing max by min, we arrive at the concept of *quasiconcave function.* Every convex (concave) function is also quasiconvex (quasiconcave). The converse works for positively homogeneous functions of degree 1, taking positive values. See Section 3.4, Lemma 3.4.2. The function $f(x_1, x_2) = x_1 x_2$ is quasiconcave on $\mathbb{R}^2_+$ but not concave.

In applications it is often useful to consider extended real-valued functions, defined on the whole ambient real linear space E.

3.2.5 Definition *An extended-value function $f \colon E \to \overline{\mathbb{R}}$ is said to be* convex *if its epigraph,*

$$\operatorname{epi}(f) = \{(\mathbf{x}, \alpha) : \mathbf{x} \in E,\ \alpha \in \mathbb{R} \text{ and } f(\mathbf{x}) \leq \alpha\},$$

is a convex subset of $E \times \mathbb{R}$.

A characterization of these functions using inequalities is offered by Exercise 2.

The *effective domain* of a convex function $f \colon E \to \overline{\mathbb{R}}$ is the set

$$\operatorname{dom} f = \{\mathbf{x} : f(\mathbf{x}) < \infty\}.$$

Since $\mathbf{x} \in \operatorname{dom}\ f$ is equivalent to the existence of some $\alpha \in \mathbb{R}$ such that $(\mathbf{x}, \alpha) \in \operatorname{epi}(f)$, the effective domain of a convex function is a convex set.

In what follows we shall only deal with *proper functions*, that is, with functions $f \colon E \to \overline{\mathbb{R}}$ which never take the value $-\infty$ and are not identically ∞. In their case, the property of convexity can be reformulated in more familiar terms:

$$f((1-\lambda)\mathbf{x} + \lambda\mathbf{y}) \leq (1-\lambda)f(\mathbf{x}) + \lambda f(\mathbf{y})$$

for all $\mathbf{x}, \mathbf{y} \in E$ and all $\lambda \in (0,1)$.

A function $f \colon E \to \mathbb{R} \cup \{-\infty\}$ is said to be a *proper concave function* if $-f$ is proper convex.

The related notions of proper quasiconvex/quasiconcave functions can be introduced in a similar way.

The restriction of a proper convex function to its effective domain is a convex function in the sense of Definition 3.1.1.

If U is a convex subset of a linear space E, then every convex function $f \colon U \to \mathbb{R}$ extends to a proper convex function $\widetilde{f}$ on E, letting $\widetilde{f}(\mathbf{x}) = \infty$ for $\mathbf{x} \in E \backslash U$. A basic example is provided by the indicator function. The *indicator function* of a *nonempty* subset A of E is defined by the formula

$$\iota_A(x) = \begin{cases} 0 & \text{if } x \in A, \\ \infty & \text{if } x \in E \backslash A. \end{cases}$$

Clearly, A is convex if and only if ι_A is a proper convex function.

A discussion of the topological nature of level sets needs the framework of lower semicontinuity.

3.2.6 Definition *An extended real-valued function f defined on a Hausdorff topological space X is called* lower semicontinuous *if*

$$f(\mathbf{x}) \leq \liminf_{\mathbf{y}\to\mathbf{x}} f(\mathbf{y}) \quad \textit{for all } \mathbf{x} \in X.$$

In the same framework, a function g is called *upper semicontinuous* if $-g$ is lower semicontinuous.

When X is a metrizable space (in particular, a normed linear space), one can formulate the condition of lower semicontinuity in terms of sequences:

$$\mathbf{x}_n \to \mathbf{x} \text{ in } X \text{ implies } f(\mathbf{x}) \leq \liminf_{n\to\infty} f(\mathbf{x}_n).$$

Lower semicontinuity is related to the topological nature of the epigraph. Assuming E a Banach space, the epigraph of any extended real-valued function f defined on a subset of E is endowed with the topology of subspace of the product Banach space $E \times \mathbb{R}$, associated to the norm

$$\|(\mathbf{x}, \alpha)\| = \left(\|\mathbf{x}\|^2 + |\alpha|^2\right)^{1/2}.$$

As in the case of the space $\mathbb{R}^N$, this norm is equivalent to each of the norms $\|(\mathbf{x}, \alpha)\|_p = (\|\mathbf{x}\|^p + |\alpha|^p)^{1/p}$ (for $1 \leq p < \infty$) and $\|(\mathbf{x}, \alpha)\|_\infty = \max\{\|\mathbf{x}\|, |\alpha|\}$, since they produce the same topology. The dual space of $E \times \mathbb{R}$ is $E^* \times \mathbb{R}$, the duality being defined by the formula

$$\langle(\mathbf{x}, \alpha), (x^*, \beta)\rangle = x^*(\mathbf{x}) + \alpha\beta.$$

3.2.7 Theorem *For a function $f : E \to \mathbb{R} \cup \{\infty\}$ the following statements are equivalent:*

(a) *The epigraph of f is closed in $E \times \mathbb{R}$, that is, for every sequence of points $(\mathbf{x}_n, y_n) \in \operatorname{epi}(f)$ converging to some point $(\mathbf{x}, y) \in E \times \mathbb{R}$ we have $(\mathbf{x}, y) \in \operatorname{epi}(f)$;*

(b) *Every level set of f is closed;*

(c) *f is lower semicontinuous on E.*

This result motivates the alternative terminology of *closed functions* for lower semicontinuous functions.

Proof. (a) $\Rightarrow$ (b) Let α be any real number. If $\operatorname{level}_\alpha(f)$ is the empty set, then it is a closed set. If $\operatorname{level}_\alpha(f)$ is nonempty, then choose $(\mathbf{x}_n)_n$ a sequence of points of $\operatorname{level}_\alpha(f)$ such that $\mathbf{x}_n \to \mathbf{x}$ in E. We have $f(\mathbf{x}_n) \leq \alpha$ for all n, whence $(\mathbf{x}_n, \alpha) \in \operatorname{epi}(f)$ for all n. Since $(\mathbf{x}_n, \alpha) \to (\mathbf{x}, \alpha)$ and the epigraph of f is closed, it follows that $(\mathbf{x}, \alpha) \in \operatorname{epi}(f)$, hence $\mathbf{x} \in \operatorname{level}_\alpha(f)$.

(b) $\Rightarrow$ (c) We reason by contradiction. If f were not lower semicontinuous, then it would exist $\mathbf{x} \in E$ and a sequence $(\mathbf{x}_n)_n$ such that $\mathbf{x}_n \to \mathbf{x}$ in E and

$$f(\mathbf{x}) > \liminf_{n\to\infty} f(\mathbf{x}_n).$$

This yields a number c and a subsequence $(\mathbf{x}_{k(n)})_n$ such that $f(\mathbf{x}) > c \geq f(\mathbf{x}_{k(n)})$. Then $\mathbf{x}_{k(n)} \in \text{level}_c(f)$ for all n. Since $\mathbf{x}_{k(n)} \to \mathbf{x}$ and every level set of f is closed it follows that $\mathbf{x} \in \text{level}_c(f)$. Hence $f(\mathbf{x}) \leq c$, a contradiction.

(c) $\Rightarrow$ (a) Assume that epi(f) is not closed. This implies the existence of a sequence of elements $(\mathbf{x}_n, \mathbf{y}_n) \in \text{epi}(f)$ such that $(\mathbf{x}_n, \mathbf{y}_n) \to (\mathbf{x}, \mathbf{y})$ in E and $(\mathbf{x}, \mathbf{y}) \notin \text{epi}(f)$. Since $f(\mathbf{x}_n) \leq \mathbf{y}_n$ for all n, we infer that

$$\liminf_{n\to\infty} f(\mathbf{x}_n) \leq \lim_{n\to\infty} \mathbf{y}_n = \mathbf{y}.$$

On the other hand, $\mathbf{y} < f(\mathbf{x})$ because $(\mathbf{x}, \mathbf{y}) \notin \text{epi}(f)$. Hence $\liminf_{n\to\infty} f(\mathbf{x}_n) < f(\mathbf{x})$, a fact that contradicts the lower semicontinuity of f. In conclusion, epi(f) must be a closed set. ■

3.2.8 Corollary *The pointwise supremum of any family $(f_i)_{i\in I}$ of lower semicontinuous convex functions defined on a normed linear space E is a function of the same nature.*

Proof. Indeed, the function $f(\mathbf{x}) = \sup\{f_i(\mathbf{x}) : i \in I\}$ verifies the equality $\text{epi}(f) = \cap_{i\in I}\, \text{epi}(f_i)$. ■

If the effective domain of a proper convex function f is closed and f is continuous relative to dom f, then f is lower semicontinuous. However, f can be lower semicontinuous under more general circumstances. The following function,

$$\varphi(x, y) = \begin{cases} y^2/2x & \text{if } x > 0, \\ \alpha & \text{if } x = y = 0, \\ \infty & \text{otherwise,} \end{cases}$$

is illustrative on what can happen at the boundary points of the effective domain. In fact, f is a proper convex function for each $\alpha \in [0, \infty]$. All points of its effective domain are points of continuity except the origin, where the limit does not exist. However the function φ is lower semicontinuous for $\alpha = 0$.

The possibility of modifying the values of a proper convex function at the boundary of its effective domain to become lower semicontinuous is discussed in Exercise 9.

3.2.9 Theorem *Every lower semicontinuous proper convex function f defined on a Banach space is continuous at every point of $U = \text{int}(\text{dom } f)$.*

Notice that if $\mathbf{a} \in \text{dom } f$ is a point of continuity, then f is bounded above on a neighborhood of $\mathbf{a}$ and this fact implies that $\mathbf{a} \in \text{int}(\text{dom } f)$.

Proof. According to Proposition 3.1.11, it suffices to prove that f is locally bounded above in the neighborhood of one point in U. We will use the fact that every nonempty open subset of a complete metric space is a Baire space. See J. Munkres [342], Lemma 48.4, p. 297. In our case, U is the union of the closed sets $U_n = \{\mathbf{x} \in U : f(\mathbf{x}) \leq n\}$ for $n \in \mathbb{N}$. Consequently there is a natural number n such that $\text{int}\, U_n \neq \emptyset$ and the proof is done. ■

A function f as in the statement of Theorem 3.2.9 can be continuous at certain boundary points a of dom f (as a function taking values in the metric space $\mathbb{R} \cup \{\infty\}$). See the case of the function

$$f(x) = \begin{cases} 1/x & \text{for } x > 0 \\ \infty & \text{for } x \in (-\infty, 0] \end{cases}$$

and the point $a = 0$.

An example of lower semicontinuous proper convex function f with no continuity points is provided by the indicator of the set $[0,1] \times \{0\}$ in $\mathbb{R}^2$.

A classical notation in convex analysis for the set of all lower semicontinuous proper convex functions $f : E \to \mathbb{R} \cup \{\infty\}$ is $\Gamma(E)$. Its theory will be presented in the next sections.

3.2.10 Remark (The extension of Jensen's criterion of convexity) *The restriction of a lower semicontinuous convex function to any closed, open, or half-open line segment is continuous. Consequently, a lower semicontinuous proper function $f : U \to \mathbb{R} \cup \{\infty\}$ is convex if and only if it is midpoint convex, that is,*

$$f\left(\frac{\mathbf{x}+\mathbf{y}}{2}\right) \leq \frac{f(\mathbf{x}) + f(\mathbf{y})}{2}$$

for all $\mathbf{x}, \mathbf{y} \in U$.

3.2.11 Remark *Every separately convex proper function $f : \mathbb{R}^N \to \mathbb{R} \cup \{\infty\}$ is locally Lipschitz, hence continuous on* int(dom f). *See B. Dacorogna* [119], *Theorem* 2.31, *p.* 47.

Exercises

1. Prove that the multiplicative inverse of a strictly positive convex function is a quasiconcave function, while the multiplicative inverse of a strictly positive concave function is a convex function.

2. Prove that $f : E \to \overline{\mathbb{R}}$ is a convex function if and only if

$$f((1-\lambda)\mathbf{x} + \lambda\mathbf{y}) \leq (1-\lambda)\alpha + \lambda\beta$$

whenever $f(x) < \alpha$, $f(y) < \beta$ and $\lambda \in (0,1)$. An important consequence is the fact that for any nonempty convex subset C of $E \times \mathbb{R}$, the function

$$f(\mathbf{x}) = \inf\{\alpha : (\mathbf{x}, \alpha) \in C \text{ for some } \mathbf{x} \in E\}$$

is convex.

3. Let C be a set in a Banach space. Prove that the indicator ι_C of C is lower semicontinuous if and only if C is closed.

4. Prove that if a lower semicontinuous convex function takes the value $-\infty$, then it is nowhere finite.

5. The Laplace transform of a Borel probability measure μ defined on $\mathbb{R}^N$ is defined by the formula

$$\mathcal{L}(\mu)(\xi) = \int_{\mathbb{R}^N} \mathrm{e}^{\langle \mathbf{x}, \xi \rangle} \mathrm{d}\mu(\mathbf{x}).$$

   Prove that $\mathcal{L}(\mu)$ is a lower semicontinuous proper convex function on $\mathbb{R}^N$.

   [*Hint:* Lower semicontinuity follows from Fatou's lemma. See [107].]

6. The *closure* $\operatorname{cl} f$ of a proper function $f : E \to \mathbb{R} \cup \{\infty\}$ is the largest lower semicontinuous minorant of f. Prove that $\operatorname{epi}(\operatorname{cl} f) = \overline{\operatorname{epi}(f)}$ and

$$(\operatorname{cl} f)(\mathbf{x}) = \liminf_{\mathbf{y} \to \mathbf{x}} f(\mathbf{y}).$$

7. Compute the closure of the function f which is 0 on the open unit disc in $\mathbb{R}^2$, arbitrary on its boundary and infinite at all points $\mathbf{x} \in \mathbb{R}^2$ with $\|\mathbf{x}\| > 1$.

8. Prove that the closure of every proper convex function is convex too.

9. (Construction of lower semicontinuous proper convex functions) Let $f : E \to \mathbb{R} \cup \{\infty\}$ be a proper convex function whose effective domain $U = \operatorname{dom} f$ is an open set. Assuming f is continuous on U, prove that

$$(\operatorname{cl} f)(\mathbf{x}) = \begin{cases} f(\mathbf{x}) & \text{if } x \in U \\ \liminf_{\mathbf{y} \to \mathbf{x}} f(\mathbf{y}) & \text{if } x \in \partial U \\ \infty & \text{if } x \in E \backslash \overline{U}. \end{cases}$$

10. Let D be a bounded closed convex subset of the real plane. Prove that D can be always represented as

$$D = \{(x, y) : f(x) \leq y \leq g(x),\, x \in [a, b]\}$$

   for suitable functions $f \colon [a, b] \to \mathbb{R}$ convex, and $g \colon [a, b] \to \mathbb{R}$ concave.

   Infer that the boundary of D is smooth except possibly for an enumerable subset.

11. (The convexification of a function) By definition, the *convex hull* of a function $f : E \to \overline{\mathbb{R}}$ is the largest convex minorant $\operatorname{conv}(f)$ of f. Suppose that $f : E \to \mathbb{R} \cup \{+\infty\}$ is a proper function for which there exists a continuous affine function $h : E \to \mathbb{R}$ such that $h \leq f$. Prove that:

   (a) The convex hull of f equals the pointwise supremum of all continuous affine minorants of f and $\operatorname{conv}(f) = \operatorname{cl} f$.

   (b) $\operatorname{epi}(\operatorname{conv}(f)) = \operatorname{conv}(\operatorname{epi}(f))$.

(c) $\operatorname{conv}(f)$ verifies the formula

$$(\operatorname{conv}(f))(\mathbf{x}) = \inf\{t : (\mathbf{x}, t) \in \operatorname{conv}(\operatorname{epi}(f))\}$$
$$= \inf\left\{\sum_{k=1}^{n} \lambda_k f(\mathbf{x}_k) : \mathbf{x} = \sum_{k=1}^{n} \lambda_k \mathbf{x}_k \text{ convex representation}\right\}.$$

(d) The functions f and $\operatorname{conv}(f)$ have the same global minimum (if any).

Remark. The points x for which $f(\mathbf{x}) = (\operatorname{conv}(f))(\mathbf{x})$ are called *points of convexity*. According to Exercise 11, Jensen's discrete inequality remains valid for any probability measure with barycenter at a point of convexity (though the function f under attention might not be convex). See also the comments after Exercise 10, Section 1.7.

3.3 The Subdifferential

A deep consequence of the characterization of convex functions through their epigraphs (see Theorem 3.2.1 above) is the existence of affine supports for convex functions.

3.3.1 Theorem *Let $f : E \to \mathbb{R} \cup \{\infty\}$ be a lower semicontinuous proper convex function defined on the linear normed space E. Then, for every $\mathbf{a} \in \operatorname{int}(\operatorname{dom} f)$ there is a continuous linear functional $x^* \in E^*$ such that*

$$f(\mathbf{x}) \geq f(\mathbf{a}) + x^*(\mathbf{x} - \mathbf{a}) \quad \textit{for all } \mathbf{x} \in E.$$

The affine minorant $h(\mathbf{x}) = f(\mathbf{a}) + x^*(\mathbf{x} - \mathbf{a})$ is said to be a *support* of f at $\mathbf{a}$. The terminology is motivated by the fact that the set $\{(\mathbf{x}, h(\mathbf{x})) : \mathbf{x} \in E\}$ is a hyperplane supporting the convex set $\operatorname{epi} f$ at the point $(\mathbf{a}, f(\mathbf{a}))$.

Proof. By Theorem 3.2.1 and Theorem 3.2.7, the epigraph of f is a closed convex set in $E \times \mathbb{R}$. It is also nonempty because f is proper. If $\mathbf{a} \in \operatorname{int}(\operatorname{dom} f)$, then $\mathbf{a}$ is a point of continuity of f; see Theorem 3.2.9. This implies that $(\mathbf{a}, f(\mathbf{a}) + \varepsilon) \in \operatorname{int}(\operatorname{epi}(f))$ for every $\varepsilon > 0$ and $(\mathbf{a}, f(\mathbf{a}))$ is a boundary point of $\operatorname{epi}(f)$. An appeal to the separation Theorem 2.3.1 yields the existence of a nonzero element in the dual of $E \times \mathbb{R}$ (that is, of a pair $(y^*, \lambda) \in E^* \times \mathbb{R}$ with $(y^*, \lambda) \neq (0, 0)$) such that

$$y^*(\mathbf{x}) + \lambda r \leq y^*(\mathbf{a}) + \lambda f(\mathbf{a})$$

for all $(\mathbf{x}, r) \in \operatorname{epi} f$. Since $\mathbf{a} \in \operatorname{int}(\operatorname{dom} f)$, it follows that $\lambda \neq 0$. On the other hand, letting $r \to \infty$ we infer that $\lambda \leq 0$. Therefore $\lambda < 0$, which yields

$$f(\mathbf{x}) \geq f(\mathbf{a}) + \left(\frac{-1}{\lambda} y^*\right)(\mathbf{x} - \mathbf{a}).$$

for all $\mathbf{x} \in E$. The proof ends by choosing $x^* = -y^*/\lambda$. ■

It is worth noticing that the affine minorants allow the reconstruction of lower semicontinuous proper convex functions.

3.3.2 Theorem *Every lower semicontinuous proper convex function $f : E \to \mathbb{R} \cup \{\infty\}$ is the pointwise supremum of the family of its continuous and affine minorants.*

Proof. We will show that for every $\mathbf{x}_0 \in E$ and every $\alpha \in \mathbb{R}$ such that $\alpha < f(\mathbf{x}_0)$ there exists a continuous affine minorant h of f for which $\alpha \leq h(\mathbf{x}_0) \leq f(\mathbf{x}_0)$. The details are similar to those used in Theorem 3.3.1. Since $\operatorname{epi}(f)$ is a closed convex set in $E \times \mathbb{R}$ and $(\mathbf{x}_0, \alpha) \notin \operatorname{epi}(f)$, the separation Theorem 2.3.2 yields the existence of an element $(x^*, \lambda) \in E^* \times \mathbb{R}$ and of a number $\varepsilon > 0$ such that

$$x^*(\mathbf{x}) + \lambda r < x^*(\mathbf{x}_0) + \lambda \alpha - \varepsilon \tag{3.5}$$

whenever $(\mathbf{x}, r) \in \operatorname{epi}(f)$ (that is, for all $x \in \operatorname{dom}\ f$ and $r \geq f(\mathbf{x})$).

If $(\mathbf{x}_0, f(\mathbf{x}_0)) \in \operatorname{epi}(f)$, then the inequality (3.5) imposes $\lambda < 0$ and in this case

$$f(\mathbf{x}) \geq h(\mathbf{x}) = \left(\frac{1}{-\lambda}\right) x^*(\mathbf{x}) + \alpha + \frac{1}{\lambda}\left(x^*(\mathbf{x}_0) - \varepsilon\right)$$

for all $\mathbf{x} \in \operatorname{dom}\ f$. Clearly, the function h is continuous and affine, $h(\mathbf{x}_0) > \alpha$ and $f(\mathbf{x}) \geq h(\mathbf{x})$ for every $\mathbf{x} \in E$.

If $f(\mathbf{x}_0) = \infty$ and $\lambda \neq 0$, letting $r \to \infty$ in (3.5) we infer that $\lambda < 0$ and the proof ends as in the preceding case.

If $f(\mathbf{x}_0) = \infty$ and $\lambda = 0$, then the inequality (3.5) shows that the continuous affine function

$$p(\mathbf{x}) = x^*(\mathbf{x}) - x^*(\mathbf{x}_0) + \varepsilon$$

is strictly negative at all $\mathbf{x} \in \operatorname{dom}\ f$.

Since $\operatorname{dom}\ f$ is nonempty, the discussion above yields the existence of a continuous affine minorant q for f. Therefore

$$h_n(\mathbf{x}) = q(\mathbf{x}) + n\left[x^*(\mathbf{x}) - x^*(\mathbf{x}_0) + \varepsilon\right],$$

for n large enough, provides an example of continuous affine minorant which takes at x_0 a value greater than α. The proof is complete. ∎

Theorem 3.3.1 imposes the following definition:

3.3.3 Definition *A functional $x^* \in E^*$ is said to be a* subgradient *of the function f at the point $\mathbf{a} \in \operatorname{dom}\ f$ if*

$$f(\mathbf{x}) \geq f(\mathbf{a}) + x^*(\mathbf{x} - \mathbf{a}) \quad \text{for all } \mathbf{x}; \tag{3.6}$$

we will refer to this inequality as the subgradient inequality.

The set of all subgradients of f at the point a is called the *subdifferential* of f at $\mathbf{a}$ and is denoted by $\partial f(\mathbf{a})$. The set-valued map $\partial f : \mathbf{a} \to \partial f(\mathbf{a})$ is called the *subdifferential* of f. In general, $\partial f(\mathbf{a})$ may be empty, a singleton or it may consist of more points. If $\partial f(\mathbf{a})$ is not empty, then f is said to be *subdifferentiable* at $\mathbf{a}$.

We denote by $\operatorname{dom} \partial f$ the *domain* of ∂f, that is, the set of all points $\mathbf{a} \in \operatorname{dom} f$ such that $\partial f(\mathbf{a}) \neq \emptyset$. According to Theorem 3.3.1,

$$\operatorname{int}(\operatorname{dom} f) \subset \operatorname{dom} \partial f \subset \operatorname{dom} f. \tag{3.7}$$

Both inclusions can be strict. Also, $\operatorname{dom} \partial f$ is not necessarily a convex set. See Exercise 3.

When E is a real Hilbert space, the subgradients $\mathbf{x}^*$ can be viewed as vectors in E and the subgradient inequality becomes

$$f(\mathbf{x}) \geq f(\mathbf{a}) + \langle \mathbf{x} - \mathbf{a}, \mathbf{x}^* \rangle \quad \text{for all } \mathbf{x}.$$

3.3.4 Remark *When E is finite dimensional, Theorem 3.3.1 works for all $\mathbf{a} \in \operatorname{ri}(\operatorname{dom} f)$. See Remark 2.3.7. As a consequence, in this case the inclusions (3.7) can be relaxed:*

$$\operatorname{ri}(\operatorname{dom} f) \subset \operatorname{dom} \partial f \subset \operatorname{dom} f.$$

3.3.5 Remark *The subdifferential at a point $\mathbf{a}$ is always a weak-star closed convex set. Recall that the weak-star topology of $E^\star$ is the topology of pointwise convergence. See Appendix B.*

If U is an open convex set in the Banach space E and $f : U \to \mathbb{R}$ is a continuous convex function, then the map $\mathbf{x} \to \partial f(\mathbf{x})$ is locally bounded. Indeed, according to Proposition 3.1.11, for each $x \in U$ there are an open ball $B_r(\mathbf{x})$ and a positive number L such that

$$|f(\mathbf{u}) - f(\mathbf{v})| \leq L \|\mathbf{u} - \mathbf{v}\|$$

for all $\mathbf{u}, \mathbf{v} \in B_r(\mathbf{x})$. Fix $\mathbf{y} \in B_r(\mathbf{x})$. Then $B_r(\mathbf{x})$ includes a closed ball $\overline{B}_\varepsilon(\mathbf{y})$ centered at $\mathbf{y}$, which yields

$$x^*(\mathbf{v}) \leq f(\mathbf{y} + \mathbf{v}) - f(\mathbf{y}) \leq L \|\mathbf{v}\|$$

for all $x^* \in \partial f(\mathbf{y})$ and all $\mathbf{v} \in E$ with $\|\mathbf{v}\| \leq \varepsilon$. Therefore $\|x^*\| \leq L$ for all $x^* \in \bigcup_{\mathbf{y} \in B_r(\mathbf{x})} \partial f(\mathbf{y})$.

Convex functions are the only functions that can be subdifferentiable at all points in their effective domains.

3.3.6 Theorem *Let E be a normed linear space and let $f : E \to \mathbb{R} \cup \{\infty\}$ be a function not identically ∞ that is subdifferentiable at any point $\mathbf{x} \in \operatorname{dom} f$. Then f is convex and lower semicontinuous.*

Proof. Let $x \in \operatorname{dom} f$ arbitrarily fixed and $x^* \in \partial f(\mathbf{x})$. Then for every two points $\mathbf{x}_1$ and $\mathbf{x}_2$ in $\operatorname{dom} f$ we have

$$f(\mathbf{x}_1) \geq f(\mathbf{x}) + x^*(\mathbf{x}_1 - \mathbf{x}) \quad \text{and} \quad f(\mathbf{x}_2) \geq f(\mathbf{x}) + x^*(\mathbf{x}_2 - \mathbf{x}).$$

Multiplying the first inequality by $\lambda \in (0,1)$ and the second one by $1-\lambda$, their sum yields

$$\lambda f(\mathbf{x}_1) + (1-\lambda) f(\mathbf{x}_2) \geq f(\mathbf{x}) + x^* \left(\lambda \mathbf{x}_1 + (1-\lambda)\mathbf{x}_2 - \mathbf{x}\right)$$

and the convexity of f follows by choosing $\mathbf{x} = \lambda \mathbf{x}_1 + (1-\lambda)\mathbf{x}_2$.

To prove that f is also lower semicontinuous, suppose that $\mathbf{x}_n \to \mathbf{x}$ in E and $x^* \in \partial f(\mathbf{x})$. Then $f(\mathbf{x}_n) \geq f(\mathbf{x}) + x^*(\mathbf{x}_n - \mathbf{x})$ for all n, which implies $\liminf_{n\to\infty} f(\mathbf{x}_n) \geq f(\mathbf{x})$. Hence f is also lower semicontinuous. ■

In Section 3.6 we will discuss the connection between subdifferentiability and directional derivatives as well as the calculus rules with subdifferentials.

The *subdifferential* of a lower semicontinuous proper convex function $f : \mathbb{R}^N \to \mathbb{R} \cup \{\infty\}$ is defined as the set-valued map ∂f which associates to each $\mathbf{x} \in \mathbb{R}^N$ the subset $\partial f(\mathbf{x}) \subset \mathbb{R}^N$. Equivalently, ∂f may be seen as a graph in $\mathbb{R}^N \times \mathbb{R}^N$.

Given a set-valued map $u \colon \mathbb{R}^N \to \mathcal{P}(\mathbb{R}^N)$, we define:

domain of u, $\operatorname{dom} u = \{\mathbf{x} : u(\mathbf{x}) \neq \emptyset\}$;

graph of u, $\operatorname{graph} u = \{(\mathbf{x}, \mathbf{y}) : \mathbf{x} \in \operatorname{dom}\ u \text{ and } \mathbf{y} \in u(\mathbf{x})\}$;

inverse of u, $u^{-1}(\mathbf{y}) = \{\mathbf{x} : \mathbf{y} \in u(\mathbf{x})\}$.

If $u, v \colon \mathbb{R}^N \to \mathcal{P}(\mathbb{R}^N)$ are set-valued maps, we write $u \subset v$, if the graph of u is contained in the graph of v.

3.3.7 Definition *A set-valued map* $u \colon \mathbb{R}^N \to \mathcal{P}(\mathbb{R}^N)$ *is said to be* monotone *if it verifies*

$$\langle \mathbf{x}_1 - \mathbf{x}_2, \mathbf{y}_1 - \mathbf{y}_2 \rangle \geq 0$$

for all $\mathbf{x}_1, \mathbf{x}_2 \in \mathbb{R}^N$ *and all* $\mathbf{y}_1 \in u(\mathbf{x}_1)$, $\mathbf{y}_2 \in u(\mathbf{x}_2)$. *A monotone function* u *is called* maximal monotone *when it is maximal with respect to inclusion in the class of monotone functions, that is, if the following implication holds:*

$$v \supset u \text{ and } v \text{ monotone} \implies v = u.$$

The subdifferential ∂f of every lower semicontinuous proper convex function $f \colon \mathbb{R}^N \to \mathbb{R} \cup \{\infty\}$ is monotone. This is a straightforward consequence of the subgradient inequality (3.6). As will be shown in Theorem 3.3.12, the subdifferential ∂f is actually a maximal monotone map.

Zorn's lemma easily yields the fact that every monotone set-valued map u admits a maximal monotone extension $\widetilde{u}$.

3.3.8 Remark *The graph of any maximal monotone map* $u \colon \mathbb{R}^N \to \mathcal{P}(\mathbb{R}^N)$ *is closed and thus it verifies the following conditions of upper semicontinuity:*

$$\mathbf{x}_k \to \mathbf{x},\ \mathbf{y}_k \to \mathbf{y}, \text{ and } \mathbf{y}_k \in u(\mathbf{x}_k) \text{ for all } k \in \mathbb{N} \implies \mathbf{y} \in u(\mathbf{x}).$$

We shall prove the existence of a one-to-one correspondence between the graphs of maximal monotone maps and graphs of nonexpansive functions. Recall that a function $h\colon \mathbb{R}^N \to \mathbb{R}^N$ is called *nonexpansive* if its Lipschitz constant verifies $\operatorname{Lip}(h) \leq 1$.

We shall need the following result concerning the extension of Lipschitz functions:

3.3.9 Theorem (M. D. Kirszbraun) *Suppose that A is a subset of $\mathbb{R}^N$ and $f\colon A \to \mathbb{R}^m$ is a Lipschitz function. Then there exists a Lipschitz function $\widetilde{f}\colon \mathbb{R}^N \to \mathbb{R}^m$ such that $\widetilde{f} = f$ on A and $\operatorname{Lip}(\widetilde{f}) = \operatorname{Lip}(f)$. Moreover, we may choose $\widetilde{f}$ convex, when A and f are also convex.*

Proof. When $m = 1$, we may choose

$$\widetilde{f}(\mathbf{x}) = \inf_{\mathbf{y} \in A} \{f(\mathbf{y}) + \operatorname{Lip}(f) \cdot \|\mathbf{x} - \mathbf{y}\|\}.$$

In the general case, a direct application of this remark at the level of components of f leads to an extension $\widetilde{f}$ with $\operatorname{Lip}(\widetilde{f}) \leq \sqrt{m}\operatorname{Lip}(f)$. The existence of an extension with the same Lipschitz constant is described in [159, Section 2.10.43, p. 201]. ■

The aforementioned correspondence between graphs is realized by the *Cayley transform*, that is, by the linear isometry

$$\Phi\colon \mathbb{R}^N \times \mathbb{R}^N \to \mathbb{R}^N \times \mathbb{R}^N, \quad \Phi(\mathbf{x}, \mathbf{y}) = \frac{1}{\sqrt{2}}(\mathbf{x} + \mathbf{y}, -\mathbf{x} + \mathbf{y}).$$

When $N = 1$, the Cayley transform represents a clockwise rotation of angle $\pi/4$. The precise statement of this correspondence is as follows:

3.3.10 Theorem (G. Minty [326]) *Let $u\colon \mathbb{R}^N \to \mathcal{P}(\mathbb{R}^N)$ be a maximal monotone map. Then $J = (\mathrm{I} + u)^{-1}$ is defined on the whole $\mathbb{R}^N$ and $\Phi(\operatorname{graph} u)$ is the graph of a nonexpansive function $v\colon \mathbb{R}^N \to \mathbb{R}^N$, given by*

$$v(\mathbf{x}) = \mathbf{x} - \sqrt{2}\,(\mathrm{I} + u)^{-1}(\sqrt{2}\,\mathbf{x}). \tag{3.8}$$

Conversely, if $v\colon \mathbb{R}^N \to \mathbb{R}^N$ is a nonexpansive function, then the inverse image of graph v *under Φ is the graph of a maximal monotone map on $\mathbb{R}^N$.*

Here I denotes the identity map of $\mathbb{R}^N$.

Proof. Let u be a monotone map and let v be the set-valued function whose graph is $\Phi(\operatorname{graph}\ u)$. We shall show that v is nonexpansive in its domain (and thus single-valued). In fact, given $\mathbf{x} \in \mathbb{R}^N$ we have

$$\mathbf{y} \in v(\mathbf{x}) \quad \text{if and only if} \quad \frac{\mathbf{x} + \mathbf{y}}{\sqrt{2}} \in u\Big(\frac{\mathbf{x} - \mathbf{y}}{\sqrt{2}}\Big) \tag{3.9}$$

and this yields $\mathbf{y} \in \mathbf{x} - \sqrt{2}\,(\mathrm{I}{+}u)^{-1}(\sqrt{2}\mathbf{x})$ for all $\mathbf{y} \in v(\mathbf{x})$. Now, if $\mathbf{x}_k \in \mathbb{R}^N$ and $\mathbf{y}_k \in v(\mathbf{x}_k)$ for $k = 1, 2$, then we infer from (3.9) that

$$\langle (\mathbf{x}_1 - \mathbf{y}_1) - (\mathbf{x}_2 - \mathbf{y}_2), (\mathbf{x}_1 + \mathbf{y}_1) - (\mathbf{x}_2 + \mathbf{y}_2)\rangle \geq 0,$$

hence $\|\mathbf{y}_1-\mathbf{y}_2\|^2 \le \|\mathbf{x}_1-\mathbf{x}_2\|^2$. This shows that v is indeed nonexpansive. The same argument shows that Φ^{-1} maps graphs of nonexpansive functions into graphs of monotone functions. Assuming that u is maximal monotone, we shall show that the domain of v is $\mathbb{R}^N$. In fact, if the contrary was true, we could apply Theorem 3.3.9 to extend v to a nonexpansive function $\widetilde{v}$ defined on the whole $\mathbb{R}^N$, and then $\Phi^{-1}(\text{graph } \widetilde{v})$ provides a monotone extension of u, which contradicts the maximality of u. ■

3.3.11 Corollary *Let $u\colon \mathbb{R}^N \to \mathcal{P}(\mathbb{R}^N)$ be a maximal monotone map. Then $J = (\mathrm{I}+u)^{-1}$ is a nonexpansive map of $\mathbb{R}^N$ into itself.*

Proof. It is easy to see that I+u (and thus $(\mathrm{I}+u)^{-1}$) is monotone. By Theorem 3.3.10, the maximality of u yields the surjectivity of I+u, hence $\operatorname{dom}(\mathrm{I}+u)^{-1} = \mathbb{R}^N$. In order to prove that $(\mathrm{I}+u)^{-1}$ is also a nonexpansive function, let us consider points $\mathbf{x}_k \in \mathbb{R}^N$ and $\mathbf{y}_k \in u(\mathbf{x}_k)$ (for $k = 1, 2$). Then

$$\begin{aligned}\|\mathbf{x}_1-\mathbf{x}_2\|^2 &\le \langle \mathbf{x}_1-\mathbf{x}_2, \mathbf{x}_1-\mathbf{x}_2+\mathbf{y}_1-\mathbf{y}_2\rangle \\ &\le \|\mathbf{x}_1-\mathbf{x}_2\| \cdot \|\mathbf{x}_1+\mathbf{y}_1-(\mathbf{x}_2+\mathbf{y}_2)\|,\end{aligned} \tag{3.10}$$

which yields $\|\mathbf{x}_1-\mathbf{x}_2\| \le \|(\mathbf{x}_1+\mathbf{y}_1)-(\mathbf{x}_2+\mathbf{y}_2)\|$. Particularly, if $\mathbf{x}_1+\mathbf{y}_1 = \mathbf{x}_2+\mathbf{y}_2$, then $\mathbf{x}_1 = \mathbf{x}_2$, and this shows that $(\mathrm{I}+u)^{-1}$ is single-valued. Consequently, $(\mathrm{I}+u)^{-1}(\mathbf{x}_k+\mathbf{y}_k) = \mathbf{x}_k$ for $k \in \{1, 2\}$ and thus (3.10) yields the nonexpansivity of $(\mathrm{I}+u)^{-1}$. ■

An important class of maximal monotone maps is provided by the subdifferentials of convex functions.

3.3.12 Theorem *If $f\colon \mathbb{R}^N \to \mathbb{R}\cup\{\infty\}$ is a lower semicontinuous proper convex function, then ∂f is a maximal monotone map.*

Proof. We already noticed that ∂f is a monotone map. According to Theorem 3.3.10, the maximality of ∂f is equivalent to the surjectivity of ∂f+I. A simple argument based on the optimality of convex functions follows from Corollary 3.10.5. ■

An example of maximal monotone set-valued function is the *duality map* of a Banach space E

$$J(\mathbf{x}) = \left\{x^* \in E^* : x^*(\mathbf{x}) = \|\mathbf{x}\|^2 = \|x^*\|^2\right\};$$

indeed, $J(\mathbf{x}) = \partial f(\mathbf{x})$, where f is the function defined by $f(\mathbf{x}) = \|\mathbf{x}\|^2/2$ for $\mathbf{x} \in E$.

Let E be one of the spaces $L^p(\mathbb{R}^N)$ with $p \in (1, \infty)$, and let A be a maximal monotone operator from E to E^*. Let $f : E \to \mathbb{R}\cup\{\infty\}$ be a lower semicontinuous proper convex function such that at least one of the following conditions is verified:

$$\operatorname{dom} A \cap \operatorname{int}(\operatorname{dom} f) \ne \emptyset \text{ or } \operatorname{dom} f \cap \operatorname{int}(\operatorname{dom} A) \ne \emptyset.$$

Then $A + \partial f$ is a maximal monotone operator. The details are quite involving, but this example is important in the theory of partial differential equations. See V. Barbu and T. Precupanu [29], Theorem 2.4, pp. 124–128.

The subdifferential of a lower semicontinuous proper convex function on $\mathbb{R}^N$ plays yet another special property that will be discussed in what follows.

3.3.13 Definition *A cyclically monotone map is any set-valued function* $u\colon \mathbb{R}^N \to \mathcal{P}(\mathbb{R}^N)$ *such that*

$$\langle \mathbf{x}_2 - \mathbf{x}_1, \mathbf{y}_1 \rangle + \langle \mathbf{x}_3 - \mathbf{x}_2, \mathbf{y}_2 \rangle + \cdots + \langle \mathbf{x}_1 - \mathbf{x}_m, \mathbf{y}_m \rangle \leq 0$$

for all finite families of points $(\mathbf{x}_k, \mathbf{y}_k) \in \mathbb{R}^N \times \mathbb{R}^N$ *such that* $\mathbf{y}_k \in u(\mathbf{x}_k)$ *for* $k \in \{1, \ldots, m\}$.

By the subgradient inequality (3.6), the subdifferential of a lower semicontinuous proper convex function on $\mathbb{R}^N$ is cyclically monotone. The converse also holds, as is shown by the following integrability result:

3.3.14 Theorem (R. T. Rockafellar [421, pp. 238–239]) *For any cyclically monotone map* $u\colon \mathbb{R}^N \to \mathcal{P}(\mathbb{R}^N)$ *there exists a lower semicontinuous proper convex function* $f : \mathbb{R}^N \to \mathbb{R} \cup \{\infty\}$ *such that* $u \subset \partial f$.

Proof. Put $\Gamma = \left\{ (\mathbf{x}, \mathbf{y}) \in \mathbb{R}^N \times \mathbb{R}^N : \mathbf{y} \in u(\mathbf{x}) \right\}$ and chose arbitrarily a point $(\mathbf{x}_0, \mathbf{y}_0) \in \Gamma$. We define the function f by the formula

$$f(\mathbf{x}) = \sup \left[\langle \mathbf{x}_1 - \mathbf{x}_0, \mathbf{y}_0 \rangle + \langle \mathbf{x}_2 - \mathbf{x}_1, \mathbf{y}_1 \rangle + \cdots + \langle \mathbf{x} - \mathbf{x}_m, \mathbf{y}_m \rangle \right],$$

where the supremum is taken over all finite families of points $(\mathbf{x}_1, \mathbf{y}_1), \ldots,$ $(\mathbf{x}_m, \mathbf{y}_m)$ in Γ. Since f is a supremum of continuous affine functions, it is lower semicontinuous and convex. Moreover, $f(\mathbf{x}_0) = 0$ due to the property of cyclic monotonicity of u. Hence f is also a proper function. It remains to prove that $\Gamma \subset \partial f$, that is, for every $(\mathbf{x}, \mathbf{y}) \in \Gamma$ we have

$$f(\mathbf{z}) \geq f(\mathbf{x}) + \langle \mathbf{z} - \mathbf{x}, \mathbf{y} \rangle$$

whenever $\mathbf{z} \in \mathbb{R}^N$. For this, it suffices to show that for all $\alpha < f(\mathbf{x})$ and $\mathbf{z} \in \mathbb{R}^N$ we have

$$f(\mathbf{z}) \geq \alpha + \langle \mathbf{z} - \mathbf{x}, \mathbf{y} \rangle.$$

Because of the definition of $f(\mathbf{x})$, the fact that $\alpha < f(\mathbf{x})$ implies the existence of a finite family $(\mathbf{x}_1, \mathbf{y}_1), \ldots, (\mathbf{x}_m, \mathbf{y}_m)$ of points in Γ such that

$$\alpha \leq \langle \mathbf{x}_1 - \mathbf{x}_0, \mathbf{y}_0 \rangle + \langle \mathbf{x}_2 - \mathbf{x}_1, \mathbf{y}_1 \rangle + \cdots + \langle \mathbf{x} - \mathbf{x}_m, \mathbf{y}_m \rangle.$$

Then

$$\begin{aligned} \alpha + \langle \mathbf{z} - \mathbf{x}, \mathbf{y} \rangle &\leq \langle \mathbf{x}_1 - \mathbf{x}_0, \mathbf{y}_0 \rangle + \langle \mathbf{x}_2 - \mathbf{x}_1, \mathbf{y}_1 \rangle + \cdots + \langle \mathbf{x} - \mathbf{x}_m, \mathbf{y}_m \rangle + \langle \mathbf{z} - \mathbf{x}, \mathbf{y} \rangle \\ &\leq f(x), \end{aligned}$$

and the proof is done. ■

Exercises

1. Consider the norm function $f(\mathbf{x}) = \|\mathbf{x}\|$ in an arbitrary real Banach space E. Prove that:
$$\partial f(\mathbf{x}) = \begin{cases} \{x^* \in E^* : \|x^*\| \leq 1\} & \text{if } \mathbf{x} = 0 \\ \{x^* \in E^* : \|x^*\| = 1,\ x^*(\mathbf{x}) = \|\mathbf{x}\|\} & \text{if } \mathbf{x} \neq 0. \end{cases}$$
Compare to the case of absolute value on the real line.

2. Suppose that $f : \mathbb{R} \to \mathbb{R}$ is a convex function. Show that the subdifferential of the function $F(x, y) = f(x) - y$ is given by the formula
$$\partial F(x, y) = \{(\lambda, -1) : \lambda \in \partial f(x)\}.$$

3. (Domain of the subdifferential) Consider the function f defined on $\mathbb{R}^2$ as follows:
$$f(x) = \begin{cases} \max\{1 - \sqrt{x}, |y|\} & \text{if } x \geq 0 \\ \infty & \text{if } x < 0. \end{cases}$$
Prove that f is a convex function and that the domain of ∂f is
$$\{(x, y) : x \geq 0\} \setminus \{(0, y) : |y| < 1)\},$$
which is not a convex set.

4. Suppose that $f : \mathbb{R}^N \to \mathbb{R} \cup \{+\infty\}$ is a lower semicontinuous proper convex function. Prove that:

   (a) $\partial f(\mathbf{a})$ is a closed convex set for every $\mathbf{a} \in \operatorname{dom} f$;

   (b) if $\mathbf{a}$ belongs to the relative interior of $\operatorname{dom} f$, then $\partial f(\mathbf{a})$ is a nonempty compact set;

   (c) the graph of the subdifferential is closed in the sense that $\mathbf{x}_n \to \mathbf{x}$ in $\operatorname{dom} f$ and $\mathbf{y}_n \to \mathbf{y}$ in $\mathbb{R}^N$ with $\mathbf{y}_n \in \partial f(\mathbf{x}_n)$ for all n imply that $\mathbf{y} \in \partial f(\mathbf{x})$.

   [*Hint:* (b) It suffices to prove that $\partial f(\mathbf{a})$ is a bounded set. Use Corollary 3.1.9.]

5. Suppose that E is a Banach space and $f : E \to \mathbb{R}$ is a Lipschitz convex function. Adapt the argument of Remark 3.3.5 to prove that $\partial f(A)$ is bounded, whenever A is a bounded subset of E.

6. (Subdifferentiability and strict convexity) Let U be an open convex subset of $\mathbb{R}^N$ and let $f : U \to \mathbb{R}$ be a convex function. Prove that f is strictly convex if, and only if, it verifies one of the following equivalent conditions:

   (a) $\mathbf{x}, \mathbf{y} \in U$, $\mathbf{x} \neq \mathbf{y}$ and $\mathbf{z} \in \partial f(\mathbf{x})$ implies $f(\mathbf{y}) - f(\mathbf{x}) > \langle \mathbf{y} - \mathbf{x}, \mathbf{z} \rangle$;

   (b) $\mathbf{x}, \mathbf{y} \in U$, $\partial f(\mathbf{x}) \cap \partial f(\mathbf{y}) \neq \emptyset$ implies $\mathbf{x} = y$.

7. Consider the function $f(\mathbf{x}) = -\left(1 - \|\mathbf{x}\|^2\right)^{1/2}$ for $\mathbf{x}$ in the closed unit ball $\overline{B}$ of $\mathbb{R}^N$ and $f(\mathbf{x}) = \infty$ on $\mathbb{R}^N \setminus \overline{B}$. Prove that f is convex and $\partial f(\mathbf{x}) = \emptyset$ at all points $\mathbf{x}$ with $\|\mathbf{x}\| = 1$.

3.4 Positively Homogeneous Convex Functions

A function f defined on a convex cone C is said to be *positively homogeneous* (of degree 1) if

$$f(\lambda \mathbf{x}) = \lambda f(\mathbf{x}) \quad \text{for all } \mathbf{x} \in C \text{ and } \lambda > 0.$$

Two simple examples of positively homogeneous functions defined on a real normed linear space E are the norm function $\|\cdot\|$ and the function $\mathbf{x} \to |x^*(\mathbf{x})|$, associated to a continuous linear functional $x^* \in E^*$.

Convexity of positively homogeneous functions can be related to other important properties like subadditivity and sublinearity. Recall that a function $p\colon E \to \mathbb{R}$ is called *subadditive* if $p(\mathbf{x}+\mathbf{y}) \leq p(\mathbf{x}) + p(\mathbf{y})$ for all $\mathbf{x}, y \in E$ and *superadditive* if $-p$ is additive; p is called *sublinear* if it is both positively homogeneous and subadditive. A sublinear function p is a *seminorm* if $p(\lambda \mathbf{x}) = |\lambda| p(\mathbf{x})$ for all $\lambda \in \mathbb{R}$. A seminorm p is a *norm* if

$$p(\mathbf{x}) = 0 \implies \mathbf{x} = 0.$$

Seminorms are instrumental in the theory of locally convex spaces.

3.4.1 Lemma *If f is a positively homogeneous function defined on a convex cone C, then f is convex (concave) if and only if f is subadditive (superadditive).*

Proof. Suppose that f is convex and $\mathbf{x}, \mathbf{y} \in C$. Then

$$\frac{1}{2} f(\mathbf{x}+\mathbf{y}) = f\left(\frac{\mathbf{x}+\mathbf{y}}{2}\right) \leq \frac{1}{2}\left(f(\mathbf{x}) + f(\mathbf{y})\right)$$

and so $f(\mathbf{x}+y) \leq f(\mathbf{x}) + f(y)$. Conversely, suppose that f is subadditive. Then

$$f((1-\lambda)\mathbf{x} + \lambda \mathbf{y}) \leq f((1-\lambda)\mathbf{x}) + f(\lambda \mathbf{y}) = (1-\lambda) f(\mathbf{x}) + \lambda f(\mathbf{y})$$

for all $\mathbf{x}, \mathbf{y} \in C$ and $\lambda \in (0,1)$, which shows that f is convex. ■

For concave functions f defined on $[0.\infty)$, the property of subadditivity is equivalent to the fact that $f(0) \geq 0$. See Exercise 7.

3.4.2 Lemma *Let f be a positively homogeneous function defined on a convex cone C and taking positive values. If the level set $\{\mathbf{x} \in C : f(\mathbf{x}) \leq 1\}$ is convex, then f is a convex function.*

In other words, every positively homogeneous, quasiconvex, and positive function is convex.

Proof. According to Lemma 3.4.1, it suffices to show that f is subadditive. For that, let $\mathbf{x}, \mathbf{y} \in C$ and choose scalars α and β such that $\alpha > f(\mathbf{x})$, and $\beta > f(\mathbf{y})$. Since f is positive and positively homogeneous, $f(\mathbf{x}/\alpha) \leq 1$ and

$f(\mathbf{y}/\beta) \leq 1$. Thus $\mathbf{x}/\alpha$ and $\mathbf{y}/\beta$ both lie in the level set of f at height 1. The assumed convexity of this level set shows that

$$\frac{1}{\alpha+\beta} f(\mathbf{x}+\mathbf{y}) = f\Big(\frac{\mathbf{x}+\mathbf{y}}{\alpha+\beta}\Big) = f\Big(\frac{\alpha}{\alpha+\beta}\cdot\frac{\mathbf{x}}{\alpha} + \frac{\beta}{\alpha+\beta}\cdot\frac{\mathbf{y}}{\beta}\Big) \leq 1,$$

that is, $f(\mathbf{x}+\mathbf{y}) \leq \alpha+\beta$ whenever $\alpha > f(\mathbf{x})$, $\beta > f(\mathbf{y})$. Hence $f(\mathbf{x}+\mathbf{y}) \leq f(\mathbf{x}) + f(\mathbf{y})$, which shows that f is subadditive. ∎

An example of how the last lemma yields the convexity of some functions is as follows. Let $p \geq 1$ and consider the function f given on the positive orthant $\mathbb{R}_+^N$ by the formula

$$f(x_1, \ldots, x_N) = (x_1^p + \cdots + x_N^p)^{1/p}.$$

Clearly, f is positive and positively homogeneous, and f^p is convex as a sum of convex functions. Hence the level set

$$\{\mathbf{x} \in X : f(\mathbf{x}) \leq 1\} = \{\mathbf{x} \in X : f^p(\mathbf{x}) \leq 1\}$$

is convex and this implies that f is a convex function. By Lemma 3.4.1 we conclude that f is subadditive, a fact which is equivalent with the Minkowski inequality.

We finish this section by introducing and investigating two important examples of positively homogeneous functions that can be attached to certain convex sets.

The *support function* of a nonempty convex set C in a real normed linear space E is defined by the formula

$$\sigma_C(u^*) = \sup_{\mathbf{x}\in C} u^*(\mathbf{x}), \quad u^* \in E^*.$$

This is a a lower semicontinuous sublinear function, taking values in $\mathbb{R} \cup \{\infty\}$. The domain of a support function is a convex cone; the domain is the entire space E if and only if the set C is bounded.

When $E = \mathbb{R}^N$, the dual space can be identified with E, so

$$\sigma_C(\mathbf{u}) = \sup_{\mathbf{x}\in C} \langle \mathbf{x}, \mathbf{u}\rangle, \quad \mathbf{u} \in \mathbb{R}^N.$$

Moreover, if $\|\mathbf{u}\| = 1$, then the sets $H_\alpha = \{\mathbf{x} \in \mathbb{R}^N : \langle \mathbf{x}, \mathbf{u}\rangle = \alpha\}$ represent parallel hyperplanes, each having $\mathbf{u}$ as a normal vector; $\alpha = \sigma_C(\mathbf{u})$ is the value for which H_α supports C, and C is contained in the half-space H_α^-.

The support function of a singleton $C = \{\mathbf{a}\}$ in $\mathbb{R}^N$ is $h_C(\mathbf{u}) = \langle \mathbf{u}, \mathbf{a}\rangle$, while in the case of the closed unit ball, $h_C(\mathbf{u}) = \|\mathbf{u}\|$.

More on the correspondence between convex sets and sublinear functions is offered by the following result:

3.4.3 Theorem (L. Hörmander [226])

(a) *The support function* σ_C *of a nonempty compact convex set* C *in* $\mathbb{R}^N$ *is positively homogeneous and convex. Moreover,*

$$C = \{\mathbf{x} \in \mathbb{R}^N : \langle \mathbf{x}, \mathbf{u} \rangle \leq \sigma_C(\mathbf{u}) \text{ for every } \mathbf{u} \in \mathbb{R}^N\},$$

which shows that C *is the intersection of all half-spaces that contain it.*

(b) *Conversely, let* $h\colon \mathbb{R}^N \to \mathbb{R}$ *be a positively homogeneous convex function. Then the set*

$$C = \{\mathbf{x} \in \mathbb{R}^N : \langle \mathbf{x}, \mathbf{u} \rangle \leq h(\mathbf{u}) \text{ for every } \mathbf{u} \in \mathbb{R}^N\}$$

is nonempty, compact, convex, and its support function is h.

Proof. (a) Clearly, $C \subset \{\mathbf{x} \in \mathbb{R}^N : \langle \mathbf{x}, \mathbf{u} \rangle \leq \sigma_C(\mathbf{u})$ for every $\mathbf{u} \in \mathbb{R}^N\}$. If a vector $\mathbf{z}$ verifies $\langle \mathbf{z}, \mathbf{u} \rangle \leq \sigma_C(\mathbf{u})$ for every $\mathbf{u} \in \mathbb{R}^N$ and $\mathbf{z} \notin C$, then by the basic separation theorem we infer the existence of a vector $\mathbf{u}$ such that $\sigma_C(\mathbf{u}) = \sup\{\langle \mathbf{x}, \mathbf{u} \rangle : \mathbf{x} \in C\} < \langle \mathbf{z}, \mathbf{u} \rangle$, a contradiction. Therefore $\mathbf{z} \in C$.

(b) By Theorem 3.3.1, $\partial h(\mathbf{v}) \neq \emptyset$ for every $\mathbf{v} \in \mathbb{R}^N$. If $\mathbf{x} \in \partial h(\mathbf{v})$, then $h(\mathbf{u}) \geq h(\mathbf{v}) + \langle \mathbf{u} - \mathbf{v}, \mathbf{x} \rangle$ for all $\mathbf{u}$. Replacing $\mathbf{u}$ by $\lambda \mathbf{u}$ (for $\lambda > 0$ arbitrarily fixed) and taking into account that h is positively homogeneous, we infer that $h(\mathbf{u}) \geq h(\mathbf{v})/\lambda + \langle \mathbf{u} - v/\lambda, \mathbf{x} \rangle$, whence, letting $\lambda \to \infty$, we infer that $h(\mathbf{u}) \geq \langle \mathbf{u}, \mathbf{x} \rangle$ for all $\mathbf{u} \in \mathbb{R}^N$. Therefore C contains the union of all subdifferentials $\partial f(\mathbf{v})$. In particular, it is a nonempty set. C is also a closed convex set being an intersection of closed subspaces.

Let $\mathbf{e}_1, ..., \mathbf{e}_N$ be the natural basis of $\mathbb{R}^N$. For each $\mathbf{x} \in C$ we have $-h(-\mathbf{e}_k) \leq -\langle \mathbf{x}, -\mathbf{e}_k \rangle = \langle \mathbf{x}, \mathbf{e}_k \rangle \leq h(\mathbf{e}_k)$ for all k, which yields the boundedness of C. Thus C is a compact set.

Clearly $\sigma_C \leq h$. For the other inequality, choose arbitrarily $\mathbf{u} \in \mathbb{R}^N$ and $\mathbf{z} \in \partial h(\mathbf{u})$ and notice that $0 = h(\mathbf{0}) \geq h(\mathbf{u}) + \langle \mathbf{0} - \mathbf{u}, \mathbf{z} \rangle$, that is, $h(\mathbf{u}) \leq \langle \mathbf{u}, \mathbf{z} \rangle$. Since $\mathbf{z} \in C$ we conclude that $h(\mathbf{u}) \leq \sigma_C(\mathbf{u})$. ■

3.4.4 Corollary *If two nonempty, compact, and convex sets in* $\mathbb{R}^N$ *have the same support, then they coincide.*

See Exercise 9 for an alternative proof.

Let C be a nonempty convex set in a normed linear space E. The *gauge function* (also called the *Minkowski functional*) associated to C is the function

$$\gamma_C\colon E \to \mathbb{R} \cup \{\infty\}, \quad \gamma_C(\mathbf{x}) = \inf\{\lambda > 0 : \mathbf{x} \in \lambda C\},$$

with the convention that $\gamma_C(\mathbf{x}) = \infty$ if $\mathbf{x} \in \lambda C$ for no $\lambda > 0$.

3.4.5 Proposition *If the convex set* C *is closed and contains the origin, then:*

(a) *the gauge function* γ_C *is a lower semicontinuous, sublinear, and positive function;*

(b) $C = \{\mathbf{x} \in E : \gamma_C(\mathbf{x}) \leq 1\}$;

(c) γ_C *is real-valued and continuous if the origin lies in the interior of* C.

Proof. Clearly, $\gamma_C \geq 0$ and $\gamma_C(0) = 0$. The positive homogeneity of γ_C results as follows:

$$\begin{aligned}\gamma_C(\lambda \mathbf{x}) &= \inf\{\mu > 0 : \lambda \mathbf{x} \in \mu C\} \\ &= \inf\{\mu > 0 : \mathbf{x} \in \lambda^{-1}\mu C\} \\ &= \lambda \inf\{\lambda^{-1}\mu > 0 : \mathbf{x} \in \lambda^{-1}\mu C\} \\ &= \lambda \inf\{\tau > 0 : \mathbf{x} \in \tau C\} = \lambda \gamma_C(\mathbf{x}),\end{aligned}$$

for all $\lambda > 0$ and $\mathbf{x} \in \operatorname{dom} \gamma_C$. We have

$$\begin{aligned}\{\mathbf{x} \in E : \gamma_C(\mathbf{x}) \leq 1\} &= \{\mathbf{x} \in E : \inf\{\lambda > 0 : \lambda^{-1}\mathbf{x} \in C\} \leq 1\} \\ &= \{\mathbf{x} \in E : \lambda^{-1}\mathbf{x} \in C \text{ for all } \lambda > 1\} \\ &= \bigcap\nolimits_{\lambda > 1} \lambda C = C,\end{aligned}$$

whence, according to Lemma 3.4.2, it follows that γ_c is a convex function. Taking into account Lemma 3.4.1, we infer that γ_C is actually sublinear. On the other hand, due to the positive homogeneity of γ_C, we have $\{\mathbf{x} \in E : \gamma_C(\mathbf{x}) \leq r\} = rC$ for all $r > 0$, a fact which implies that γ_C is lower semicontinuous.

If 0 is interior to C, then C includes a ball $\overline{B}_\varepsilon(0)$ for some $\varepsilon > 0$. Then every point $\mathbf{x} \in E$, $\mathbf{x} \neq 0$, verifies the formula

$$\gamma_C(\mathbf{x}) = \gamma_C\left(\frac{\|\mathbf{x}\|}{\varepsilon} \cdot \frac{\varepsilon \mathbf{x}}{\|\mathbf{x}\|}\right) = \frac{\|\mathbf{x}\|}{\varepsilon}\gamma_C\left(\frac{\varepsilon \mathbf{x}}{\|\mathbf{x}\|}\right) \leq \frac{1}{\varepsilon}\|\mathbf{x}\| < \infty,$$

so γ_C is finite everywhere. ■

3.4.6 Corollary *If C is a closed, bounded, and convex subset of a real normed linear space E and* int $C \neq \emptyset$, *then C is homeomorphic to the closed unit ball $B = \{\mathbf{x} \in E : \|\mathbf{x}\| \leq 1\}$.*

Proof. Via a translation, we may assume that $0 \in \operatorname{int} C$. Then there is $r > 0$ such that C contains all $\mathbf{x} \in E$ with $\|\mathbf{x}\| \leq r$, whence $\gamma_C(\mathbf{x}) \leq \|\mathbf{x}\|/r$ for all $\mathbf{x} \in E$. Since C is bounded, there is $R > 0$ such that $\|\mathbf{x}\| \leq R$ for all $\mathbf{x} \in C$. Therefore, if $\mathbf{x} \in \lambda C$, then $\|\mathbf{x}/\lambda\| \leq R$, whence $\gamma_C(\mathbf{x}) \geq \|\mathbf{x}\|/R$ for all $\mathbf{x} \in E$.

An immediate consequence is the continuity of the gauge function γ_C. Indeed,

$$|\gamma_C(\mathbf{x}) - \gamma_C(\mathbf{y})| \leq \max\{\gamma_C(\mathbf{x} - \mathbf{y}), \gamma_C(\mathbf{y} - \mathbf{x})\} \leq \frac{1}{r}\|\mathbf{x} - \mathbf{y}\|$$

for all $\mathbf{x}, y \in E$. The proof ends by noticing that the map

$$T : E \to E, \quad T(\mathbf{x}) = \begin{cases} \gamma_C(\mathbf{x})\frac{\mathbf{x}}{\|\mathbf{x}\|} & \text{if } \mathbf{x} \neq 0 \\ 0 & \text{if } \mathbf{x} = 0 \end{cases}$$

is continuous, with continuous inverse and $T(C) = B$. ■

We shall show in Section 6.1 that the support function and the gauge function are related through the concept of Legendre–Fenchel duality. A particular case makes the object of Exercise 6.

Exercises

1. Let $p : E \to \mathbb{R}$ be a sublinear function on a normed linear space. Prove that p is continuous if and only if there is $M > 0$ such that $p(\mathbf{x}) \leq M \|\mathbf{x}\|$ for all $\mathbf{x}$.

2. (B. Dacorogna [119]) Let $f : \mathbb{R}^2 \to \mathbb{R}$ be separately convex and positively homogeneous of degree one. Prove that f is convex. Show, by examples that this result does not work for functions of three (or more) variables.

3. Let f be a positive function defined on a cone C such that for some $\alpha > 0$ it verifies the condition $f(\lambda \mathbf{x}) = \lambda^\alpha f(\mathbf{x})$ for all $\lambda \geq 0$ and $\mathbf{x} \in C$ (that is, f is positively homogeneous of degree α).
   (a) Assume $\alpha \geq 1$. Prove that f is convex if and only if it is quasiconvex. Infer the convexity of the function $f(x_1, \dots, x_N) = \left(\sum_{k=1}^N x_k^\alpha\right)^p$ for all $p \geq 1/\alpha$.
   (b) If $\alpha < 1$, then prove that f is concave if and only if it is quasiconcave.

4. The *Cobb–Douglas function* is defined as
$$f(\mathbf{x}) = A x_1^{r_1} \cdots x_N^{r_n}, \quad (x_1, \dots, x_n) \in \mathbb{R}_{++}^N,$$
   where $A > 0$ and all exponents r_k are strictly positive.
   (a) Prove that $\log f$ is a concave function and infer that $f = e^{\log f}$ is quasiconcave.
   (b) Prove that f is concave if and only if $\sum_{k=1}^N r_k \leq 1$.

5. The *Leontief function* is defined by $f(\mathbf{x}) = \left(\min\left\{\frac{x_k}{\alpha_k} : 1 \leq k \leq N\right\}\right)^\alpha$ for $\mathbf{x} \in \mathbb{R}_{++}^N$, where $\alpha > 0$ and all parameters α_k are strictly positive. Prove that this function is quasiconcave and that it is concave if and only if $\alpha \in (0, 1]$.

6. (Polar duality) Let C be a closed convex set containing the origin. Prove that:
   (a) The Minkowski functional of C is the support function of the polar set C°, and the Minkowski functional of C° is the support function of C, that is,
$$p_C = \sigma_{C^\circ} \text{ and } p_{C^\circ} = \sigma_C.$$
   (b) C° is bounded if and only if $0 \in \text{int } C$ (so, by the bipolar theorem, C is bounded if and only if $0 \in \text{int } C^\circ$). Infer that the Minkowski functional of C is real-valued if $0 \in \text{int } C$.

7. Find the support and the gauge function of each of the following sets:

   (a) $\{\mathbf{x} \in \mathbb{R}^N : \sup\{|x_1|, \ldots, |x_N|\} \leq 1\}$;

   (b) $\{\mathbf{x} \in \mathbb{R}^N : |x_1| + \cdots + |x_N| \leq 1\}$;

   (c) $\{\mathbf{x} \in \mathbb{R}^N : x_1^2 + \cdots + x_N^2 \leq 1\}$.

8. (Subadditivity of concave functions) Suppose C is a convex set that contains the origin and $f : C \to \mathbb{R}$ is a concave function such that $f(0) \geq 0$.

   (a) Prove that $f(t\mathbf{x}) \geq tf(\mathbf{x})$ for all $\mathbf{x} \in C$ and $t \geq 0$.

   (b) Infer that f is subadditive.

   This exercise also shows that every convex function $f : [0, \infty) \to \mathbb{R}$ such that $f(0) \leq 0$ is superadditive.

9. Suppose that C_1 and C_2 are two nonempty compact convex subsets in $\mathbb{R}^N$. Prove that

$$C_1 \subset C_2 \text{ if and only if } \sigma_{C_1} \leq \sigma_{C_2}.$$

   This shows that $C_1 = C_2$ if and only if $\sigma_{C_1} = \sigma_{C_2}$.

10. If C is a nonempty convex set in $\mathbb{R}^N$, prove that

$$\operatorname{ri}(C) = \{\mathbf{x} : \langle \mathbf{x}, \mathbf{v} \rangle < \sigma_C(\mathbf{v}) \text{ for all } \mathbf{v} \neq 0\}.$$

11. (A converse of Proposition 3.4.5) Prove that any positively homogeneous, subadditive, and continuous function $p : E \to [0, \infty)$ is a gauge function.

3.5 Inequalities Associated to Perspective Functions

Many basic inequalities can be easily derived via the following functional operation generating new convex functions from old ones.

Assuming C a convex cone in a linear space, the *perspective* of a function $f : C \to \mathbb{R}$ is the positively homogeneous function

$$\widetilde{f} : C \times (0, \infty) \to \mathbb{R}, \quad \widetilde{f}(\mathbf{x}, t) = tf\left(\frac{\mathbf{x}}{t}\right).$$

3.5.1 Lemma *The perspective of a convex/concave function is a function of the same nature.*

Proof. Indeed, assuming (to make a choice) that f is convex, then for all $(\mathbf{x}, s), (y, t) \in C \times (0, \infty)$ and $\lambda \in [0, 1]$ we have

$$f\left(\frac{(1-\lambda)\mathbf{x}+\lambda y}{(1-\lambda)s+\lambda t}\right) = f\left(\frac{(1-\lambda)s}{(1-\lambda)s+\lambda t}\cdot\frac{\mathbf{x}}{s}+\frac{\lambda t}{(1-\lambda)s+\lambda t}\cdot\frac{y}{t}\right)$$
$$\leq \frac{(1-\lambda)s}{(1-\lambda)s+\lambda t} f\left(\frac{\mathbf{x}}{s}\right)+\frac{\lambda t}{(1-\lambda)s+\lambda t} f\left(\frac{y}{t}\right), \tag{3.11}$$

that is,

$$\widetilde{f}((1-\lambda)\mathbf{x}+\lambda y, (1-\lambda)s+\lambda t) \leq (1-\lambda)\widetilde{f}(\mathbf{x}, s)+\lambda\widetilde{f}(y, t). \tag{3.12}$$

■

3.5.2 Corollary *The following functions are convex on the indicated domains:*
(a) $x \log x - x \log t$ *on* $(0, \infty) \times (0, \infty)$;
(b) $\|\mathbf{x}\|^2 / t$ *on* $\mathbb{R}^N \times (0, \infty)$;
(c) $(\langle \mathbf{c}, \mathbf{x}\rangle + d) f\left(\frac{A\mathbf{x}+b}{\langle \mathbf{c}, \mathbf{x}\rangle+d}\right)$ *on* $\left\{\mathbf{x} : \langle \mathbf{c}, \mathbf{x}\rangle + d > 0 \text{ and } \frac{A\mathbf{x}+b}{\langle \mathbf{c}, \mathbf{x}\rangle+d} \in \operatorname{dom} f\right\}$,
where $A \in \mathrm{M}_N(\mathbb{R})$, $b \in \mathbb{R}^N$ *and* f *is a convex function defined on a convex set in* $\mathbb{R}^N$.

The convexity of the function in case (a) easily yields the convexity of the Kullback–Leibler divergence.

According to Lemma 3.4.1, the perspective of a convex (respectively concave) function is subadditive (respectively superadditive). Exercise 1 addresses the case of strictly concave functions and indicates at this basis short proofs for some classical inequalities.

Deeper results can be obtained via a generalization of Jensen's inequality that works for positively homogeneous continuous convex functions $\Phi\colon \mathbb{R}^2_+ \to \mathbb{R}$. Since

$$\Phi(u, v) = \begin{cases} v\Phi(u/v, 1) & \text{if } u \geq 0 \text{ and } v > 0 \\ u\Phi(1, 0) & \text{if } u \geq 0 \text{ and } v = 0, \end{cases}$$

it follows that Φ is the perspective function associated to the continuous convex function $\varphi(u) = \Phi(u, 1)$, $u \geq 0$.

Since Φ is continuous, $\Phi(0, 0) = 0$. Notice also that

$$\sup_{u>0} \varphi'_+(u) = \lim_{u\to\infty} \frac{\varphi(u)}{u} = \Phi(1, 0). \tag{3.13}$$

The details are left as Exercise 4. According to Theorem 1.6.5, the function φ is the pointwise supremum of a family of affine functions, precisely,

$$\varphi(u) = \sup\{a + bu : (a, b) \in G\}, \text{ for all } u \geq 0,$$

where $G = \{(\varphi(s) - s\varphi'_+(s), \varphi'_+(s)) : s \in (0, \infty)\}$. Therefore for $u, v \in \mathbb{R}^2_+$ with $v > 0$ we have

$$\Phi(u, v) = \sup\{av + bu : (a, b) \in G\}, \tag{3.14}$$

but the last equality also works when $v = 0$, due to formula (3.13).

The next result extends the basic ingredient used by O. Hanner [198] to prove his famous inequalities. See Exercise 7.

3.5.3 Theorem (T.-C. Lim [283]) *Let $\Phi\colon \mathbb{R}^2_+ \to \mathbb{R}$ be a positively homogeneous continuous convex function. Then for every measure space (X, Σ, μ) and every μ-integrable function $f\colon X \to \mathbb{R}^2_+$ we have the inequality*

$$\Phi\Big(\int_X f\,d\mu\Big) \leq \int_X \Phi \circ f\,d\mu. \tag{3.15}$$

If Φ is concave, then the inequality holds in the reverse way.

Put $f = (f_1, f_2)$. Since Φ is continuous, the inverse image of every Borel set is a Borel set, so $\Phi \circ f$ is measurable. On the other hand, $\Phi \circ f$ is majorized in absolute value by a multiple of $f_1 + f_2$. Indeed, for all x such that $f_1(x) + f_2(x) \neq 0$ we have

$$\begin{aligned}
|(\Phi \circ f)(x)| &= |\Phi(f_1(x), f_2(x))| \\
&= (f_1(x) + f_2(x)) \left|\Phi\left(\frac{f(x)}{f_1(x)+f_2(x)}, \frac{f_2(x)}{f_1(x)+f_2(x)}\right)\right| \\
&\leq \max_{0\leq x\leq 1} |\Phi(x, 1-x)|\,(f_1(x) + f_2(x)),
\end{aligned}$$

while for all other values of x we have $\Phi(f_1(x), f_2(x)) = \Phi(0,0) = 0$. Therefore $\Phi \circ f$ is an integrable function. According to formula (3.14),

$$\begin{aligned}
\int_X (\Phi \circ f)(x)\,\mathrm{d}\mu &= \int_X \sup_{(a,b)\in G} (af_2 + bf_1)\,\mathrm{d}\mu \\
&\geq \sup_{(a,b)\in G} \left(a\int_X f_2\,\mathrm{d}\mu + b\int_X f_1\,\mathrm{d}\mu\right) = \Phi\Big(\int_X f\,\mathrm{d}\mu\Big).
\end{aligned}$$

See Exercise 5 for a converse. Also, the role of $\mathbb{R}^2_+$ can be taken by every convex cone in $\mathbb{R}^N$ $(N \geq 2)$. Indeed, the only change is to use Theorem 3.3.2 instead of Theorem 1.6.5.

In applications we shall need the following remark to derive the convexity of Φ :

3.5.4 Remark *If $\Phi : \mathbb{R}^2_+ \to \mathbb{R}$ is a positively homogeneous continuous function, then its convexity is equivalent to the convexity of the restriction of Φ to $(0,\infty)\times(0,\infty)$, which in turn is equivalent to the convexity of the function $\varphi(u) = \Phi(u, 1)$ on $(0, \infty)$.*

3.5.5 Theorem (Minkowski's inequality) *For $p \in (-\infty, 0) \cup [1, \infty)$ and $f, g \in L^p(\mu)$ we have*

$$\|f + g\|_{L^p} \leq \|f\|_{L^p} + \|g\|_{L^p}, \tag{3.16}$$

while for $0 < p < 1$ the inequality works in the reverse sense,

$$\|f + g\|_{L^p} \geq \|f\|_{L^p} + \|g\|_{L^p}. \tag{3.17}$$

If f is not 0 almost everywhere, then we have equality if and only if $g = \lambda f$ almost everywhere, for some $\lambda \geq 0$.

Proof. Apply Theorem 3.5.3 for $f(x) = (|u(x)|^p, |v(x)|^p)$ and

$$\Phi(u,v) = (u^{1/p} + v^{1/p})^p \quad (p \in \mathbb{R},\ p \neq 0)$$

The convexity of Φ follows from that of $\varphi(u) = (1 + u^{1/p})^p$ for $u > 0$. Using the second derivative test we obtain that φ is strictly convex for $0 < p < 1$ and strictly concave for $p \in (-\infty, 0) \cup (1, \infty)$. ■

There is an integral Minkowski-type inequality even in the case $p = 0$. In fact, letting $p \to 0+$ in (3.17), and taking into account Exercise 1 (d) in Section 1.7, we obtain the following result:

3.5.6 Theorem (Minkowski's inequality for $p = 0$) *Assume that (Ω, Σ, μ) is a finite measure space. For every $f, g \in L^1(\mu)$, $f, g \geq 0$, we have*

$$\exp\left(\frac{1}{\mu(X)}\int_X \log(f(x) + g(x))\,\mathrm{d}\mu\right) \geq \exp\left(\frac{1}{\mu(X)}\int_X \log f(x)\,\mathrm{d}\mu\right) + \exp\left(\frac{1}{\mu(X)}\int_X \log g(x)\,\mathrm{d}\mu\right).$$

A simple proof in the discrete case was detailed in Example 3.1.5.

Exercises

1. Prove that the functions $xy/(x + y)$ and $(xy + yz + zx)/(x + y + z)$ are concave for $x, y, z > 0$. For more examples of rational concave functions, see Appendix C.

2. (E. K. Godunova [183] and G. J. Woeginger [490]) Consider the perspective function $\widetilde{f}(x, y) = yf(x/y)$, associated to a strictly concave function $f : [0, \infty) \to \mathbb{R}$. Prove that

$$\sum_{k=1}^{n} \tilde{f}(x_k, y_k) \leq \tilde{f}\left(\sum_{k=1}^{n} x_k, \sum_{k=1}^{n} y_k\right)$$

whenever $x_1, ..., x_n$ and $y_1, ..., y_n$ are strictly positive numbers. The inequality works in the reverse way if f is strictly convex. In either case the equality holds if and only if there exists a number $t > 0$ such that $x_k = ty_k$ for all k.

This result encompasses many classical inequalities such as those of Cauchy, Rogers–Hölder, Minkowski and Gibbs (for $f(x) = x^{1/2}$, $x^{1/p}$, $(x^{1/p} + 1)^p$ and $\log x$ respectively).

3. (E. K. Godunova [183]) Suppose that $M(x, y)$ is a concave function of two real variables, increasing with x for constant y and decreasing with y for constant x. Prove that:

(a) $M(f(x), g(y))$ is a concave function whenever $f : U \to \mathbb{R}$ is concave and $g : V \to \mathbb{R}$ is convex.

(b) A similar result holds if the words "convex" and "concave" are interchanged.

If $g(y) = y$ for all y, then the condition that M decreases with y for constant x is not necessary. Infer that the function $y \log f(x) - y \log y$ is concave on $U \times (0, \infty)$, whenever $f : U \to \mathbb{R}$ is a strictly positive concave function.

4. Prove the formula (3.13), that is, the fact $\sup_{u>0} \varphi'_+(u) = \lim_{u\to\infty} \frac{\varphi(u)}{u} = \Phi(1,0)$.

5. Prove the following converse of Theorem 3.5.3: If Φ is continuous and the inequality (3.15) holds for every μ-integrable function $f : X \to \mathbb{R}^2_+$ for which $\Phi \circ f$ is also μ-integrable, then Φ is positively homogeneous and convex.

6. (More on the Rogers–Hölder inequality) Extend the result of Exercise 3, Section 1.2, to the general context of measure spaces.

7. (Hanner's inequalities [198]) If $f, g \in L^p(\mu)$ and $2 \leq p < \infty$, then

$$\|f + g\|_{L^p}^p + \|f - g\|_{L^p}^p \leq (\|f\|_{L^p} + \|g\|_{L^p})^p + \left|\|f\|_{L^p} - \|g\|_{L^p}\right|^p,$$

equivalently (by making the replacements $f \to f + g$ and $g \to f - g$),

$$(\|f+g\|_{L^p} + \|f-g\|_{L^p})^p + \left|\|f+g\|_{L^p} - \|f-g\|_{L^p}\right|^p \geq 2^p(\|f\|_{L^p}^p + \|g\|_{L^p}^p).$$

In case $1 < p \leq 2$, the above inequalities are reversed.

[*Hint*: Apply Theorem 3.5.3 for $f(x) = (|u(x)|^p, |v(x)|^p)$ and $\Phi(u, v) = (u^{1/p} + v^{1/p})^p + |u^{1/p} - v^{1/p}|^p$. The function Φ is convex for $1 < p \leq 2$ and concave for $2 \leq p < \infty$. To check this, notice that the second derivative of the function $\varphi(u) = \Phi(u, 1) = (u^{1/p} + 1)^p + |u^{1/p} - 1|^p$ is

$$\varphi''(u) = \frac{1-p}{p} x^{-(p+1)/p} \left[\left(x^{-1/p} + 1\right)^{p-2} - \left|x^{-1/p} - 1\right|^{p-2} \right].]$$

Remark. The *modulus of convexity* of a Banach space E is the function $\delta_E : (0, 2] \to [0, 1]$ defined by

$$\delta_E(\varepsilon) = \inf\{1 - \left\|\frac{\mathbf{x} + \mathbf{y}}{2}\right\| : \|\mathbf{x}\| \leq 1, \|\mathbf{y}\| \leq 1 \text{ and } \|\mathbf{x} - \mathbf{y}\| \geq \varepsilon\}.$$

Hanner's inequalities provide exact values for the moduli of convexity in the case of L^p-spaces. Precisely,

$$\left(1 - \delta_{L^p}(\varepsilon) + \frac{\varepsilon}{2}\right)^p + \left|1 - \delta_{L^p}(\varepsilon) - \frac{\varepsilon}{2}\right|^p = 2 \text{ for } p \in (1, 2)$$

and $\delta_{L^p}(\varepsilon) = 1 - (1 - (\varepsilon/2)^p)^{1/p}$ for p in $[2, \infty)$. As a consequence, $\delta_{L^p}(\varepsilon) > 0$ for all $\varepsilon \in (0, 2]$, which shows that the L^p-spaces with $p \in (1, \infty)$ are uniformly convex.

8. (P. Maréchal [303]) Notice that the function $f(x_1, ..., x_N) = \prod_{k=1}^{N} x_k^{-\alpha_k}$ is log-convex on $\mathbb{R}_{++}^N$ whenever $\alpha_1, ..., \alpha_N > 0$. Infer that the function

$$g(x_1, ..., x_N, z) = |z|^\gamma \prod_{k=1}^{N} x_k^{-\alpha_k}$$

is convex on $\mathbb{R}_{++}^N \times \mathbb{R}$ provided that $\gamma \geq 1 + \sum_{k=1}^{N} \alpha_k$.

[*Hint:* In the case of function g, use the properties of perspective functions.]

9. (The contractive version of Jensen's inequality) Suppose that C is a convex set that contains the origin and $f : C \to \mathbb{R} \cup \{\infty\}$ is a proper convex function such that $f(0) \leq 0$. Prove that

$$f\left(\sum_{k=1}^{n} \lambda_k x_k\right) \leq \sum_{k=1}^{n} \lambda_k f(x_k)$$

for all finite families $\{x_1, ..., x_n\} \subset C$ and $\{\lambda_1, ..., \lambda_n\} \subset [0,1]$ with $\sum_{k=1}^{n} \lambda_k \leq 1$. Infer that

$$\sum_{k=1}^{n} \lambda_k x_k \geq \prod_{k=1}^{n} x_k^{\lambda_k}$$

for all $x_1, ..., x_n > 0$ and all $\lambda_1, ..., \lambda_n \in [0,1]$ with $\sum_{k=1}^{n} \lambda_k \leq 1$.

10. (A generalization of the concept of perspective function due to P. Maréchal [303]) Let C be a cone in a linear E, and let K be convex set in a linear space F. The perspective function $f \bigtriangleup g$, associated to a pair of functions $f : C \to \mathbb{R}$ and $g : K \to (0, \infty)$, is defined by the formula

$$(f \bigtriangleup g)(\mathbf{x}, \mathbf{y}) = g(\mathbf{y}) f\left(\frac{\mathbf{x}}{g(\mathbf{y})}\right) \quad \text{for } (\mathbf{x}, \mathbf{y}) \in C \times K.$$

Prove that:

(a) $f \bigtriangleup g$ is convex provided that f is a convex function such that $f(\mathbf{0}) \leq 0$ and g is concave;

(b) $f \bigtriangleup g$ is concave when f and g are concave and $f(\mathbf{0}) \geq 0$.

A noncommutative generalization of the notion of perspective function is described in Section 5.2.

3.6 Directional Derivatives

Subdifferentiability can be expressed in terms of directional derivatives.

3.6.1 Definition *Let $f : E \to \mathbb{R} \cup \{\infty\}$ be a proper function defined on a normed linear space E. The* right-hand directional derivative *of f at $\mathbf{a} \in \operatorname{int}(\operatorname{dom} f)$ relative to $\mathbf{v} \in E$ is defined via the formula*

$$f'_+(\mathbf{a}\,;\mathbf{v}) = \lim_{t\to 0+} \frac{f(\mathbf{a}+t\mathbf{v}) - f(\mathbf{a})}{t}.$$

The main features of the right-hand directional derivative are summarized in the following lemma.

3.6.2 Lemma *Suppose $f : E \to \mathbb{R} \cup \{\infty\}$ is a proper convex function and $\mathbf{a} \in \operatorname{int}(\operatorname{dom} f)$. Then:*

(a) *the right-hand directional derivative $f'_+(\mathbf{a}\,;\mathbf{v})$ is a real-valued positively homogeneous sublinear function of $\mathbf{v} \in E$;*

(b) *$f'_+(\mathbf{a}\,;\mathbf{0}) = 0$ and*

$$f'_+(\mathbf{a}\,;\mathbf{v}) \geq -f'_+(\mathbf{a}\,;-\mathbf{v}) \quad \text{for all } \mathbf{v} \in E;$$

(c) *if $\mathbf{a}$ is a point of continuity for f, then the directional derivative $f'_+(\mathbf{a}\,;\mathbf{v})$ is a continuous function in $\mathbf{v}$.*

Proof. (a) & (b) Since $\mathbf{a} \in \operatorname{int}(\operatorname{dom} f)$, there is $\varepsilon > 0$ such that the function $t \to f(\mathbf{a}+t\mathbf{v})$ is convex on an interval $[-\varepsilon, \varepsilon]$ whenever $\mathbf{v} \in E$. According to the three chords inequality (1.16), the ratio $[f(\mathbf{a}+t\mathbf{v}) - f(\mathbf{a})]/t$ is an increasing function of t, which yields

$$\frac{f(\mathbf{a}-\varepsilon\mathbf{v}) - f(\mathbf{a})}{-\varepsilon} \leq -\frac{f(\mathbf{a}-t\mathbf{v}) - f(\mathbf{a})}{t} \leq \frac{f(\mathbf{a}+t\mathbf{v}) - f(\mathbf{a})}{t} \leq \frac{f(\mathbf{a}+\varepsilon\mathbf{v}) - f(\mathbf{a})}{\varepsilon}$$

for all $t \in (0, \varepsilon]$. This easily implies the existence and finiteness of $f'_+(\mathbf{a}\,;\mathbf{v})$ and $f'_+(\mathbf{a}\,;-\mathbf{v})$ and also the inequality $-f'_+(\mathbf{a}\,;-\mathbf{v}) \leq f'_+(\mathbf{a}\,;\mathbf{v})$. Notice that

$$f'_+(\mathbf{a}\,;\mathbf{v}) = \inf_{t>0} \frac{f(\mathbf{a}+t\mathbf{v}) - f(\mathbf{a})}{t} \tag{3.18}$$

and $f'_+(\mathbf{a}\,;\mathbf{0}) = 0$. For the positive homogeneity of $f'_+(\mathbf{a}\,;\cdot)$ notice that

$$\begin{aligned} f'_+(\mathbf{a}\,;\alpha\mathbf{v}) &= \lim_{t\to 0+} \frac{f(\mathbf{a}+t\alpha\mathbf{v}) - f(\mathbf{a})}{t} = \alpha \lim_{t\to 0+} \frac{f(\mathbf{a}+\alpha t\mathbf{v}) - f(\mathbf{a})}{\alpha t} \\ &= \alpha \lim_{s\to 0+} \frac{f(\mathbf{a}+s\mathbf{v}) - f(\mathbf{a})}{s} = \alpha\, f'(\mathbf{a}\,;\mathbf{v}) \end{aligned}$$

for all $\alpha > 0$. The fact that $f'_+(\mathbf{a}\,;\cdot)$ is subadditive (and thus sublinear) is left as an exercise. See Exercise 1.

(c) If $\mathbf{a}$ is a point of continuity for f, then f is Lipschitz in a ball $\overline{B}_R(\mathbf{a})$. See Proposition 3.1.11. Therefore, there is $M > 0$ such that for all $\mathbf{v} \in E$, $\mathbf{v} \neq 0$,

$$\left| \frac{f(\mathbf{a} + t\mathbf{v}) - f(\mathbf{a})}{t} \right| \leq M \|\mathbf{v}\|$$

for all $t \in (0, R/\|v\|)$. This implies $f'_+(\mathbf{a}; \mathbf{v}) \leq M \|\mathbf{v}\|$ for all $\mathbf{v} \in E$ and the continuity of $f'_+(\mathbf{a}; \mathbf{v})$ follows. See Exercise 1, Section 3.4. ■

Next we will discuss the connection between subgradients and right-hand directional derivatives.

3.6.3 Lemma *Suppose that $f : E \to \mathbb{R} \cup \{\infty\}$ is a proper convex function and $\mathbf{a} \in \operatorname{int}(\operatorname{dom} f)$. Then $x^* \in E^*$ is a subgradient of f at $\mathbf{a}$ if and only if*

$$x^*(\mathbf{v}) \leq f'_+(\mathbf{a}; \mathbf{v}) \quad \text{for all} \quad \mathbf{v} \in E.$$

Proof. If $x^* \in \partial f(\mathbf{a})$, then for every $\mathbf{v} \in E$ and every $t > 0$ we have

$$x^*(\mathbf{v}) \leq \frac{f(\mathbf{a} + t\mathbf{v}) - f(\mathbf{a})}{t},$$

whence, by taking limits as $t \to 0+$, we infer that $x^*(\mathbf{v}) \leq f'_+(\mathbf{a}; \mathbf{v})$.

Conversely, if $x^*(\mathbf{v}) \leq f'_+(\mathbf{a}; \mathbf{v})$ for all $\mathbf{v} \in E$, then the formula (3.18) yields

$$x^*(\mathbf{v}) \leq f'_+(\mathbf{a}; \mathbf{v}) \leq \frac{f(\mathbf{a} + t\mathbf{v}) - f(\mathbf{a})}{t} \quad \text{for all } \mathbf{v} \in E \text{ and } t > 0.$$

In particular, for $\mathbf{v} = \mathbf{x} - \mathbf{a}$ and $t = 1$, we obtain $x^*(\mathbf{x} - \mathbf{a}) \leq f(\mathbf{x}) - f(\mathbf{a})$, that is, $x^* \in \partial f(\mathbf{a})$. The proof is complete. ■

3.6.4 Theorem (Moreau's max formula) *If $f : E \to \mathbb{R} \cup \{\infty\}$ is a proper convex function and a is a point of continuity for f, then $\partial f(\mathbf{a})$ is nonempty and*

$$f'_+(\mathbf{a}; \mathbf{v}) = \max \{x^*(\mathbf{v}) : x^* \in \partial f(\mathbf{a})\}.$$

In other words, $f'_+(\mathbf{a}; \cdot)$ is the support function of $\partial f(\mathbf{a})$.

Proof. According to Lemma 3.6.3, we have to prove that for $\mathbf{v} \in E$ arbitrarily fixed there is a subgradient $x^* \in E^*$ such that $x^*(\mathbf{v}) = f'_+(\mathbf{a}; \mathbf{v})$. This follows from the Hahn–Banach theorem. Indeed, by Lemma 3.6.2 (c), the sublinear function $\mathbf{x} \to f'_+(\mathbf{a}; \mathbf{x})$ is continuous, and an appeal to Corollary B.1.2 shows the existence of a linear functional $x^* : E \to \mathbb{R}$ such that $x^*(\mathbf{v}) = f'_+(\mathbf{a}; \mathbf{v})$ and $x^*(\mathbf{x}) \leq f'_+(\mathbf{a}; \mathbf{x})$ for all $\mathbf{x}$ in E. It remains to prove the continuity of x^* at the origin. We have $\limsup_{\mathbf{x} \to 0} x^*(\mathbf{x}) \leq \lim_{\mathbf{x} \to 0} f'_+(\mathbf{a}; \mathbf{x}) = 0$ and

$$\limsup_{\mathbf{x} \to 0} x^*(-\mathbf{x}) = -\liminf_{\mathbf{x} \to 0} x^*(\mathbf{x}) \leq \lim_{\mathbf{x} \to 0} f'_+(\mathbf{a}; -\mathbf{x}) = 0,$$

so

$$0 \leq \liminf_{\mathbf{x} \to 0} x^*(\mathbf{x}) \leq \limsup_{\mathbf{x} \to 0} x^*(\mathbf{x}) \leq 0,$$

that is, $\lim_{\mathbf{x}\to 0} x^*(\mathbf{x}) = 0$. ∎

In connection with Moreau's max formula notice that for every nonempty convex compact C in $\mathbb{R}^N$ we have

$$\partial\sigma_C(0) = C \text{ and } (\sigma_C)'(0;\cdot) = \sigma_C.$$

Some applications of Moreau's max formula are in order.

3.6.5 Theorem (Subdifferential of the max function) *Suppose that $f_1, \dots, f_n$ are real-valued convex functions on $\mathbb{R}^N$ and denote*

$$f = \max\{f_1, \dots, f_n\}.$$

For $\mathbf{a} \in \mathbb{R}^N$, let $J(\mathbf{a}) = \{j : f_j(\mathbf{a}) = f(\mathbf{a})\}$ be the active index set. Then

$$\partial f(\mathbf{a}) = \operatorname{conv}\left(\cup_{j\in J(\mathbf{a})}\partial f_j(\mathbf{a})\right).$$

More general results, as well as applications to the other subdifferential calculus rules, can be found in the paper of A. Hantoute, M. A. López, and C. Zălinescu [204].

Proof. The inclusion $\supset$ follows from the fact that $\partial f(\mathbf{a})$ is a closed convex set and $\partial f(\mathbf{a}) \supset \partial f_j(\mathbf{a})$ for all $j \in J(\mathbf{a})$. Since the functions f_k are continuous, there is $\varepsilon > 0$ such that $f_k(\mathbf{x}) < f(\mathbf{x})$ for all $k \notin J(\mathbf{a})$ and $\mathbf{x} \in B_\varepsilon(\mathbf{a})$. Then for every $\mathbf{v} \in \mathbb{R}^N \backslash \{\mathbf{0}\}$ arbitrarily fixed and $0 < t < \varepsilon / \|v\|$ we have $f(\mathbf{a} + t\mathbf{v}) = \max_{j\in J(\mathbf{a})} f(\mathbf{a} + t\mathbf{v})$. Therefore

$$\begin{aligned} f'_+(\mathbf{a};\mathbf{v}) &= \lim_{t\to 0+} \frac{f(\mathbf{a}+t\mathbf{v}) - f(\mathbf{a})}{t} = \lim_{t\to 0+} \max_{j\in J(\mathbf{a})} \frac{f_j(\mathbf{a}+t\mathbf{v}) - f(\mathbf{a})}{t} \\ &= \lim_{t\to 0+} \max_{j\in J(\mathbf{a})} \frac{f_j(\mathbf{a}+t\mathbf{v}) - f_j(\mathbf{a})}{t} = \max_{j\in J(a)} f'_j(\mathbf{a},\mathbf{v}) \\ &= \max_{j\in J(\mathbf{a})} \{\langle \mathbf{v}, x_j^*\rangle : x_j^* \in \partial f_j(\mathbf{a})\} \\ &= \max\left\{\langle \mathbf{v}, x^*\rangle : x^* \in \bigcup\nolimits_{j\in J(\mathbf{a})} \partial f_j(\mathbf{a})\right\} \\ &= \max\left\{\langle \mathbf{v}, x^*\rangle : x^* \in \operatorname{conv}\bigcup\nolimits_{j\in J(\mathbf{a})} \partial f_j(\mathbf{a})\right\}, \end{aligned}$$

which shows that the function $\mathbf{v} \to f'_+(\mathbf{x};\mathbf{v})$ is the support of the convex set $C = \operatorname{conv}\left(\cup_{j\in J(\mathbf{x})}\partial f_j(\mathbf{x})\right)$. By Theorem 3.6.4, this function is also the support of $\partial f(\mathbf{x})$, so by Theorem 3.4.3 we conclude that these two sets coincide. ∎

An application of Theorem 3.6.5 is indicated in Exercise 3.

3.6.6 Theorem (Subdifferential sum rule) *Suppose that f_1 and f_2 are two real-valued convex functions defined on $\mathbb{R}^N$ and $t_1, t_2 > 0$. Then, for every $\mathbf{x} \in \mathbb{R}^N$,*

$$\partial(t_1 f_1 + t_2 f_2)(\mathbf{x}) = t_1\partial f_1(\mathbf{x}) + t_2\partial f_2(\mathbf{x}).$$

In the general setting of proper convex functions, only the inclusion $\supset$ works. The equality needs additional assumptions, for example, $\mathrm{ri}(\mathrm{dom}\ f_1) \cap \mathrm{ri}(\mathrm{dom}\ f_2) \neq \emptyset$. A proof based on the Fenchel–Moreau duality is available in [421], Theorem 23.8, p. 223.

Proof. Indeed, $t_1\partial f_1(\mathbf{x}) + t_2\partial f_2(\mathbf{x})$ is a compact convex set whose support function is $t_1 f_1'(\mathbf{x};\cdot) + t_2 f_2'(\mathbf{x};\cdot)$. On the other hand, the support function of $\partial(t_1 f_1 + t_2 f_2)(\mathbf{x})$ is

$$(t_1 f_1 + t_2 f_2)'(\mathbf{x};\cdot) = t_1 f_1'(\mathbf{x};\cdot) + t_2 f_2'(\mathbf{x};\cdot).$$

Therefore the two compact convex sets $t_1\partial f_1(\mathbf{x}) + t_2\partial f_2(\mathbf{x})$ and $\partial(t_1 f_1 + t_2 f_2)(\mathbf{x})$ have the same support. According to Corollary 3.4.4, they coincide. ■

3.6.7 Theorem (Subdifferential of a composition) *Let f be a real-valued convex function on $\mathbb{R}^N$ and let A be a linear transformation from $\mathbb{R}^M$ to $\mathbb{R}^N$. Then, for every $\mathbf{x} \in \mathbb{R}^M$,*

$$\partial(f \circ A)(\mathbf{x}) = A^*\partial f(A\mathbf{x}).$$

In the general setting of proper convex functions, only the inclusion $\partial(f \circ A)(\mathbf{x}) \supset A^*\partial f(A\mathbf{x})$ works. The equality needs additional assumptions, for example, it works when the range of A contains a point of $\mathrm{ri}(\mathrm{dom}\ f)$. See [421], Theorem 23.9, p. 225, for a proof based on Fenchel–Moreau duality.

Proof. Using the definition of directional derivatives and Moreau's max formula, one can show that $(f\circ A)'(\mathbf{x};\mathbf{v}) = f'(A\mathbf{x}, A\mathbf{v})$. Thus the support function of the compact convex set $\partial(f \circ A)(\mathbf{x})$ verifies the formula

$$\sigma_{\partial(f\circ A)(\mathbf{x})}(\mathbf{v}) = f'(A\mathbf{x}, A\mathbf{v}).$$

On the other hand,

$$\sigma_{A^*\partial f(A\mathbf{x})}(\mathbf{v}) = \sup_{\mathbf{y}\in\partial f(A\mathbf{x})} \langle A^*\mathbf{y}, \mathbf{v}\rangle = \sup_{\mathbf{y}\in\partial f(A\mathbf{x})} \langle \mathbf{y}, A\mathbf{v}\rangle = f'(A\mathbf{x}, A\mathbf{v}).$$

Therefore the compact convex sets $\partial(f \circ A)(\mathbf{x})$ and $A^*\partial f(A\mathbf{x})$ have the same support. Hence, according to Corollary 3.4.4, they coincide. ■

Subdifferentiability can be seen as a generalization of Gâteaux differentiability.

3.6.8 Definition *Suppose that E and F are two real Banach spaces and U is an open subset of E. A function $f : U \to F$ is said to be Gâteaux differentiable at a point $\mathbf{a} \in U$ if the directional derivatives*

$$f'(\mathbf{a};\mathbf{v}) = \lim_{t\to 0} \frac{f(\mathbf{a}+t\mathbf{v}) - f(\mathbf{a})}{t}.$$

exist for every $\mathbf{v} \in E$ and define a continuous linear operator $f'(\mathbf{a})\colon \mathbf{v} \to f'(\mathbf{a};\mathbf{v})$ from E to F (called the Gâteaux derivative of f at $\mathbf{a}$ in the direction $\mathbf{v}$).

In the case of real-valued functions f defined on $\mathbb{R}^N$, the Gâteaux differentiability means the existence of all partial derivatives

$$\frac{\partial f}{\partial x_k}(\mathbf{a}) = f'(\mathbf{a})(\mathbf{e}_k) \quad \text{for } k = 1, ..., N,$$

where $\mathbf{e}_1 = (1, 0, ..., 0), ..., \mathbf{e}_N = (0, 0, ..., 1)$ represent the natural basis of $\mathbb{R}^N$. Thus

$$f'(\mathbf{a})(\mathbf{v}) = \langle \nabla f(\mathbf{a}), \mathbf{v} \rangle \quad \text{for all } \mathbf{v} \in \mathbb{R}^N,$$

where

$$\nabla f(\mathbf{a}) = \sum_{k=1}^{n} \frac{\partial f}{\partial x_k}(\mathbf{a})\mathbf{e}_k$$

denotes the *gradient* of f at $\mathbf{a}$.

In the case of a proper convex function $f : E \to \mathbb{R} \cup \{\infty\}$, the definition of Gâteaux differentiability applies to points $\mathbf{a} \in \operatorname{int}(\operatorname{dom} f)$ (which corresponds to the choice $U = \operatorname{int}(\operatorname{dom} f)$ and $F = \mathbb{R}$ in Definition 3.6.8). According to Lemma 3.6.2, the function f is Gâteaux differentiable at a point $\mathbf{a} \in \operatorname{int}(\operatorname{dom} f)$, of continuity for f, if and only if

$$f'_+(\mathbf{a}; \mathbf{v}) = -f'_+(\mathbf{a}; -\mathbf{v}) \quad \text{for all } \mathbf{v} \in E.$$

Indeed, a sublinear function p is linear if and only if $p(-\mathbf{x}) = -p(\mathbf{x})$ for all x. The condition of continuity is automatically fulfilled when E is finite dimensional. See Proposition 3.1.8.

3.6.9 Proposition *If a proper convex function $f : E \to \mathbb{R} \cup \{\infty\}$ is Gâteaux differentiable at a point $a \in \operatorname{int}(\operatorname{dom} f)$, then*

$$\partial f(\mathbf{a}) = \{f'(\mathbf{a})\}$$

and the subgradient inequality becomes

$$f(\mathbf{x}) \geq f(\mathbf{a}) + f'(\mathbf{a})(\mathbf{x} - \mathbf{a}) \quad \textit{for all } \mathbf{x} \in E.$$

Proof. Clearly, $f'(\mathbf{a}) \in \partial f(\mathbf{a})$. If $x^* \in \partial f(\mathbf{a})$, then $x^*(\mathbf{v}) \leq f'(\mathbf{a})(\mathbf{v})$ for all $\mathbf{v} \in E$. Replacing $\mathbf{v}$ by $-\mathbf{v}$, we infer the converse inequality. Therefore $x^* = f'(\mathbf{a})$. ∎

Gâteaux differentiability is equivalent to the uniqueness of the support function.

3.6.10 Corollary *Let $f : E \to \mathbb{R} \cup \{\infty\}$ be a proper convex function which is continuous at a point $\mathbf{a} \in \operatorname{int}(\operatorname{dom} f)$. Then f is Gâteaux differentiable at $\mathbf{a}$ if and only if f has a unique support at $\mathbf{a}$.*

Proof. Suppose that $\partial f(\mathbf{a}) = \{x^*\}$. Then, by Moreau's max formula, $f'_+(\mathbf{a}; \mathbf{v}) = x^*(\mathbf{v})$ and $f'_+(\mathbf{a}; -\mathbf{v}) = -x^*(\mathbf{v})$ for all $\mathbf{v} \in E$. According to the comments after Definition 3.6.8, this implies that f is Gâteaux differentiable at a. The other implication follows from Proposition 3.6.9. ∎

If E is finite dimensional, then the condition on continuity in Corollary 3.6.10 is automatic. See Proposition 3.1.8. In infinite dimensions things are different and without the continuity hypothesis the assertion of Corollary 3.6.10 may fail. An example is provided by the indicator function ι_C of the *Hilbert cube*,

$$C = \left\{(x_n)_{n=1}^{\infty} \in \ell^2 : |x_n| \leq 1/n \text{ for all } n\right\}.$$

The function ι_C is lower semicontinuous because C is a weakly compact convex set and weakly closed sets are also closed in the norm topology. See Exercise 5. Since int $C = \emptyset$, the function ι_C is nowhere continuous. The subdifferential of ι_C at the origin of ℓ^2 is a singleton, but ι_C is nowhere Gâteaux differentiable.

As we already noticed, the subdifferentiability is related to the existence of directional derivatives. However, while a derivative is a local property, the subgradient definition (3.3.3) describes a global property. An illustration of this fact is the following theorem that plays an important role in optimization theory.

3.6.11 Theorem (Subdifferential at optimality) *Let E be a Banach space and let $f : E \to \mathbb{R} \cup \{\infty\}$ be a proper convex function. Then a point $\mathbf{a} \in \operatorname{dom} f$ is a global minimizer of f if and only if f is subdifferentiable at $\mathbf{a}$ and*

$$\mathbf{0} \in \partial f(\mathbf{a}).$$

If $\mathbf{a} \in \operatorname{int}(\operatorname{dom} f)$ and f is differentiable at $\mathbf{a}$, then the condition $\mathbf{0} \in \partial f(\mathbf{a})$ reduces to $f'(\mathbf{a}) = 0$.

The proof of Theorem 3.6.11 is straightforward.

Theorem 3.6.11 represents the analogue of *Fermat's rule* for finding points of extrema in the context of convex functions.

IMPORTANT. In the general framework of differentiable functions a critical point is not necessarily a point of extremum. However, when $f : E \to \mathbb{R}$ is differentiable and convex, the condition $f'(\mathbf{a}) = 0$ is equivalent to the fact that $\mathbf{a}$ is a global minimizer!

A detailed discussion on the extremal properties of convex functions makes the objective of Section 3.10.

Exercises

1. (Subadditivity of the directional derivatives) Suppose that f is a convex function (on an open convex set U in a normed linear space E). For $\mathbf{a} \in U$, $\mathbf{u}, \mathbf{v} \in E$ and $t > 0$ small enough, show that
$$\frac{f(\mathbf{a} + t(\mathbf{u} + \mathbf{v})) - f(\mathbf{a})}{t} \leq \frac{f(\mathbf{a} + 2t\mathbf{u}) - f(\mathbf{a})}{2t} + \frac{f(\mathbf{a} + 2t\mathbf{v}) - f(\mathbf{a})}{2t}$$
and conclude that $f'_+(\mathbf{a}\,; \mathbf{u} + \mathbf{v}) \leq f'_+(\mathbf{a}\,; \mathbf{u}) + f'_+(\mathbf{a}\,; \mathbf{v})$.

2. Let $f, g : \mathbb{R} \to \mathbb{R}$ be two convex functions. Infer from Theorem 3.6.6 that the subdifferential of the function $h(x, y) = f(x) + g(\mathbf{y})$ is given by the formula $\partial h(x, \mathbf{y}) = \partial f(x) \times \partial g(y)$.

3. Consider the space $\mathbb{R}^N$ endowed with the norm $\|\mathbf{x}\|_1 = \sum_{n=1}^N |x_n|$. Prove that the subdifferential of the function $f(\mathbf{x}) = \|\mathbf{x}\|_1$ is

$$\partial f(\mathbf{x}) = \left\{ \sigma \in \mathbb{R}^N : \sup_{1\leq n\leq N} |\sigma_n| \leq 1,\ \sum_{n=1}^{N} \sigma_n \mathbf{x}_n = \|\mathbf{x}\|_1 \right\}.$$

[*Hint:* Apply Theorem 3.6.5, taking into account that f can be expressed as the maximum of $2N$ linear and differentiable functions,

$$f(\mathbf{x}) = \max\left\{ \langle \mathbf{x}, \sigma\rangle = \sum\nolimits_{n=1}^{N} \sigma_n x_n : \sigma_n \in \{-1,1\} \text{ for all } n \right\},$$

and the subgradient of each map $\mathbf{x} \to \langle \mathbf{x}, \sigma\rangle$ is σ.]

4. (Mean value theorem) Let $f : U \to \mathbb{R}$ be a convex function defined on an open convex subset U of the Banach space E and let $\mathbf{x}$ and $\mathbf{y}$ be two distinct points in E. Prove that there is a number $t \in (0,1)$ such that

$$f(\mathbf{x}) - f(\mathbf{y}) \in \langle \mathbf{x} - \mathbf{y}, \partial f((1-t)\mathbf{x} + t\mathbf{y})\rangle.$$

[*Hint:* Consider a local extremum for the convex function $\varphi : [0,1] \to \mathbb{R}$ defined by $\varphi(t) = f((1-t)\mathbf{x} + t\mathbf{y})$.]

Remark. The mean value theorem easily yields the following characterization of the Lipschitz property: a convex function $f : U \to \mathbb{R}$ is Lipschitz with $\mathrm{Lip}(f) = K$ if, and only if, $\|x^*\| \leq K$ for all $x^* \in \partial f(\mathbf{x})$ and $\mathbf{x} \in U$.

5. Prove that the Hilbert cube is a weakly compact convex subset of ℓ^2.

[*Hint:* Since ℓ^2 is a reflexive Banach space, it suffices to prove that the Hilbert cube is a closed, bounded, and convex set. See the Banach–Alaoglu theorem (Theorem B.1.6, Appendix B).]

6. Find convex functions $f, g : \mathbb{R} \to \mathbb{R} \cup \{\infty\}$ with $\partial f(0) + \partial g(0) \neq \partial(f+g)(0)$.

7. Give an example of convex function $g : \mathbb{R}^2 \to \mathbb{R} \cup \{\infty\}$ and of a linear map $A : \mathbb{R} \to \mathbb{R}^2$ with $\partial(g \circ A)(0) \neq A^* \partial g(0)$.

3.7 Differentiability of Convex Functions

A stronger condition than Gâteaux differentiability is Fréchet differentiability, which is crucial for some basic results in nonlinear analysis such as the local inversion theorem and the implicit function theorem.

3.7.1 Definition *Suppose that E and F are two real Banach spaces and f is a function defined on an open subset U of E and taking values in F. We say that f is* Fréchet differentiable *at a point $\mathbf{a} \in U$ if there exists a continuous linear operator $df(\mathbf{a}) \in L(E,F)$ such that*

$$\lim_{\mathbf{x}\to\mathbf{a}} \frac{|f(\mathbf{x}) - f(\mathbf{a}) - df(\mathbf{a})(\mathbf{x}-\mathbf{a})|}{\|\mathbf{x}-\mathbf{a}\|} = 0.$$

Equivalently, using Landau's little-o notation,

$$f(\mathbf{a}+\mathbf{v}) = f(\mathbf{a}) + df(\mathbf{a})(\mathbf{v}) + o(\mathbf{v}), \tag{3.19}$$

which means the existence of a ball $B_\varepsilon(\mathbf{0}) \subset E$ centered at $\mathbf{0}$ and a function $\omega : B_\varepsilon(\mathbf{0}) \to \mathbb{R}$ such that $\omega(\mathbf{0}) = \lim_{\mathbf{v}\to\mathbf{0}} \omega(\mathbf{v}) = 0$ and

$$f(\mathbf{a}+\mathbf{v}) = f(\mathbf{a}) + df(\mathbf{a})(\mathbf{v}) + \omega(\mathbf{v})\,\|\mathbf{v}\|$$

for all $\mathbf{v} \in B_\varepsilon(\mathbf{0})$.

The operator $df(\mathbf{a})$, when it exists, is unique. It is called the *Fréchet derivative* of f at $\mathbf{a}$.

Notice the following analogue of formula (3.19) in the case of Gâteaux differentiability:

$$f(\mathbf{a}+\varepsilon\mathbf{v}) = f(\mathbf{a}) + \varepsilon f'(\mathbf{a})(\mathbf{v}) + o(\mathbf{v}). \tag{3.20}$$

Clearly, Fréchet differentiability implies Gâteaux differentiability and also the equality of the respective derivatives,

$$f'(\mathbf{a}) = df(\mathbf{a}).$$

A simple test of Fréchet differentiability for functions defined on open sets in $\mathbb{R}^N$ is the membership to the class C^1 (that is, the existence and continuity of all partial derivatives of first order). This result, as well as many more on Fréchet differentiability, can be found in the book by R. Coleman [112], Theorem 2.2, p. 45.

Remarkably, the Gâteaux and Fréchet differentiability agree in the case of convex functions defined on open subsets of $\mathbb{R}^N$.

3.7.2 Theorem *Suppose that a convex function f defined on an open convex set U in $\mathbb{R}^N$ possesses all its partial derivatives $\frac{\partial f}{\partial x_1}, \dots, \frac{\partial f}{\partial x_N}$ at some point $\mathbf{a} \in U$. Then f is Fréchet differentiable at $\mathbf{a}$.*

Proof. Since U is open, there is an $r > 0$ such that $B_r(\mathbf{a}) \subset U$. We have to prove that the function

$$g(\mathbf{u}) = f(\mathbf{a}+\mathbf{u}) - f(\mathbf{a}) - \sum_{k=1}^{N} \frac{\partial f}{\partial x_k}(\mathbf{a}) u_k,$$

defined for all vectors $\mathbf{u} = (u_1, \dots, u_N)$ with $\|\mathbf{u}\| < r$, verifies

$$\lim_{\|\mathbf{u}\|\to\mathbf{0}} g(\mathbf{u})/\|\mathbf{u}\| = 0.$$

Clearly, the function g is convex. Then

$$0 = g(\mathbf{0}) = g\Big(\frac{\mathbf{u}+(-\mathbf{u})}{2}\Big) \le \frac{1}{2}\,(g(\mathbf{u}) + g(-\mathbf{u})),$$

which yields $g(\mathbf{u}) \geq -g(-\mathbf{u})$. On the other hand, for each $\mathbf{u}$ with $N\|\mathbf{u}\| < r$, we have

$$g(\mathbf{u}) = g\Big(\frac{1}{N}\sum_{k=1}^{N} N u_k \mathbf{e}_k\Big) \leq \frac{1}{N}\sum_{k=1}^{N} g(N u_k \mathbf{e}_k)$$
$$= \sum_{\{k:u_k \neq 0\}} u_k \frac{g(N u_k \mathbf{e}_k)}{N u_k} \leq \|\mathbf{u}\| \sum_{\{k:u_k \neq 0\}} \left|\frac{g(N u_k \mathbf{e}_k)}{N u_k}\right|,$$

where $\mathbf{e}_1, ..., \mathbf{e}_N$ denotes the natural basis of $\mathbb{R}^N$. Similarly,

$$g(-\mathbf{u}) \leq \|\mathbf{u}\| \sum_{\{k:u_k \neq 0\}} \left|\frac{g(-N u_k \mathbf{e}_k)}{N u_k}\right|.$$

Then

$$-\|\mathbf{u}\| \sum_{\{k:u_k \neq 0\}} \left|\frac{g(-N u_k \mathbf{e}_k)}{N u_k}\right| \leq -g(-\mathbf{u}) \leq g(\mathbf{u}) \leq \|\mathbf{u}\| \sum_{\{k:u_k \neq 0\}} \left|\frac{g(N u_k \mathbf{e}_k)}{N u_k}\right|,$$

and to complete the proof it remains to notice that from the definition of the partial derivative we have $g(N u_k \mathbf{e}_k)/(N u_k) \to 0$ as $u_k \to 0$. ■

The analogue of Theorem 3.7.2 for differentiability of higher order does not work. See Exercise 9 and the Comments at the end of this chapter.

In the context of several variables, the set of points where a convex function is not differentiable is negligible (though it can be uncountable). This fact can be seen as a consequence of Rademacher's theorem on almost everywhere differentiability of Lipschitz functions, but we prefer to present here a direct argument.

3.7.3 Theorem *Suppose that f is a convex function on an open subset U of $\mathbb{R}^N$. Then f is differentiable almost everywhere in U.*

Proof. Consider first the case when U is also bounded. According to Theorem 3.7.2 we must show that each of the sets

$$E_k = \Big\{\mathbf{x} \in U \;\Big|\; \frac{\partial f}{\partial x_k}(\mathbf{x}) \text{ does not exist}\Big\}, \quad k = 1, ..., N,$$

is a set of zero Lebesgue measure. The measurability of E_k is a consequence of the fact that the limit of a pointwise converging sequence of measurable functions is measurable too. In fact, the formula

$$f'_+(\mathbf{x}, \mathbf{e}_k) = \lim_{j \to \infty} \frac{f(\mathbf{x} + \mathbf{e}_k/j) - f(\mathbf{x})}{1/j}$$

motivates the measurability of the one-sided directional derivative $f'_+(\mathbf{x}, \mathbf{e}_k)$ and a similar argument applies for $f'_-(\mathbf{x}, \mathbf{e}_k)$. Consequently the set

$$E_k = \{\mathbf{x} \in U : f'_+(\mathbf{x}, \mathbf{e}_k) - f'_-(\mathbf{x}, \mathbf{e}_k) > 0\}$$

is measurable. Being bounded, it is also integrable. By Fubini's theorem,

$$\begin{aligned}\mathcal{L}(E_k) &= \int_{\mathbb{R}^N} \chi_{E_k}\,\mathrm{d}\mathbf{x} \\ &= \int_{\mathbb{R}} \cdots \left(\int_{\mathbb{R}} \chi_{E_k}\,\mathrm{d}x_k \right) \mathrm{d}x_1 \cdots \mathrm{d}x_{k-1}\,\mathrm{d}x_{k+1} \cdots \mathrm{d}x_N\end{aligned}$$

and the interior integral is zero since f is convex as a function of x_k (and thus differentiable except at a countable set of points). If U is arbitrary, the argument above shows that all the sets $E_k \cap B_n(\mathbf{0})$ are negligible. Or, $E_k = \bigcup_{n=1}^{\infty} (E_k \cap B_n(\mathbf{0}))$ and a countable union of negligible sets is negligible too. ■

The function $f(x, y) = \sup\{x, 0\}$ is convex on $\mathbb{R}^2$ and nondifferentiable at the points of the y-axis (which constitutes an uncountable set).

The coincidence of Gâteaux and Fréchet differentiability is no longer true in the context of infinite-dimensional spaces. See Exercises 3–6.

An example of a real-valued continuous convex function which is not Gâteaux differentiable at any point is provided in Exercise 5. However, according to Theorem 3.3.1, such a function is everywhere subdifferentiable.

3.7.4 Theorem *Let E be a Banach space such that for each continuous convex function $f\colon E \to \mathbb{R}$, every point of Gâteaux differentiability is also a point of Fréchet differentiability. Then E is finite dimensional.*

The proof depends on a deep result in Banach space theory:

3.7.5 Theorem (The Josephson–Nissenzweig theorem [236], [382]) *If E is a Banach space such that*

$$x_n^\star \to 0 \text{ in the weak-star topology of } E^\star \quad \text{implies } \|x_n^\star\| \to 0,$$

then E is finite dimensional.

Proof. (of Theorem 3.7.4). Consider a sequence $(x_n^*)_n$ of norm-1 functionals in E^* and a sequence $(\alpha_n)_n$ of real numbers such that $\alpha_n \downarrow 0$. Then the function

$$f(\mathbf{x}) = \sup_n [\langle \mathbf{x}, x_n^* \rangle - \alpha_n]$$

is convex and continuous and, moreover,

$$\begin{aligned} f \text{ is Gâteaux differentiable at } 0 &\iff x_n^*(\mathbf{x}) \to 0 \text{ for all } x \in E \\ f \text{ is Frechet differentiable at } 0 &\iff \|x_n^*\| \to 0. \end{aligned}$$

The proof ends by applying the Josephson–Nissenzweig theorem. ■

3.7.6 Remark *A result due to S. Mazur asserts that if E is a separable Banach space and f is a continuous convex function defined on a convex open subset U of E, then the set of points where f is Gâteaux differentiable is always a dense subset of U (even more, it is the intersection of a sequence of dense open subsets). See R. R. Phelps [397], Theorem 1.20, p. 12. A similar result in the* *case of Fréchet differentiability is known to work in the framework of reflexive Banach spaces. See Theorem* 3.12.1 *in the Comments section at the end of this chapter.*

We next discuss an important geometric property of L^p-spaces with $1 < p < \infty$, the smoothness of the unit sphere.

3.7.7 Definition *A Banach space E is called* smooth *if its norm is Gâteaux differentiable at every nonzero point.*

Smoothness and strict convexity are related by duality. If E^* is strictly convex (respectively, smooth), then E is smooth (respectively, strictly convex). The converse is true for reflexive spaces, but not in general.

3.7.8 Proposition *Let (Ω, Σ, μ) be a measure space. Then all Lebesgue spaces $L^p(\mu)$ with $1 < p < \infty$ are smooth and the Gâteaux derivative of the norm function*

$$\mathrm{N}(u) = \|u\|_p = \left(\int_\Omega |u(x)|^p \, \mathrm{d}\mu\right)^{1/p}$$

at a point $u \in L^p(\mu)$, $u \neq 0$, is

$$\mathrm{N}'(u)(v) = \|u\|^{1-p} \int_\Omega |u(x)|^{p-2} u(x)v(x)\mathrm{d}\mu \quad \textit{for all } v \in L^p(\mu).$$

Proof. We prove first that the function $G(u) = \|u\|_p^p$ is Gâteaux differentiable on $L^p(\mu)$ and

$$G'(u)(v) = p \int_\Omega |u(x)|^{p-2} u(x)v(x)\mathrm{d}\mu \quad \text{for all } v.$$

Indeed, for every two real numbers a, b we have

$$\lim_{t \to 0} \frac{|a+tb|^p - |a|^p}{t} = p\,|a|^{p-2} ab; \tag{3.21}$$

the case $a = 0$ is not excluded because the function $|a|^{p-2} a$ has the limit 0 at $a = 0$ when $p > 1$. The function $|a+tb|^p$ is convex (being the composition of the convex and increasing function x^p on $[0,\infty)$ and the convex function $|a+tb|$). Therefore

$$|a|^p - |a-b|^p \leq \frac{|a+tb|^p - |a|^p}{t} \leq |a+b|^p - |a|^p, \tag{3.22}$$

according to the three chords inequality. The double estimate (3.22) yields the integrability of the function

$$\frac{|u(x)+tv(x)|^p - |u(x)|^p}{t}.$$

Taking into account the formula (3.21) and Lebesgue's dominated convergence theorem we infer that

$$G'(u)(v) = \lim_{t\to 0} \frac{\|u+tv\|^p - \|u\|^p}{t} = p \int_\Omega |u(x)|^{p-2} u(x)v(x)\mathrm{d}\mu.$$

Hence, for $u \neq 0$,

$$N'(u)(v) = \frac{d}{d\varepsilon} G^{1/p}(u+\varepsilon v)\bigg|_{\varepsilon=0} = G^{(1-p)/p}(u) \int_\Omega |u(x)|^{p-2}\, u(x)v(x)\mathrm{d}\mu.$$

■

If one considers the squared norm $\mathrm{N}^2(\mathbf{x}) = \|\mathbf{x}\|^2$ in a real Hilbert space H, then the identity

$$\|\mathbf{x}+\mathbf{v}\|^2 = \|\mathbf{x}\|^2 + 2\langle \mathbf{x}, \mathbf{v}\rangle + \|\mathbf{v}\|^2$$

easily yields the fact that this function is Fréchet differentiable everywhere and its Fréchet derivative verifies the formula

$$d\mathrm{N}^2(\mathbf{x})(\mathbf{v}) = 2\langle \mathbf{x}, \mathbf{v}\rangle \quad \text{for all } \mathbf{x}, \mathbf{v} \in H.$$

The Fréchet differentiability of the norm function $\mathrm{N}(\mathbf{x}) = \|\mathbf{x}\| = \sqrt{\mathrm{N}^2(\mathbf{x})}$ for all $\mathbf{x} \in H$, $\mathbf{x} \neq 0$, results via the chain rule. We have

$$d\mathrm{N}(\mathbf{x})(\mathbf{v}) = \langle \mathbf{v}, \frac{\mathbf{x}}{\|\mathbf{x}\|}\rangle \quad \text{whenever } \mathbf{x}, \mathbf{v} \in H,\ \mathbf{x} \neq 0.$$

One can prove that all L^p-norms are Fréchet differentiable at every nonzero point of $L^p(\mu)$ when $1 < p < \infty$. See K. Sundaresan and S. Swaminathan [467].

Exercises

1. Consider a Lipschitz function f defined on an open set in $\mathbb{R}^N$ and taking values in a Banach space E. Prove that f is Gâteaux differentiable if and only if it is Fréchet differentiable.

2. Suppose that U is a convex open subset of a Banach space E and $f : U \to \mathbb{R}$ is a differentiable convex function. Prove that f is actually a continuously differentiable function.

   [*Hint:* Notice that for any $\mathbf{x} \in U$ and any $\varepsilon > 0$ there exists $\delta > 0$ such that

   $$\|\mathbf{y} - \mathbf{x}\| < \delta \text{ implies } \partial f(\mathbf{y}) \subset \partial f(\mathbf{x}) + \varepsilon B,$$

   where B denotes the open unit ball of E.]

3. Consider the space ℓ^1, of all absolutely summing sequences of real numbers, endowed with its natural norm, $f(\mathbf{x}) = \|\mathbf{x}\|_1 = \sum_{n=1}^\infty |x_n|$. The dual of this space is ℓ^∞, the space of all bounded sequences $\sigma = (\sigma_n)_n$ of real numbers, endowed with the sup norm, $\|\sigma\|_\infty = \sup |\sigma_n|$. See R. Bhatia [51], p. 50. Prove that the norm in ℓ^1 is Gâteaux differentiable only at the points $\mathbf{x} = (x_n)_n$ in ℓ^1 whose all entries x_n are nonzero.

4. Prove that the norm $\|\cdot\|_1$ in ℓ^1 is not Fréchet differentiable at any point.

5. Prove that the natural norm in the space $\ell^1(A)$, of all absolutely summing families $\mathbf{x} = (x_a)_{a \in A}$ of real numbers,

$$\|\mathbf{x}\|_1 = \sum_{F \subset A,\ F \text{ finite}} |x_a|,$$

is not Gâteaux differentiable at any point if A is uncountable.

6. Prove that the norm function on $C([0,1])$,

$$\|x\| = \sup\{|x(t)| : t \in [0,1]\},$$

is not Fréchet differentiable at any point, but it is Gâteaux differentiable at those points x of the unit sphere for which $|x(t_0)| = 1$ is attained for only one value t_0.

7. (An example due to M. Sova; see [180]) Consider the maps $f : L^1([0,\pi]) \to \mathbb{R},\ f(x) = \int_0^\pi \sin(x(t))dt$, and $g : L^2([0,\pi]) \to \mathbb{R},\ g(x) = \int_0^\pi \sin(x(t))dt$. Prove that:

   (a) f is everywhere Gâteaux differentiable, with $f'(x) = \cos x$, but nowhere Fréchet differentiable;

   (b) g is Fréchet differentiable everywhere.

8. Infer from the Josephson–Nissenzweig theorem that any Banach space E on which all continuous convex functions are bounded on bounded subsets is finite dimensional.

   [*Hint*: Consider a sequence $(x_n^*)_n$ of norm-1 functionals in E^* and the convex function $f(\mathbf{x}) = \sum_n n(|x_n^*(\mathbf{x})| - 1/2)^+$. Then notice that f is finite and continuous if and only if $(x_n^*)_n$ is weak-star convergent to 0.]

9. (J. Stoer and C. Witzgall [460], p. 152) There is no analogue of Theorem 3.7.2 for the second-order partial derivatives. Consider the function

$$f(x,y) = \begin{cases} \frac{xy(x^2-y^2)}{x^2+y^2} + 13x^2 + 13y^2 & \text{if } (x,y) \in \mathbb{R}^2 \setminus \{(0,0)\} \\ 0 & \text{if } x = y = 0. \end{cases}$$

   (a) Prove that f is convex.

   (b) Prove that all second-order partial derivatives of f exist, but $\frac{\partial^2 f}{\partial x \partial y}(0,0)$ differs from $\frac{\partial^2 f}{\partial y \partial x}(0,0)$. Therefore f is not twice differentiable despite the fact that its Hessian matrix exists.

3.8 Differential Criteria of Convexity

The following characterization of convexity within the class of Gâteaux differentiable functions combines Theorem 3.3.6 and Proposition 3.6.9:

3.8.1 Theorem (First derivative test of convexity) *Suppose that U is an open convex set in a real Banach space. Then a Gâteaux differentiable function $f : U \to \mathbb{R}$ is convex if and only if*

$$f(\mathbf{x}) \geq f(\mathbf{a}) + f'(\mathbf{a})(\mathbf{x} - \mathbf{a}) \quad \text{for every } \mathbf{x}, \mathbf{a} \in U. \tag{3.23}$$

The function f is strictly convex if and only if the inequality (3.23) *is strict for $\mathbf{x} \neq \mathbf{a}$.*

The case of strongly convex functions makes the objective of Exercise 5.

On intervals, a differentiable function is convex if and only if its derivative is increasing. The higher dimensional analogue of this result is due to I. R. Kachurovskii [239] and illustrates the concept of monotone map.

3.8.2 Definition *Suppose that C is a nonempty set in a real Banach space E. A map $V : C \to E^*$ is called monotone if*

$$(V(\mathbf{x}) - V(\mathbf{y}))\,(\mathbf{x} - \mathbf{y}) \geq 0$$

for all $\mathbf{x}, \mathbf{y} \in C$ and strictly monotone if this inequality is strict for $\mathbf{x} \neq \mathbf{y}$.

This definition is motivated by the case where $E = \mathbb{R}$ and C is an interval in $\mathbb{R}$. In this case the monotonicity of a function $V : C \to \mathbb{R}$ reduces to the fact that $(V(x) - V(y)) \cdot (x - y)) \geq 0$, equivalently,

$$x \leq y \text{ implies } V(x) \leq V(y).$$

3.8.3 Theorem *Suppose that f is Gâteaux differentiable on the open convex set U in a real Banach space. Then f is (strictly) convex if and only if its Gâteaux differential is a (strictly) monotone map from U to E^*.*

Proof. If f is convex, then for $\mathbf{x}$ and $\mathbf{y}$ in U and $0 < t < 1$ we have

$$\frac{f(\mathbf{y} + t(\mathbf{x} - \mathbf{y})) - f(\mathbf{y})}{t} \leq f(\mathbf{x}) - f(\mathbf{y}),$$

so by letting $t \to 0+$ we obtain that $f'(\mathbf{y})(\mathbf{x} - \mathbf{y}) \leq f(\mathbf{x}) - f(\mathbf{y})$. Interchanging $\mathbf{x}$ and $\mathbf{y}$, we also have $f'(\mathbf{x})(\mathbf{y} - \mathbf{x}) \leq f(\mathbf{y}) - f(\mathbf{x})$. By adding these two inequalities we obtain the monotonicity of f', that is,

$$(f'(\mathbf{x}) - f'(\mathbf{y}))\,(\mathbf{x} - \mathbf{y}) \geq 0 \quad \text{for all } \mathbf{x}, \mathbf{y} \in U. \tag{3.24}$$

Suppose now that (3.24) holds. Let $\mathbf{x}, \mathbf{y} \in U$ and consider the function $g(\lambda) = f((1 - \lambda)\mathbf{x} + \lambda\mathbf{y})$, $\lambda \in [0, 1]$. One can easily verify that

$$\lambda_1 \leq \lambda_2 \text{ implies } g'(\lambda_1) \leq g'(\lambda_2),$$

which shows that g is convex. Then

$$\begin{aligned} f((1 - \lambda)\mathbf{x} + \lambda\mathbf{y}) &= g(\lambda) = g(\lambda \cdot 1 + (1 - \lambda) \cdot 0) \\ &\leq \lambda g(1) + (1 - \lambda) g(0) = (1 - \lambda) f(\mathbf{x}) + \lambda f(\mathbf{y}), \end{aligned}$$

which proves that f is a convex function. The case of strictly convex functions can be treated in the same manner. ■

When the ambient space is a real Hilbert space (in particular, an Euclidean space), the inequality (3.24) defining the monotonicity of the differential becomes

$$\langle \nabla f(\mathbf{x}) - \nabla f(\mathbf{y}), \mathbf{x} - \mathbf{y} \rangle \geq 0. \tag{3.25}$$

The analogue of Theorem 3.8.3 in the context of subdifferentiability was presented in Section 3.3.

Our next goal is an important characterization of convexity under the presence of twice Gâteaux differentiability. For convenience, we recall here some basic facts.

Given a Gâteaux differentiable function $f\colon U \to \mathbb{R}$, the limit

$$f''(\mathbf{a}; \mathbf{v}, \mathbf{w}) = \lim_{\lambda \to 0} \frac{f'(\mathbf{a} + \lambda \mathbf{w})(\mathbf{v}) - f'(\mathbf{a})(\mathbf{v})}{\lambda}$$

(when it exists in $\mathbb{R}$) is called the *second directional derivative* of f at the point $\mathbf{a}$ in the directions $\mathbf{v}$ and $\mathbf{w}$. We say that f is *twice Gâteaux differentiable* at $\mathbf{a}$ if the second directional derivatives $f''(\mathbf{a}; \mathbf{v}, \mathbf{w})$ exist for all vectors $\mathbf{v}, \mathbf{w}$ in the ambient Banach space E and the map $f''(\mathbf{a}) : (\mathbf{v}, \mathbf{w}) \to f''(\mathbf{a}; \mathbf{v}, \mathbf{w})$ is a continuous bilinear form on $E \times E$; in this case, $f''(\mathbf{a})$ represents the *second Gâteaux differential* of f at $\mathbf{a}$.

When $E = \mathbb{R}^N$, the twice Gâteaux differentiability simply means the existence of the *Hessian matrix* of f at $\mathbf{a}$,

$$\operatorname{Hess}_{\mathbf{a}} f = \Big(\frac{\partial^2 f}{\partial x_i \partial x_j}(\mathbf{a}) \Big)_{i,j=1}^{N}$$

for all points $\mathbf{a} \in U$. We have

$$f''(\mathbf{a})(\mathbf{v}, \mathbf{w}) = \langle (\operatorname{Hess}_{\mathbf{a}} f)\mathbf{v}, \mathbf{w} \rangle. \tag{3.26}$$

The Hessian matrix $\operatorname{Hess}_{\mathbf{a}} f$ is symmetric when all second-order derivatives of f are continuous at $\mathbf{a}$.

The twice Fréchet differentiability of a function $f\colon U \to \mathbb{R}$ means the differentiability of both f and $\mathrm{d}f$. The *second Fréchet differential* of a function $f\colon U \to \mathbb{R}$ at $\mathbf{a} \in U$ is defined as the differential at $\mathbf{a}$ of the first differential, that is,

$$\mathrm{d}^2 f(\mathbf{a}) = \mathrm{d}\,(\mathrm{d}f(\mathbf{x}))\,(\mathbf{a}).$$

This imposes the existence of the first differential at least on a neighborhood of $\mathbf{a}$. The Fréchet differential of second order belongs to the space $L(E, L(E, \mathbb{R}))$. This space can be identified with the space $\mathcal{B}(E \times E, \mathbb{R})$ (of all continuous bilinear maps from $E \times E$ to $\mathbb{R}$) via the linear isomorphism

$$\Phi : L(E, L(E, \mathbb{R})) \to \mathcal{B}(E \times E, \mathbb{R}),$$

that associates to each operator $T \in L(E, L(E, \mathbb{R}))$ the bilinear map $B = \Phi(T)$ given by $B(\mathbf{v}, \mathbf{w}) = (T(\mathbf{v}))(\mathbf{w})$.

All functions defined on open sets in $\mathbb{R}^N$ that admit continuous partial derivatives up to second order (that is, of class C^2) are twice Fréchet differentiable. See R. Coleman [112], Proposition 4.9, p. 99.

3.8.4 Lemma *If $f\colon U \to \mathbb{R}$ is twice Fréchet differentiable, then it is also twice Gâteaux differentiable and*

$$\mathrm{d}^2 f(\mathbf{a})(\mathbf{v}, \mathbf{w}) = f''(\mathbf{a}\,; \mathbf{v}, \mathbf{w}) \tag{3.27}$$

for all $\mathbf{a} \in U$ and $\mathbf{v}, \mathbf{w} \in E$.

Our next goal is to establish the analogue of Taylor's formula in the context of Gâteaux differentiability and to infer from it an important characterization of convexity under the presence of twice Gâteaux differentiability.

3.8.5 Theorem (*Taylor's formula*) *If f is twice Gâteaux differentiable at all points of the linear segment $[\mathbf{a}, \mathbf{a} + \mathbf{v}]$ relative to the pair $(\mathbf{v}, \mathbf{v})$, then there exists a number $\theta \in (0, 1)$ such that*

$$f(\mathbf{a} + \mathbf{v}) = f(\mathbf{a}) + f'(\mathbf{a}\,; \mathbf{v}) + \frac{1}{2} f''(\mathbf{a} + \theta\mathbf{v}\,; \mathbf{v}, \mathbf{v}). \tag{3.28}$$

Proof. Consider the function $g(t) = f(\mathbf{a} + t\mathbf{v})$, for $t \in [0, 1]$. Its first derivative is

$$\begin{aligned} g'(t) &= \lim_{\varepsilon \to 0} \frac{g(t + \varepsilon) - g(t)}{\varepsilon} \\ &= \lim_{\varepsilon \to 0} \frac{f(\mathbf{a} + t\mathbf{v} + \varepsilon\mathbf{v}) - f(\mathbf{a} + t\mathbf{v})}{\varepsilon} = f'(\mathbf{a} + t\mathbf{v}\,; \mathbf{v}) \end{aligned}$$

and similarly we infer that $g''(t) = f''(\mathbf{a} + t\mathbf{v}\,; \mathbf{v}, \mathbf{v})$. By usual Taylor's formula we infer the existence of a number $\theta \in (0, 1)$ such that

$$g(1) = g(0) + g'(0) + \frac{1}{2}\, g''(\theta),$$

which in turn yields the formula (3.28). ■

3.8.6 Corollary (The second derivative test of convexity) *Suppose that f is twice Gâteaux differentiable on an open convex subset U of the real Banach space E.*

(a) *If*

$$f''(\mathbf{x}\,; \mathbf{v}, \mathbf{v}) \geq 0 \quad \textit{for all } \mathbf{x} \in U,\ \mathbf{v} \in E, \tag{3.29}$$

then f is convex on U.

(b) *If the above inequality is strict for $\mathbf{v} \neq \mathbf{0}$, then f is strictly convex.*

(c) *Every convex function f verifies the inequality (3.29).*

Proof. Combine Taylor's formula (3.28) and Theorem 3.8.1. ■

In the context of finite-dimensional Banach spaces, Corollary 3.8.6 shows that the positivity (strict positivity) of the Hessian matrix at all points of U guarantees the convexity (strict convexity) of f. In connection with this fact it is important to recall here the following result from linear algebra:

3.8.7 Theorem (Sylvester's criterion of positivity of a symmetric matrix) *Suppose that $A \in \mathrm{Sym}(N, \mathbb{R})$. Then:*

(a) *$A > 0$ if and only if all the leading principal minors of A are strictly positive;*

(b) *$A \geq 0$ if and only if all principal minors of A are positive.*

For details, see the book of C. D. Meyer [319], p. 559 and p. 566. Another simple proof of the assertion (a) makes the objective of Exercise 4, Section 4.4.

Simple examples show that the strict positivity of the Hessian matrix is a sufficient condition but not necessary for strict convexity of a function. See Section 3.7, Exercise 9.

In practice, handling second-order differentials is not always very comfortable and the methods exposed in sections 3.1–3.3 for recognizing convexity are much more convenient.

3.8.8 Example *For $A \in \mathrm{Sym}^{++}(N, \mathbb{R})$ and $\mathbf{b} \in \mathbb{R}^N$, the function*

$$F(\mathbf{x}) = \frac{1}{2}\langle A\mathbf{x}, \mathbf{x}\rangle - \langle \mathbf{x}, \mathbf{b}\rangle$$

is Fréchet differentiable of class C^∞ and its first two differentials are

$$\begin{aligned} \mathrm{d}F(\mathbf{x})\mathbf{v} &= \langle A\mathbf{x}, \mathbf{v}\rangle - \langle \mathbf{v}, \mathbf{b}\rangle \\ \mathrm{d}^2F(\mathbf{x})(\mathbf{v}, \mathbf{v}) &= \langle A\mathbf{v}, \mathbf{v}\rangle. \end{aligned}$$

This implies that F is strictly convex (and even strongly convex, according to Exercise 5).

3.8.9 Example *According to Example 3.1.4, the following particular variant of the log-sum-exp function,*

$$f(x) = \log\left(\sum_{k=1}^{N} e^{x_k}\right)$$

is convex on $\mathbb{R}^N$. An alternative argument based on Corollary 3.8.6 starts by noticing that this function has continuous partial derivatives of order 2. Its Hessian matrix $\mathrm{Hess}_x f$ equals

$$\frac{1}{\left(\sum_{k=1}^{N} e^{x_k}\right)^2}\left[\left(\sum_{k=1}^{N} e^{x_k}\right)\mathrm{diag}\,(e^{x_1}, ..., e^{x_N}) - \begin{pmatrix} e^{x_1} \\ \vdots \\ e^{x_N} \end{pmatrix}(e^{x_1}, ..., e^{x_N})\right],$$

so by taking into account the Cauchy–Bunyakovsky–Schwarz inequality we infer that

$$\langle (\mathrm{Hess}_{\mathbf{x}}\, f)\, \mathbf{z}, \mathbf{z}\rangle = \left(\sum_{k=1}^{N} e^{x_k} z_k^2\right)\left(\sum_{k=1}^{N} e^{x_k}\right) - \left(\sum_{k=1}^{N} e^{x_k} z_k\right)^2 \geq 0,$$

for all $\mathbf{z} \in \mathbb{R}^N$. *Consequently the log-sum-exp function* f *is convex (and not strictly convex).*

3.8.10 Example *According to Example* 3.1.5, *the function* $f(A) = (\det A)^{1/N}$ *is concave on* $\mathrm{Sym}^{+}(N,\mathbb{R})$. *Since* f *is continuous and* $\mathrm{Sym}^{++}(N,\mathbb{R})$ *is a convex open set dense in* $\mathrm{Sym}^{+}(N,\mathbb{R})$, *one can reduce the proof of concavity of* f *to the proof of concavity of its restriction to* $\mathrm{Sym}^{++}(N,\mathbb{R})$. *The later will be done via Corollary* 3.8.6. *The ambient space is* $\mathrm{Sym}(N,\mathbb{R})$ *and we have*

$$\mathrm{d}f(A)(X) = \frac{1}{N}(\det A)^{1/N}\, \mathrm{trace}(A^{-1}X)$$

and

$$\begin{aligned}\mathrm{d}^2 f(A)(X,Y) = \frac{1}{N^2}(\det A)^{1/N}\, &\mathrm{trace}(A^{-1}X)\, \mathrm{trace}(A^{-1}Y)\\ &- \frac{1}{N}(\det A)^{1/N}\, \mathrm{trace}(A^{-1}XA^{-1}Y).\end{aligned}$$

See J. R. Magnus and H. Neudecker [293] *for the two formulas concerning the matrix differential calculus. We have*

$$\begin{aligned}\mathrm{d}^2 f(A)(X,X) &= \frac{1}{N^2}(\det A)^{1/N}\left(\mathrm{trace}(A^{-1}X)\right)^2\\ &\qquad - \frac{1}{N}(\det A)^{1/N}\, \mathrm{trace}(A^{-1}XA^{-1}X)\\ &= \frac{1}{N^2}(\det A)^{1/N}\left[\left(\mathrm{trace}(A^{-1/2}XA^{-1/2})\right)^2 - N\, \mathrm{trace}\left((A^{-1/2}XA^{-1/2})^2\right)\right]\end{aligned}$$

and the fact that $\mathrm{d}^2 f(A)(X,X) \leq 0$ *for all* $A \in \mathrm{Sym}^{++}(N,\mathbb{R})$ *and* $X \in \mathrm{Sym}(N,\mathbb{R})$ *is a consequence of the matrix form of the Cauchy–Bunyakovsky–Schwarz inequality applied to the inner product*

$$\langle A, B\rangle = \mathrm{trace}(AB), \quad \text{for } A, B \in \mathrm{Sym}(N,\mathbb{R}).$$

Surprisingly, things are simpler in the case of the differentiable function $g(A) = \log\det(A)$ *(defined on* $\mathrm{Sym}^{++}(N,\mathbb{R})$*). Indeed, in this case*

$$\mathrm{d}g(A) = A^{-1} \quad \text{and} \quad \mathrm{d}^2 g(A)(X,Y) = -\,\mathrm{trace}(A^{-1}XA^{-1}Y),$$

which yields $\mathrm{d}^2 g(A)(X,X) \leq 0$ *for all* $X \in \mathrm{Sym}(N,\mathbb{R})$ *(and thus the fact that* g *is concave).*

3.8.11 Example *One can prove that the following function defined on $\mathbb{R}^N_{++}$ is convex:*

$$f(x_1, ..., x_N) = \sum_{i=1}^{N} \frac{1}{x_i} - \sum_{1 \le i < j \le N}^{N} \frac{1}{x_i + x_j} + \sum_{1 \le i < j < k \le N}^{N} \frac{1}{x_i + x_j + x_k} - \cdots + (-1)^{N-1} \frac{1}{x_1 + x_2 + \cdots + x_N},$$

See J. M. Borwein, D. Bailey, and R. Girgensohn [70], *p.* 36, *for details and history of this example. For $N = 2$, or $N = 3$, the Hessian method works well, but for large N this method is certainly not helpful.*

3.8.12 Remark *Differential criteria apply to functions defined on open sets. In the context of convex functions, they can be applied to the relative interior of the domain, taking into account the following simple consequence of the Accessibility Lemma* (see Exercise 6, Section 2.1)*: Suppose that C is a convex set in a Banach space E and $f : C \to \mathbb{R}$ is a continuous function. If f is convex on $\operatorname{ri}(C)$, then f is convex on C.*

Exercises

1. Prove the strict convexity on $\mathbb{R}^N$ of the function

$$f(x_1, \ldots, x_n) = \sum_{1 \le i < j \le N} c_{ij}(x_i - x_j)^2,$$

where the coefficients c_{ij} are strictly positive.

2. Consider the open set $A = \{(x, y, z) \in \mathbb{R}^3 : x, y > 0, xy > z^2\}$. Prove that A is convex and the function

$$f: A \to \mathbb{R}, \quad f(x, y, z) = \frac{1}{xy - z^2}$$

is strictly convex. Then, infer the inequality

$$\frac{8}{(x_1 + x_2)(y_1 + y_2) - (z_1 + z_2)^2} < \frac{1}{x_1 y_1 - z_1^2} + \frac{1}{x_2 y_2 - z_2^2},$$

which works for every pair of distinct points (x_1, y_1, z_1) and (x_2, y_2, z_2) of the set A.

[*Hint*: Compute the Hessian of f.]

3. (A generalization of Minkowski's inequality for $p = 0$) Use calculus to prove that the function

$$\Pi: [0, \infty)^n \to \mathbb{R}, \quad \Pi(x_1, \ldots, x_n) = x_1^{\alpha_1} \cdots x_n^{\alpha_n}$$

is concave if $\alpha_1, ..., \alpha_n > 0$ and $\alpha_1 + \cdots + \alpha_n \leq 1$.

[*Hint*: It suffices to prove the positivity of $H(\mathbf{x})$, the Hessian of $\Pi(\mathbf{x})$ at $\mathbf{x}$, divided by $\Pi(\mathbf{x})$. For every vector $\mathbf{v} \in \mathbb{R}^n$ we have

$$\langle H(\mathbf{x})\mathbf{v}, \mathbf{v}\rangle = \left(\sum_{j=1}^{n} \frac{\alpha_j v_j}{x_j}\right)^2 - \sum_{j=1}^{n} \frac{1}{\alpha_j}\left(\frac{\alpha_j v_j}{x_j}\right)^2 \leq 0,$$

according to the *Cauchy–Bunyakovsky–Schwarz* inequality.]

4. Prove that the function $f(x, y) = y^2/(1 - |x|)$ is convex and bounded on the open unit disc $D_1(0) = \{(x, y) \in \mathbb{R}^2 : x^2 + y^2 < 1\}$.

   [*Hint*: The function f can be represented as the maximum of two convex functions, $y^2/(1 - x)$ and $y^2/(1 + x)$.]

5. (The case of strongly convex functions) Suppose that f is a Gâteaux differentiable function defined on an open convex set U in $\mathbb{R}^N$. Prove that the following assertions are equivalent:

   (a) f is strongly convex (with parameter $C > 0$);

   (b) $f(\mathbf{y}) \geq f(\mathbf{x}) + f'(\mathbf{x})(\mathbf{y} - \mathbf{x}) + \frac{C}{2}\|\mathbf{x} - \mathbf{y}\|^2$ for all $\mathbf{x}, \mathbf{y} \in U$;

   (c) $(\nabla f(\mathbf{x}) - \nabla f(\mathbf{y}))(\mathbf{x} - \mathbf{y}) \geq C\|\mathbf{x} - \mathbf{y}\|^2$ for all $\mathbf{x}, \mathbf{y} \in U$;

   (d) $\langle(\text{Hess}_{\mathbf{x}} f)\mathbf{v}, \mathbf{v}\rangle \geq C\|\mathbf{v}\|^2$ for all $\mathbf{x} \in U$ and $\mathbf{v} \in \mathbb{R}^N$.

6. Suppose that f is a twice Gâteaux differentiable function defined on an open convex set U in $\mathbb{R}^N$. Prove that f is strongly convex (with constant $C > 0$) if and only if $f''(\mathbf{x}; \mathbf{v}, \mathbf{v}) \geq C\|\mathbf{v}\|^2$ for all $\mathbf{x} \in U$, $\mathbf{v} \in E$.

7. Suppose that f is a twice continuously differentiable function defined on an open convex set U in $\mathbb{R}^N$. Prove that

$$\langle\nabla f(\mathbf{x}) - \nabla f(\mathbf{y}), \mathbf{x} - \mathbf{y}\rangle = \int_0^1 \langle\left(\text{Hess}_{(1-t)\mathbf{x}+t\mathbf{y}} f\right)(\mathbf{x} - \mathbf{y}), \mathbf{x} - \mathbf{y}\rangle\, dt.$$

8. Let f be a continuous function on the closure of an open convex subset D of $\mathbb{R}^N$ and quasiconvex on the interior of D. Prove that f is quasiconvex on $\overline{D}$.

9. Suppose that f is continuously differentiable on an open convex subset D of $\mathbb{R}^N$. Prove that f is quasiconvex if, and only if,

$$f(\mathbf{y}) \leq f(\mathbf{x}) \text{ implies } \langle\mathbf{y} - \mathbf{x}, \nabla f(\mathbf{x})\rangle \leq 0.$$

   Infer that the function $g(x, y) = -2x^2 - 2xy$ is quasiconvex on $\mathbb{R}^2_+$.

3.9 Jensen's Integral Inequality in the Context of Several Variables

Given an arbitrary measure space (Ω, Σ, μ), the integrability of a vector-valued function $f : \Omega \to \mathbb{R}^N$, $f = (f_1, ..., f_N)$ means the integrability of its components, so that its integral is defined via the formula

$$\int_\Omega f(x) d\mu(x) = \left(\int_\Omega f_1(x) d\mu(x), ..., \int_\Omega f_N(x) d\mu(x) \right).$$

Moreover, for every linear functional L on $\mathbb{R}^N$ we have

$$L\left(\int_\Omega f(x) d\mu(x) \right) = \int_\Omega L(f(x)) d\mu(x).$$

As in the scalar case, we call the integral of a function f with respect to a probability measure, the *expectation* (or *expected value*, or *arithmetic mean*) of f and we will denote it also by $E(f)$.

In this section we are especially interested in the case where Ω is a convex subset of $\mathbb{R}^N$. According to Remark 2.1.11, the convex subsets C of $\mathbb{R}^N$ are not necessarily Borel sets if $N \geq 2$. Therefore, in order to ensure that the inclusion of C into $\mathbb{R}^N$ is measurable, we will endow C with its *relative Borel algebra*,

$$\mathcal{B}(C) = \left\{ B \cap C : B = \text{Borel subset of } \mathbb{R}^N \right\},$$

and we will call every probability measure $\mu : \mathcal{B}(C) \to [0,1]$ a *Borel probability measure* on C. The concept of barycenter is attached to all Borel probability measures on C that belong to the class

$$\mathcal{P}^1(C) = \left\{ \mu : \mu \text{ Borel probability measure on } C \text{ and } \int_C \|\mathbf{x}\| \, d\mu(\mathbf{x}) < \infty \right\}.$$

This class includes all Borel probability measures if C is bounded.

3.9.1 Definition *The barycenter of a probability measure $\mu \in \mathcal{P}^1(C)$ is the point*

$$\text{bar}(\mu) = \int_C \mathbf{x} d\mu(\mathbf{x}).$$

Clearly,

$$x^*(\text{bar}(\mu)) = \int_C x^*(\mathbf{x}) d\mu(\mathbf{x})$$

for all linear functionals x^* on $\mathbb{R}^N$. This fact serves as a model for introducing the concept of barycenter for any Borel probability measure defined on an arbitrary compact convex set. See Chapter 7.2.

The barycenter of the discrete probability measure $\lambda = \sum_{k=1}^n \lambda_k \delta_{\mathbf{x}_k}$ concentrated at the points $\mathbf{x}_1, ..., \mathbf{x}_n \in C$ is

$$\text{bar}(\lambda) = \int_{\mathbb{R}^N} \mathbf{x} d\lambda(\mathbf{x}) = \sum_{k=1}^n \lambda_k \mathbf{x}_k,$$

which clearly belongs to the convex set C. The membership of the barycenter to the convex set on which the probability measure is defined is valid in all finite-dimensional spaces.

3.9.2 Proposition (H. Rubin and O. Wesler [428]) *If C is a nonempty convex set in $\mathbb{R}^N$, then it contains the barycenter of every probability measure $\mu \in \mathcal{P}^1(C)$.*

Proof. By mathematical induction with respect to $n = \dim C$. For $n = 0$, C consists of one single point, $\mathbf{x}_0$, which implies that $\mathcal{P}^1(C) = \{\delta_{\mathbf{x}_0}\}$. Clearly, $\operatorname{bar}(\delta_{\mathbf{x}_0}) = \mathbf{x}_0$.

Suppose now that $\dim C = n \leq N$ and the statement holds for all convex sets of dimension m with $m < n$. Since C is nonempty, its relative interior is nonempty, and we can assume without loss of generality that $\mathbf{0} \in \operatorname{ri}(C)$. In this case $\operatorname{aff}(C)$ coincides with $L = \operatorname{span}(C)$ and $\mathbf{0}$ is an interior point of C relative to the ambient space L.

We have to prove that the barycenter of every probability measure $\mu \in \mathcal{P}^1(C)$ lies in C. For this, notice first that $\operatorname{bar}(\mu) \in L$. Indeed, if $\operatorname{bar}(\mu) \notin L$, then there would exist a linear functional $x^* : \mathbb{R}^N \to \mathbb{R}$ such that $x^*(\operatorname{bar}(\mu)) > 0$ and $x^*|_L = 0$. But this is impossible because

$$0 = x^*(\operatorname{bar}(\mu)) - x^*\left(\int_C \mathbf{x} \mathrm{d}\mu(\mathbf{x})\right) = x^*(\operatorname{bar}(\mu)) - \int_C x^*(\mathbf{x})\mathrm{d}\mu(\mathbf{x}) = x^*(\operatorname{bar}(\mu)).$$

Therefore $\operatorname{bar}(\mu) \in L$. Now consider the case $\operatorname{bar}(\mu) \in L \backslash C$. According to the Separation Theorem 2.3.2, there exists a nonzero linear functional $x^* : L \to \mathbb{R}$ such that

$$x^*(\operatorname{bar}(\mu)) > \sup\{x^*(\mathbf{c}) : \mathbf{c} \in C\}.$$

This functional can be extended to a linear functional on $\mathbb{R}^N$ (also denoted x^*) by letting $x^*(\mathbf{x}) = 0$ for $\mathbf{x}$ in the orthogonal complement of L with respect to $\mathbb{R}^N$. We have

$$\begin{aligned}\int_C [x^*(\operatorname{bar}(\mu)) - x^*(\mathbf{x})]\, \mathrm{d}\mu(\mathbf{x}) &= x^*(\operatorname{bar}(\mu)) - \int_C x^*(\mathbf{x})\mathrm{d}\mu(\mathbf{x}) \\ &= x^*(\operatorname{bar}(\mu)) - x^*\left(\int_C \mathbf{x}\mathrm{d}\mu(\mathbf{x})\right) = 0,\end{aligned}$$

and since the function under square brackets is positive, it follows that μ is concentrated on the set $C_0 = C \cap H$, where $H = \{\mathbf{x} \in L : x^*(\mathbf{x}) = x^*(\operatorname{bar}(\mu))\}$ is a hyperplane in L. Thus $\dim C_0 \leq n-1$ and $\mu_0 = \mu|_{C_0} \in \mathcal{P}^1(C_0)$. According to our induction assumption, $\operatorname{bar}(\mu) = \operatorname{bar}(\mu_0) \in C_0 \subset C$, and this contradiction completes the proof. ■

A special method to construct new probability measure spaces from old ones is described in what follows.

Let (Ω_1, Σ_1) and (Ω_2, Σ_2) be two measurable spaces and let $g : \Omega_1 \to \Omega_2$ be a measurable map (that is, $A \in \Sigma_2$ implies $g^{-1}(A) \in \Sigma_1$). Then, for every probability measure $\mu_1 : \Sigma_1 \to [0,1]$, the formula

$$\mu_2(A) = \mu_1(g^{-1}(A)), \quad A \in \Sigma_2,$$

defines a probability measure μ_2 on Σ_2 called the *push-forward measure* (or the *image* of μ_1 through g). Usually, the measure μ_2 is denoted $g\#\mu_1$. The main feature of this measure is the following *change-of-variables formula,*

$$\int_{\Omega_2} f(y)\mathrm{d}\left(g\#\mu_1\right)(y) = \int_{\Omega_1} f(g(x))\mathrm{d}\mu_1(x), \tag{3.30}$$

valid for all μ_2-integrable functions f.

When $\Omega_2 = \mathbb{R}^N$, $\Sigma_2 = \mathcal{B}\left(\mathbb{R}^N\right)$, $f =$ the identity of $\mathbb{R}^N$, and $g \in L^1(\mu_1)$, this formula yields the existence of the barycenter of $g\#\mu_1$ and also the fact that

$$\mathrm{bar}(g\#\mu_1) = \int_{\Omega_1} g(x)\mathrm{d}\mu_1(x) = E(g). \tag{3.31}$$

When f is an arbitrary convex functions, the formulas (3.30) and (3.31) are connected by Jensen's inequality:

3.9.3 Theorem (The integral form of Jensen's inequality) *Let (Ω, Σ, μ) be a probability space and let $g\colon \Omega \to \mathbb{R}^N$ be a μ-integrable function. If f is a lower semicontinuous proper convex function whose effective domain C includes the image of g, then $E(g) \in C$ and*

$$f(E(g)) \leq \int_{\Omega} f(g(x))\mathrm{d}\mu(x). \tag{3.32}$$

Notice that the right-hand side integral always exists but it might be ∞ if the μ-integrability of $f \circ g$ is not expressly asked. When both functions g and $f \circ g$ are μ-integrable, then the above inequality becomes

$$f(E(g)) \leq E(f \circ g).$$

Moreover, if in addition f is strictly convex and $\partial f\left(E(g)\right) \neq \emptyset$, then this inequality becomes an equality if and only if g is constant μ-almost everywhere.

Proof. The membership of $E(g)$ to C follows from Proposition 3.9.2 and the fact that $E(g) = \mathrm{bar}(g\#\mu)$.

The function f is $\mathcal{B}(C)$-measurable since the level sets of f are relatively closed subsets of C. As a consequence, the function $f \circ g$ is μ-measurable (and the same is true for $(f \circ g)^+$ and $(f \circ g)^-$).

According to Theorem 3.3.2, for every $\varepsilon > 0$ there is a continuous affine function $A(\mathbf{y}) = \langle \mathbf{y}, \mathbf{v}\rangle + a$ on $\mathbb{R}^N$ such that

$$A(\mathbf{y}) \leq f(\mathbf{y}) \text{ for all } \mathbf{y} \in C \text{ and } A(E(g)) > f(E(g)) - \varepsilon.$$

Then

$$f(g(x)) \geq A(g(x)) = \langle g(x), \mathbf{v}\rangle + a \quad \text{for all } \ x \in \Omega, \tag{3.33}$$

whence we infer that $(f(g(x)))^-$ is a μ-integrable function (being dominated by the μ-integrable function $\|g(x)\|\,\|\mathbf{v}\| + a$). Since every positive measurable function admits an integral (possibly $+\infty$) it follows that the integral in the right-hand side of inequality (3.32) exists. This inequality holds because $\varepsilon > 0$ was arbitrarily chosen and

$$\begin{aligned}\int_\Omega f(g(x))\,\mathrm{d}\mu(x) &\geq \int_\Omega A(g(x))\,\mathrm{d}\mu(x)\\ &= A\left(\int_\Omega g(x)\mathrm{d}\mu\right) = A(E(g)) > f(E(g)) - \varepsilon.\end{aligned}$$

The last assertion, concerning the equality case when f is strictly convex, is a consequence of the inequality

$$f(\mathbf{y}) > f(E(g)) + \langle \mathbf{y} - E(g), \mathbf{z}\rangle, \quad \text{for all } \mathbf{y} \in C,\ \mathbf{y} \neq E(g),$$

where $\mathbf{z} \in \partial f\,(E(g))$. This makes the equality $E(f \circ g) = f(E(g))$ possible if, and only if, $g = E(g)$ μ-almost everywhere. ■

The condition $\partial f\,(E(g)) \neq \emptyset$ in Theorem 3.9.3 is fulfilled when $E(g) \in \operatorname{ri}(C)$. See Remark 3.3.4.

3.9.4 Corollary *If f is a lower semicontinuous proper convex function defined on a convex subset C of $\mathbb{R}^N$ and $\mu \in \mathcal{P}^1(C)$, then $\operatorname{bar}(\mu) \in C$ and*

$$f\,(\operatorname{bar}(\mu)) \leq \int_C f\mathrm{d}\mu.$$

Theorem 3.9.3 and Corollary 3.9.4 are actually equivalent (due to the technique of push-forward measure).

The derivation of the discrete case of Jensen's inequality from Theorem 3.9.3 can be done in two ways:

(*a*) by considering the probability measure space (Ω, Σ, μ), where

$$\Omega = \{1, ..., n\}\,,\ \Sigma = \mathcal{P}(\Omega) \text{ and } \mu = \sum_{k=1}^n \lambda_k \delta_k, \tag{3.34}$$

and the function $g : \{1, ..., n\} \to \mathbb{R}$ defined by $g(k) = \mathbf{x}_k$ for all k. In this case the inequality $f(E(g)) \leq E(f \circ g)$ becomes

$$f\left(\sum_{k=1}^n \lambda_k \mathbf{x}_k\right) \leq \sum_{k=1}^n \lambda_k f(\mathbf{x}_k).$$

(*b*) by considering the probability measure space $(\mathbb{R}^N, \mathcal{B}(\mathbb{R}^N), \lambda)$, corresponding to the discrete probability measure $\lambda = \sum_{k=1}^n \lambda_k \delta_{\mathbf{x}_k}$ concentrated at the points $\mathbf{x}_1, ..., \mathbf{x}_n \in \mathbb{R}^N$ and choosing as g the identity of $\mathbb{R}^N$. In this variant, Theorem 1.7.3 yields the same inequality, written as

$$f(\operatorname{bar}(\lambda)) \leq E(f).$$

Surprisingly, Jensen's integral inequality can be deduced from the discrete case via an approximation argument. See Lemma 7.2.6.

Exercises

1. (Convex series closed sets) A subset C of a real Banach space E is called convex series closed if every point $\mathbf{x} \in E$ that can be represented as the sum of a convergent series $\sum_{k=1}^{\infty} \lambda_k \mathbf{x}_k$ with $\lambda_k \geq 0$, $\sum_{k=1}^{\infty} \lambda_k = 1$ and $\mathbf{x}_k \in C$, must be in C. Prove that all open convex subsets (as well as all closed convex subsets) of E are convex series closed.

   [*Hint:* Use the separation theorems.]

2. (Jensen's inequality for series) Suppose that $f : \mathbb{R}^N \to \mathbb{R}$ is a convex function, $(\mathbf{x}_n)_n$ is a sequence in $\mathbb{R}^N$ and $(\lambda_n)_n$ is a sequence in $[0, 1]$ such that $\sum_{n=1}^{\infty} \lambda_n = 1$ and the series $\sum_{n=1}^{\infty} \lambda_n \mathbf{x}_n$ is absolutely convergent. Prove that the series $\sum_{n=1}^{\infty} \lambda_n f(\mathbf{x}_n)$ is either convergent or has infinite sum, and in both cases,

$$f\left(\sum_{n=1}^{\infty} \lambda_n \mathbf{x}_n\right) \leq \sum_{n=1}^{\infty} \lambda_n f(\mathbf{x}_n).$$

3. (An estimate of Jensen gap) Let (Ω, Σ, μ) be a probability measure space and let $g : \Omega \to \mathbb{R}^N$, $g = (g_1, ..., g_N)$, be a function such that $\|g(\cdot)\| \in L^2(\mu)$. Consider also a twice differentiable convex function f defined on an open convex set $U \subset \mathbb{R}^N$, that includes the image of g and verifies the double inequality

$$\alpha \|\mathbf{v}\|^2 \leq d^2 f(\mathbf{x})(\mathbf{v}, \mathbf{v}) \leq \beta \|\mathbf{v}\|^2 \quad \text{for all } \mathbf{v} \in \mathbb{R}^N,$$

   where α and β are strictly positive constants. Prove that

$$\alpha \|\mathrm{var}(g)\|_{HS} \leq E(f \circ g) - f(E(g)) \leq \beta \|\mathrm{var}(g)\|_{HS} \cdot$$

   Here $\mathrm{cov}(g) = (\mathrm{cov}(g_i, g_j))_{i,j}$ denotes the *covariance matrix* of g.

   [*Hint*: Apply Taylor's formula 3.8.5.]

3.10 Extrema of Convex Functions

Convex functions exhibit a series of nice properties concerning their extrema, which make them successful in numerous applications of mathematics.

3.10.1 Theorem *Suppose that U is a convex subset of a real Banach space E and $f : U \to \mathbb{R}$ is a convex function. If a point $\mathbf{a} \in U$ is a local minimizer for f, then it is also a global minimizer. In other words, if*

$$f(\mathbf{x}) \geq f(\mathbf{a}) \quad \textit{in a neighborhood of } \mathbf{a},$$

then

$$f(\mathbf{x}) \geq f(\mathbf{a}) \quad \text{for all } \mathbf{x} \in U.$$

Moreover, the set of global minimizers of f is a convex set. If f is strictly convex, then this set is either empty or a singleton.

It is worth noticing that, according to Theorem 3.6.11, if $f\colon U \to \mathbb{R}$ is a differentiable convex function, then every point $\mathbf{a} \in \operatorname{int} U$ that verifies the condition $f'(\mathbf{a}) = 0$ is a global minimizer.

Proof. Let $r > 0$ such that $\mathbf{a}$ is a minimizer of $f|_{B_r(\mathbf{a})}$ and let $\mathbf{x} \in U$. Since U is convex, the line segment $[\mathbf{a}, \mathbf{x}]$ is contained in U and for some $\varepsilon \in (0,1)$, the point $(1-\varepsilon)\mathbf{a} + \varepsilon\mathbf{x} = \mathbf{a} + \varepsilon(\mathbf{x} - \mathbf{a})$ must be in $B_r(\mathbf{a}) \cap U$. By the convexity of f,

$$\begin{aligned} f(\mathbf{a}) &\leq f(\mathbf{a} + \varepsilon(\mathbf{x} - \mathbf{a})) = f((1-\varepsilon)\mathbf{a} + \varepsilon\mathbf{x}) \\ &\leq (1-\varepsilon) f(\mathbf{a}) + \varepsilon f(\mathbf{x}), \end{aligned} \tag{3.35}$$

which yields $f(\mathbf{a}) \leq f(\mathbf{x})$. Hence a is a global minimizer. If f is strictly convex in a neighborhood of $\mathbf{a}$, then the last inequality in (3.35) is strict and the conclusion becomes $f(\mathbf{x}) > f(\mathbf{a})$ for all $\mathbf{x} \in U$, $\mathbf{x} \neq \mathbf{a}$. The convexity of the set of global minimizers of f is immediate. ■

It is customary in optimization theory to denote

$$\mathbf{a} \in \arg\min_{\mathbf{x} \in U} f(\mathbf{x})$$

for the fact that $\mathbf{a}$ is a minimizer of f over U.

An archetype for the existence of a global minimum is the following classical result:

3.10.2 Theorem (K. Weierstrass) *Assume that $f\colon \mathbb{R}^N \to \mathbb{R} \cup \{\infty\}$ is a lower semicontinuous proper function whose level sets are bounded. Then f has a global minimum.*

Proof. Choose arbitrarily $\bar{\mathbf{x}} \in \operatorname{dom} f$ and notice that

$$m = \inf\left\{f(\mathbf{x}) : \mathbf{x} \in \mathbb{R}^N\right\} = \inf\left\{f(\mathbf{x}) : \mathbf{x} \in \operatorname{level}_{f(\bar{\mathbf{x}})}\right\}.$$

According to our hypotheses, the level set $\operatorname{level}_{f(\bar{\mathbf{x}})}$ is bounded and closed in $\mathbb{R}^N$ and thus compact. By the definition of m, there exists a sequence $(\mathbf{x}_n)_n$ in $\operatorname{level}_{f(\bar{\mathbf{x}})}$ such that $f(\mathbf{x}_n) \to m$. Passing to a subsequence (if necessary) we may assume that $(\mathbf{x}_n)_n$ is convergent to a point $\mathbf{a}$. Necessarily, $\mathbf{a} \in \operatorname{level}_{f(\bar{\mathbf{x}})}$. Since f is lower semicontinuous, we have

$$f(\mathbf{a}) \leq \liminf_{n \to \infty} f(\mathbf{x}_n) = m,$$

which shows that f has a global minimum at $\mathbf{a}$. ■

The condition of having bounded level sets is a growth condition.

3.10.3 Lemma *Let E be a Banach space and $f : E \to \mathbb{R} \cup \{\infty\}$ be a lower semicontinuous proper function. Then, the following conditions are equivalent:*

(a) *all levels sets of f are bounded;*

(b) *the function f is coercive in the sense that*

$$\lim_{\|\mathbf{x}\| \to \infty} f(\mathbf{x}) = \infty; \tag{3.36}$$

(c) *there exist $\alpha > 0$ and $\beta \in \mathbb{R}$ such that $f(\mathbf{x}) \geq \alpha \|\mathbf{x}\| + \beta$ for all $\mathbf{x} \in E$;*

(d) *we have*

$$\liminf_{\|\mathbf{x}\| \to \infty} \frac{f(\mathbf{x})}{\|\mathbf{x}\|} > 0.$$

Proof. The only nontrivial implication is (b) $\implies$ (c) This will be done noticing that we may assume (replacing f by a translate of it) that $\mathbf{0} \in \operatorname{dom} f$ and $f(\mathbf{0}) = 0$. Since f is coercive, there is a number $R > 0$ such that $f(\mathbf{x}) \geq 1$ if $\|\mathbf{x}\| \geq R$. Then for all such $\mathbf{x}$ we have

$$1 \leq f\left(\frac{R\mathbf{x}}{\|\mathbf{x}\|}\right) \leq \frac{R}{\|\mathbf{x}\|} f(\mathbf{x}) + \left(1 - \frac{R}{\|\mathbf{x}\|}\right) f(\mathbf{0}) = \frac{R}{\|\mathbf{x}\|} f(\mathbf{x}),$$

whence $f(\mathbf{x}) \geq \|\mathbf{x}\| / R$. Since f is lower semicontinuous at the origin, the set $\{\mathbf{x} : f(\mathbf{x}) > -1\}$ is open and thus there is $\varepsilon \in (0, R)$ such that $\|\mathbf{x}\| \leq \varepsilon$ implies $f(\mathbf{x}) > -1$. Then for every $\mathbf{x}$ with $\|\mathbf{x}\| \leq R$ we have

$$-1 < f\left(\frac{\varepsilon}{R}\mathbf{x}\right) \leq \frac{\varepsilon}{R} f(\mathbf{x}) + \left(1 - \frac{\varepsilon}{R}\right) f(\mathbf{0}) = \frac{\varepsilon}{R} f(\mathbf{x}).$$

Therefore $f(\mathbf{x}) \geq \|\mathbf{x}\| / R - R/\varepsilon$ for all $\mathbf{x}$, which shows that assertion (c) works for $\alpha = 1/R$ and $\beta = -R/\varepsilon$. ■

Combining Lemma 3.10.3 with Theorem 3.10.1 and Theorem 3.10.2 one obtains the following important consequence concerning the minimization of convex functions defined on an Euclidean space.

3.10.4 Corollary *Every lower semicontinuous, proper, convex, and coercive function $f : \mathbb{R}^N \to \mathbb{R} \cup \{\infty\}$ has a global minimizer.*

The following particular case of Corollary 3.10.4 is useful to establish the maximality of ∂f; see Theorem 3.3.12.

3.10.5 Corollary *If $f : \mathbb{R}^N \to \mathbb{R} \cup \{\infty\}$ is a lower semicontinuous proper convex function, then for every $\mathbf{y} \in \mathbb{R}^N$ there is $\mathbf{a} \in \mathbb{R}^N$ such that $\mathbf{y} \in (\mathrm{I} + \partial f)(\mathbf{a})$.*

Proof. The function

$$g : \mathbf{x} \to f(\mathbf{x}) + \frac{1}{2} \|\mathbf{x}\|^2 - \langle \mathbf{x}, \mathbf{y} \rangle$$

is lower semicontinuous, strictly convex, and coercive, so it admits a unique minimizer $\mathbf{x} = \mathbf{a}$. Then $\mathbf{0} \in \partial g(\mathbf{a})$, whence $\mathbf{y} \in (\partial f + \mathrm{I})(\mathbf{a})$ since $\partial(f + \|\cdot\|^2/2) = \partial f + \mathrm{I}$. ■

A nice consequence of Theorem 3.10.2 and Lemma 3.10.3 is the existence of "approximate" critical points:

3.10.6 Proposition *If $f : \mathbb{R}^N \to \mathbb{R}$ is differentiable and bounded below, then for every $\varepsilon > 0$ there exist points $\mathbf{x}_\varepsilon$ such that $\|\nabla f(\mathbf{x}_\varepsilon)\| \leq \varepsilon$.*

Proof. The function $f + \varepsilon \|\mathbf{x}\|$ is coercive, so by Theorem 3.10.2 it has a global minimizer $\mathbf{x}_\varepsilon$. We will show that $\mathbf{v} = \nabla f(\mathbf{x}_\varepsilon)$ verifies $\|\mathbf{v}\| \leq \varepsilon$. Indeed, assuming that $\|\mathbf{v}\| > \varepsilon$, the formula

$$\lim_{t \to 0+} \frac{f(\mathbf{x}_\varepsilon - t\mathbf{v}) - f(\mathbf{x}_\varepsilon)}{t} = -\langle \nabla f(\mathbf{x}_\varepsilon), \mathbf{v} \rangle = -\|\mathbf{v}\|^2 < -\varepsilon \|\mathbf{v}\|$$

implies for $t > 0$ small enough that

$$\begin{aligned} -t\varepsilon \|\mathbf{v}\| &> f(\mathbf{x}_\varepsilon - t\mathbf{v}) - f(\mathbf{x}_\varepsilon) \\ &= [f(\mathbf{x}_\varepsilon - t\mathbf{v}) + \varepsilon \|\mathbf{x}_\varepsilon - tv\|] - [f(\mathbf{x}_\varepsilon) + \varepsilon \|\mathbf{x}_\varepsilon\|] \\ &+ \varepsilon [\|\mathbf{x}_\varepsilon\| - \|\mathbf{x}_\varepsilon - t\mathbf{v}\|] \\ &> -\varepsilon t \|\mathbf{v}\|, \end{aligned}$$

which is a contradiction. Therefore, $\|\mathbf{v}\| \leq \varepsilon$ and the proof is complete. ■

Given a finite family of vectors $\mathbf{v}_0, ..., \mathbf{v}_m$ in $\mathbb{R}^N$, their convex hull may contain the origin or not. Using the separation Theorem 2.3.2, this dichotomy can be reformulated as follows: *Either*

$$\textit{there exist } \lambda_0, ..., \lambda_m \geq 0 \textit{ such that } \sum_{k=0}^{m} \lambda_k = 1 \textit{ and } \sum_{k=0}^{m} \lambda_k \mathbf{v}_k = 0, \tag{3.37}$$

or

$$\textit{there exists } \mathbf{x} \in \mathbb{R}^N \textit{such that} \langle \mathbf{v}_k, \mathbf{x} \rangle < 0 \quad \textit{for } k = 0, ..., m. \tag{3.38}$$

Taking into account Proposition 3.10.6, one obtains the following surprising consequence:

3.10.7 Corollary *If $\mathbf{v}, \mathbf{v}_0, ..., \mathbf{v}_m \in \mathbb{R}^N$, then $\mathbf{v} \in \operatorname{conv} \{\mathbf{v}_0, ..., \mathbf{v}_m\}$ if and only if the function*

$$f(\mathbf{x}) = \log \left(\sum_{k=0}^{m} \mathrm{e}^{\langle \mathbf{v}_k - \mathbf{v}, \mathbf{x} \rangle} \right)$$

is bounded below on $\mathbb{R}^N$.

Proof. Replacing $\mathbf{v}_0, ..., \mathbf{v}_m$ by $\mathbf{v}_0 - \mathbf{v}, ..., \mathbf{v}_m - \mathbf{v}$, we may restrict ourselves to the case $\mathbf{v} = 0$. If $\mathbf{0} \in \operatorname{conv} \{\mathbf{v}_0, ..., \mathbf{v}_m\}$, then the alternative (3.38) fails and clearly the function f is bounded below.

Conversely, if the function f is bounded below, then Proposition 3.10.6 applies, which yields the existence of a sequence $(\mathbf{x}_n)_n$ in $\mathbb{R}^N$ such that $\|\nabla f(\mathbf{x}_n)\| \leq 1/n$ for all n. We have

$$\nabla f(\mathbf{x}_n) = \sum_{k=0}^{m} \lambda_{nk} \mathbf{v}_k,$$

where the scalars

$$\lambda_{nk} = \mathrm{e}^{\langle \mathbf{v}_k, \mathbf{x}_n \rangle} / \sum_{k=0}^{m} \mathrm{e}^{\langle \mathbf{v}_k, \mathbf{x}_n \rangle}$$

are positive and verify $\sum_{k=0}^{m} \lambda_{nk} = 1$. The sequences $(\lambda_{n0})_n, ..., (\lambda_{nm})_n$ are bounded so, by the Bolzano–Weierstrass theorem, we may assume that the limits $\lim_{n\to\infty} \lambda_{nk} = \lambda_k$ exist in $\mathbb{R}$ for $k = 0, ..., m$. This provides a solution for the system (3.37) and the proof is complete. ■

Corollary 3.10.7 can be used to prove Birkhoff's theorem 2.3.6. See the book of J. M. Borwein and Q. J. Zhu [74], Theorem 2.4.10, pp. 25–26.

A stronger condition than coercivity is *supercoercivity*, which means that

$$\lim_{\|\mathbf{x}\|\to\infty} \frac{f(\mathbf{x})}{\|\mathbf{x}\|} = \infty. \tag{3.39}$$

3.10.8 Proposition *Every lower semicontinuous, proper, strongly convex function f defined on a real Hilbert space H is supercoercive (and thus has exactly one minimizer in H).*

Proof. Indeed, if f is strongly convex with constant $C > 0$, then $g(\mathbf{x}) = f(\mathbf{x}) - \frac{C}{2}\|\mathbf{x}\|^2$ is a lower semicontinuous, proper convex function. According to Theorem 3.3.1, there is $\mathbf{x}_0 \in H$ such that $\partial g(\mathbf{x}_0) \neq \emptyset$. Therefore, for $\mathbf{z}$ arbitrarily fixed in $\partial g(\mathbf{x}_0)$, we have

$$f(\mathbf{x}) \geq f(\mathbf{x}_0) + \frac{C}{2}\left(\|\mathbf{x}\|^2 - \|\mathbf{x}_0\|^2\right) + \langle \mathbf{x} - \mathbf{x}_0, \mathbf{z}\rangle,$$

whenever $\mathbf{x} \in H$. By the Cauchy–Bunyakovsky–Schwarz inequality, $\langle \mathbf{x}-\mathbf{x}_0, \mathbf{z}\rangle \geq -\|\mathbf{x}\|\,\|\mathbf{z}\| - \langle \mathbf{x}_0, \mathbf{z}\rangle$, and the fact that $\lim_{\|\mathbf{x}\|\to\infty} f(\mathbf{x})/\|\mathbf{x}\| = \infty$ is now clear. ■

The converse of Proposition 3.10.8 is not true. For example, the function $f(x) = (x^2-1)^+$ is convex, continuous, and supercoercive on $\mathbb{R}$, but it is not strongly convex.

If $A \in \mathrm{Sym}^{++}(N, \mathbb{R})$ and $\mathbf{b} \in \mathbb{R}^N$, then the function

$$F(\mathbf{x}) = \frac{1}{2}\langle A\mathbf{x}, \mathbf{x}\rangle - \langle \mathbf{x}, \mathbf{b}\rangle$$

is strongly convex and differentiable, of class C^∞ on $\mathbb{R}^N$. By Proposition 3.10.8, it is also supercoercive and admits a (unique) global minimizer $\bar{\mathbf{x}}$. Necessarily,

$$\nabla F(\bar{\mathbf{x}}) = A\bar{\mathbf{x}} - \mathbf{b} = 0,$$

and the subgradient inequality assures the sufficiency of this condition. Therefore

$$\bar{\mathbf{x}} = \arg\min_{\mathbf{x}\in\mathbb{R}^N} \left[\frac{1}{2}\langle A\mathbf{x}, \mathbf{x}\rangle - \langle \mathbf{x}, \mathbf{b}\rangle\right] \iff A\bar{\mathbf{x}} = \mathbf{b}.$$

The above reasoning suggests the valuable idea to solve certain systems of equations by finding the minimum of suitable functionals. Put in the general framework of Hilbert spaces, it constitutes the basis of a very useful tool in partial differential equations. See Appendix E.

3.10.9 Remark (Barycenters as minimizers) *Suppose that C is a convex set in $\mathbb{R}^N$ and fix arbitrarily $\mathbf{y} \in \mathbb{R}^N$. Then the barycenter* $\operatorname{bar}(\mu)$ *of every probability measure $\mu \in \mathcal{P}^1(C)$ (as defined in Section 3.9) is the unique point that minimizes the strongly convex continuous function*

$$F(\mathbf{x}) = F_{\mathbf{y}}(\mathbf{x}) = \int_C \left[\|\mathbf{x}-\mathbf{u}\|^2 - \|\mathbf{y}-\mathbf{u}\|^2 \right] \mathrm{d}\mu(\mathbf{u}), \quad \mathbf{x} \in \mathbb{R}^N. \tag{3.40}$$

Notice first that F is well defined. Indeed, the integrand is continuous and its absolute value is majorized by an integrable function since

$$\begin{aligned} \left| \|\mathbf{x}-\mathbf{u}\|^2 - \|\mathbf{y}-\mathbf{u}\|^2 \right| &\leq \left| \|\mathbf{x}-\mathbf{u}\| - \|\mathbf{y}-\mathbf{u}\| \right| \left| \|\mathbf{x}-\mathbf{u}\| + \|\mathbf{y}-\mathbf{u}\| \right| \\ &\leq \|\mathbf{x}-\mathbf{y}\| \left(\|\mathbf{x}-\mathbf{u}\| + \|\mathbf{y}-u\| \right) \\ &\leq \|\mathbf{x}-\mathbf{y}\| \left(\|\mathbf{x}\| + \|\mathbf{y}\| + 2\|\mathbf{u}\| \right) \end{aligned}$$

and $\mu \in \mathcal{P}^1(C)$. The function $F(\mathbf{x})$ is continuous, being Lipschitz on every bounded ball.

Let $\mathbf{x}', \mathbf{x}'' \in C$ and $\lambda \in [0,1]$. Since the function $\mathbf{u} \to \|\mathbf{x}-\mathbf{u}\|^2$ is strongly convex, an easy computation yields

$$\begin{aligned} F((1-\lambda)\mathbf{x}' + \lambda\mathbf{x}'') &= \int_C \left[\|(1-\lambda)\mathbf{x}' + \lambda\mathbf{x}'' - \mathbf{u}\|^2 - \|\mathbf{y}-\mathbf{u}\|^2 \right] \mathrm{d}\mu(\mathbf{u}) \\ &\leq (1-\lambda) \int_C \left[\|\mathbf{x}'-\mathbf{u}\|^2 - \|\mathbf{y}-\mathbf{u}\|^2 \right] \mathrm{d}\mu(\mathbf{u}) \\ &\quad + \lambda \int_C \left[\|\mathbf{x}''-u\|^2 - \|\mathbf{y}-\mathbf{u}\|^2 \right] \mathrm{d}\mu(\mathbf{u}) \\ &\qquad - \frac{1}{2}\lambda(1-\lambda) \|\mathbf{x}'-\mathbf{x}''\|^2, \end{aligned}$$

so $F(\mathbf{x})$ is also strongly convex. Consequently the function $F(\mathbf{x})$ admits a unique minimizer, say $\mathbf{z}_\mu$. Since

$$\begin{aligned} F'(\mathbf{x})(\mathbf{v}) &= \lim_{t\to 0} \frac{F(\mathbf{x}+t\mathbf{v}) - F(\mathbf{x})}{t} = 2\int_C \langle \mathbf{x}-\mathbf{u}, \mathbf{v} \rangle d\mu(\mathbf{u}) \\ &= 2\langle \mathbf{x} - \operatorname{bar}(\mu), \mathbf{v} \rangle, \end{aligned}$$

for all $\mathbf{x}$ and $\mathbf{v}$ in $\mathbb{R}^N$, we infer that F is differentiable, with $F'(\mathbf{x}) = \mathbf{x} - \operatorname{bar}(\mu)$. According to Theorem 3.6.11, $\mathbf{z}_\mu$ is a solution of the equation $F'(\mathbf{x}) = 0$, whence

$$\mathbf{z}_\mu = \operatorname{bar}(\mu).$$

The discussion above shows $F_{\operatorname{bar}(\mu)}(\mathbf{x}) \geq 0$, that is,

$$\int_C \|\mathbf{x}-\mathbf{u}\|^2 \,\mathrm{d}\mu(\mathbf{u}) \geq \int_C \|\operatorname{bar}(\mu) - \mathbf{u}\|^2 \,\mathrm{d}\mu(\mathbf{u}).$$

When μ represents a mass distribution and $\|\mathbf{x}-\mathbf{u}\|^2$ represents the transportation cost per unit of mass from $\mathbf{u}$ to $\mathbf{x}$, then the last inequality asserts that the transportation cost for moving the whole mass at a point $\mathbf{x}$ is minimum for $\mathbf{x} = \operatorname{bar}(\mu)$.

Another simple example illustrating how supercoercivity yields the existence of solutions of certain nonlinear equations is offered in Exercise 1.

The main ingredient of Theorem 3.10.2 is the fact that all bounded and closed sets in $\mathbb{R}^N$ are compact sets. This does not work in the context of infinite-dimensional Banach spaces, but the reflexive Banach spaces exhibit a valuable substitute, the sequential weak compactness of their bounded closed and convex subsets. See Appendix E for applications to calculus of variations.

Some of the results above concerning the minimization of convex functions can be extended to the case of quasiconvex functions defined on reflexive Banach spaces. See Exercise 6.

Convex functions attain their maxima at the relative boundary:

3.10.10 Theorem (*The maximum principle*) *Let U be a convex subset of $\mathbb{R}^N$. If a convex function $f\colon U \to \mathbb{R}$ attains its maximum on U at a point $\mathbf{a}$ from the relative interior of U, then f is constant on U.*

Proof. We should prove that $f(\mathbf{x}) = f(\mathbf{a})$ for $\mathbf{x} \in U$, $\mathbf{x} \neq \mathbf{a}$. Since $\mathbf{a} \in \operatorname{ri} U$, there exists a point $\mathbf{x}' \in U$ such that $\mathbf{a}$ is an interior point of the segment $[\mathbf{x}, \mathbf{x}']$, that is, $\mathbf{a} = (1-\lambda)\mathbf{x} + \lambda\mathbf{x}'$ for some $\lambda \in (0,1)$. From the definition of convexity,

$$f(\mathbf{a}) \leq (1-\lambda)f(\mathbf{x}) + \lambda f(\mathbf{x}').$$

Since both $f(\mathbf{x})$ and $f(\mathbf{x}')$ do not exceed $f(\mathbf{a})$ and both weights $1-\lambda$ and λ are strictly positive, the last inequality forces $f(\mathbf{a}) = f(\mathbf{x}) = f(\mathbf{x}')$. The proof is complete. ■

A generalization of the maximum principle is given in Corollary B.3.5 (Appendix B). We end this section with an important consequence of Minkowski's Theorem 2.3.5.

3.10.11 Theorem *If f is a continuous convex function on a compact convex subset K of $\mathbb{R}^N$, then f attains a global maximum at an extreme point.*

Proof. Assume that f attains its global maximum at $\mathbf{a} \in K$. By Theorem 2.3.5, the point $\mathbf{a}$ can be represented as a convex combination of extreme points, say $\mathbf{a} = \sum_{k=1}^{m} \lambda_k \mathbf{e}_k$. Then $f(\mathbf{a}) \leq \sum_{k=1}^{m} \lambda_k f(\mathbf{e}_k) \leq \sup_k f(\mathbf{e}_k)$, which forces $f(\mathbf{a}) = f(\mathbf{e}_k)$ for some k. ■

For functions defined on N-dimensional intervals $[a_1, b_1] \times \cdots \times [a_N, b_N]$ in $\mathbb{R}^N$, Theorem 3.10.11 extends to the case of continuous functions which are convex in each variable (when the others are kept fixed). This fact can be proved by one-variable means (taking into account that convexity is equivalent to convexity on segment lines).

In the infinite-dimensional setting, it is difficult to state fairly general results on maximum-attaining. Besides, the deep results of Banach space theory appear to be crucial in answering questions which at first glance may look simple. Here is an example. By the Alaoglu–Bourbaki theorem (see Theorem B.1.6, Appendix B) it follows that each continuous linear functional on a reflexive Banach space E achieves its norm on the unit ball. Surprisingly, these are the *only* Banach spaces for which the norm-attaining phenomenon occurs. This was proved by R. C. James [229].

Exercises

1. (Surjectivity of derivatives) Suppose that $f : \mathbb{R}^N \to \mathbb{R}$ is a Fréchet differentiable supercoercive function. Prove that its differential $df : \mathbb{R}^N \to \mathbb{R}^N$, $df(\mathbf{x}) = \nabla f(\mathbf{x})$, is onto.

   [*Hint:* For $\mathbf{z} \in \mathbb{R}^N$ arbitrarily fixed, the function $g(\mathbf{x}) = f(\mathbf{x}) - \langle \mathbf{x}, \mathbf{z} \rangle$ admits a global minimum.]

2. (First-order condition of minimization) Let C be a nonempty convex subset of $\mathbb{R}^N$ and let $f : C \to \mathbb{R}$ be a convex and differentiable function. Prove that $\bar{\mathbf{x}}$ is a minimizer of f on C if and only if

   $$\langle \nabla f(\bar{\mathbf{x}}), \mathbf{x} - \bar{\mathbf{x}} \rangle \geq 0 \text{ for all } \mathbf{x} \in C.$$

   Infer that the projection $P_C(\mathbf{x})$ of a point $\mathbf{x} \in \mathbb{R}^N$ onto a nonempty closed convex set C in $\mathbb{R}^N$ verifies the condition

   $$\langle \mathbf{x} - P_C(\mathbf{x}), \mathbf{y} - P_C(\mathbf{x}) \rangle \leq 0 \quad \text{for all } \mathbf{y} \in C.$$

3. Suppose that $f : E \to \mathbb{R}$ is convex and C is a nonempty subset of E. Prove that

   $$\sup_{\mathbf{x} \in C} f(\mathbf{x}) = \sup_{\mathbf{x} \in \operatorname{conv}(C)} f(\mathbf{x}).$$

4. Suppose that E is a smooth Banach space. Prove that the Gâteaux differential of the norm function $\mathrm{N}(\mathbf{x}) = \|\mathbf{x}\|$ at $\mathbf{x} \neq 0$ is a norm-1 functional that verifies the formula

   $$\mathrm{N}'(\mathbf{x})(\mathbf{x}) = \|\mathbf{x}\| = \sup \{x^*(\mathbf{x}) : x^* \in E^*, \ \|x^*\| = 1\}.$$

5. Suppose that E is a uniformly convex Banach space and $x^* \in E^*$, $x^* \neq 0$. Prove that x^* attains its norm at an element $\mathbf{u} \in E$ with $\|\mathbf{u}\| = 1$.

   Remark. According to the preceding exercise, the element $\mathbf{u}$ is unique if E is also a smooth space.

6. (Minimum of quasiconvex functions) Let E be a real Banach space.

   (a) Prove that every lower semicontinuous quasiconvex proper function $f : E \to \mathbb{R} \cup \{\infty\}$ is also (sequentially) weakly lower semicontinuous, that is,

   $$\mathbf{u}_n \to \mathbf{u} \text{ weakly in } E \text{ implies } f(\mathbf{u}) \leq \liminf_{n \to \infty} f(\mathbf{u}_n).$$

   (b) Suppose in addition that E is an L^p-space (or, more generally, a reflexive Banach space) and that $f : E \to \mathbb{R} \cup \{\infty\}$ is a lower semicontinuous, quasiconvex, coercive, and proper function. Prove that f attains its infimum.

   (c) A function $f : E \to \mathbb{R} \cup \{\infty\}$ is said to be *semistrictly quasiconvex* if it is quasiconvex and for any $t \in (0, 1)$ and any $\mathbf{x}_0, \mathbf{x}_1 \in E$ with $f(\mathbf{x}_0) \neq f(\mathbf{x}_1)$ we have $f((1 - \lambda)\mathbf{x}_0 + \lambda \mathbf{x}_1) < \max \{f(\mathbf{x}_0), f(\mathbf{x}_1)\}$. Prove that any superposition $g \circ f$ of a strictly increasing function $g : \mathbb{R} \to \mathbb{R}$ and a strictly convex function $f : E \to \mathbb{R}$ is semistrictly quasiconvex. Indicate concrete examples when $E = \mathbb{R}^N$.

(d) Prove that for every semistrictly quasiconvex function any local minimizer is also a global minimizer.

(e) Give an example of quasiconvex function on $\mathbb{R}$ for which not every local minimizer is a global minimizer.

[*Hint:* For (a), see Corollary B.2.6, Appendix B. For (b), adapt the argument of Theorem 3.10.2 by using the Eberlein–Šmulyan compactness theorem (Theorem B.1.8, Appendix B).]

The following set of exercises concerns the constrained optimization.

7. Suppose that C is a convex subset of $\mathbb{R}^N$ and $\mathbf{a} \in C$. The *tangent cone* to C at $\mathbf{a}$ is the closed cone $T_C(\mathbf{a}) = \overline{\mathbb{R}_+(C - \mathbf{a})}$, and the *normal cone* to C at $\mathbf{a}$ is the closed cone $N_C(\mathbf{a}) = \{\mathbf{v} \in \mathbb{R}^N : \langle \mathbf{v}, \mathbf{x} - \mathbf{a} \rangle \leq 0 \text{ for all } \mathbf{x} \in C\}$.

   (a) Compute $T_C(\mathbf{a})$ and $N_C(\mathbf{a})$ when C is the unit disc.

   (b) Prove that the polar set of $T_C(\mathbf{a})$ is $N_C(\mathbf{a})$ (and vice versa).

8. Let $f \colon \mathbb{R}^N \to \mathbb{R}$ be a convex function and $C = \{\mathbf{x} : f(\mathbf{x}) \leq 0\}$. Assume that there exists a point $\mathbf{x}$ such that $f(\mathbf{x}) < 0$. Prove that

$$T_C(\mathbf{a}) = \{\mathbf{v} : f'(\mathbf{a}; \mathbf{v}) \leq 0\} \quad \text{and} \quad N_C(\mathbf{a}) = \mathbb{R}_+ \partial f(\mathbf{a})$$

   for all $\mathbf{a} \in \mathbb{R}^N$ such that $f(\mathbf{a}) = 0$.

9. (Self-dual cones) Suppose that C is one of the following cones: $\mathbb{R}^N$, $\mathrm{Sym}^+(n, \mathbb{R})$ or $\{\mathbf{x} \in \mathbb{R}^N : x_1^2 \geq x_2^2 + \cdots + x_n^2\}$. Prove that $N_C(\mathbf{0}) = -C$.

10. Suppose that C is a convex subset of $\mathbb{R}^N$ and that $f \colon \mathbb{R}^N \to \mathbb{R}$ is a convex function. Prove that the following assertions are equivalent for $\mathbf{a} \in C$:

    (a) $\mathbf{a}$ is a minimizer for $f|_C$;

    (b) $f'(\mathbf{x}; \mathbf{v}) \geq 0$ for all $\mathbf{v} \in T_C(\mathbf{a})$;

    (c) $\mathbf{0} \in \partial f(\mathbf{a}) + N_C(\mathbf{a})$.

3.11 The Prékopa–Leindler Inequality

The aim of this section is to prove that the property of log-concavity is preserved by integration. The basic ingredient is an inequality whose roots lie in the geometric theory of convexity.

3.11.1 Theorem (Prékopa–Leindler inequality) *Let $0 < \lambda < 1$ and let f, g, and h be positive integrable functions on $\mathbb{R}^N$ satisfying*

$$h((1 - \lambda)\mathbf{u} + \lambda \mathbf{v}) \geq f(\mathbf{u})^{1-\lambda} g(\mathbf{v})^{\lambda}$$

for all $\mathbf{u}, \mathbf{v} \in \mathbb{R}^N$. *Then*

$$\int_{\mathbb{R}^N} h(\mathbf{x})\,\mathrm{d}\mathbf{x} \geq \left(\int_{\mathbb{R}^N} f(\mathbf{x})\,\mathrm{d}\mathbf{x}\right)^{1-\lambda} \left(\int_{\mathbb{R}^N} g(\mathbf{x})\,\mathrm{d}\mathbf{x}\right)^{\lambda}.$$

The equality condition for this inequality can be found in the paper by S. Dubuc [136].

The Prékopa–Leindler inequality is just the particular case of a much more general result, the Borell–Brascamp–Lieb inequality. See Theorem 3.11.6. Before entering into details, we mention several immediate consequences of the Prékopa–Leindler inequality.

3.11.2 Corollary (Preservation of log-concavity by integration) *If*

$$F : \mathbb{R}^M \times \mathbb{R}^N \to [0, \infty), \quad F = F(\mathbf{x}, \mathbf{y}),$$

is an integrable log-*concave function, then the function*

$$\varphi(\mathbf{x}) = \int_{\mathbb{R}^N} F(\mathbf{x}, \mathbf{y})\mathrm{d}\mathbf{y}, \quad \mathbf{x} \in \mathbb{R}^M,$$

is also log-*concave.*

Proof. Fix arbitrarily $\mathbf{x}_1, \mathbf{x}_2 \in \mathbb{R}^M$ and $\lambda \in (0,1)$. The functions $h(\mathbf{y}) = F((1-\lambda)\mathbf{x}_1 + \lambda\mathbf{x}_2, \mathbf{y})$, $f(\mathbf{y}) = F(\mathbf{x}_1, \mathbf{y})$, and $g(\mathbf{y}) = F(\mathbf{x}_2, \mathbf{y})$ verify the hypotheses of Theorem 3.11.1, which implies that

$$\begin{aligned} \varphi\left((1-\lambda)\mathbf{x}_1 + \lambda\mathbf{x}_2\right) &= \int_{\mathbb{R}^N} F((1-\lambda)\mathbf{x}_1 + \lambda\mathbf{x}_2, \mathbf{y})\mathrm{d}\mathbf{y} \\ &\geq \left(\int_{\mathbb{R}^N} F(\mathbf{x}_1, \mathbf{y})\mathrm{d}\mathbf{y}\right)^{1-\lambda} \left(\int_{\mathbb{R}^N} F(\mathbf{x}_2, \mathbf{y})\mathrm{d}\mathbf{y}\right)^{\lambda} \\ &= \left(\varphi(\mathbf{x}_1)\right)^{1-\lambda} \left(\varphi(\mathbf{x}_2)\right)^{\lambda}. \end{aligned}$$

Therefore φ is log-concave. ■

3.11.3 Corollary (Preservation of log-concavity by convolution product) *The convolution product of two* log-*concave functions defined on* $\mathbb{R}^N$ *is also a* log-*concave function.*

Proof. Indeed, if $f_1, f_2 : \mathbb{R}^N \to [0, \infty)$ are log-concave, so is the function $F(\mathbf{x}, \mathbf{y}) = f_1(\mathbf{x} - \mathbf{y}) f_2(\mathbf{y})$ on $\mathbb{R}^N \times \mathbb{R}^N$. The proof ends by taking into account Corollary 3.11.2. ■

It is worth noticing that the Prékopa–Leindler inequality represents the functional analogue of another important inequality related to the isoperimetric problem in Euclidean spaces:

3.11.4 Theorem (Lusternik's general Brunn–Minkowski inequality) *Let $s, t > 0$ and let A and B be two nonempty bounded measurable sets in $\mathbb{R}^N$ such that $sA + tB$ is also measurable. Then*

$$\operatorname{Vol}_N(sA + tB)^{1/N} \geq s \operatorname{Vol}_N(A)^{1/N} + t \operatorname{Vol}_N(B)^{1/N}.$$

Here Vol_N denotes the N-dimensional Lebesgue measure.

Proof. Since the Lebesgue measure Vol_N is positively homogeneous of degree N (that is, $\operatorname{Vol}_N(\alpha A) = \alpha^N \operatorname{Vol}_N(A)$ for every Borel set A and every $\alpha \geq 0$), we may restrict to the case where $s = 1 - \lambda$ and $t = \lambda$ for some $\lambda \in (0, 1)$. Then we apply the Prékopa–Leindler inequality for $f = \chi_A$, $g = \chi_B$ and $h = \chi_{(1-\lambda)A+\lambda B}$, which yields

$$\begin{aligned}\operatorname{Vol}_N((1-\lambda)A + \lambda B) &= \int_{\mathbb{R}^N} \chi_{(1-\lambda)A+\lambda B}(\mathbf{x})\, dx \\ &\geq \left(\int_{\mathbb{R}^N} \chi_A(\mathbf{x})\, d\mathbf{x}\right)^{1-\lambda} \left(\int_{\mathbb{R}^N} \chi_B(\mathbf{x})\, d\mathbf{x}\right)^{\lambda} \\ &= \operatorname{Vol}_N(A)^{1-\lambda} \operatorname{Vol}_N(B)^{\lambda}.\end{aligned}$$

Applying this inequality for A replaced by $\operatorname{Vol}_N(A)^{-1/N} A$, B replaced by $\operatorname{Vol}_N(B)^{-1/N} B$, and λ replaced by

$$\frac{\lambda \operatorname{Vol}_N(B)^{1/N}}{(1-\lambda) \operatorname{Vol}_N(A)^{1/N} + \lambda \operatorname{Vol}_N(B)^{1/N}},$$

we obtain

$$\operatorname{Vol}_N((1-\lambda)A + \lambda B)^{1/N} \geq (1-\lambda) \operatorname{Vol}_N(A)^{1/N} + \lambda \operatorname{Vol}_N(B)^{1/N},$$

which ends the proof. ∎

The hypothesis on the measurability of $sA + tB$ cannot be deduced from the measurability of A and B. A counterexample can be found in a paper by W. Sierpiński [444].

The classical Brunn–Minkowski inequality represents the particular case of Theorem 3.11.4 for convex bodies. A *convex body* is understood as a compact convex set in $\mathbb{R}^N$, with nonempty interior. In this case the measurability of the sets $sX + tY$ is automatic.

3.11.5 Theorem (The Brunn–Minkowski inequality) *Let $\lambda \in (0, 1)$ and let K and L be two convex bodies in $\mathbb{R}^N$. Then*

$$\operatorname{Vol}_N((1-\lambda)K + \lambda L)^{1/N} \geq (1-\lambda) \operatorname{Vol}_N(K)^{1/N} + \lambda \operatorname{Vol}_N(L)^{1/N}.$$

Equality holds precisely when K and L are equal up to translation and dilation.

The connection of the Brunn–Minkowski inequality with the isoperimetric problems is briefly described in the Comments at the end of this chapter.

The Prékopa–Leindler inequality represents the case $p = 0$ of the following general result:

3.11.6 Theorem (The Borell–Brascamp–Lieb inequality) *Suppose that* $0 < \lambda < 1$, $-1/N \leq p \leq \infty$, *and* f, g, *and* h *are positive integrable functions on* $\mathbb{R}^N$ *satisfying*

$$h((1-\lambda)\mathbf{x} + \lambda\mathbf{y}) \geq M_p(f(\mathbf{x}), g(\mathbf{y}); 1-\lambda, \lambda),$$

for all $\mathbf{x}, \mathbf{y} \in \mathbb{R}^N$. *Then*

$$\int_{\mathbb{R}^N} h(\mathbf{x})\,\mathrm{d}\mathbf{x} \geq M_{p/(Np+1)}\Big(\int_{\mathbb{R}^N} f(\mathbf{x})\,\mathrm{d}\mathbf{x}, \int_{\mathbb{R}^N} g(\mathbf{x})\,\mathrm{d}\mathbf{x}; 1-\lambda, \lambda\Big).$$

Here $p/(Np+1)$ *means* $-\infty$, *if* $p = -1/N$, *and* $1/N$, *if* $p = \infty$.

Proof. We start with the case $N = 1$. Without loss of generality we may assume that

$$\int_{\mathbb{R}} f(x)\,\mathrm{d}x = A > 0 \quad \text{and} \quad \int_{\mathbb{R}} g(x)\,\mathrm{d}x = B > 0.$$

We define $u, v\colon [0,1] \to \overline{\mathbb{R}}$ such that $u(t)$ and $v(t)$ are the smallest numbers satisfying

$$\frac{1}{A}\int_{-\infty}^{u(t)} f(x)\,\mathrm{d}x = \frac{1}{B}\int_{-\infty}^{v(t)} g(x)\,\mathrm{d}x = t.$$

Clearly, the two functions are increasing and thus they are differentiable almost everywhere. This yields

$$\frac{f(u(t))u'(t)}{A} = \frac{g(v(t))v'(t)}{B} = 1 \qquad \text{almost everywhere,}$$

so that $w(t) = (1-\lambda)u(t) + \lambda v(t)$ verifies

$$\begin{aligned} w'(t) &= (1-\lambda)u'(t) + \lambda v'(t) \\ &= (1-\lambda)\frac{A}{f(u(t))} + \lambda\frac{B}{g(v(t))} \end{aligned}$$

at every t with $f(u(t)) > 0$ and $g(v(t)) > 0$. Or,

$$\int_{\mathbb{R}} h(x)\,\mathrm{d}x \geq \int_0^1 h(w(t))w'(t)\,\mathrm{d}t$$

and the last inequality can be continued as

$$\begin{aligned} &\geq \int_0^1 M_p\big(f(u(t)), g(v(t)); 1-\lambda, \lambda\big) M_1\Big(\frac{A}{f(u(t))}, \frac{B}{g(v(t))}; 1-\lambda, \lambda\Big)\,\mathrm{d}t \\ &\geq \int_0^1 M_{p/(p+1)}(A, B; 1-\lambda, \lambda)\,\mathrm{d}t \\ &= M_{p/(p+1)}(A, B; 1-\lambda, \lambda), \end{aligned}$$

by a generalization of the discrete Rogers–Hölder inequality (provided by Section 1.2, Exercise 3 (a), for $N = 2$, $q = 1$, and $p + q \geq 0$).

The general case follows by induction. Suppose that it is true for all natural numbers less than N. For each $s \in \mathbb{R}$, attach to f, g, and h section functions f_s, g_s, and h_s, following the model

$$f_s \colon \mathbb{R}^{N-1} \to \mathbb{R}, \quad f_s(\mathbf{z}) = f(\mathbf{z}, s).$$

Let $\mathbf{x}, \mathbf{y} \in \mathbb{R}^{N-1}$, let $a, b \in \mathbb{R}$ and put $c = (1-\lambda)a + \lambda b$. Then

$$\begin{aligned} h_c((1-\lambda)\mathbf{x} + \lambda\mathbf{y}) &= h((1-\lambda)\mathbf{x} + \lambda\mathbf{y}, (1-\lambda)a + \lambda b) \\ &= h((1-\lambda)(\mathbf{x}, a) + \lambda(\mathbf{y}, b)) \\ &\geq M_p(f(\mathbf{x}, a), g(\mathbf{y}, b); 1-\lambda, \lambda) \\ &= M_p(f_a(\mathbf{x}), g_b(\mathbf{y}); 1-\lambda, \lambda) \end{aligned}$$

and thus, by our inductive hypothesis,

$$\int_{\mathbb{R}^{N-1}} h_c(\mathbf{x})\, \mathrm{d}\mathbf{x} \geq M_{p/((N-1)p+1)}\left(\int_{\mathbb{R}^{N-1}} f_a(\mathbf{x})\, \mathrm{d}\mathbf{x}, \int_{\mathbb{R}^{N-1}} g_b(\mathbf{x})\, \mathrm{d}\mathbf{x}; 1-\lambda, \lambda\right).$$

Letting

$$H(c) = \int_{\mathbb{R}^{N-1}} h_c(\mathbf{x})\, \mathrm{d}x, \quad F(a) = \int_{\mathbb{R}^{N-1}} f_a(\mathbf{x})\, \mathrm{d}\mathbf{x}, \quad G(b) = \int_{\mathbb{R}^{N-1}} g_b(\mathbf{x})\, \mathrm{d}\mathbf{x},$$

we have

$$H(c) = H((1-\lambda)a + \lambda b) \geq M_r(F(a), G(b); 1-\lambda, \lambda),$$

where $r = p/((N-1)p+1)$, so by Fubini's theorem and our inductive hypothesis we conclude that

$$\begin{aligned} \int_{\mathbb{R}^N} h(\mathbf{x})\, \mathrm{d}\mathbf{x} &\geq \int_{\mathbb{R}}\int_{\mathbb{R}^{N-1}} h_c(\mathbf{z})\, \mathrm{d}\mathbf{z}\, \mathrm{d}c = \int_{\mathbb{R}} H(c)\, \mathrm{d}c \\ &\geq M_{r/(r+1)}\left(\int_{\mathbb{R}} F(a)\, \mathrm{d}a, \int_{\mathbb{R}} G(b)\, \mathrm{d}b; 1-\lambda, \lambda\right) \\ &= M_{p/(Np+1)}\left(\int_{\mathbb{R}^N} f(\mathbf{x})\, \mathrm{d}\mathbf{x}, \int_{\mathbb{R}^N} g(\mathbf{x})\, \mathrm{d}\mathbf{x}; 1-\lambda, \lambda\right). \end{aligned}$$

■

The above argument of Theorem 3.11.6 goes back to R. Henstock and A. M. Macbeath [212] (when $N = 1$) and illustrates a powerful tool of convex analysis: the *Brenier map*. See the Comments at the end of this chapter. Basically the same argument (plus some computation that makes the objective of Exercise 10, Section 6.1) led F. Barthe [31], [32] to a simplified approach of the best constants in some famous inequalities like the Young convolution inequality and the reverse Young convolution inequality.

Exercises

1. Settle the equality case in the Brunn–Minkowski inequality (as stated in Theorem 3.11.5).

2. (Log-concavity of marginals) Let $f = f(\mathbf{x}, \mathbf{y})$ be an integrable log-concave function defined on an open convex subset Ω of $\mathbb{R}^m \times \mathbb{R}^n$. For each $\mathbf{x}$ in the orthogonal projection $\mathrm{pr}_1 \Omega$, of Ω onto $\mathbb{R}^m$, put

$$F(\mathbf{x}) = \int_{\Omega(\mathbf{x})} f(\mathbf{x}, \mathbf{y})\, d\mathbf{y},$$

where $\Omega(\mathbf{x}) = \{\mathbf{y} \in \mathbb{R}^n : (\mathbf{x}, \mathbf{y}) \in \Omega\}$. Infer from the Prékopa–Leindler inequality that the function $F(\mathbf{x})$ is log-concave on $\mathrm{pr}_1 \Omega$.

[*Hint*: Suppose that $\mathbf{x}_k \in \mathrm{pr}_1 \Omega$ and $\mathbf{y}_k \in \Omega(\mathbf{x}_k)$ for $k = 1, 2$ and $\lambda \in (0, 1)$. Then

$$\Omega((1-\lambda)\mathbf{x}_1 + \lambda \mathbf{x}_2) \supset (1-\lambda)\Omega(\mathbf{x}_1) + \lambda \Omega(\mathbf{x}_2)$$

and

$$f((1-\lambda)\mathbf{x}_1 + \lambda \mathbf{x}_2, (1-\lambda)\mathbf{y}_1 + \lambda \mathbf{y}_2) \geq f(\mathbf{x}_1, \mathbf{y}_1)^{1-\lambda} f(\mathbf{x}_2, \mathbf{y}_2)^{\lambda}.]$$

3. Consider a convex body K in $\mathbb{R}^3$ and a vector $\mathbf{v} \in \mathbb{R}^3$ with $\|\mathbf{v}\| \neq 0$. Prove that the function

$$A(t) = \sqrt{Area(K \cap \{\mathbf{x} : \langle \mathbf{x}, \mathbf{v} \rangle = t\})}$$

is concave on the interval consisting of all t for which

$$K \cap \{\mathbf{x} : \langle \mathbf{x}, \mathbf{v} \rangle = t\} \neq \emptyset.$$

In other words, the square root of the area of the cross section of K by parallel hyperplanes is a concave function.

4. (The essential form of the Prékopa–Leindler inequality; see H. J. Brascamp and E. H. Lieb [76]) Let $f, g \in L^1(\mathbb{R}^N)$ be two positive functions and let $\lambda \in (0, 1)$. The function

$$S(\mathbf{x}) = \operatorname*{ess\,sup}_{y} f\left(\frac{\mathbf{x} - \mathbf{y}}{1-\lambda}\right)^{1-\lambda} g\left(\frac{\mathbf{y}}{\lambda}\right)^{\lambda}$$

is measurable since

$$S(\mathbf{x}) = \sup_n \int_{\mathbb{R}^N} f\left(\frac{\mathbf{x} - \mathbf{y}}{1-\lambda}\right)^{1-\lambda} g\left(\frac{\mathbf{y}}{\lambda}\right)^{\lambda} \varphi_n(\mathbf{y})\, d\mathbf{y}$$

for every sequence $(\varphi_n)_n$, dense in the unit ball of $L^1(\mathbb{R}^N)$. Prove that

$$\|S(\mathbf{x})\|_{L^1} \geq \|f\|_{L^1}^{1-\lambda} \|g\|_{L^1}^{\lambda} \tag{3.41}$$

and derive from this result the classical Prékopa–Leindler inequality.

Remark. As was noticed in [76], the essential form of the Prékopa–Leindler inequality represents the limiting case as $r \to 0$ of the following reverse Young convolution inequality with sharp constants: *Let* $0 < p, q, r \leq 1$ *with* $1/p + 1/q = 1 + 1/r$, *and let* $f \in L^p(\mathbb{R}^N)$ *and* $g \in L^q(\mathbb{R}^N)$ *be positive functions. Then*

$$\|f * g\|_{L^r} \geq C(p, q, r, N)\|f\|_{L^p}\|g\|_{L^q}.$$

5. A positive regular measure μ defined on the (Lebesgue) measurable subsets of $\mathbb{R}^N$ is called M_p-*concave* (for some $p \in \overline{\mathbb{R}}$) if

$$\mu((1-\lambda)X + \lambda Y) \geq M_p(\mu(X), \mu(Y); 1-\lambda, \lambda)$$

for all measurable sets X and Y in $\mathbb{R}^N$ and all $\lambda \in (0,1)$ such that the set $(1-\lambda)X + \lambda Y$ is measurable. When $p = 0$, an M_p-concave measure is also called log-*concave*. By the Prékopa–Leindler inequality, the Lebesgue measure is $M_{1/N}$-concave. Suppose that $-1/N \leq p \leq \infty$, and let f be a positive integrable function which is M_p-concave on an open convex set C in $\mathbb{R}^N$. Prove that the measure $\mu(X) = \int_{C \cap X} f(\mathbf{x})\,\mathrm{d}\mathbf{x}$ is $M_{p/(Np+1)}$-concave. Infer that the standard *Gaussian measure* in $\mathbb{R}^N$,

$$\mathrm{d}\gamma_N = (2\pi)^{-N/2}\,\mathrm{e}^{-\|\mathbf{x}\|^2/2}\mathrm{d}\mathbf{x},$$

is log-concave.

6. (S. Dancs and B. Uhrin [121]) Extend Theorem 3.11.6 by replacing the Lebesgue measure by an M_q-concave measure, for some $-\infty \leq q \leq 1/N$.

3.12 Comments

The theory of convex functions of one real variable can be generalized to several variables in many different ways. A long time ago, P. Montel [335] pointed out the alternative of subharmonic functions. The question when the superharmonic functions are concave is discussed by B. Kawohl [244]. Nowadays, many other alternatives are known. An authoritative monograph on this subject has been published by L. Hörmander [227].

The first modern exposition on convexity in $\mathbb{R}^N$ was written by W. Fenchel [161]. He used the framework of lower semicontinuous proper convex functions to provide a valuable extension of the classical theory. At the beginning of the 1960s, R. T. Rockafellar and J.-J. Moreau initiated a systematic study of the field of convex analysis, that paved the way to many deep applications in convex optimization, variational analysis and partial differential equations, mathematical economics, etc.

The notion of perspective function has a long history with deep roots in statistics. The term “perspective function” was coined by C. Lemaréchal ca. 1987–1988, and first appeared in print in [217], Section IV.2.2. A nice account on the present-day research based on this notion is available in the paper of P. L. Combettes [114].

A nice boundary behavior is available for continuous convex functions defined on polytopes. If K is a polytope in $\mathbb{R}^N$, then every bounded convex function defined on the relative interior of K has a unique extension to a continuous convex function on K. D. Gale, V. Klee, and R. T. Rockafellar [172] have noticed that this property actually characterizes the polytopes among the convex sets in $\mathbb{R}^N$.

Support functions, originally defined by H. Minkowski in the case of bounded convex sets, have been studied for general convex sets in $\mathbb{R}^N$ by W. Fenchel [160], [161] and, in infinite-dimensional spaces, by L. Hörmander [226].

The almost everywhere differentiability of convex functions defined on $\mathbb{R}^N$ is generalized by a famous result due to H. Rademacher concerning locally Lipschitz functions. See Theorem D.1.1, Appendix D.

As we noticed in Section 3.8, the second derivative test of convexity needs some good experience with matrix differential calculus when applied to functions whose variables are the minors of a given matrix. *The Matrix Cookbook* of K. B. Petersen and M. S. Pedersen, available on Internet, might be helpful in this context.

A Banach space E is said to be a *strong differentiability space* if, for each convex open set U in E and each continuous convex function $f: U \to \mathbb{R}$, the set of points of Fréchet differentiability of f contains a dense G_δ subset of E. A characterization of such spaces is given by the following result:

3.12.1 Theorem (E. Asplund [23], I. Namioka and R. R. Phelps [343]) *A Banach space E is a strong differentiability space if and only if the dual of every separable subspace of E is separable.*

In particular, every reflexive Banach space is a strong differentiability space. See R. R. Phelps [397] for details and a survey on the differentiability properties of convex functions on a Banach space.

A survey on the problem of renorming Banach spaces to improve their smoothness properties is available in the book by M. M. Day [128].

The convex functions can be characterized in terms of distributional derivatives:

3.12.2 Theorem (G. Alberti and L. Ambrosio [7]) *If Ω is an open convex subset of $\mathbb{R}^n$, and $f: \Omega \to \mathbb{R}$ is a convex function, then Df is monotone, and $D^2 f$ is a positive and symmetric (matrix-valued and locally bounded) measure. Conversely, if f is locally integrable and $D^2 f$ is a positive (matrix-valued) distribution on Ω, then f agrees almost everywhere on Ω with a convex function g such that $\Omega \subset \operatorname{dom} g$.*

If $f: \mathbb{R}^N \to \mathbb{R}$ is a convex function, then it admits distributional derivatives of second order. More precisely, for all $i, j \in \{1, \ldots, N\}$ there exist signed

Radon measures μ_{ij} (with $\mu_{ij} = \mu_{ji}$) such that

$$\int_{\mathbb{R}^N} f(\mathbf{x}) \frac{\partial^2 \varphi}{\partial x_i \partial x_j}(\mathbf{x})\,\mathrm{d}\mathbf{x} = \int_{\mathbb{R}^N} \varphi(\mathbf{x}) \mathrm{d}\mu_{ij} \quad \text{for every } \varphi \in C_c^2(\mathbb{R}^N).$$

In addition, the measures μ_{ii} are positive. For details, see L. C. Evans and R. Gariepy [152], Theorem 6.8, p. 271. Yu. G. Reshetnyak [418] proved that a locally integrable function f defined on an open subset of $\mathbb{R}^N$ is equal almost everywhere to a convex function if and only if for every $\xi_1, ..., \xi_N \in \mathbb{R}$, $\sum_{i,j=1}^N \xi_i \xi_j \frac{\partial^2 [f]}{\partial x_i \partial x_j}$ is a positive Radon measure; here $[f]$ represents the distribution associated to f.

In Appendix D we will prove (following the argument of M. G. Crandall, H. Ishii and P.-L. Lions [118]) the following theorem of A. D. Alexandrov [9] concerning the twice differentiability of convex functions defined on $\mathbb{R}^N$:

3.12.3 Theorem *Every convex function f on $\mathbb{R}^N$ is twice differentiable almost everywhere in the following sense: f is twice differentiable at $\mathbf{a}$, with Alexandrov Hessian $\nabla^2 f(\mathbf{a})$ in $\mathrm{Sym}^+(N, \mathbb{R})$, if $\nabla f(\mathbf{a})$ exists, and if for every $\varepsilon > 0$ there exists $\delta > 0$ such that*

$$\|\mathbf{x} - \mathbf{a}\| < \delta \text{ implies } \sup_{\mathbf{y} \in \partial f(\mathbf{x})} \|\mathbf{y} - \nabla f(\mathbf{a}) - \nabla^2 f(\mathbf{a})(\mathbf{x} - \mathbf{a})\| \leq \varepsilon \|\mathbf{x} - \mathbf{a}\|.$$

Moreover, if $\mathbf{a}$ is such a point, then

$$\lim_{\mathbf{h} \to 0} \frac{f(\mathbf{a} + \mathbf{h}) - f(\mathbf{a}) - \langle \nabla f(\mathbf{a}), \mathbf{h} \rangle - \frac{1}{2} \langle \nabla^2 f(\mathbf{a}) \mathbf{h}, \mathbf{h} \rangle}{\|\mathbf{h}\|^2} = 0.$$

Convolution by smooth functions provides a powerful technique for approximating locally integrable functions by C^∞ functions. Particularly, this applies to the convex functions.

Let φ be a *mollifier*, that is, a positive function in $C_c^\infty(\mathbb{R}^N)$ such that

$$\int_{\mathbb{R}^N} \varphi \,\mathrm{d}\mathbf{x} = 1 \quad \text{and} \quad \operatorname{supp} \varphi \subset \overline{B}_1(\mathbf{0}).$$

The standard example of such a function is

$$\varphi(\mathbf{x}) = \begin{cases} C \exp(-1/(1 - \|\mathbf{x}\|^2)) & \text{if } \|\mathbf{x}\| < 1, \\ 0 & \text{if } \|\mathbf{x}\| \geq 1, \end{cases}$$

where C is chosen such that $\int_{\mathbb{R}^N} \varphi \,\mathrm{d}\mathbf{x} = 1$. Each mollifier φ gives rise to an one-parameter family of positive functions

$$\varphi_\varepsilon(\mathbf{x}) = \frac{1}{\varepsilon^N} \varphi\left(\frac{\mathbf{x}}{\varepsilon}\right), \quad \varepsilon > 0$$

with similar properties:

$$\varphi_\varepsilon \in C_c^\infty(\mathbb{R}^N), \quad \operatorname{supp} \varphi_\varepsilon \subset \overline{B}_\varepsilon(\mathbf{0}) \quad \text{and} \quad \int_{\mathbb{R}^N} \varphi_\varepsilon \,\mathrm{d}\mathbf{x} = 1.$$

The following result is standard and its proof is available from many sources. See, for example [152], Theorem 4.1, p. 146.

3.12.4 Theorem *Suppose that* $f \in L^1_{\text{loc}}(\mathbb{R}^N)$ *and* $(\varphi_\varepsilon)_{\varepsilon>0}$ *is the one-parameter family of functions associated to a mollifier* φ. *Then:*

(a) *The functions* $f_\varepsilon = \varphi_\varepsilon * f$ *belong to* $C^\infty(\mathbb{R}^N)$ *and*

$$D^\alpha f_\varepsilon = D^\alpha \varphi_\varepsilon * f \quad \text{for every multi-index } \alpha \in \mathbb{N}^N;$$

(b) $f_\varepsilon(\mathbf{x}) \to f(\mathbf{x})$ *whenever* $\mathbf{x}$ *is a point of continuity of* f. *If* f *is continuous on an open subset* U, *then* f_ε *converges uniformly to* f *on each compact subset of* U;

(c) *If* $f \in L^p(\mathbb{R}^N)$ *(for some* $p \in [1, \infty)$*), then* $f_\varepsilon \in L^p(\mathbb{R}^N)$, $\|f_\varepsilon\|_{L^p} \leq \|f\|_{L^p}$ *and* $\lim_{\varepsilon \to 0} \|f_\varepsilon - f\|_{L^p} = 0$;

(d) *If* f *is a convex function on an open convex subset* U *of* $\mathbb{R}^N$, *then* f_ε *is convex too.*

In Section 6.5 we will discuss the approximation of convex functions via infimal convolution, that provides convex and C^1 smooth approximants. The following result due to D. Azagra and C. Mudarra [27], shows that the uniform approximation of continuous convex functions in Banach spaces is possible only in some special cases.

3.12.5 Theorem *For a separable Banach space* E, *the following statements are equivalent:*

(a) E^* *is separable;*

(b) *For every* $U \subset E$ *open and convex, every continuous convex function* $f : U \to \mathbb{R}$ *and every* $\varepsilon > 0$, *there exists a convex function* $g \in C^1(U)$ *such that* $f - \varepsilon \leq g \leq f$ *on* U.

In section 3.10 we noticed the role played by lower semicontinuity to the problem of finding the minimum of a convex function defined on $\mathbb{R}^N$. For the applications of the calculus of variations to solving partial differentiable equations it is necessary to consider also the variant of sequential lower semicontinuity. See Appendix E. The ubiquity of convexity in this connection was noticed by a number of mathematicians starting with L. Tonelli. For simplicity, we will consider here only the case of functionals of the form

$$I = I(u) = \int_\Omega f(u(\mathbf{x}))d\mathbf{x}, \quad u \in L^p(\Omega),$$

where Ω is a bounded open subset of $\mathbb{R}^n$ with smooth boundary, $p \in (1, \infty)$ and $f : \mathbb{R} \to \mathbb{R}$ is a continuous function bounded below.

3.12.6 Theorem (L. Tonelli) *The functional* I *is sequentially weakly lower semicontinuous, that is*

$$u_n \to u \text{ weakly in } L^p(\Omega) \text{ implies } I(u) \leq \liminf_{n\to\infty} I(u_n),$$

if, and only if, the function f *is convex.*

For details and much more general results (involving integrands of the form $f = f(\mathbf{x}, u(\mathbf{x}), \nabla u(\mathbf{x}))$ see the books of L. C. Evans [150] and B. Dacorogna [119]. A proof of Tonelli's theorem is also available in the book of M. Renardy and R. C. Rogers [416], pp. 347–350.

In some cases (for example in nonlinear elasticity) the assumption of convexity is physically unreasonable. Then, a substitute of lower semicontinuity (known as Morrey quasiconvexity) is used and that is obtained via the property of polyconvexity. All three aforementioned books detail this important matter.

The quasiconvex functions (introduced by B. De Finetti around 1949) play many special properties similar to those for convex functions. For example, any lower semicontinuous quasiconvex function is weakly lower semicontinuous. Thus, if the ambient space is reflexive and if moreover the function is coercive, it attains its infimum. See Exercise 6, Section 3.10. Meanwhile, there are serious differences. The sums of quasiconvex functions may be not quasiconvex. In fact, as observed by J.-P. Crouzeix, a real-valued function f defined on a normed linear space E is convex if, and only if, for each $x^* \in E^*$ the function $f + x^*$ is quasiconvex. The quasiconvex functions defined on open convex sets in $\mathbb{R}^N$ may be continuous only almost everywhere and the existence of the one-sided partial derivatives takes place only almost everywhere.

Our book includes some basic information on quasiconvex functions, most of it presented as exercises (accompanied by hints when necessary). A systematic treatment of this subject can be found in the books of M. Avriel, W. E. Diewert, S. Schaible and I. Zang [26], and of A. Cambini and L. Martein [88]. The very enjoyable surveys by H. J. Greenberg and W. P. Pierskalla [187], and J.-P. Penot [392] describe the evolution of our understanding of quasiconvexity by the 1960s and the 2000s, respectively.

The Brunn–Minkowski inequality is among the most important results in the theory of convex bodies. R. J. Gardner [175], [176] wrote a nice survey about it, including historical remarks and a description of links to other inequalities and various extensions. Nowadays, a dozen of Brunn–Minkowski-type inequalities are already known for the first eigenvalue of the Laplacian, the Newton capacity, the torsional rigidity, etc. See the paper of A. Colesanti [113] and references therein.

The Brunn–Minkowski inequality was primarily motivated by the isoperimetric problem in Euclidean spaces. The clue is provided by the formula

$$S = \frac{\mathrm{d}V}{\mathrm{d}R},$$

that relates the volume V of a ball $B_R(0)$ in $\mathbb{R}^3$ and the area S of its surface. This led H. Minkowski to define the *surface area* of a convex body K in $\mathbb{R}^N$ by the formula

$$S_{N-1}(K) = \lim_{\varepsilon \to 0+} \frac{\mathrm{Vol}_N(K + \varepsilon B) - \mathrm{Vol}_N(K)}{\varepsilon},$$

where B denotes the closed unit ball of $\mathbb{R}^N$. The agreement of this definition with the usual definition of the surface of a smooth surface is discussed in books by H. Federer [159] and Y. D. Burago and V. A. Zalgaller [85].

3.12.7 Theorem (The isoperimetric inequality for convex bodies in $\mathbb{R}^N$) *Let K be a convex body in $\mathbb{R}^N$ and let B denote the closed unit ball of this space. Then*

$$\left(\frac{\mathrm{Vol}_N(K)}{\mathrm{Vol}_N(B)}\right)^{1/N} \leq \left(\frac{S_{N-1}(K)}{S_{N-1}(B)}\right)^{1/(N-1)},$$

with equality if and only if K is a ball.

For $N = 2$, this means that the area A of a compact convex domain in plane, bounded by a curve of length L, verifies the inequality $4\pi A \leq L^2$, the equality taking place only for discs.

Proof. In fact, by the Brunn–Minkowski inequality,

$$\begin{aligned} S_{N-1}(K) &= \lim_{\varepsilon \to 0+} \frac{\mathrm{Vol}_N(K + \varepsilon B) - \mathrm{Vol}_N(K)}{\varepsilon} \\ &\geq \lim_{\varepsilon \to 0+} \frac{(\mathrm{Vol}_N(K)^{1/N} + \varepsilon \, \mathrm{Vol}_N(B)^{1/N})^N - \mathrm{Vol}_N(K)}{\varepsilon} \\ &= N \, \mathrm{Vol}_N(K)^{(N-1)/N} \, \mathrm{Vol}(B)^{1/N}, \end{aligned}$$

and it remains to notice that $S_{N-1}(B) = N \, \mathrm{Vol}_N(B)$. ∎

A survey of the convexity properties of solutions of partial differential equations may be found in the notes of B. Kawohl [245]. We shall mention here one application of the Prékopa–Leindler inequality, which refers to the diffusion equation

$$\frac{\partial u}{\partial t} = \frac{1}{2} \Delta u - V(\mathbf{x})u \quad \text{for } (\mathbf{x}, t) \in \Omega \times (0, \infty)$$

with zero Dirichlet boundary conditions (that is, $\lim_{\mathbf{x} \to \partial\Omega} u(\mathbf{x}, t) = 0$ for each t). Here Ω is an open convex set in $\mathbb{R}^N$, $u : \Omega \times (0, \infty) \to \mathbb{R}$, $u = u(\mathbf{x}, t)$, is the unknown function, and V is a positive continuous function defined on Ω. When $\Omega = \mathbb{R}^N$ and $V = 0$, the fundamental solution is given by formula $f(\mathbf{x}, \mathbf{y}, t) = (2\pi t)^{-n/2} \mathrm{e}^{-\|\mathbf{x}-\mathbf{y}\|^2/2t}$, which is log-concave on $\mathbb{R}^N \times \mathbb{R}^N$ for all $t > 0$. H. J. Brascamp and E. H. Lieb [76] have proved, based on the Prékopa–Leindler inequality, that in general the fundamental solution $f(\mathbf{x}, \mathbf{y}, t)$ of the above Dirichlet problem is log-concave on $\Omega \times \Omega$, whenever V is a convex function. The idea is to show that $f(\mathbf{x}, \mathbf{y}, t)$ is a pointwise limit of convolutions of log-concave functions. Later on, Ch. Borell [65] considered potentials that depend on a parameter, and this fact led him to more general results and a Brownian motion treatment of the Brunn–Minkowski inequality.

The Borell–Brascamp–Lieb inequality was first stated and proved in full generality by Ch. Borell [64] and, independently, by H. J. Brascamp and E. H. Lieb [76].

Chapter 4

Convexity and Majorization

This chapter is aimed to offer a glimpse on the majorization theory and the beautiful inequalities associated to it. Introduced by G. H. Hardy, J. E. Littlewood, and G. Pólya [208] in 1929, and popularized by their celebrated book on *Inequalities* [209], the relation of majorization has attracted along the time a big deal of attention not only from the mathematicians, but also from people working in various other fields such as statistics, economics, physics, signal processing, data mining, etc. Part of this research activity is summarized in the 900 pages of the recent book by A. W. Marshall, I. Olkin, and B. Arnold [305].

The discussion in this chapter is limited to some very basic aspects. However, due to the importance of the subject of majorization, we will come back to it in Chapter 7, by presenting a series of integral inequalities involving signed measures instead of positive measures.

4.1 The Hardy–Littlewood–Pólya Theory of Majorization

In what follows, for any vector $\mathbf{x} = (x_1, \ldots, x_N) \in \mathbb{R}^N$, we will denote by

$$x_1^{\downarrow} \geq \cdots \geq x_N^{\downarrow}$$

the components of $\mathbf{x}$ in decreasing order.

4.1.1 Definition *Given two vectors* $\mathbf{x}$ *and* $\mathbf{y}$ *in* $\mathbb{R}^N$, *we say that* $\mathbf{x}$ *is weakly* majorized *by* $\mathbf{y}$ (*denoted* $\mathbf{x} \prec_{HLPw} \mathbf{y}$) *if*

$$\sum_{i=1}^{k} x_i^{\downarrow} \leq \sum_{i=1}^{k} y_i^{\downarrow} \quad \textit{for } k = 1, \ldots, N$$

C. P. Niculescu and L.-E. Persson, *Convex Functions and Their Applications*, CMS Books in Mathematics, https://doi.org/10.1007/978-3-319-78337-6_4

and that $\mathbf{x}$ *is majorized by* $\mathbf{y}$ *(denoted* $\mathbf{x} \prec_{HLP} \mathbf{y}$*) if in addition*

$$\sum_{i=1}^{N} x_i^{\downarrow} = \sum_{i=1}^{N} y_i^{\downarrow}.$$

4.1.2 Remark *As was noticed by G. Pólya [401], if*

$$(x_1, \ldots, x_N) \prec_{HLPw} (y_1, \ldots, y_N),$$

then there exist real numbers x_{N+1} *and* y_{N+1} *such that*

$$(x_1, \ldots, x_N, x_{N+1}) \prec_{HLP} (y_1, \ldots, y_N, y_{N+1}).$$

To check this, choose

$$x_{N+1} = \min\{x_1, \ldots, x_N, y_1, \ldots, y_N\} \text{ and } y_{N+1} = \sum_{k=1}^{N+1} x_k - \sum_{k=1}^{N} y_k.$$

W. Arveson and R. Kadison have found another reduction of weak majorization to majorization. Let $x_1 \geq \cdots \geq x_N$ *and* $y_1 \geq \cdots \geq y_N$ *be two decreasing sequences of positive real numbers such that*

$$x_1 + \cdots + x_k \leq y_1 + \cdots + y_k \quad \text{for } k = 1, ..., N.$$

Then there is a decreasing sequence $\bar{y}_1 \geq \cdots \geq \bar{y}_N$ *such that* $0 \leq \bar{y}_k \leq y_k$ *for all* k *and*

$$(x_1, \ldots, x_N) \prec_{HLP} (\bar{y}_1, \ldots, \bar{y}_N).$$

See [22], Lemma 4.3.

The basic result relating majorization and convexity is the *Hardy–Littlewood–Pólya inequality of majorization:*

4.1.3 Theorem *If* $\mathbf{x}$ *and* $\mathbf{y}$ *belong to* $\mathbb{R}^N$ *and* $\mathbf{x} \prec_{HLP} \mathbf{y}$, *then*

$$\sum_{k=1}^{N} f(x_k) \leq \sum_{k=1}^{N} f(y_k) \tag{4.1}$$

for every continuous convex function f *whose domain of definition is an interval that contains the components of* $\mathbf{x}$ *and* $\mathbf{y}$.

Conversely, if the inequality (4.1) *holds for every continuous convex function whose domain includes the components of* $\mathbf{x}$ *and* $\mathbf{y}$, *then* $\mathbf{x} \prec_{HLP} \mathbf{y}$.

Proof. For the first part, notice that we may assume that $x_k \neq y_k$ for all indices k. Then, according to Abel's partial summation formula, we have

$$\sum_{k=1}^{N} f(y_k^{\downarrow}) - \sum_{k=1}^{N} f(x_k^{\downarrow}) = \sum_{k=1}^{N} \left[\left(y_k^{\downarrow} - x_k^{\downarrow} \right) \frac{f(y_k^{\downarrow}) - f(x_k^{\downarrow})}{y_k^{\downarrow} - x_k^{\downarrow}} \right]$$

$$= \sum_{k=1}^{N-1} \Big(\frac{f(y_k^{\downarrow}) - f(x_k^{\downarrow})}{y_k^{\downarrow} - x_k^{\downarrow}} - \frac{f(y_{k+1}^{\downarrow}) - f(x_{k+1}^{\downarrow})}{y_{k+1}^{\downarrow} - x_{k+1}^{\downarrow}} \Big) \Big(\sum_{i=1}^{k} y_i^{\downarrow} - \sum_{i=1}^{k} x_i^{\downarrow} \Big)$$

and the proof of (4.1) ends by taking into account the three chords inequality.

For the converse, since the identity and its opposite are convex functions, we infer from (4.1) that $\sum_{i=1}^N x_i^{\downarrow} = \sum_{i=1}^N y_i^{\downarrow}$. Also, using the convexity of $f = (x - y_k^{\downarrow})^+$, we obtain

$$x_1^{\downarrow} + \cdots + x_k^{\downarrow} - ky_k^{\downarrow} \leq \sum_{j=1}^N f(x_j^{\downarrow}) \leq \sum_{j=1}^N f(y_j^{\downarrow}) \leq y_1^{\downarrow} + \cdots + y_k^{\downarrow} - ky_k^{\downarrow},$$

that is, $x_1^{\downarrow} + \cdots + x_k^{\downarrow} \leq y_1^{\downarrow} + \cdots + y_k^{\downarrow}$. The proof is done. ■

The simplest example of majorization is

$$\underbrace{(\overline{x}, ..., \overline{x})}_{N \text{ times}} \prec_{HLP} (x_1, \ldots, x_N), \quad \text{where } \overline{x} = \frac{1}{N}\sum_{k=1}^N x_k. \tag{4.2}$$

This yields, via Theorem 4.1.3, the discrete Jensen inequality. Notice also that

$$(y_1, \ldots, y_N) \prec_{HLP} (x_1, \ldots, x_N) \text{ implies } (\overline{x}, ..., \overline{x}) \prec_{HLP} (y_1, \ldots, y_N),$$

which shows that $(\overline{x}, ..., \overline{x})$ is the smallest vector (in the sense of majorization) that is majorized by $(x_1, \ldots, x_N)$. For other examples and applications see exercises at the end of this section.

The Hardy–Littlewood–Pólya inequality of majorization admits the following companion in the case of weak majorization:

4.1.4 Corollary (M. Tomić [470] and H. Weyl [485]) *Let f be an increasing convex function defined on a non-empty interval I. If $(a_k)_{k=1}^n$ and $(b_k)_{k=1}^n$ are two families of numbers in I with $a_1 \geq \cdots \geq a_n$ and*

$$\sum_{k=1}^m a_k \leq \sum_{k=1}^m b_k \quad \text{for } m = 1, \ldots, n,$$

then

$$\sum_{k=1}^n f(a_k) \leq \sum_{k=1}^n f(b_k).$$

Proof. Apply Theorem 4.1.3, taking into account Remark 4.1.2. ■

The aforementioned book of Hardy, Littlewood, and Pólya also includes a characterization of the relation of majorization using the doubly stochastic matrices. Recall that a matrix $P \in \mathrm{M}_N(\mathbb{R})$ is doubly stochastic if P has positive entries and each row and each column sums to unity. A special class of doubly stochastic matrices is that of *T-transformations*. They have the form

$$T = \lambda I + (1 - \lambda)Q$$

where $0 \leq \lambda \leq 1$ and Q is a permutation mapping which interchanges two coordinates, that is,

$$Tx = \big(x_1, \ldots, x_{j-1}, \lambda x_j + (1-\lambda)x_k, x_{j+1}, \ldots, x_{k-1}, \\ \lambda x_k + (1-\lambda)x_j, x_{k+1}, \ldots, x_N\big).$$

4.1.5 Theorem *Let* $\mathbf{x}, \mathbf{y} \in \mathbb{R}^N$. *Then the following assertions are equivalent:*

(a) $\mathbf{x} \prec_{HLP} \mathbf{y}$;

(b) $\mathbf{x} = P\mathbf{y}$ *for a suitable doubly stochastic matrix* $P \in \mathrm{M}_N(\mathbb{R})$;

(c) $\mathbf{x}$ *can be obtained from* $\mathbf{y}$ *by successive applications of finitely many* T*-transformations.*

The implication (b) $\Rightarrow$ (a) is due to I. Schur [439] and constituted the starting point for this theorem.

Proof. (c) $\Rightarrow$ (b) Since T-transformations are doubly stochastic, the product of T-transformations is a doubly stochastic transformation.

The implication (b) $\Rightarrow$ (a) is a consequence of Theorem 4.1.3. Assume that $P = (p_{jk})_{j,k=1}^N$ and consider an arbitrary continuous convex function f defined on an interval including the components of $\mathbf{x}$ and $\mathbf{y}$. Since $x_k = \sum_j y_j p_{jk}$, and $\sum_j p_{jk} = 1$ for all indices k, it follows from Jensen's inequality that

$$f(x_k) \leq \sum_{j=1}^N p_{jk} f(y_j).$$

Then

$$\sum_{k=1}^N f(x_k) \leq \sum_{k=1}^N \left(\sum_{j=1}^N p_{jk} f(y_j) \right) = \sum_{j=1}^N \left(\sum_{k=1}^N p_{jk} f(y_j) \right) = \sum_{j=1}^N f(y_j)$$

and Theorem 4.1.3 applies.

(a) $\Rightarrow$ (c) Let $\mathbf{x}$ and $\mathbf{y}$ be two distinct vectors in $\mathbb{R}^N$ such that $\mathbf{x} \prec_{HLP} \mathbf{y}$. Since permutations are T-transformations, we may assume that their components verify the conditions

$$x_1 \geq x_2 \geq \cdots \geq x_N \quad \text{and} \quad y_1 \geq y_2 \geq \cdots \geq y_N.$$

Let j be the largest index such that $x_j < y_j$ and let k be the smallest index such that $k > j$ and $x_k > y_k$. The existence of such a pair of indices is motivated by the fact that the largest index i with $x_i \neq y_i$ verifies $x_i > y_i$. Then

$$y_j > x_j \geq x_k > y_k.$$

Put $\varepsilon = \min\{y_j - x_j, x_k - y_k\}$, $\lambda = 1 - \varepsilon/(y_j - y_k)$ and

$$\mathbf{y}^* = (y_1, \ldots, y_{j-1}, y_j - \varepsilon, y_{j+1}, \ldots, y_{k-1}, y_k + \varepsilon, y_{k+1}, \ldots, y_N).$$

Clearly, $\lambda \in (0,1)$. Denoting by Q the permutation matrix which interchanges the components of order j and k, we see that $\mathbf{y}^* = T\mathbf{y}$ for the representation

$$T = \lambda I + (1 - \lambda) Q.$$

According to the implication $(b) \Rightarrow (a)$, it follows that $\mathbf{y}^* \prec_{HLP} \mathbf{y}$. On the other hand, $\mathbf{x} \prec_{HLP} \mathbf{y}^*$. In fact,

$$\sum_{r=1}^{s} y_r^* = \sum_{r=1}^{s} y_r \geq \sum_{r=1}^{s} x_r \quad \text{for } s = 1, \ldots, j-1,$$

$$y_j^* \geq x_j \quad \text{and} \quad y_r^* = y_r \quad \text{for } r = j+1, \ldots, k-1,$$

$$\sum_{r=1}^{s} y_r^* = \sum_{r=1}^{s} y_r \geq \sum_{r=1}^{s} x_r \quad \text{for } s = k+1, \ldots, N$$

and

$$\sum_{r=1}^{N} y_r^* = \sum_{r=1}^{N} y_r = \sum_{r=1}^{N} x_r.$$

Letting $d(\mathbf{u}, \mathbf{v})$ be the number of indices r such that $u_r \neq v_r$, it is clear that $d(\mathbf{x}, \mathbf{y}^*) \leq d(\mathbf{x}, \mathbf{y}) - 1$, so by repeating the above algorithm (at most) $N-1$ times, we arrive at $\mathbf{x}$. ■

The next two results need some preparation concerning the action of the permutation group on the Euclidean space.

The permutation group of order N is the group $\Pi(N)$ of all bijective functions from $\{1, ..., N\}$ onto itself. This group acts on $\mathbb{R}^N$ via the map $\Psi : \Pi(N) \times \mathbb{R}^N \to \mathbb{R}^N$, defined by the formula

$$\Psi(\pi, \mathbf{x}) = \pi\mathbf{x} = (x_{\pi(1)}, ..., x_{\pi(N)}).$$

The orbits of this action, that is, the sets of the form $\mathcal{O}(\mathbf{x}) = \{\pi\mathbf{x} : \pi \in \Pi(N)\}$, play an important role in majorization theory.

4.1.6 Definition *A subset C of $\mathbb{R}^N$ is called* invariant under permutations *(or $\Pi(N)$-invariant) if $\pi\mathbf{x} \in C$ whenever $\pi \in \Pi(N)$ and $\mathbf{x} \in C$. Accordingly, a function F defined on a $\Pi(N)$-invariant subset C is called* $\Pi(N)$-invariant *(or* invariant under permutations*) if $F(\pi\mathbf{x}) = F(\mathbf{x})$ whenever $\pi \in \Pi(N)$ and $\mathbf{x} \in C$.*

For example, all elementary symmetric functions (as well as all norms of index $p \in [1, \infty]$) are invariant under permutations. Moreover, to every convex function $\varphi : \mathbb{R}^N \to \mathbb{R}$ one can attach a convex function $\varphi_\Pi : \mathbb{R}^N \to \mathbb{R}$ invariant under permutations via the formula

$$\varphi_\Pi(\mathbf{x}) = \sum_{\pi \in \Pi(N)} \varphi(\pi\mathbf{x}).$$

A geometric insight into majorization was revealed by R. Rado, who noticed that $\mathbf{x} \prec_{HLP} \mathbf{y}$ means that the components of $\mathbf{x}$ spread out less than those of $\mathbf{y}$ in the sense that $\mathbf{x}$ lies in the convex hull of the $N!$ permutations of y.

4.1.7 Theorem (R. Rado [412]) *$\mathbf{x} \prec_{HLP} \mathbf{y}$ in $\mathbb{R}^N$ if and only if $\mathbf{x}$ belongs to the convex hull of the $N!$ permutations of $\mathbf{y}$. Therefore,*

$$\{\mathbf{x} : \mathbf{x} \prec_{HLP} \mathbf{y}\} = \operatorname{conv}\{\pi\mathbf{y} : \pi \in \Pi(N)\}.$$

Proof. According to Theorem 4.1.5, if $\mathbf{x} \prec_{HLP} \mathbf{y}$, then $\mathbf{x} = P\mathbf{y}$ for some doubly stochastic matrix. Taking into account Birkhoff's Theorem 2.3.6, P can be represented as a convex combination $P = \sum_\pi \lambda_\pi P_\pi$ of the $N!$ permutation matrices P_π. Then

$$\mathbf{x} = \sum_\pi \lambda_\pi P_\pi(\mathbf{y}) \in \operatorname{conv}\{P_\pi(\mathbf{y}) : \pi \in \Pi(N)\}.$$

Conversely, if $\mathbf{x} \in \operatorname{conv}\{P_\pi(\mathbf{y}) : \pi \in \Pi(N)\}$, then $\mathbf{x}$ admits a convex representation of the form $\mathbf{x} = \sum_\pi \lambda_\pi P_\pi(\mathbf{y})$, whence $\mathbf{x} = \left(\sum_\pi \lambda_\pi P_\pi\right)(\mathbf{y})$. ■

Remarkably, the relation of majorization gives rise to inequalities of the type (4.1) not only for the continuous convex functions of the form

$$F(x_1, ..., x_N) = \sum_{k=1}^N f(x_k),$$

but also for all quasiconvex functions $F(x_1, ..., x_N)$ which are invariant under the action of the permutation group.

4.1.8 Theorem (I. Schur [439]) *If C is a convex set in $\mathbb{R}^N$ invariant under permutations and $F : C \to \mathbb{R}$ is a quasiconvex function invariant under permutations, then*

$$\mathbf{x} \prec_{HLP} \mathbf{y} \text{ implies } F(\mathbf{x}) \leq F(\mathbf{y}).$$

Proof. Indeed, according to Theorem 4.1.7,

$$\begin{aligned} F(\mathbf{x}) &\leq \sup\{F(\mathbf{u}) : \mathbf{u} \in \operatorname{conv}\{\pi\mathbf{y} : \pi \in \Pi(N)\}\} \\ &= \sup\{F(\pi\mathbf{y}) : \pi \in \Pi(N)\} = F(\mathbf{y}). \end{aligned}$$

■

Some simple examples of positive quasiconvex functions to which this remark applies are:

$$\begin{aligned} &f(x, y) = (x \log x)/y \quad \text{for } x \geq 1,\ y > 0 \\ &f(x, y) = 1/\sqrt{xy} \quad \text{for } x, y > 0 \\ &f(x, y) = (ax^\alpha + by^\alpha)^{1/\alpha} \quad \text{for } x, y \geq 0 \ (a, b > 0,\ \alpha > 0). \end{aligned}$$

An illustration of Theorem 4.1.8 is offered by the following result due to R. F. Muirhead [341]:

4.1.9 Theorem (Muirhead's inequality) *If $\mathbf{x}$ and $\mathbf{y}$ belong to $\mathbb{R}^N$ and $\mathbf{x} \prec_{HLP} \mathbf{y}$, then*

$$\sum_{\pi \in \Pi(N)} \alpha_{\pi(1)}^{x_1} \cdots \alpha_{\pi(N)}^{x_N} \leq \sum_{\pi \in \Pi(N)} \alpha_{\pi(1)}^{y_1} \cdots \alpha_{\pi(N)}^{y_N}, \tag{4.3}$$

for every $\alpha_1, \ldots, \alpha_N > 0$, the sum being taken over all permutations π of the set $\{1, \ldots, N\}$.

Actually, Muirhead has considered only the case where $\mathbf{x}$ and $\mathbf{y}$ have positive integer components; the extension to the case of real components is due to G. H. Hardy, J. E. Littlewood, and G. Pólya [209].

Proof. Put $\mathbf{w} = (\log a_1, ..., \log a_N)$. Then we have to prove that

$$\sum_{\pi \in \Pi(N)} \mathrm{e}^{\langle \mathbf{x}, \pi \mathbf{w} \rangle} \leq \sum_{\pi \in \Pi(N)} \mathrm{e}^{\langle \mathbf{y}, \pi \mathbf{w} \rangle}.$$

This follows from Theorem 4.1.8 because the function $\mathbf{u} \to \sum_{\pi \in \Pi(N)} \mathrm{e}^{\langle \mathbf{u}, \pi \mathbf{w} \rangle}$ is convex and invariant under permutations. ∎

The converse of Theorem 4.1.9 also works: *If the inequality* (4.3) *is valid for all* $\alpha_1, \ldots, \alpha_N > 0$, *then* $\mathbf{x} \prec_{HLP} \mathbf{y}$. Indeed, the case where $\alpha_1 = \cdots = \alpha_N > 0$ gives us

$$\alpha_1^{\sum_{k=1}^N x_k} \leq \alpha_1^{\sum_{k=1}^N y_k},$$

so that $\sum_{k=1}^N x_k = \sum_{k=1}^N y_k$ since $\alpha_1 > 0$ is arbitrary. Denote by $\mathcal{P}$ the set of all subsets of $\{1, \ldots, N\}$ of size k and take $\alpha_1 = \cdots = \alpha_k > 1$ and $\alpha_{k+1} = \cdots = \alpha_N = 1$. By our hypotheses,

$$\sum_{S \in \mathcal{P}} \alpha_1^{\sum_{i \in S} x_i} \leq \sum_{S \in \mathcal{P}} \alpha_1^{\sum_{i \in S} y_i}.$$

If $\sum_{j=1}^k x_j^{\downarrow} > \sum_{j=1}^k y_j^{\downarrow}$, this leads to a contradiction for α_1 large enough. Thus $\mathbf{x} \prec_{HLP} \mathbf{y}$.

See Exercise 5, for an application of Muirhead's inequality.

We end this section with a short discussion about the following multiplicative analogue of the Hardy–Littlewood–Pólya inequality of majorization:

4.1.10 Theorem *Suppose that* $x_1 \geq x_2 \geq \cdots \geq x_n$ *and* $y_1 \geq y_2 \geq \cdots \geq y_n$ *are two families of numbers in a subinterval* I *of* $(0, \infty)$ *such that*

$$\begin{aligned} x_1 &\leq y_1 \\ x_1 x_2 &\leq y_1 y_2 \\ &\cdots \\ x_1 x_2 \cdots x_{n-1} &\leq y_1 y_2 \cdots y_{n-1} \\ x_1 x_2 \cdots x_n &= y_1 y_2 \cdots y_n. \end{aligned}$$

Then

$$f(x_1) + f(x_2) + \cdots + f(x_n) \leq f(y_1) + f(y_2) + \cdots + f(y_n)$$

for every continuous convex function $f : I \to \mathbb{R}$.

Weyl's inequality for singular numbers, stated below, gives us the basic example of a pair of sequences satisfying the hypothesis of Theorem 4.1.10; the fact that all examples come this way was noted by A. Horn [221].

4.1.11 Theorem (H. Weyl [485]) *Consider a matrix $A \in M_N(\mathbb{C})$ having the eigenvalues $\lambda_1, \ldots, \lambda_N$ and the singular numbers $s_1, \ldots, s_N$, and assume that they are rearranged such that $|\lambda_1| \geq \cdots \geq |\lambda_N|$, and $s_1 \geq \cdots \geq s_N$. Then:*

$$\prod_{k=1}^{m} |\lambda_k| \leq \prod_{k=1}^{m} s_k \quad \text{for } m = 1, \ldots, N \quad \text{and} \quad \prod_{k=1}^{N} |\lambda_k| = \prod_{k=1}^{N} s_k.$$

Proof. The fact that $|\lambda_1| \leq \|A\| = s_1$ is clear. The other inequalities follow by applying this argument to the antisymmetric tensor powers $\wedge^k A$. See R. Bhatia [49] , for details. As concerns the equality $\prod_{k=1}^{N} |\lambda_k| = \prod_{k=1}^{N} s_k$, notice that each side equals $(\det(A^*A))^{1/2}$. ■

An immediate consequence of Theorem 4.1.11, also due to H. Weyl, is as follows:

4.1.12 Corollary *Let $A \in M_N(\mathbb{C})$ be a matrix having the eigenvalues $\lambda_1, \ldots, \lambda_N$ and the singular numbers $s_1, \ldots, s_N$, listed such that $|\lambda_1| \geq \cdots \geq |\lambda_N|$ and $s_1 \geq \cdots \geq s_N$. Then*

$$\sum_{k=1}^{m} \varphi(|\lambda_k|) \leq \sum_{k=1}^{m} \varphi(s_k)$$

for every number $m \in \{1, ..., N\}$ and every continuous function $\varphi : \mathbb{R}_+ \to \mathbb{R}_+$ such that $\varphi(\mathrm{e}^t)$ is increasing and convex in t.

In particular,

$$\sum_{k=1}^{m} |\lambda_k|^p \leq \sum_{k=1}^{m} s_k^p \quad \text{for all } p > 0 \text{ and } m \in \{1, ..., N\}.$$

Proof. Combine Weyl's inequality for singular numbers with Corollary 4.1.4. ■

Exercises

1. Infer from the Hardy–Littlewood–Pólya inequality of majorization the following inequality due to M. Petrović: if f is a convex function defined on $[0, \infty)$, then

$$\sum_{k=1}^{N} f(x_k) \leq f\left(\sum_{k=1}^{N} x_k\right) + (N-1)f(0)$$

for all $x_1, \ldots, x_N \geq 0$. As an application show that

$$\sum_{1 \leq i < j \leq N} x_i x_j \left(x_i^2 + x_j^2\right) \leq \frac{1}{8}\left(\sum_{k=1}^{N} x_k\right)^4,$$

whenever $x_1, x_2, ..., x_n \geq 0$ with $n \geq 2$.

2. (R. Bellman) Suppose that $a_1 \geq a_2 \geq \cdots \geq a_N > 0$ and f is a convex function defined on $[0, a_1]$ such that $\left((-1)^N + 1\right) f(0) \leq 0$. Prove that

$$\sum_{k=1}^{N}(-1)^{k-1}f(a_k) \geq f\Big(\sum_{k=1}^{N}(-1)^{k-1}a_k\Big).$$

3. Use the relation $(1/N, 1/N, \ldots, 1/N) \prec_{HLP} (1, 0, \ldots, 0)$ to infer the *AM-GM* inequality from the Muirhead inequality of majorization.

4. Prove that a positive matrix $P \in \mathrm{M}_N(\mathbb{R})$ is doubly stochastic if, and only if, $P\mathbf{x} \prec_{HLP} \mathbf{x}$ for every $\mathbf{x} \in \mathbb{R}^N$.

5. Use Muirhead's inequality of majorization to prove that every completely monotonic function f verifies *Fink's inequalities* [163],

$$(-1)^{nk}(f^{(k)}(x))^n \leq (-1)^{nk}(f^{(n)}(x))^k(f(x))^{n-k}$$

for all $x > 0$ and all integers n, k with $n \geq k \geq 0$.

[*Hint:* See the integral representation (1.13) of completely monotonic functions.]

Remark. A. M. Fink [163] also noticed that this inequality yields similar inequalities for moments: if μ is a positive Borel measure that admits moments of all orders,

$$\mu_n = \int_0^1 x^n \mathrm{d}\mu(x), \quad n = 0, 1, 2, \ldots,$$

then $\prod_{k=1}^N \mu_{\alpha_k} \leq \prod_{k=1}^N \mu_{\beta_k}$, provided that $(\alpha_1, \ldots, \alpha_N) \prec_{HLP} (\beta_1, \ldots, \beta_N)$.

6. (An order-free characterization of majorization). Adapt the proof of the Hardy–Littlewood–Pólya inequality of majorization to show that $(x_1, \ldots, x_N) \prec_{HLP} (y_1, \ldots, y_N)$ if and only if $\sum_{k=1}^N x_k = \sum_{k=1}^N y_k$ and

$$\sum_{k=1}^{N}(x_k - t)^+ \leq \sum_{k=1}^{N}(y_k - t)^+ \quad \text{for all } t \in \mathbb{R}.$$

Here the positive part can be replaced by the absolute value function.

7. (Q. J. Zhu; see [74]) Prove that $(x_1, \ldots, x_N) \prec_{HLP} (y_1, \ldots, y_N)$ if and only if for every $(z_1, \ldots, z_N) \in \mathbb{R}^N$ we have

$$\sum_{k=1}^{N} x_k^{\downarrow} z_k^{\downarrow} \leq \sum_{k=1}^{N} y_k^{\downarrow} z_k^{\downarrow}.$$

[*Hint:* Use Abel's partial summation formula.]

8. (S. M. Malamud [297]) The convex hull of a set A of real numbers (denoted by $\operatorname{conv}(A)$) is the line segment joining the minimal and maximal elements of A. Prove that $(x_1, \dots, x_N) \prec_{HLP} (y_1, \dots, y_N)$ if and only if, for $j = 1, \dots, N$,

$$\operatorname{conv}\{x_{i_1} + \cdots + x_{i_j} : 1 \leq i_1 \leq \cdots \leq i_j \leq n\} \\ \subset \operatorname{conv}\{y_{i_1} + \cdots + y_{i_j} : 1 \leq i_1 \leq \cdots \leq i_j \leq N\}.$$

This characterization of majorization extends Corollary 1.4.6.

9. (L. Fuchs' generalization of the Hardy–Littlewood–Pólya inequality [170]) Let f be a convex function defined on an interval I and consider points $x_1, \dots, x_N, y_1, \dots, y_N \in I$ and weights $p_1, \dots, p_N \in \mathbb{R}$ such that:

   (*a*) $x_1 \geq \cdots \geq x_N$ and $y_1 \geq \cdots \geq y_N$;

   (*b*) $\sum_{k=1}^{r} p_k x_k \leq \sum_{k=1}^{r} p_k y_k$ for every $r = 1, \dots, N$;

   (*c*) $\sum_{k=1}^{N} p_k x_k = \sum_{k=1}^{N} p_k y_k$.

   Prove that
$$\sum_{k=1}^{N} p_k f(x_k) \leq \sum_{k=1}^{N} p_k f(y_k).$$

   If $x_1 \leq \cdots \leq x_N$ and $y_1 \leq \cdots \leq y_N$, then the conclusion works in the reversed way.

   [*Hint:* Adapt the argument used for the proof of Hardy–Littlewood–Pólya inequality of majorization.]

10. Infer from Theorem 4.1.10 and Theorem 4.1.11 that for every $A \in M_N(\mathbb{C})$ and every integer number $n \geq 1$, one has
$$|\operatorname{trace} A^n| \leq \operatorname{trace} |A|^n ;$$
    here $|A| = (A^*A)^{1/2}$ denotes the modulus of A.

11. (Convex ordering) Given two real random variables X and Y we say that X is smaller than Y in *convex order* (abbreviated, $X \leq_{cv} Y$) if
$$E\varphi(X) \leq E\varphi(Y)$$
    for all continuous convex functions φ such that the integrals defining expectation exist; if this inequality holds for all continuous, increasing and convex functions, we say that X is smaller than Y in *increasing convex order* (abbreviated, $X \leq_{icv} Y$). Prove that:

    (a) $X \leq_{icv} Y$ if and only if $E((X - t)_+) \leq E((Y - t)_+)$ for all $t \in \mathbb{R}$.

    (b) If $X \leq_{icv} Y$ and $E(X) = E(Y)$, then $X \leq_{cv} Y$.

(c) X is smaller than Y in convex order if $E(X) = E(Y)$ and for some $t_0 \in \mathbb{R}$,

$$F(t) \leq G(t) \text{ for } t < t_0 \text{ and } F(t) \geq G(t) \text{ for } t > t_0;$$

here F and G are respectively the cumulative distributions of X and Y.

[*Hint:* (a) use the fact that the increasing convex functions defined on an interval $[a, b]$ can be uniformly approximated by linear combinations (with positive coefficients) of functions $(x - t)_+$; see Lemma 1.9.1 and Lemma 1.9.2. For (*c*), use the assertion (b) and the formula

$$E\left((X - t)_+\right) = \int_t^\infty (1 - F(s))ds.]$$

The assertion (c) of Exercise 11 is the *cut criterion* due to S. Karlin and A. Novikoff [242]. Important applications of convex ordering to mathematical risk analysis are available in the book of L. Rüschendorf [431].

4.2 The Schur–Horn Theorem

In this section we are interested in determining the possible diagonal entries of self-adjoint matrices with a fixed set of real eigenvalues. The answer will be given by the following result:

4.2.1 Theorem (The Schur–Horn theorem) *Suppose that*

$$\mathbf{d} = (d_1, ..., d_N) \text{ and } \lambda = (\lambda_1, ..., \lambda_N)$$

are two vectors in $\mathbb{R}^N$. *Then there is a real symmetric matrix with diagonal entries* $d_1, ..., d_N$ *and eigenvalues* $\lambda_1, ..., \lambda_N$ *if and only if* $\mathbf{d} \prec_{HLP} \lambda$.

Schur's contribution was the striking remark concerning the implication of majorization to this matter.

4.2.2 Lemma (I. Schur [439]) *Let* $A \in \mathrm{M}_N(\mathbb{R})$ *be a self-adjoint matrix with diagonal elements* $a_{11}, \ldots, a_{NN}$ *and eigenvalues* $\lambda_1, \ldots, \lambda_N$. *Then*

$$(a_{11}, \ldots, a_{NN}) \prec_{HLP} (\lambda_1, \ldots, \lambda_N).$$

Proof. By the spectral decomposition theorem, $A = UDU^*$, where $U = (u_{kj})_{k,j}$ is orthogonal and D is diagonal, with diagonal entries $\lambda_1, \ldots, \lambda_N$. The diagonal elements of A are

$$a_{kk} = \langle A\mathbf{e}_k, \mathbf{e}_k \rangle = \sum_{j=1}^{N} \lambda_j u_{kj}^2 = \sum_{j=1}^{N} p_{kj} \lambda_j,$$

where $p_{kj} = u_{kj}^2$; as usually, $\mathbf{e}_1, ..., \mathbf{e}_N$ denote the natural basis of $\mathbb{R}^N$. Since U is orthogonal, the matrix $P = (p_{kj})_{k,j}$ is doubly stochastic and Theorem 4.1.5 applies. ■

Since the function log is concave, from Theorem 4.1.3 and Lemma 4.2.2 we infer the following inequality:

4.2.3 Corollary (Hadamard's determinant inequality) *If A is an $N\times N$-dimensional positive matrix with diagonal elements $a_{11},\dots,a_{NN}$ and eigenvalues $\lambda_1,\dots,\lambda_N$, then*

$$\prod_{k=1}^{N} a_{kk} \geq \prod_{k=1}^{N} \lambda_k.$$

See Exercise 1 for a more general result.

A. Horn [220] has proved a converse to Lemma 4.2.2, which led to the statement of Theorem 4.2.1:

4.2.4 Lemma *If $\mathbf{x}$ and $\mathbf{y}$ are two vectors in $\mathbb{R}^N$ such that $\mathbf{x} \prec_{HLP} \mathbf{y}$, then there exists a symmetric matrix A such that the entries of $\mathbf{x}$ are the diagonal elements of A and the entries of $\mathbf{y}$ are the eigenvalues of A.*

Proof. We follow here the argument of W. Arveson and R. Kadison [22].

Step 1: If $A=(a_{ij})_{i,j}\in \mathrm{M}_N(\mathbb{R})$ is a symmetric matrix with diagonal $\mathbf{d}$, then for every T-transform T there exists a unitary matrix U such that UAU^* has diagonal $T\mathbf{d}$.

Indeed, suppose that $T=(1-\cos^2\theta)I+\left(\sin^2\theta\right)P_\pi$, where π is a permutation that interchanges i_0 and j_0. Then the matrix $U=(u_{ij})_{i,j}$, obtained by modifying four entries of the identity matrix as follows

$$\begin{aligned} u_{i_0i_0} &= \mathrm{i}\sin\theta, \quad u_{i_0j_0} = -\cos\theta,\\ u_{j_0i_0} &= \mathrm{i}\cos\theta, \quad u_{j_0j_0} = \sin\theta, \end{aligned}$$

is unitary and a straightforward computation shows that the diagonal of UAU^* is equal to $(1-\cos^2\theta)d+\left(\sin^2\theta\right)P_\pi(d)$.

Step 2: Let $\Lambda=\mathrm{Diag}(\mathbf{y})$. By Theorem 4.1.5, $\mathbf{x}$ can be obtained from $\mathbf{y}$ by successive applications of finitely many T-transformations,

$$\mathbf{x}=T_mT_{m-1}\cdots T_1\mathbf{y}.$$

By Step 1, there is a unitary matrix U_1 such that $U_1\Lambda U_1^*$ has diagonal $T_1\mathbf{y}$. Similarly, there is a unitary matrix U_2 such that $U_2\left(U_1\Lambda U_1^*\right)U_2^*$ has diagonal $T_2(T_1\mathbf{y})$. Iterating this argument, we obtain a self-adjoint matrix

$$A=(U_mU_{m-1}\cdots U_1)\,\Lambda\,(U_mU_{m-1}\cdots U_1)^*$$

whose diagonal elements are the entries of $\mathbf{x}$ and the eigenvalues are the entries of $\mathbf{y}$. ∎

The Schur–Horn theorem (when combined with Theorem 4.1.7) offers an illustration of Kostant's convexity theorem. See [258]. Indeed, if we fix an N-tuple α of real numbers and denote by $\mathcal{O}_\alpha$ the set of all symmetric matrices in $\mathrm{M}_N(\mathbb{R})$ with eigenvalues α, then the range of the map $\Phi\colon \mathcal{O}_\alpha\to\mathbb{R}^N$, that takes a matrix to its diagonal, is a convex polyhedron (whose vertices are the $N!$ permutations of α). See M. F. Atiyah [24] for a large generalization and a surprising link between mechanics, Lie group theory, and spectra of matrices.

Exercises

1. (Hadamard's determinant inequality) *Suppose that* $A = (a_{jk})_{j,k}$ *is an* $N \times N$*-dimensional complex matrix. Prove that*

$$|\det A| \leq \left(\prod_{j=1}^{N} \sum_{k=1}^{N} |a_{jk}|^2 \right)^{1/2}.$$

[*Hint:* Use Corollary 4.2.3 and the fact that $|\det A|^2 = \det(AA^*)$.]

2. (I. Schur [437]) Consider a matrix $A = (a_{ij})_{i,j=1}^N$, whose eigenvalues are $\lambda_1, \ldots, \lambda_N$. Prove that

$$\sum_{k=1}^{N} |\lambda_k|^2 \leq \sum_{j,k=1}^{N} |a_{jk}|^2 = \|A\|_{HS}^2 .$$

3. Apply the result of the preceding exercise to derive the *AM–GM* inequality.

[*Hint*: Write down an $N \times N$-dimensional matrix whose nonzero entries are $x_1, \ldots, x_N > 0$ and whose characteristic polynomial is $x^N - \prod_{k=1}^N x_k$.]

4.3 Schur-Convexity

According to Theorem 4.1.8, the quasiconvex functions $f : \mathbb{R}^N \to \mathbb{R}$ invariant under permutations are isotonic with respect to the relation of majorization, that is,

$$\mathbf{x} \prec_{HLP} \mathbf{y} \text{ implies } f(\mathbf{x}) \leq f(\mathbf{y}). \tag{4.4}$$

Simple examples (such as $f(x_1, x_2) = -x_1 x_2$ on $\mathbb{R}^2$) show that the implication (4.4) holds true beyond the framework of quasiconvex functions invariant under permutations. This led I. Schur [439] to initiate a systematic study of the functions that verify the property (4.4).

4.3.1 Definition *A function* $f : C \to \mathbb{R}$ *defined on a set invariant under permutations is called Schur-convex if*

$$\mathbf{x} \prec_{HLP} \mathbf{y} \textit{ implies } f(\mathbf{x}) \leq f(\mathbf{y});$$

if in addition $f(\mathbf{x}) < f(\mathbf{y})$ *whenever* $\mathbf{x} \prec_{HLP} \mathbf{y}$ *but* $\mathbf{x}$ *is not a permutation of* $\mathbf{y}$*, then* f *is said to be strictly Schur-convex.*

We call the function f *(strictly) Schur-concave if* $-f$ *is (strictly) Schur-convex.*

Every Schur-convex (as well as every Schur-concave) function defined on a set C invariant under permutations is a function invariant under permutations. This is a consequence of the fact that $\mathbf{x} \prec_{HLP} P_\pi(\mathbf{x})$ and $P_\pi(\mathbf{x}) \prec_{HLP} \mathbf{x}$ for every vector $\mathbf{x} \in \mathbb{R}^N$ and every permutation matrix $P_\pi \in \mathrm{M}_N(\mathbb{R})$.

The Schur-convex functions encompass a large variety of examples.

4.3.2 Example *If* $H : \mathbb{R}^N \to \mathbb{R}$ *is a function (strictly) increasing in each variable and* $f_1, ..., f_N$ *are (strictly) Schur-convex functions on* $\mathbb{R}^n$*, then the function* $h(\mathbf{x}) = H(f_1(\mathbf{x}), ..., f_N(\mathbf{x}))$ *is (strictly) Schur-convex on* $\mathbb{R}^n$*. Some particular examples of strictly Schur-convex functions are:*

$$\max\{x_1, ..., x_N\} \text{ and } \log\left(\sum\nolimits_{k=1}^N x_k^2\right) \text{ on } \mathbb{R}^N;$$

$$-\prod\nolimits_{k=1}^N x_k \text{ on } (0,\infty)^N.$$

4.3.3 Theorem (Schur–Ostrowski criterion of Schur-convexity) *Let* I *be a nonempty open interval. A differentiable function* $f : I^N \to \mathbb{R}$ *is Schur-convex if and only if it fulfils the following two conditions:*

(a) f *is invariant under permutations;*

(b) *for every* $\mathbf{x} \in I^N$ *and* $i, j \in \{1, \ldots, N\}$ *we have*

$$(x_i - x_j)\left(\frac{\partial f}{\partial x_i}(\mathbf{x}) - \frac{\partial f}{\partial x_j}(\mathbf{x})\right) \geq 0.$$

Proof. The necessity. For (a), see the comments after Definition 4.3.1. This reduces the verification of (b) to the case where $i = 1$ and $j = 2$. Fix arbitrarily $\mathbf{x} \in I^N$ and choose $\varepsilon > 0$ sufficiently small such that

$$\mathbf{x}(t) = ((1-t)x_1 + tx_2, tx_1 + (1-t)x_2, x_3, ..., x_N) \in D \tag{4.5}$$

for $t \in (0, \varepsilon]$. Then $\mathbf{x}(t) \prec_{HLP} \mathbf{x}$, which yields $f(\mathbf{x}(t)) \leq f(\mathbf{x})$. Therefore

$$0 \geq \lim_{t\to 0} \frac{f(\mathbf{x}(t)) - f(\mathbf{x})}{t} = \left.\frac{\mathrm{d}f(\mathbf{x}(t))}{\mathrm{d}t}\right|_{t=0} = -(x_1 - x_2)\left(\frac{\partial f}{\partial x_1}(\mathbf{x}) - \frac{\partial f}{\partial x_2}(\mathbf{x})\right).$$

The sufficiency. We have to prove that $\mathbf{y} \prec_{HLP} \mathbf{x}$ implies $f(\mathbf{y}) \leq f(\mathbf{x})$. According to Theorem 4.1.5, it suffices to consider the case where

$$\mathbf{y} = ((1-s)\,x_1 + sx_2, sx_1 + (1-s)x_2, x_3, ..., x_N).$$

for some $s \in [0, 1/2]$. Consider $x(t)$ as in formula (4.5). Then

$$\begin{aligned} f(\mathbf{y}) - f(\mathbf{x}) &= \int_0^s \frac{\mathrm{d}}{\mathrm{d}t} f(\mathbf{x}(t))\mathrm{d}t \\ &= -\int_0^s (x_1 - x_2)\left(\frac{\partial f}{\partial x_1}(\mathbf{x}(t)) - \frac{\partial f}{\partial x_2}(\mathbf{x}(t))\right)\mathrm{d}t \\ &= -\int_0^s \frac{\mathrm{pr}_1\,\mathbf{x}(t) - \mathrm{pr}_2\,\mathbf{x}(t)}{1-2t}\left(\frac{\partial f}{\partial x_1}(\mathbf{x}(t)) - \frac{\partial f}{\partial x_2}(\mathbf{x}(t))\right)\mathrm{d}t, \end{aligned}$$

where pr_k denotes the projection to the k-th coordinate. According to condition (b), $f(\mathbf{y}) - f(\mathbf{x}) \leq 0$, and the proof is done. ■

The elementary symmetric functions of N variables are defined by the formulas

$$\begin{aligned}
e_0(x_1, x_2, \ldots, x_N) &= 1 \\
e_1(x_1, x_2, \ldots, x_N) &= x_1 + x_2 + \cdots + x_N \\
&\vdots \\
e_k(x_1, x_2, \ldots, x_N) &= \sum_{1 \leq i_1 < \cdots < i_k \leq N} x_{i_1} \cdots x_{i_k} \\
&\vdots \\
e_N(x_1, x_2, \ldots, x_N) &= x_1 x_2 \cdots x_N.
\end{aligned}$$

Clearly, they are invariant under permutations. A small computation shows that

$$\frac{\partial}{\partial x_i} e_k(x_1, \ldots, x_N) = e_{k-1}(x_1, ..., \widehat{x_i}, \ldots, x_N)$$

and

$$\frac{\partial}{\partial x_i} e_k(x_1, \ldots, x_N) - \frac{\partial}{\partial x_j} e_k(x_1, \ldots, x_N) = -(x_i - x_j) e_{k-2}(x_1, ..., \widehat{x_i}, \ldots, \widehat{x_j}, ..., x_N),$$

where the cup indicates the omission of the coordinate underneath. This leads us to the following consequence of Theorem 4.3.3:

4.3.4 Corollary (I. Schur) *All elementary symmetric functions $e_k(x_1, x_2, \ldots, x_N)$ of N variables are Schur-concave on $\mathbb{R}_+^N$.*

Since

$$\left(\frac{x_1 + \cdots + x_N}{N}, \ldots, \frac{x_1 + \cdots + x_N}{N} \right) \prec_{HLP} (x_1, \ldots, x_N)$$

for every $(x_1, \ldots, x_N) \in \mathbb{R}_+^N$, we infer from Corollary 4.3.4 that

$$e_k(x_1, x_2, \ldots, x_N) = \sum_{1 \leq i_1 < \cdots < i_k \leq N} x_{i_1} \cdots x_{i_k} \leq \binom{N}{k} \left(\frac{x_1 + \cdots + x_N}{N} \right)^k;$$

for $k = N$ we retrieve the *AM–GM* inequality. A nice application of these inequalities is mentioned in Exercise 7.

More information about the elementary symmetric functions can be found in Appendix C.

4.3.5 Remark *The relation of majorization is closely related to the duality of cones, more precisely, to the fact that the dual cone of the monotone cone* $\mathbb{R}^N_{\geq} = \{(x_1, \ldots, x_N) \in \mathbb{R}^N : x_1 \geq \cdots \geq x_N\}$ *is the cone*

$$\left(\mathbb{R}^N_{\geq}\right)^* = \left\{\mathbf{y} \in \mathbb{R}^N : \sum_{k=1}^{m} y_k \geq 0 \text{ for } m = 1, 2, ..., N-1, \text{ and } \sum_{k=1}^{N} y_k = 0\right\}.$$

See Exercise 11, Section 2.3. Indeed, if $\mathbf{x}, \mathbf{y} \in \mathbb{R}^N_{\geq}$, *then*

$$\mathbf{x} \prec_{HLP} \mathbf{y} \quad \text{if, and only if,} \quad \mathbf{y} - \mathbf{x} \in \left(\mathbb{R}^N_{\geq}\right)^*.$$

This leads to the more general concept of majorization with respect to a convex cone C *in a real vector space* V,

$$\mathbf{x} \prec_C \mathbf{y} \quad \text{if, and only if,} \quad \mathbf{y} - \mathbf{x} \in C^*,$$

and, implicitly, to a generalization of Schur convexity. In the case of self-dual cones C *(like* $\mathbb{R}^N_+$ *and* $\mathrm{Sym}^+(n, \mathbb{R})$*), the corresponding concept of Schur convexity coincides with that of a function which is monotone increasing on* C.

Exercises

1. Suppose that $f : [0, \infty) \to (0, \infty)$ is a log-convex function. Prove that $f(x) + f(y)$ and $f(x)f(y)$ are Schur-convex functions on $[0, \infty) \times [0, \infty)$. Infer that for every $0 < \varepsilon < x \leq y$, we have

$$f(x)f(y) \leq f(x - \varepsilon)f(y + \varepsilon) \leq f(0)f(x + y).$$

2. Prove that the area of a triangle is a Schur-concave function of its sides. Also, the radius of the circumscribed circle of a triangle is a Schur-convex function of its sides.

3. Suppose that f is a real-valued convex function defined on an interval I.

   (a) Prove that the function

$$F : I^N \to \mathbb{R}^N, \quad F(x_1, ..., x_N) = (f(x_1), ..., f(x_N))$$

   verifies the condition

$$\mathbf{x} \prec_{HLP} \mathbf{y} \text{ implies } F(\mathbf{x}) \prec_{HLPw} F(\mathbf{y}).$$

   (b) Suppose that f is convex and increasing. Prove that in this case $\mathbf{x} \prec_{HLP} \mathbf{y}$ implies $F(\mathbf{x}) \prec_{HLP} F(\mathbf{y})$.

   (c) Illustrate with examples the assertions (a) and (b).

4. Consider the Laplace transform of a Borel probability measure μ on $\mathbb{R}$,

$$\mathcal{L}(s) = \int_{\mathbb{R}^N} \mathrm{e}^{sx} \mathrm{d}\mu(x).$$

Prove that the function $f(s_1, ..., s_N) = \prod_{k=1}^{N} \mathcal{L}(s_k)$ is Schur-convex on $[0, \infty)^N$.

5. (D. B. Hunter) The complete homogeneous symmetric polynomials of N variables are defined by the formulas

$$h_0(x_1, .., x_N) = 0$$
$$h_d(x_1, \ldots, x_N) = \sum_{1 \leq i_1 \leq \cdots \leq i_d \leq N} x_{i_1} \cdots x_{i_d} \quad \text{for } d \geq 1.$$

Prove that all such polynomials of even degree d are Schur-convex and positive on $\mathbb{R}^N$.

[*Hint:* Use mathematical induction and the identity

$$(x_i - x_j) \left[\frac{\partial h_d}{\partial x_i}(x_1, \ldots, x_N) - \frac{\partial h_d}{\partial x_j}(x_1, \ldots, x_N) \right]$$
$$= (x_i - x_j)^2 h_{d-2}(x_1, \ldots, x_N).\,]$$

6. (A. W. Marshall and I. Olkin) Suppose that φ and ψ are Schur-concave functions defined on $\mathbb{R}^N$. Prove that their convolution product

$$(\varphi * \psi)(\mathbf{x}) = \int_{\mathbb{R}^N} \varphi(\mathbf{x} - \mathbf{y}) \psi(y) \mathrm{d}\mathbf{y}$$

is Schur-concave (whenever the integral exists).

7. (The birthday inequality) Let the n-tuple $\mathbf{p} = (p_1, ..., p_n)$ be any probability vector of birthdays for the $n = 365$ days of the year. Given $\mathbf{p}$, let $P_k(\mathbf{p})$ be the probability of at least one match among k people. Prove that

$$P_k(\mathbf{p}) \geq 1 - \frac{n(n-1)\cdots(n-k+1)}{n^k},$$

along the following steps:

(a) Let $Q_k(\mathbf{p}) = 1 - P_k(\mathbf{p})$ be the probability of the complementary event that all k birthdays are different. Show that $Q_k(\mathbf{p}) = \sum p_{i_1} \cdots p_{i_k}$, where the sum is taken over all strings $i_1, ..., i_k$ of distinct integers in $\{1, ..., n\}$.

(b) Prove that $Q_k(\mathbf{p}) = k! e_k(p_1, ..., p_n)$.

(c) Compute $P_k(\mathbf{u})$, where $\mathbf{u} = (1/n, ..., 1/n)$ is the uniform distribution of the birthdays.

(d) Conclude the proof by using Corollary 4.3.4.

8. (D. E. Wulbert) Suppose that f is a continuous function defined on an interval I. Prove that the symmetric function

$$F(x,y) = \frac{1}{y-x}\int_x^y f(t)dt \text{ for } x,y \in I,\ x \neq y$$
$$F(x,y) = f(x) \text{ for } x = y \in I$$

is convex if and only if the function f is convex.

9. (An example of Schur-convex like function due to V. Senderov) Let $f : [0,\infty)^N \to \mathbb{R}$ be a function such that

$$f(s/N, ..., s/N) \leq f(x_1, \dots, x_N) \leq f(s, 0, ..., 0)$$

whenever $s \geq 0$ and $x_1 + \cdots + x_N = s$. Every Schur-convex function f verifies this property, but the converse is not true in general. Illustrate this fact using the polynomial function $4x_1^3 + 4x_2^3 + 4x_3^3 + 15x_1x_2x_3$.

4.4 Eigenvalue Inequalities

According to the spectral decomposition, for every matrix $A \in \mathrm{Sym}(N,\mathbb{R})$ there is an orthonormal basis $(\mathbf{u}_k)_k$ of $\mathbb{R}^N$ consisting of eigenvectors of A. Assuming $A\mathbf{u}_k = \lambda_k(A)\mathbf{u}_k$ for each index k, this fact gives rise to a representation of the form

$$A = \sum_{k=1}^N \lambda_k(A)\langle \cdot, \mathbf{u}_k\rangle \mathbf{u}_k. \tag{4.6}$$

In what follows the downwards/upwards rearrangements of the eigenvalues $\lambda_k(A)$ will be denoted respectively by $\lambda_k^{\downarrow}(A)$ and $\lambda_k^{\uparrow}(A)$. Of a special interest is the *eigenvalues map,*

$$\Lambda : \mathrm{Sym}(N,\mathbb{R}) \to \mathbb{R}^N, \quad \Lambda(A) = \lambda^{\downarrow}(A) = \left(\lambda_1^{\downarrow}(A), ..., \lambda_N^{\downarrow}(A)\right).$$

The representation (4.6) easily yields the equalities

$$\lambda_1^{\downarrow}(A) = \max_{\|\mathbf{x}\|=1} \langle A\mathbf{x}, \mathbf{x}\rangle \tag{4.7}$$

and

$$\lambda_N^{\downarrow}(A) = \min_{\|\mathbf{x}\|=1} \langle A\mathbf{x}, \mathbf{x}\rangle. \tag{4.8}$$

Therefore

$$\{\langle A\mathbf{x}, \mathbf{x}\rangle : \mathbf{x} \in \mathbb{R}^N, \|\mathbf{x}\| = 1\} = [\lambda_N^{\downarrow}(A), \lambda_1^{\downarrow}(A)], \tag{4.9}$$

due to the fact that the function $\mathbf{x} \to \langle A\mathbf{x}, \mathbf{x}\rangle$ is continuous and the unit sphere is compact and connected.

The equalities (4.7) and (4.8) also imply the following result:

4.4.1 Proposition *If A and B belong to* $\mathrm{Sym}(N,\mathbb{R})$, *then*

$$\lambda_1^{\downarrow}(A+B) \leq \lambda_1^{\downarrow}(A) + \lambda_1^{\downarrow}(B)$$

and

$$\lambda_N^{\downarrow}(A+B) \geq \lambda_N^{\downarrow}(A) + \lambda_N^{\downarrow}(B).$$

Thus the function $\lambda_1^{\downarrow}(A)$ is positively homogeneous and subadditive (in particular, convex), while the function $\lambda_N^{\downarrow}(A)$ is positively homogeneous and superadditive (in particular, concave). These two conclusions are equivalent since

$$\lambda_k^{\downarrow}(-A) = -\lambda_{N-k+1}^{\downarrow}(A) = -\lambda_k^{\uparrow}(A). \tag{4.10}$$

The formulas (4.7) and (4.8) admit analogues for all indices k:

4.4.2 Theorem (Courant–Fischer minimax principle) *If $A \in \mathrm{Sym}(N,\mathbb{R})$, then its eigenvalues can be computed by the formulas*

$$\lambda_k^{\downarrow}(A) = \max_{\substack{V\subset\mathbb{R}^N \\ \dim V=k}} \min_{\substack{\mathbf{x}\in V \\ \|\mathbf{x}\|=1}} \langle A\mathbf{x},\mathbf{x}\rangle = \min_{\substack{V\subset\mathbb{R}^N \\ \dim V=N-k+1}} \max_{\substack{\mathbf{x}\in V \\ \|\mathbf{x}\|=1}} \langle A\mathbf{x},\mathbf{x}\rangle.$$

Equivalently, taking into account (4.10),

$$\lambda_k^{\uparrow}(A) = \min_{\substack{V\subset\mathbb{R}^N \\ \dim V=k}} \max_{\substack{\mathbf{x}\in V \\ \|\mathbf{x}\|=1}} \langle A\mathbf{x},\mathbf{x}\rangle = \max_{\substack{V\subset\mathbb{R}^N \\ \dim V=N-k+1}} \min_{\substack{\mathbf{x}\in V \\ \|\mathbf{x}\|=1}} \langle A\mathbf{x},\mathbf{x}\rangle.$$

Proof. Let $\mathbf{u}_1,\ldots,\mathbf{u}_N$ be the orthonormal basis which appears in the spectral representation (4.6) of A. The vector space $W = \mathrm{span}\{\mathbf{u}_k,\mathbf{u}_{k+1},\ldots,\mathbf{u}_N\}$ is $(N-k+1)$-dimensional and thus every k-dimensional vector subspace $V \subset \mathbb{R}^N$ will contain a point $\mathbf{z} \in W \cap V$ with $\|\mathbf{z}\| = 1$. According to (4.9),

$$\langle A\mathbf{z},\mathbf{z}\rangle \in [\lambda_N^{\downarrow}(A), \lambda_k^{\downarrow}(A)],$$

from which it follows that

$$\min_{\substack{\mathbf{x}\in V \\ \|\mathbf{x}\|=1}} \langle A\mathbf{x},\mathbf{x}\rangle \leq \lambda_k^{\downarrow}(A).$$

Finally, note that equality occurs for $V = \mathrm{span}\{\mathbf{u}_1,\ldots,\mathbf{u}_k\}$. ∎

4.4.3 Corollary (Weyl's monotonicity principle) *If A and B belong to* $\mathrm{Sym}(N,\mathbb{R})$ *and $A \leq B$, then $\lambda_k^{\downarrow}(A) \leq \lambda_k^{\downarrow}(B)$ for every index k.*

If $A \in \mathrm{Sym}(N,\mathbb{R})$ and P_W is the orthogonal projection of $\mathbb{R}^N$ onto the m-dimensional vector subspace W, we call

$$B = P_W A P_W$$

the *compression* of A to W. For example, the principal submatrix obtained from A by deleting the last row and last column is a compression of A, corresponding to the linear span W of the first $N-1$ vectors of the canonical basis of $\mathbb{R}^N$.

4.4.4 Corollary (Cauchy's interlace theorem) *Let $A \in \mathrm{Sym}(N,\mathbb{R})$ be a symmetric matrix and let $B = P_W A P_W$ be a compression of it. If $m = \dim W$, then*

$$\lambda_k^{\uparrow}(A) \le \lambda_k^{\uparrow}(B) \le \lambda_{N-m+k}^{\uparrow}(A) \quad \textit{for every } k \in \{1,...,m\}. \tag{4.11}$$

Proof. Denote by $\mathbf{w}_1,...,\mathbf{w}_m$ an orthonormal basis of W consisting of eigenvectors of B such that $B\mathbf{w}_i = \lambda_i^{\uparrow}(B)\mathbf{w}_i$ for every i. Put $W_k = \mathrm{span}\,\{w_1,...,w_k\}$. According to the Courant–Fischer minimax theorem,

$$\begin{aligned}\lambda_k^{\uparrow}(B) &= \max_{\mathbf{x}\in W_k,\ \|\mathbf{x}\|=1} \langle B\mathbf{x},\mathbf{x}\rangle = \max_{\mathbf{x}\in W_k,\ \|\mathbf{x}\|=1} \langle AP_W\mathbf{x}, P_W\mathbf{x}\rangle \\ &= \max_{\mathbf{x}\in W_k,\ \|\mathbf{x}\|=1} \langle A\mathbf{x},\mathbf{x}\rangle \ge \lambda_k^{\uparrow}(A).\end{aligned}$$

In a similar way, by considering the space $\widetilde{W}_k = \mathrm{span}\,\{w_k,...,w_m\}$,

$$\begin{aligned}\lambda_k^{\uparrow}(B) &= \min_{\mathbf{x}\in \widetilde{W}_k,\ \|\mathbf{x}\|=1} \langle B\mathbf{x},\mathbf{x}\rangle = \min_{\mathbf{x}\in \widetilde{W}_k,\ \|\mathbf{x}\|=1} \langle AP_W\mathbf{x}, P_W\mathbf{x}\rangle \\ &= \min_{\mathbf{x}\in \widetilde{W}_k,\ \|\mathbf{x}\|=1} \langle A\mathbf{x},\mathbf{x}\rangle \le \lambda_{N-m+k}^{\uparrow}(A).\end{aligned}$$

■

One can prove (see [223]) that the converse of Corollary 4.4.4 is also true, that is, for any two sequences $(\lambda_k)_{k=1}^N$ and $(\mu_k)_{k=1}^{N-1}$ of real numbers, satisfying

$$\lambda_1 \le \mu_1 \le \lambda_2 \le \mu_2 \le \cdots \le \lambda_{N-1} \le \mu_{N-1} \le \lambda_N,$$

there exists a matrix $A \in \mathrm{Sym}(N,\mathbb{R})$ such that $\sigma(A) = (\lambda_k)_{k=1}^N$ and $\sigma(A_{N-1}) = (\mu_k)_{k=1}^{N-1}$, where A_{N-1} is the principal submatrix obtained from A by deleting the last row and column. We say that such a matrix A solves the inverse spectral problem for these sequences.

4.4.5 Theorem (Weyl's inequalities) *If A and B belong to* $\mathrm{Sym}(N,\mathbb{R})$, *then*

$$\lambda_{i+j-1}^{\downarrow}(A+B) \le \lambda_i^{\downarrow}(A) + \lambda_j^{\downarrow}(B) \quad \textit{if } i+j-1 \le N.$$

Proof. Suppose that A, B, and $A+B$ have the spectral representations

$$A = \sum_{k=1}^N \lambda_k^{\downarrow}(A)\langle\cdot,\mathbf{u}_k\rangle\mathbf{u}_k,\ B = \sum_{k=1}^N \lambda_k^{\downarrow}(B)\langle\cdot,\mathbf{v}_k\rangle\mathbf{v}_k,\ A+B = \sum_{k=1}^N \lambda_k^{\downarrow}(C)\langle\cdot,\mathbf{w}_k\rangle\mathbf{w}_k.$$

Since

$$\dim\mathrm{span}\{\mathbf{u}_i,\dots,\mathbf{u}_N\} + \dim\mathrm{span}\{\mathbf{v}_j,\dots,\mathbf{v}_N\} + \dim\mathrm{span}\{\mathbf{w}_1,\dots,\mathbf{w}_{i+j-1}\}$$

equals $(N-i+1)+(N-j+1)+(i+j-1) = 2N+1$, the spaces

$$\mathrm{span}\{\mathbf{u}_i,\dots,\mathbf{u}_N\},\ \mathrm{span}\{\mathbf{v}_j,\dots,\mathbf{v}_N\} \text{ and } \mathrm{span}\{\mathbf{w}_1,\dots,\mathbf{w}_{i+j-1}\}$$

must have in common a vector $\mathbf{x}$ with $\|\mathbf{x}\| = 1$. Then, according to (4.7) and (4.8),

$$\langle A\mathbf{x}, \mathbf{x}\rangle \le \lambda_i^{\downarrow}(A), \quad \langle B\mathbf{x}, \mathbf{x}\rangle \le \lambda_j^{\downarrow}(B), \quad \langle (A+B)\mathbf{x}, \mathbf{x}\rangle \ge \lambda_{i+j-1}^{\downarrow}(C)$$

and the proof is complete. ■

4.4.6 Corollary *The following inequalities hold:*

$$\lambda_i^{\downarrow}(A) + \lambda_N^{\downarrow}(B) \le \lambda_i^{\downarrow}(A+B) \le \lambda_i^{\downarrow}(A) + \lambda_1^{\downarrow}(B) \quad \text{for } i = 1, ..., N.$$

4.4.7 Theorem (Weyl's perturbation theorem) *For every pair of matrices A, B in* $\mathrm{Sym}(N, \mathbb{R})$, *we have*

$$\max_{1 \le k \le n} |\lambda_k(A) - \lambda_k(B)| \le \|A - B\|.$$

In particular, the eigenvalues $\lambda_k(A)$ are continuous functions of A.

Proof. In fact, for every symmetric matrix A we have

$$\|A\| = \sup_{\|\mathbf{x}\|=1} |\langle A\mathbf{x}, \mathbf{x}\rangle| = \max\{|\lambda_1^{\downarrow}(A)|, |\lambda_n^{\downarrow}(A)|\}.$$

Consequently, by applying Corollary 4.4.6 to A and $B - A$, we get

$$\lambda_k(A) - \|B - A\| \le \lambda_k(B) \le \lambda_k(A) + \|B - A\|.$$

■

Weyl's perturbation theorem has applications to numerical analysis.

According to Lemma 4.2.2, if A is a symmetric matrix with diagonal elements $a_{11}, \ldots, a_{NN}$ and eigenvalues $\lambda_1, \ldots, \lambda_N$, then

$$(a_{11}, \ldots, a_{NN}) \prec_{HLP} (\lambda_1, \ldots, \lambda_N).$$

Since the spectrum is invariant under unitary equivalence of matrices, this result yields *Ky Fan's maximum principle*:

$$\sum_{k=1}^{r} \lambda_k^{\downarrow}(A) = \max_{\substack{(\mathbf{x}_k)_{k=1}^r \\ \text{orthonormal} \\ \text{family}}} \sum_{k=1}^{r} \langle A\mathbf{x}_k, \mathbf{x}_k\rangle \quad \text{for } r = 1, \ldots, N. \tag{4.12}$$

As a consequence, the sums $\sum_{k=1}^{r} \lambda_k^{\downarrow}(A)$ are *convex* functions of A and the following set of inequalities (known as *Ky Fan's inequalities*) hold:

$$\sum_{k=1}^{r} \lambda_k^{\downarrow}(A+B) \le \sum_{k=1}^{r} \lambda_k^{\downarrow}(A) + \sum_{k=1}^{r} \lambda_k^{\downarrow}(B), \quad \text{for } r = 1, \ldots, N. \tag{4.13}$$

In other words,

$$\lambda^{\downarrow}(A+B) \prec_{HLP} \lambda^{\downarrow}(A) + \lambda^{\downarrow}(B). \tag{4.14}$$

The complementary inequality,

$$\lambda^{\downarrow}(A) + \lambda^{\uparrow}(B) \prec_{HLP} \lambda(A+B), \tag{4.15}$$

also works and it was proved in an equivalent form by V. B. Lidskii [281] and later by H. Wielandt [488]:

4.4.8 Theorem (Lidskii–Wielandt inequalities) *Let A, B, C be three $N \times N$-dimensional symmetric matrices such that $C = A + B$. Then, for every $1 \leq r \leq N$ and every $1 \leq i_1 < \cdots < i_r \leq N$, we have the inequalities*

$$\sum_{k=1}^{r} \lambda_{i_k}^{\downarrow}(C) \leq \sum_{k=1}^{r} \lambda_{i_k}^{\downarrow}(A) + \sum_{k=1}^{r} \lambda_{k}^{\downarrow}(B)$$

as well as the corresponding inequalities obtained by interchanging A and B.

Proof. (C. K. Li and R. Mathias [279]) We have to prove the inequality

$$\sum_{k=1}^{r} \left[\lambda_{i_k}^{\downarrow}(A+B) - \lambda_{i_k}^{\downarrow}(A)\right] \leq \sum_{k=1}^{r} \lambda_{k}^{\downarrow}(B). \tag{4.16}$$

Without loss of generality we may assume that $\lambda_r^{\downarrow}(B) = 0$; for this, replace B by $B - \lambda_r^{\downarrow}(B) \cdot \mathrm{I}$. Let $B = B^+ - B^-$ be the canonical decomposition of B into the positive and negative part. Since $B \leq B^+$, Weyl's monotonicity principle yields $\lambda_{i_k}^{\downarrow}(A+B) \leq \lambda_{i_k}^{\downarrow}(A+B^+)$, so that the left-hand side of (4.16) is majorized by

$$\sum_{k=1}^{r} [\lambda_{i_k}^{\downarrow}(A+B^+) - \lambda_{i_k}^{\downarrow}(A)],$$

which in turn is less than or equal to

$$\sum_{k=1}^{n} [\lambda_{k}^{\downarrow}(A+B^+) - \lambda_{k}^{\downarrow}(A)] = \operatorname{trace}(B^+).$$

Or, $\operatorname{trace}(B^+) = \sum_{k=1}^{r} \lambda_k^{\downarrow}(B)$ since $\lambda_r^{\downarrow}(B) = 0$. ■

The following result is known as von Neumann's trace inequality:

4.4.9 Theorem (von Neumann [349]) *Any matrices $A, B \in \mathrm{M}_N(\mathbb{R})$ verify the inequality*

$$|\operatorname{trace}(A^* B)| \leq \langle s^{\downarrow}(A), s^{\downarrow}(B)\rangle.$$

Here $s^{\downarrow}(A) = (s_1^{\downarrow}(A), s_2^{\downarrow}(A), ..., s_N^{\downarrow}(A))$ and similarly for $s^{\downarrow}(B)$.

The proof needs the following generalization of the Hardy–Littlewood–Pólya rearrangement inequality.

4.4.10 Lemma *Suppose that* $\mathbf{x} = (x_1, ..., x_N)$ *and* $\mathbf{y} = (y_1, ..., y_N)$ *are two vectors in* $\mathbb{R}^N$ *such that* $x_1 \geq \cdots \geq x_N \geq 0$ *and* $y_1 \geq \cdots \geq y_N \geq 0$. *Then, for every doubly stochastic matrix* $D \in \mathrm{M}_N(\mathbb{R})$, *we have the inequality*

$$\langle D\mathbf{x}, \mathbf{y}\rangle \leq \langle \mathbf{x}, \mathbf{y}\rangle.$$

Proof. This is an immediate consequence of the Hardy–Littlewood–Pólya rearrangement inequality 1.9.8 and of the Birkhoff Theorem 2.3.6. ■

Proof of Theorem 4.4.9. According to the singular value decomposition theorem,

$$A = USV^* \text{ and } B = XTY^*$$

where S and T are diagonal matrices and U, V, X, Y are orthogonal matrices. Then

$$\begin{aligned}\operatorname{trace}(A^*B) &= \operatorname{trace}(VSU^*XTY^*) = \operatorname{trace}(Y^*VSU^*XT)\\ &= \operatorname{trace}(Q^*SPT),\end{aligned}$$

where $P = U^*X = (p_{ij})_{i,j}$ and $Q = V^*Y = (q_{ij})_{i,j}$ are orthogonal matrices. Thus the matrices $(p_{ij}^2)_{i,j}$ and $(q_{ij}^2)_{i,j}$ are doubly stochastic and we have

$$\begin{aligned}\operatorname{trace}(A^*B) &= \operatorname{trace}\left((SQ)^*\, PT\right)\\ &= \sum_{i,j=1}^{N} s_i(A)s_j(B)q_{ij}p_{ij}\\ &\leq \frac{1}{2}\sum_{i,j=1}^{N} s_i(A)s_j(B)q_{ij}^2 + \frac{1}{2}\sum_{i,j=1}^{N} s_i(A)s_j(B)p_{ij}^2\\ &\leq \sum_{k=1}^{N} s_k^{\downarrow}(A)s_k^{\downarrow}(B),\end{aligned}$$

the last inequality being implied by Lemma 4.4.10. ■

4.4.11 Corollary *Any matrices* $A, B \in \mathrm{Sym}(N, \mathbb{R})$ *verify the inequality.*

$$\operatorname{trace}(AB) \leq \langle \lambda^{\downarrow}(A), \lambda^{\downarrow}(B)\rangle.$$

Equality occurs if, and only if, there exists an orthogonal matrix V *such that* $V^*AV = \operatorname{Diag}\lambda^{\downarrow}(A)$ *and* $V^*BV = \operatorname{Diag}\lambda^{\downarrow}(B)$.

Proof. Choose scalars α and β such $A + \alpha\mathrm{I} \geq 0$ and $B + \beta\mathrm{I} \geq 0$ and apply Theorem 4.4.9 taking into account that $\lambda^{\downarrow}(C) = s^{\downarrow}(C)$ if $C \geq 0$. A simple proof of the equality case is available in [278], Theorem 2.2. ■

Most of the above results extend easily to the framework of compact self-adjoint operators. A good starting point in this direction is offered by the book of B. Simon [449].

Exercises

1. Recall Ky Fan's maximum principle (4.12),

$$\sum_{k=1}^{r} \lambda_k^{\downarrow}(A) = \max_P \operatorname{trace}(AP) \quad \text{for } r = 1, \ldots, N,$$

where the maximum is taken over all r-dimensional orthogonal projections P. Prove that this principle is equivalent to Schur's Lemma 4.2.2.

2. Consider a symmetric matrix $A \in \operatorname{Sym}(N, \mathbb{R})$ and its principal submatrix B, obtained from A by deleting the last row and column of A. Prove that

$$(\lambda_1^{\downarrow}(B), \ldots, \lambda_{N-1}^{\downarrow}(B), \sum\nolimits_{i=1}^{N} \lambda_1^{\downarrow}(A) - \sum\nolimits_{i=1}^{N-1} \lambda_1^{\downarrow}(B)) \prec_{HLP} (\lambda_1^{\downarrow}(A), \ldots, \lambda_N^{\downarrow}(A)).$$

3. (A determinant inequality due to M. Lin) Assuming $A, B, C \in \operatorname{Sym}^+(N, \mathbb{R})$, prove the following matrix analogue of the Hlawka inequality:

$$\det(A + B + C) + \det A + \det B + \det C \geq \det(A+B) + \det(B+C) + \det(C+A).$$

(a) Show that the inequality is equivalent to its particular case where $C = \mathrm{I}$.

(b) Prove the inequality in the case where A and B are diagonal matrices, that is,

$$\prod_{i=1}^{N}(a_i + b_i + 1) + \prod_{i=1}^{N} a_i + \prod_{i=1}^{N} b_i + 1 \geq \prod_{i=1}^{N}(a_i + b_i) + \prod_{i=1}^{N}(1 + a_i) + \prod_{i=1}^{N}(1 + b_i).$$

(c) Prove that the function $\prod_{i=1}^{N}(1 + x_i) - \prod_{i=1}^{N} x_i$ is Schur-concave on $\mathbb{R}_+^N$.

(d) Notice that

$$\det(A + B + \mathrm{I}) - \det(A + B) = \prod\nolimits_{i=1}^{N}(1 + \lambda_i(A + B)) - \prod\nolimits_{i=1}^{N} \lambda_i(A + B)$$

and conclude the proof using (in order) Ky Fan's inequality (4.14), the assertion (b) and next the assertion (a).

4. Infer from Cauchy's interlace theorem that a real symmetric matrix A is strictly positive if and only if the determinants of all its leading minors are strictly positive.

5. Consider two interlaced sequences $a_1 \geq b_1 \geq a_2 \geq \cdots \geq b_{N-1} \geq a_N$. Prove that

$$(b_1, ..., b_{N-1}, \sum\nolimits_{i=1}^{N} a_i - \sum\nolimits_{j=1}^{N-1} b_j) \prec_{HLP} (a_1, ..., a_{N-1}, a_N).$$

6. (Ky Fan) Consider a matrix $A \in \mathrm{M}_N(\mathbb{C})$ and put $\operatorname{Re} A = (A + A^*)/2$. Infer from Ky Fan's maximum principle that

$$(\operatorname{Re} \lambda_1(A), ..., \operatorname{Re} \lambda_N(A)) \prec_{HLP} (\lambda_1(\operatorname{Re} A), ..., \lambda_N(\operatorname{Re} A)).$$

4.5 Horn's Inequalities

The following problem was raised by H. Weyl [484] in 1912: *Let A,B, and C be Hermitian $N \times N$ matrices and denote the string of eigenvalues of A by α, where*

$$\alpha : \ \alpha_1 \geq \cdots \geq \alpha_N,$$

and similarly write β and γ for the spectra of B and C. What α, β and γ can be the eigenvalues of the Hermitian matrices A, B and C when $C = A + B$?

There is one obvious condition, namely that the trace of C is the sum of the traces of A and B:

$$\sum_{k=1}^{N} \gamma_k = \sum_{k=1}^{N} \alpha_k + \sum_{k=1}^{N} \beta_k. \tag{4.17}$$

H. Weyl was able to indicate supplementary additional conditions in terms of linear inequalities on the possible eigenvalues. They were presented in Section 4.4.

Weyl's problem was studied extensively by A. Horn [222] who solved it for small N and proposed a complete set of necessary inequalities to accompany (4.17) for $N \geq 5$. Horn's inequalities have the form

$$\sum_{k \in K} \gamma_k \leq \sum_{i \in I} \alpha_i + \sum_{j \in J} \beta_j, \tag{4.18}$$

where

$$I = \{i_1, \ldots, i_r\}, \quad J = \{j_1, \ldots, j_r\}, \quad K = \{k_1, \ldots, k_r\}$$

are subsets of $\{1, \ldots, N\}$ with the same cardinality $r \in \{1, \ldots, N-1\}$ in a certain finite set T_r^N. Let us call such triplets (I, J, K) *admissible*. When $r = 1$, the condition of admissibility is

$$i_1 + j_1 = k_1 + 1.$$

If $r > 1$, this condition is:

$$\sum_{i \in I} i + \sum_{j \in J} j = \sum_{k \in K} k + \binom{r+1}{2}$$

and, for all $1 \leq p \leq r-1$ and all $(U, V, W) \in T_p^r$,

$$\sum_{u \in U} i_u + \sum_{v \in V} j_v = \sum_{w \in W} k_w + \binom{p+1}{2}.$$

Notice that Horn's inequalities are defined by an inductive procedure.

4.5.1 Conjecture (Horn's Conjecture) *A triplet (α, β, γ) of elements of $\mathbb{R}^N_{\geq}$ occurs as eigenvalues of symmetric matrices $A, B, C \in \mathrm{M}_N(\mathbb{R})$, with $C = A + B$, if and only if the trace equality* (4.17) *and Horn's inequalities* (4.18) *hold for every (I, J, K) in T_r^N, and every $r < N$.*

Nowadays this conjecture is a theorem due to works by A. A. Klyachko [254] and A. Knutson and T. Tao [256]. It appeals to advanced facts from algebraic geometry and representation theory (beyond the goal of this book). A thorough introduction to the mathematical world of Horn's conjecture is offered by the paper of R. Bhatia [50]. A more technical description of the work of Klyachko, Knutson, and Tao can be found in the paper of W. Fulton [171].

Just to get a flavor of what Horn's conjecture is about we will detail here the proof in the case of 2×2 symmetric matrices. Precisely, we will prove that for all families of real numbers $\alpha_1 \geq \alpha_2$, $\beta_1 \geq \beta_2$, $\gamma_1 \geq \gamma_2$, which verify Weyl's inequalities,

$$\gamma_1 \leq \alpha_1 + \beta_1, \quad \gamma_2 \leq \alpha_2 + \beta_1, \quad \gamma_2 \leq \alpha_1 + \beta_2,$$

and the trace formula (4.17),

$$\gamma_1 + \gamma_2 = \alpha_1 + \alpha_2 + \beta_1 + \beta_2,$$

there exist symmetric matrices $A, B, C \in \mathrm{M}_2(\mathbb{R})$ with $C = A + B$, $\sigma(A) = \{\alpha_1, \alpha_2\}$, $\sigma(B) = \{\beta_1, \beta_2\}$ and $\sigma(C) = \{\gamma_1, \gamma_2\}$.

Assume, for the sake of simplicity, that the spectra of A and B are respectively $\alpha = (4, 2)$ and $\beta = (2, -2)$. Then the conditions above may be read as

$$\gamma_1 + \gamma_2 = 6, \quad \gamma_1 \geq \gamma_2 \tag{4.19}$$

$$\gamma_1 \leq 6, \quad \gamma_2 \leq 2. \tag{4.20}$$

This shows that γ has the form $\gamma = (6 - a, a)$, with $0 \leq a \leq 2$; clearly, $\gamma_1 \geq \gamma_2$. We shall prove that every pair $(6 - a, a)$ with $0 \leq a \leq 2$ can be the spectrum of a sum $A + B$.

In fact, the relations (4.19) and (4.20) lead us to consider (in the plane $0\gamma_1\gamma_2$) the line segment XY, where $X = (6, 0)$ and $Y = (4, 2)$. Starting with the matrices

$$A = \begin{pmatrix} 4 & 0 \\ 0 & 2 \end{pmatrix}$$

and

$$R_\theta^\star \begin{pmatrix} 2 & 0 \\ 0 & -2 \end{pmatrix} R_\theta,$$

where

$$R_\theta = \begin{pmatrix} \cos\theta & \sin\theta \\ -\sin\theta & \cos\theta \end{pmatrix},$$

we should remark that the spectrum $(\lambda_1^\downarrow(C_\theta), \lambda_2^\downarrow(C_\theta))$ of the matrix

$$C_\theta = \begin{pmatrix} 4 & 0 \\ 0 & 2 \end{pmatrix} + R_\theta^\star \begin{pmatrix} 2 & 0 \\ 0 & -2 \end{pmatrix} R_\theta$$

lies on the line segment XY for all $\theta \in [0, \pi/2]$. In fact, since the eigenvalues of a matrix are continuous functions on the entries of that matrix, the map

$$\theta \to (\lambda_1^\downarrow(C_\theta), \lambda_2^\downarrow(C_\theta))$$

is continuous. The trace formula shows that the image of this map is a subset of the line $\gamma_1 + \gamma_2 = 6$. The point X corresponds to $\theta = 0$, and Y corresponds to $\theta = \pi/2$. Since the image should be a line segment, we conclude that each point of the linear segment XY represents the spectrum of a matrix C_θ with $\theta \in [0, \pi/2]$.

The list of inequalities involved in the 3×3-dimensional case of Horn's conjecture is considerably larger and consists of 13 items:

- *Weyl's inequalities,*

$$\begin{array}{lll} \gamma_1 \leq \alpha_1 + \beta_1 & \gamma_2 \leq \alpha_1 + \beta_2 & \gamma_2 \leq \alpha_2 + \beta_1 \\ \gamma_3 \leq \alpha_1 + \beta_3 & \gamma_3 \leq \alpha_3 + \beta_1 & \gamma_3 \leq \alpha_2 + \beta_2; \end{array}$$

- *Ky Fan's inequality,*

$$\gamma_1 + \gamma_2 \leq \alpha_1 + \alpha_2 + \beta_1 + \beta_2;$$

- *Lidskii–Wielandt inequalities* (taking into account the symmetric role of A and B),

$$\begin{aligned} \gamma_1 + \gamma_3 &\leq \alpha_1 + \alpha_3 + \beta_1 + \beta_2 \\ \gamma_2 + \gamma_3 &\leq \alpha_2 + \alpha_3 + \beta_1 + \beta_2 \\ \gamma_1 + \gamma_3 &\leq \alpha_1 + \alpha_2 + \beta_1 + \beta_3 \\ \gamma_2 + \gamma_3 &\leq \alpha_1 + \alpha_2 + \beta_2 + \beta_3; \end{aligned}$$

- *Horn's inequality,*

$$\gamma_2 + \gamma_3 \leq \alpha_1 + \alpha_3 + \beta_1 + \beta_3; \tag{4.21}$$

- *trace identity,*

$$\gamma_1 + \gamma_2 + \gamma_3 = \alpha_1 + \alpha_2 + \alpha_3 + \beta_1 + \beta_2 + \beta_3.$$

Horn's inequality (4.21) follows from (4.15), which in the case $n = 3$ may be read as

$$(\alpha_1 + \beta_3, \alpha_2 + \beta_2, \alpha_3 + \beta_1) \prec_{HLP} (\gamma_1, \gamma_2, \gamma_3).$$

The aforementioned set of 13 relations provides necessary and sufficient conditions for the existence of three symmetric matrices $A, B, C \in \mathrm{M}_3(\mathbb{R})$, with $C = A + B$, and spectra equal respectively to

$$\alpha_1 \geq \alpha_2 \geq \alpha_3; \quad \beta_1 \geq \beta_2 \geq \beta_3; \quad \gamma_1 \geq \gamma_2 \geq \gamma_3.$$

The proof is similar to that in the case $n = 2$.

For larger n, things become much more intricate. For example, for $n = 7$, Horn's list includes 2062 inequalities, not all of them independent.

The multiplicative companion to Horn's inequalities is a by-product of the solution of Horn's conjecture. See W. Fulton [171].

4.5.2 Theorem *Let $\alpha_1 \geq \cdots \geq \alpha_N$, $\beta_1 \geq \cdots \geq \beta_N$, $\gamma_1 \geq \cdots \geq \gamma_N$, be strings of positive real numbers. Then there exist matrices A and B with singular numbers $s_k(A) = \alpha_k$, $s_k(B) = \beta_k$, $s_k(AB) = \gamma_k$, if and only if*

$$\prod_{k \in K} \gamma_k \leq \prod_{i \in I} \alpha_i \prod_{j \in J} \beta_j$$

for all admissible triplets (I, J, K).

4.6 The Way of Hyperbolic Polynomials

Motivated by the theory of hyperbolic partial differential equations, L. Gårding [177], [178] has developed the theory of hyperbolic polynomials, that connects in an unexpected way algebra with convex analysis.

In what follows, by a polynomial on a finite-dimensional real vector space V, we will mean any real-valued functions on V that can be represented as finite sum of finite products of linear functionals. Assuming that $\dim V = N$, and considering a vector basis $\mathbf{v}_1, ..., \mathbf{v}_N$ of V and its dual vector basis $\mathbf{v}_1^*, ..., \mathbf{v}_N^*$ of V^*, it follows that every such polynomial $p(\mathbf{x})$ is nothing but a usual polynomial in the coefficients of $\mathbf{x}$.

If p is a nonconstant polynomial on V and m is a strictly positive integer, then we say that p is *homogeneous* of degree m if $p(t\mathbf{x}) = t^m p(\mathbf{x})$ for all $t \in \mathbb{R}$ and every $\mathbf{x} \in V$.

4.6.1 Definition (L. Gårding [177]) *A homogeneous polynomial p is called* hyperbolic *in the direction $\mathbf{d} \in V$ if $p(\mathbf{d}) > 0$ and the one real-variable polynomial function $t \to P(\mathbf{x} - t\mathbf{d})$ has only real roots for all vectors $\mathbf{x}$.*

Some simple examples of hyperbolic polynomials are as follows.

The polynomial $p(x_1, ..., x_N) = x_1 \cdots x_N$ is hyperbolic in the direction $\mathbf{d} = (1, ..., 1)$; the roots of $p(\mathbf{x} - t\mathbf{d})$ are exactly $x_1, x_2, ..., x_N$. This polynomial is actually hyperbolic in every direction $\mathbf{d} \in \mathbb{R}^N_{++}$ such that $p(\mathbf{d}) > 0$.

The polynomial $p(x_0, x_1, ..., x_N) = x_0^2 - x_1^2 - \cdots - x_N^2$ is hyperbolic in the direction $\mathbf{d} = (1, 0, ..., 0)$; the roots of $p(\mathbf{x} - t\mathbf{d})$ are $x_0 \pm \sqrt{x_1^2 + \cdots + x_N^2}$. It is also hyperbolic in every direction $\mathbf{d}$ for which $p(\mathbf{d}) > 0$.

The function det, when restricted to $\mathrm{Sym}(N, \mathbb{R})$, can be considered as a polynomial in the entries above the diagonal. The polynomial det is hyperbolic in the direction $\mathbf{d} = \mathrm{I}$, the identity matrix. For each $X \in \mathrm{Sym}(N, \mathbb{R})$, the roots of $\det(X - t\mathrm{I})$ are precisely the eigenvalues of X; det is also hyperbolic in every direction $\mathbf{d} \in \mathrm{Sym}^{++}(N, \mathbb{R})$.

Now, consider the polynomial $p(\mathbf{x}) = \det(x_1 A_1 + \cdots + x_N A_N)$, where $A_1, ..., A_N \in \mathrm{Sym}^{++}(N, \mathbb{R})$. Then $p(x)$ is hyperbolic in the direction $\mathbf{d} = (1, 0, ..., 0)$ because

$$p(\mathbf{x} - t\mathbf{d}) = \det\left(\sum_{k=1}^{n} x_k A_k - tA_1\right) = \det A_1 \det\left(\sum_{k=1}^{n} x_k A_1^{-1/2} A_k A_1^{-1/2} - t\mathrm{I}\right).$$

4.6.2 Remark *The Helton–Vinnikov theorem [211] (previously a conjecture due to P. Lax) asserts that every hyperbolic polynomial $p(x_1, x_2, x_3)$ in three variables possesses a representation of the form*

$$p(x_1, x_2, x_3) = \det(x_1 A_1 + x_2 A_2 + x_3 A_3),$$

where A_1, A_2, A_3 are real symmetric matrices.

A simple way to generate new hyperbolic polynomials from old ones is by differentiation.

4.6.3 Proposition (L. Gårding [178]) *Suppose that p is a hyperbolic polynomial in the direction $\mathbf{d}$ and p is homogeneous, of degree $m > 1$. Then $Q(\mathbf{x}) = \langle \nabla p(\mathbf{x}), \mathbf{d} \rangle$ is also a hyperbolic polynomial in the direction $\mathbf{d}$.*

Proof. Since $Q(\mathbf{x} + t\mathbf{d}) = \langle \nabla p(\mathbf{x} + t\mathbf{d}), d \rangle = \frac{\mathrm{d}}{\mathrm{d}t} p(\mathbf{x} + t\mathbf{d})$, we infer from Rolle's theorem that the polynomial $Q(\mathbf{x} + t\mathbf{d})$ has $m - 1$ real roots, separating the roots of $p(\mathbf{x} + t\mathbf{d})$. On the other hand, by differentiating the identity $p(t\mathbf{d}) = t^m p(\mathbf{d})$ we obtain $\langle \nabla p(t\mathbf{d}), \mathbf{d} \rangle = mt^{m-1} p(\mathbf{d}) > 0$ for $t \neq 0$, whence $Q(\mathbf{d}) > 0$. ■

4.6.4 Corollary *All elementary symmetric polynomials $e_k(x_1, ..., x_n)$ of strictly positive degree are hyperbolic in the direction $\mathbf{d} = (1, ..., 1)$.*

Proof. Indeed, it was already noticed that $e_n(x_1, ..., x_n) = x_1 \cdots x_n$ is hyperbolic in the direction $\mathbf{d}$. Then $e_{n-1}(\mathbf{x}) = \langle \nabla e_n(\mathbf{x}), \mathbf{d} \rangle$ is also hyperbolic by Proposition 4.6.3 and this argument can be iterated $n - 1$ steps. ■

By applying Proposition 4.6.3 to the polynomial $p(X) = \det X$ and the direction $\mathbf{d} = \mathrm{I}$, one obtains a new family of hyperbolic polynomials defined on $\mathrm{Sym}(N, \mathbb{R})$,

$$\sigma_k(X) = e_k(\lambda_1(X), ..., \lambda_N(X)), \quad k = 1, ..., N,$$

where $\lambda_1(X), ..., \lambda_N(X)$ are the eigenvalues of X. This example can be generalized.

If $p(\mathbf{x})$ is a hyperbolic polynomial in the direction $\mathbf{d}$, of degree m, then for each $\mathbf{x} \in V$ we have

$$p(\mathbf{x}+t\mathbf{d}) = p(\mathbf{d}) \prod_{k=1}^{m} (t+\lambda_k(\mathbf{x})),$$

where $\lambda_1(\mathbf{x}), ..., \lambda_m(\mathbf{x})$ are the so called $\mathbf{d}$-*eigenvalues* of $\mathbf{x}$. The terminology is motivated by the case of the polynomial $p(X) = \det X$ and the direction $\mathbf{d} = \mathrm{I}$. These functions are continuous (and dependent on p and $\mathbf{d}$). The continuity follows from the continuous dependence of the roots of the polynomial of its coefficients. See Q. I. Rahman and G. Schmeisser [413], Theorem 1.3.1, p. 10.

Notice that

$$p(\mathbf{x}) = p(\mathbf{d}) \prod_{k=1}^{m} \lambda_k(\mathbf{x}). \tag{4.22}$$

The following immediate result implies that the $\mathbf{d}$-eigenvalues of $\mathbf{x}$ are also positively homogeneous.

4.6.5 Lemma *Denote by* $\lambda_k^{\downarrow}(\mathbf{x})$ *the* k*-th largest* $\mathbf{d}$*-eigenvalue of* $\mathbf{x}$*. Then for all* $s, t \in \mathbb{R}$ *we have*

$$\lambda_k^{\downarrow}(s\mathbf{x}+t\mathbf{d}) = \begin{cases} s\lambda_k^{\downarrow}(\mathbf{x})+t & \text{if } s \geq 0 \\ s\lambda_{m-k+1}^{\downarrow}(\mathbf{x})+t & \text{if } s < 0. \end{cases}$$

The *eigenvalues map* associated to the family of $\mathbf{d}$-eigenvalues of $\mathbf{x}$ is defined as

$$\Lambda : V \to \mathbb{R}^m, \quad \Lambda(\mathbf{x}) = \left(\lambda_1^{\downarrow}(\mathbf{x}), ..., \lambda_m^{\downarrow}(\mathbf{x})\right).$$

See Exercise 2 for an important result concerning this map.

As will be shown in what follows, the $\mathbf{d}$-eigenvalues share many of the properties of the eigenvalues of symmetric matrices.

The *hyperbolicity cone* of p is the set

$$\begin{aligned} \Gamma_{p,\mathbf{d}} &= \{\mathbf{x} \in V : p(\mathbf{x}-t\mathbf{d}) \neq 0, \text{ for all } t \leq 0\} \\ &= \{\mathbf{x} \in V : \lambda_k(\mathbf{x}) > 0, \text{ for } k = 1, ..., m\}. \end{aligned}$$

The hyperbolicity cone $\Gamma_{p,\mathbf{d}}$ associated with the polynomial $x_1 \cdots x_n$ and the direction $\mathbf{d} = (1, ..., 1)$ is the open positive orthant $\mathbb{R}^N_{++}$. In the case of the polynomial det and the direction $\mathbf{d} = \mathrm{I}$, the hyperbolicity cone is $\mathrm{Sym}^{++}(n, \mathbb{R})$.

The hyperbolicity cone plays an important role in emphasizing the convexity properties of hyperbolic polynomials.

4.6.6 Theorem (L. Gårding [177], [178]) *Suppose that* $p(\mathbf{x})$ *is a hyperbolic polynomial in the direction* $\mathbf{d}$*, of degree* m*. Then:*

(a) *The hyperbolicity cone* $\Gamma_{p,\mathbf{d}}$ *is open and convex and* $\mathbf{d} \in \Gamma_{p,\mathbf{d}}$;

(b) *The polynomial* p *is hyperbolic in every direction* $\mathbf{e} \in \Gamma_{p,\mathbf{d}}$ *and* $\Gamma_{p,d} = \Gamma_{p,\mathbf{e}}$;

(c) $p(\mathbf{x}+\mathbf{y}) \geq p(\mathbf{x})$ *whenever* $\mathbf{y} \in \Gamma_{p,\mathbf{d}}$;

(d) $\Gamma_{p,\mathbf{d}} = \{\mathbf{x} \in V : \lambda_m^{\downarrow}(\mathbf{x}) > 0\}$ *and* $\overline{\Gamma}_{p,\mathbf{d}} = \{\mathbf{x} \in V : \lambda_m^{\downarrow}(\mathbf{x}) \geq 0\}$;

(e) $p(\mathbf{x})^{1/m}$ *is a concave function on* $\Gamma_{p,\mathbf{d}}$, *which vanishes on its boundary;*

(f) *The largest eigenvalue* $\lambda_1^{\downarrow}(\mathbf{x})$ *is a convex function (while* $\lambda_m^{\downarrow}(\mathbf{x})$ *is a concave function).*

An important ingredient in the proof of Theorem 4.6.6 is the following lemma.

4.6.7 Lemma (J. Renegar [417]) *The hyperbolicity cone is the connected component of the open set* $\{\mathbf{x} : p(\mathbf{x}) \neq 0\}$, *containing* $\mathbf{d}$.

Proof. Let S denote the connected component containing $\mathbf{d}$. Since $\mathbf{x}$ has 0 as an eigenvalue only if $p(\mathbf{x}) = 0$, and since $\mathbf{d} \in \Gamma_{p,\mathbf{d}}$, it follows from the continuity of the eigenvalues that $S \subset \Gamma_{p,\mathbf{d}}$. For the other inclusion, consider $\mathbf{x} \in \Gamma_{p,\mathbf{d}}$ and let L be the line segment with endpoints $\mathbf{x}$ and $\mathbf{d}$. For sufficiently large $t^* > 0$, all $\mathbf{y} \in L$ satisfy $p(\mathbf{y} + t^*\mathbf{d}) > 0$. Also, since $\mathbf{x}, \mathbf{d} \in \Gamma_{p,\mathbf{d}}$, we know that $\mathbf{x} + t\mathbf{d}$ and $\mathbf{d} + t\mathbf{d}$ belong to $\Gamma_{p,\mathbf{d}}$ for $t \geq 0$, which implies that $p(\mathbf{x} + t\mathbf{d}) > 0$ and $p(\mathbf{d} + t\mathbf{d}) > 0$ for $t \geq 0$. Thus, the segments $\{\mathbf{x} + t\mathbf{d} : 0 \leq t \leq t^*\}$, $\{\mathbf{y} + t^*\mathbf{d} : \mathbf{y} \in L\}$, and $\{\mathbf{d} + t\mathbf{d} : 0 \leq t \leq t^*\}$ form a path from $\mathbf{x}$ to $\mathbf{d}$ on which p remains strictly positive. This shows that $\mathbf{x} \in S$ and the proof of the inclusion $\Gamma_{p,\mathbf{d}} \subset S$ is done. ∎

Proof of Theorem 4.6.6. (a) According to Lemma 4.6.7, the hyperbolicity cone $\Gamma_{p,\mathbf{d}}$ is open and contains $\mathbf{d}$. This cone is also convex. To show this, let $\mathbf{v}, \mathbf{w} \in \Gamma_{p,\mathbf{d}}$. As was noticed in the proof of Lemma 4.6.7, the line segment joining $\mathbf{v}$ and $\mathbf{d}$ lies inside $\Gamma_{p,\mathbf{d}}$. Now, if $\mathbf{e}$ is an arbitrary point in $\Gamma_{p,\mathbf{d}}$, we have $\Gamma_{p,\mathbf{d}} = \Gamma_{p,\mathbf{e}}$ because of the maximality of a connected component. Therefore $\mathbf{v} \in \Gamma_{p,\mathbf{e}}$ and the line segment joining $\mathbf{v}$ and $\mathbf{w}$ lies inside $\Gamma_{p,\mathbf{w}} = \Gamma_{p,\mathbf{e}}$ $(= \Gamma_{p,\mathbf{d}})$, which ends the proof.

(b) If $\mathbf{e} \in \Gamma_{p,\mathbf{d}}$, then $p(\mathbf{e}) > 0$. See formula (4.22). We will show that for every $\mathbf{x} \in V$, the polynomial $t \to p(\mathbf{x} + t\mathbf{e})$ has only real roots. Let $\alpha > 0$ and $s \in \mathbb{R}$ be two parameters, and consider the polynomial $t \to p(s\mathbf{x} + t\mathbf{e} + \alpha \mathrm{i}\mathbf{d})$. We claim that if $s \geq 0$, then this polynomial has only roots with strictly negative imaginary part. Clearly, this is true for $s = 0$ since $t \to p(t\mathbf{e} + \alpha \mathrm{i}\mathbf{d})$ cannot have a root at $t = 0$, since $p(\alpha \mathrm{i}\mathbf{d}) = (\alpha \mathrm{i})^m p(\mathbf{d}) = 0$. If $t \neq 0$, then $p(t\mathbf{e} + \alpha \mathrm{i}\mathbf{d}) = 0$ if, and only if, $p(\mathbf{e} + \alpha t^{-1}\mathrm{i}\mathbf{d}) = 0$ which implies $\alpha t^{-1}\mathrm{i} < 0$, and thus $t = r\mathrm{i}$ for some $r < 0$.

If for some $s > 0$ the polynomial $t \to p(s\mathbf{x} + t\mathbf{e} + \alpha \mathrm{i}\mathbf{d})$ would have a root in the upper half-plane, then there must exist s^* for which $t \to p(s^*\mathbf{x} + t\mathbf{e} + \alpha \mathrm{i}\mathbf{d})$ has a real root t^*, that is, $p(s^*\mathbf{x} + t^*\mathbf{e} + \alpha \mathrm{i}\mathbf{d}) = 0$. However, this contradicts the hyperbolicity of p, since $s^*\mathbf{x} + t^*\mathbf{e} \in V$. Thus, for all $s \geq 0$, the roots of $t \to p(s\mathbf{x} + t\mathbf{e} + \alpha \mathrm{i}\mathbf{d})$ have strictly negative imaginary parts.

The conclusion above was true for every $\alpha > 0$. Letting $\alpha \to 0$, by continuity of the roots we have that the polynomial $t \to p(s\mathbf{x} + t\mathbf{e})$ must also have only roots with negative imaginary parts. However, since it is a polynomial with real coefficients (and therefore its roots always appear in complexconjugate pairs), then all the roots must actually be real. Taking now $s = 1$, we have that $t \to p(\mathbf{x} + t\mathbf{e})$ has real roots for all $\mathbf{x}$.

The assertion (c) is a consequence of assertion (b) and Lemma 4.6.5.

(d) The proof is left to the reader as Exercise 7.

(e) We follow the ingenious argument of L. Hörmander [227], p. 63. According to Proposition 3.1.2, it suffices to prove that if $\mathbf{x} \in \Gamma$ and $\mathbf{y} \in V$, then the function $\varphi(t) = p^{1/m}(\mathbf{x} + t\mathbf{y})$ is concave on $\operatorname{dom} \varphi = \{t \in \mathbb{R} : \mathbf{x} + t\mathbf{y} \in \Gamma\}$. We know that

$$p(\mathbf{y} + t\mathbf{x}) = p(\mathbf{x}) \prod\nolimits_{k=1}^{m} (t - r_i)$$

for suitable real numbers r_i, whence

$$p(\mathbf{x} + t\mathbf{y}) = t^m p(\mathbf{y} + \mathbf{x}/m) = p(\mathbf{x}) \prod\nolimits_{k=1}^{m} (1 - tr_i);$$

we have $1 - tr_i > 0$ since $\mathbf{x} + t\mathbf{y} \in \Gamma$. Consider the function $f(t) = \log p(\mathbf{x} + t\mathbf{y})$. Then

$$f'(t) = -\sum\nolimits_{k=1}^{m} \frac{r_i}{1 - tr_i}, \quad f''(t) = -\sum\nolimits_{k=1}^{m} \frac{r_i^2}{(1 - tr_i)^2},$$

and

$$\begin{aligned} m^2 \mathrm{e}^{-f(t)/m} \frac{\mathrm{d}^2}{\mathrm{d}t^2} \mathrm{e}^{f(t)/m} &= f'(t)^2 + m f''(t) \\ &= \left(\sum\nolimits_{k=1}^{m} \frac{r_i}{1 - tr_i} \right)^2 - m \sum\nolimits_{k=1}^{m} \frac{r_i^2}{(1 - tr_i)^2} \leq 0, \end{aligned}$$

according to the Cauchy–Bunyakovsky–Schwarz inequality. This proves that the function $t \to p^{1/m}(\mathbf{x} + t\mathbf{y})$ is concave on $\Gamma_{p,\mathbf{d}}$. The fact that p vanishes on the boundary of $\Gamma_{p,\mathbf{d}}$ follows from assertion (d).

(f) Notice that $\lambda_m^{\downarrow}(\mathbf{x}) \geq \alpha$ if, and only if, $\lambda_m^{\downarrow}(\mathbf{x} - \alpha\mathbf{d}) \geq 0$, whence

$$\{\mathbf{x} : \lambda_m^{\downarrow}(\mathbf{x}) \geq \alpha\} = \alpha\mathbf{d} + \Gamma_{p,\mathbf{d}}.$$

According to assertion (a) this is a convex set. Since the function $\lambda_m^{\downarrow}(\mathbf{x})$ is positively homogeneous, we conclude from Lemma 3.4.2 that $\lambda_m^{\downarrow}(\mathbf{x})$ is also a concave function. ■

The following theorem extends the results stated in Section 4.5.

4.6.8 Theorem (D. Serre [441]) *Horn's inequalities remain valid for the eigenvalues of every hyperbolic polynomial.*

Exercises

1. (New hyperbolic polynomials from old ones)

   (a) Suppose that $p : V \to \mathbb{R}$ is a hyperbolic polynomial in the direction $\mathbf{d} \in V$ and W is a vector subspace of V that contains $\mathbf{d}$. Prove that $p_{|W}$ is also hyperbolic (in the same direction). Moreover $\Gamma_{p_{|W},\mathbf{d}} = \Gamma_{p,\mathbf{d}} \cap W$.

(b) Let p and q be two hyperbolic polynomials in the direction $\mathbf{d}$. Prove that pq is also hyperbolic in the direction $\mathbf{d}$ and $\Gamma_{pq,\mathbf{d}} = \Gamma_{p,\mathbf{d}} \cap \Gamma_{q,\mathbf{d}}$.

(c) Let p be a hyperbolic polynomial in the direction $\mathbf{d}$, of degree $m \geq 1$, and assume that $p(\mathbf{x} + t\mathbf{d}) = \sum_{k=0}^{m} p_k(\mathbf{x})t^k$. Prove that all coefficients $p_k(\mathbf{x})$ are hyperbolic polynomials in the direction $\mathbf{d}$.

[*Hint:* For (c), use the formula

$$p_k(\mathbf{x}) = \frac{1}{k!} \frac{d^k}{dt^k} p(\mathbf{x} + t\mathbf{d}) \bigg|_{t=0} = \langle \nabla^k p(\mathbf{x}), \mathbf{d} \rangle.]$$

2. (H. H. Bauschke, O. Güler, A. S. Lewis and H. S. Sendov [35]). Infer from Exercise 1 (c) the following new construction for hyperbolic polynomials. Suppose that q is a homogeneous symmetric polynomial of degree m on $\mathbb{R}^N$, hyperbolic in the direction $\mathbf{e} = (1, 1, ..., 1)$, with eigenvalue map Φ and let Λ be the eigenvalue map of a hyperbolic polynomial of degree m in the direction $\mathbf{d}$. Prove that $q \circ \Lambda$ is also a hyperbolic polynomial of degree m in the direction $\mathbf{d}$, with eigenvalue map $\Phi \circ \Lambda$.

3. Let $1 \leq k \leq N$ and $\mathbf{e} = (1, 1, ..., 1) \in \mathbb{R}^N$. Prove that the polynomial

$$q(\mathbf{u}) = \prod_{1 \leq i_1 < \cdots < i_k \leq N} \sum_{j=1}^{k} u_{i_j}$$

is hyperbolic, of degree $\binom{N}{k}$, in the direction $\mathbf{e}$, and its eigenvalues are the functions

$$\lambda_{i_1,...,i_k}(\mathbf{u}) = \frac{1}{k} \sum_{j=1}^{k} u_{i_j} \quad \text{for } 1 \leq i_1 < \cdots < i_k \leq N.$$

Notice that the largest eigenvalue is the arithmetic mean of the k largest components of $\mathbf{u}$.

4. Infer from Exercises 2 and 3 and Theorem 4.6.6 (f) that the sums of the first k largest eigenvalues of a hyperbolic polynomial of degree m

$$\sigma_k(\mathbf{x}) = \lambda_1^{\downarrow}(\mathbf{x}) + \cdots + \lambda_k^{\downarrow}(\mathbf{x}) \quad \text{for } k = 1, ..., m,$$

are positively homogeneous and sublinear (in particular, convex). Prove that $\sigma_m(\mathbf{x})$ is actually a linear function.

5. Let $\mathbf{w} \in \mathbb{R}^m$ be a vector such that $w_1 \geq w_2 \geq \cdots \geq w_m$ and let Λ be the eigenvalues map of a hyperbolic polynomial of degree m. Infer from Exercise 4 that the function $\mathbf{x} \to \langle \mathbf{w}, \Lambda(\mathbf{x}) \rangle$ is positively homogeneous and sublinear (in particular, convex). This extends the fact that the largest eigenvalue is positively homogeneous and sublinear.

[*Hint:* According to Abel's partial summation formula,

$$\langle \mathbf{w}, \Lambda(\mathbf{x})\rangle = \sum_{k=1}^{m} w_k \lambda_k(\mathbf{x}) = w_m \sigma_m(\mathbf{x}) + \sum_{k=1}^{m-1} (w_k - w_{k+1})\, \lambda_k(\mathbf{x}).]$$

6. Prove Lemma 4.6.5.

7. Prove assertion (d) of Theorem 4.6.6, that is, the formulas

$$\Gamma_{p,\mathbf{d}} = \left\{\mathbf{x} \in V : \lambda_m^{\downarrow}(\mathbf{x}) > 0\right\} \text{ and } \overline{\Gamma}_{p,\mathbf{d}} = \left\{\mathbf{x} \in V : \lambda_m^{\downarrow}(\mathbf{x}) \geq 0\right\}.$$

8. A polynomial $q \in \mathbb{R}[x_1, ..., x_N]$ of degree $m \geq 1$ is called *real stable* if for all $\mathbf{x} \in \mathbb{R}^N$ and $\mathbf{d} \in \mathbb{R}^N_{++}$, the polynomial $q(\mathbf{x} + t\mathbf{d}) \in \mathbb{R}[t]$ has only real roots. Prove that q is real stable if, and only if, its homogenization

$$p_h(x_0, x_1, ...x_N) = x_0^m p\left(\frac{x_1}{x_0}, ..., \frac{x_N}{x_0}\right)$$

is hyperbolic in the direction $(0, \mathbf{d})$ for all $\mathbf{d} \in \mathbb{R}^N_{++}$.

4.7 Vector Majorization in $\mathbb{R}^N$

The usual relation of majorization, described in Section 4.1 as a relation between strings of real numbers can be easily generalized as a relation between strings of weighted vectors in $\mathbb{R}^N$. This was done by S. Sherman [442], inspired by the equivalence of conditions (a) and (b) in Theorem 4.1.5.

4.7.1 Definition *The relation of majorization*

$$(\mathbf{x}_1, ..., \mathbf{x}_m; \lambda_1, ..., \lambda_m) \prec_{Sh} (\mathbf{y}_1, ..., \mathbf{y}_n; \mu_1, ..., \mu_n) \tag{4.23}$$

between two strings of weighted points in $\mathbb{R}^N$ is defined by asking the existence of an $m \times n$-dimensional matrix $A = (a_{ij})_{i,j}$ such that

$$a_{ij} \geq 0 \quad \text{for } (i,j) \in \{1, ..., m\} \times \{1, ..., n\} \tag{4.24}$$

$$\sum_{j=1}^{n} a_{ij} = 1 \quad \text{for } i = 1, ..., m \tag{4.25}$$

$$\mu_j = \sum_{i=1}^{m} a_{ij} \lambda_i \quad \text{for } j = 1, ..., n \tag{4.26}$$

and

$$\mathbf{x}_i = \sum_{j=1}^{n} a_{ij} \mathbf{y}_j \quad \text{for } i = 1, ..., m. \tag{4.27}$$

We assume that all weights λ_i and μ_j belong to $[0,1]$ and

$$\sum_{i=1}^{m} \lambda_i = \sum_{j=1}^{n} \mu_j = 1.$$

The matrices verifying the conditions (4.24) and (4.25) are called *stochastic on rows*. When $m = n$ and all weights λ_i and μ_j are equal to each other, then the condition (4.26) assures the *stochasticity on columns*, so in that case we deal with doubly stochastic matrices.

4.7.2 Remark *The relation of majorization introduced by Definition* 4.7.1 *can be restated as a relation between probability measures, letting*

$$\sum_{i=1}^{m} \lambda_i \delta_{\mathbf{x}_i} \prec_{Sh} \sum_{j=1}^{n} \mu_j \delta_{\mathbf{y}_j},$$

if the conditions (4.24)–(4.27) *hold. In this context, the condition* (4.27) *means that*

$$\mathbf{x}_i = \operatorname{bar}\left(\sum_{j=1}^{n} a_{ij} \delta_{\mathbf{y}_j}\right) \quad \text{for } i = 1, ..., m,$$

a fact that allows easily to define the relation of majorization between probability measures not only in $\mathbb{R}^N$, *but also in any space where the notion of barycenter of a probability measure makes sense. Notice also that the conditions* (4.25) *and* (4.27) *imply*

$$\mathbf{x}_1, ..., \mathbf{x}_m \in \operatorname{conv}\{\mathbf{y}_1, ..., \mathbf{y}_n\}.$$

The following theorem provides a large extension of the Hardy–Littlewood–Pólya inequality of majorization:

4.7.3 Theorem (S. Sherman [442]) *Suppose that* $\sum_{i=1}^{m} \lambda_i \delta_{\mathbf{x}_i}$ *and* $\sum_{j=1}^{n} \mu_j \delta_{\mathbf{y}_j}$ *are two Borel probability measures on* $\mathbb{R}^N$. *Then the following assertions are equivalent:*

(a) $\sum_{i=1}^{m} \lambda_i \delta_{\mathbf{x}_i} \prec_{Sh} \sum_{j=1}^{n} \mu_j \delta_{\mathbf{y}_j}$;

(b) $\mathbf{x}_1, ..., \mathbf{x}_m \in \operatorname{conv}\{\mathbf{y}_1, ..., \mathbf{y}_n\}$ *and every continuous convex function* f *defined on* $\operatorname{conv}\{\mathbf{y}_1, ..., \mathbf{y}_n\}$ *(or to a larger convex subset of* $\mathbb{R}^N$*) verifies the inequality*

$$\sum_{i=1}^{m} \lambda_i f(\mathbf{x}_i) \leq \sum_{j=1}^{n} \mu_j f(\mathbf{y}_j). \tag{4.28}$$

Proof. The implication (a) $\Rightarrow$ (b) is a consequence of Remark 4.7.2. Indeed, assuming the existence of an $m \times n$-dimensional matrix $A = (a_{ij})_{i,j}$ which verifies the conditions (4.24)–(4.27), we infer from Jensen's inequality that

$$f(\mathbf{x}_i) \leq \sum_{j=1}^{n} a_{ij} f(\mathbf{y}_j)$$

for every $i \in \{1, ..., m\}$. Multiplying each side by λ_i and then summing up over i we conclude that

$$\sum_{i=1}^{m} \lambda_i f(\mathbf{x}_i) \leq \sum_{i=1}^{m} \left(\lambda_i \sum_{j=1}^{n} a_{ij} f(\mathbf{y}_j) \right) = \sum_{j=1}^{n} \left(\sum_{i=1}^{m} a_{ij} \lambda_i \right) f(\mathbf{y}_j) = \sum_{j=1}^{n} \mu_j f(\mathbf{y}_j).$$

The proof of the implication (b) $\Rightarrow$ (a) follows the argument of J. Borcea [63], pp. 3212–3214.

The starting remark is the non-emptiness of the set $\mathcal{M}$, of $m \times n$-dimensional matrices $A = (a_{ij})_{i,j}$ that are stochastic on rows and verify the condition (4.27); this is a consequence of the fact that $\mathbf{x}_1, ..., \mathbf{x}_m \in \operatorname{conv}\{\mathbf{y}_1, ..., \mathbf{y}_n\}$. Thus, the proof of the implication (b) $\Rightarrow$ (a) reduces to the existence of a matrix $A \in \mathcal{M}$ such that

$$\mu = \lambda \cdot A,$$

where $\mu = (\mu_1, ..., \mu_n)$ and $\lambda = (\lambda_1, ..., \lambda_m)$. The set $\mathcal{M}$ is a closed convex set and the same is true for

$$\lambda \mathcal{M} = \{\lambda \cdot A : A \in \mathcal{M}\}.$$

By the basic separation theorem, if $\mu \notin \lambda \mathcal{M}$, then there exist $c \in \mathbb{R}$ and a $(r_1, ..., r_n) \in \mathbb{R}^n$ such that

$$\sum_{j=1}^{n} r_j \nu_j < c < \sum_{j=1}^{n} r_j \mu_j \tag{4.29}$$

for all $(\nu_1, ..., \nu_n) \in \lambda \mathcal{M}$. Consider the function $F : \operatorname{conv}\{\mathbf{y}_1, ..., \mathbf{y}_n\} \to \mathbb{R}$ defined by

$$F(\mathbf{y}) = \max \sum_{j=1}^{n} r_j c_j,$$

where the maximum is taken over the nonempty compact convex set

$$C(\mathbf{y}) = \left\{ (c_1, ..., c_n) : c_1, ..., c_n \geq 0, \sum_{j=1}^{n} c_j = 1 \text{ and } \mathbf{y} = \sum_{j=1}^{n} c_j \mathbf{y}_j \right\}.$$

One can show easily that this is a continuous concave function.

Since $C(\mathbf{y})$ is compact, we may choose $(c_1(\mathbf{y}), ..., c_n(\mathbf{y}))$ in $C(\mathbf{y})$ such that $F(\mathbf{y}) = \sum_{j=1}^{n} r_j c_j(\mathbf{y})$. Since $c_j(\mathbf{y}_i) = \delta_{ij}$, we have $F(\mathbf{y}_j) \geq r_j$ for all j, hence

$$\sum_{j=1}^{n} \mu_j F(\mathbf{y}_j) \geq \sum_{j=1}^{n} r_j \mu_j > c, \tag{4.30}$$

according to the right inequality in (4.29). On the other hand, since

$$(c_1(x_i), ..., c_n(x_i)) \in \mathcal{M}$$

for every index i, we infer from the left inequality in (4.29) that

$$c \geq \sum_{j=1}^{n} r_j \left(\sum_{i=1}^{m} \lambda_i c_j(\mathbf{x}_i) \right) = \sum_{i=1}^{m} \lambda_i \left(\sum_{j=1}^{n} r_j c_j(\mathbf{x}_i) \right) = \sum_{i=1}^{m} \lambda_i F(\mathbf{x}_i). \quad (4.31)$$

Combining (4.30) and (4.31) we get

$$-\sum_{i=1}^{m} \lambda_i F(\mathbf{x}_i) > -\sum_{j=1}^{n} \mu_j F(\mathbf{y}_j),$$

a fact that contradicts our hypothesis (b) because the function $-F$ is convex and continuous. Consequently $\mu \in \lambda\mathcal{M}$ and the proof is done. ■

Concerning the assertion (b) of Theorem 4.7.3, let us notice that if the inequality (4.28) holds for all continuous convex functions defined on convex sets containing all points $\mathbf{x}_1, ..., \mathbf{x}_m, \mathbf{y}_1, ..., \mathbf{y}_n$, then necessarily $\mathbf{x}_1, ..., \mathbf{x}_m \in \operatorname{conv}\{\mathbf{y}_1, ..., \mathbf{y}_n\}$. To check this, see the case of the function

$$f(\mathbf{x}) = d\left(\mathbf{x}, \operatorname{conv}\{\mathbf{y}_1, ..., \mathbf{y}_n\}\right).$$

We next present a connection between the majorization theory and the distribution of roots of a polynomial.

The Gauss–Lucas theorem asserts that the roots $(\mu_k)_{k=1}^{n-1}$ of the derivative P' of any complex polynomial $P \in \mathbb{C}[z]$ of degree $n \geq 2$ lie in the smallest convex polygon containing the roots $(\lambda_j)_{j=1}^{n}$ of the polynomial P. See Section 2.1. Letting $\mu_n = (1/n)\sum_{k=1}^{n} \lambda_k$, this theorem yields the existence of a stochastic on rows matrix S such that $\mu = S\lambda$, where $\mu = (\mu_1, ..., \mu_n)$ and $\lambda = (\lambda_1, ..., \lambda_n)$. See Theorem 4.1.5 and Theorem 4.1.7. S. M. Malamud [297] and, independently, R. Pereira [393], were able to prove stronger results concerning the location of the roots of a polynomial and of its derivative. In particular, they turned into a theorem a conjecture raised by N. G. de Bruijn and T. A. Springer in 1947:

4.7.4 Theorem *For any convex function $f : \mathbb{C} \to \mathbb{R}$ and any polynomial P of degree $n \geq 2$,*

$$\frac{1}{n-1} \sum_{k=1}^{n-1} f(\mu_k) \leq \frac{1}{n} \sum_{k=1}^{n} f(\lambda_k),$$

where $(\lambda_k)_{k=1}^{n}$ and $(\mu_k)_{k=1}^{n-1}$ are respectively the roots of P and P'.

The proof of Theorem 4.7.4 needs some preliminaries relating the pair of roots $(\lambda_k)_{k=1}^{n}$ and $(\mu_k)_{k=1}^{n-1}$ with the spectrum of a normal matrix and the spectrum of a compression of it. Our argument follows closely that of J. Borcea [63], pp. 3222–3223.

Recall that a matrix $A \in \mathrm{M}_n(\mathbb{C})$ is called *normal* if it commutes with its adjoint, that is, $AA^* = A^*A$. Since

$$A = \frac{A + A^*}{2} + \mathrm{i}\frac{A - A^*}{2\mathrm{i}}$$

and the self-adjoint matrices $(A + A^*)/2$ and $-\mathrm{i}(A - A^*)/2$ commute, the spectral decomposition of self-adjoint matrices easily yields the existence of a unitary matrix U such that UAU^* is diagonal.

Choose a normal matrix $A \in \mathrm{M}_n(\mathbb{C})$ whose characteristic polynomial is P (as in Theorem 4.7.4) and denote by $(\mathbf{v}_k)_{k=1}^n$ an orthonormal basis of $H = \mathbb{C}^n$ consisting of eigenvectors of A such that $A\mathbf{v}_k = \lambda_k \mathbf{v}_k$ for all k. Let

$$\mathbf{w}_n = (1/\sqrt{n}) \sum\nolimits_{k=1}^n \mathbf{v}_k$$

and let π be the projection onto the linear subspace X, orthogonal to w_n. According to Schur's unitary triangularization theorem (see R. A. Horn and C. R. Johnson [223], Theorem 2.3.1, p. 79), there exists an orthonormal basis $(\mathbf{w}_k)_{k=1}^{n-1}$ of X which triangularizes the operator

$$B = \pi A \pi|_X,$$

the compression of A to X. Since $\|\mathbf{w}_n\| = 1$, the family $\{\mathbf{w}_1, ..., \mathbf{w}_{n-1}, \mathbf{w}_n\}$ constitutes an orthonormal basis of $\mathbb{C}^n$.

4.7.5 Lemma *We have*

$$\langle (A - z\mathrm{I}_H)^{-1} \mathbf{w}_n, \mathbf{w}_n \rangle = \frac{\det(B - z\mathrm{I}_X)}{\det(A - z\mathrm{I}_H)} = -\frac{P'(z)}{nP(z)},$$

for every $z \in \mathbb{C} \setminus \sigma(A)$.

Proof. The matrix representation of B in the basis $(\mathbf{w}_k)_{k=1}^{n-1}$ of X is given by the $(n-1) \times (n-1)$-dimensional principal minor in the matrix of representation of A in the basis $(\mathbf{w}_k)_{k=1}^n$ of H. The $n \times n$ coefficient of the matrix representing $(A - z\mathrm{I}_H)^{-1}$ in this basis is

$$\langle (A - z\mathrm{I}_H)^{-1} \mathbf{w}_n, \mathbf{w}_n \rangle = \langle \sum_{k=1}^n \frac{\langle \mathbf{w}_n, \mathbf{v}_k \rangle}{\lambda_k - z} \mathbf{v}_k, \sum_{k=1}^n \langle \mathbf{w}_n, \mathbf{v}_k \rangle \mathbf{v}_k \rangle = -\frac{P'(z)}{nP(z)}$$

for every $z \in \mathbb{C} \setminus \sigma(A)$. On the other hand, by Cramer's rule, this coefficient equals the cofactor of the $n \times n$ coefficient of $A - z\mathrm{I}_H$ in the same basis. Thus

$$\langle (A - z\mathrm{I}_H)^{-1} \mathbf{w}_n, \mathbf{w}_n \rangle = \frac{\det(B - z\mathrm{I}_X)}{\det(A - z\mathrm{I}_H)}$$

for every $z \in \mathbb{C} \setminus \sigma(A)$ and the proof is done. ∎

Proof of Theorem 4.7.4. According to Lemma 4.7.5, the roots μ_k are the same with the eigenvalues of the compression $B = \pi A \pi|_X$ as defined in the comments preceding this lemma. As was noticed above $\{w_1, ..., w_{n-1}\}$ is an orthonormal basis of X which triangularizes the operator $B = \pi A \pi|_X$ (that is, the compression of A to X). We have

$$\mu_j = \langle A\mathbf{w}_j, \mathbf{w}_j \rangle = \sum\nolimits_{i=1}^n \lambda_i \left| \langle \mathbf{w}_j, \mathbf{v}_i \rangle \right|^2.$$

Let S denote the $n-1$ by n matrix with coefficients $s_{ij} = |\langle \mathbf{w}_i, \mathbf{v}_j \rangle|^2$. By Parseval's theorem,

$$\sum\nolimits_{j=1}^{n} s_{ij} = \sum\nolimits_{j=1}^{n} |\langle \mathbf{w}_i, \mathbf{v}_j \rangle|^2 = \|\mathbf{w}_i\|^2 = 1$$

for $i \in \{1, ..., n-1\}$, and

$$\sum\nolimits_{i=1}^{n-1} s_{ij} = \sum\nolimits_{i=1}^{n-1} |\langle \mathbf{w}_i, \mathbf{v}_j \rangle|^2 = \|\mathbf{v}_j\|^2 - |\langle \mathbf{v}_j, \mathbf{w}_n \rangle|^2 = 1 - 1/n,$$

for $j \in \{1, ..., n-1\}$. This shows that S is a rectangular row stochastic matrix whose columns sum to $(n-1)/n$. Its existence shows that

$$\frac{1}{n-1} \sum_{k=1}^{n-1} \delta_{\mu_k} \prec_{Sh} \frac{1}{n} \sum_{k=1}^{n} \delta_{\lambda_k},$$

so the conclusion of Theorem 4.7.4 follows now from Sherman's majorization Theorem 4.7.3. ∎

See the Comments at the end of this chapter for other deeper results connecting the roots of a polynomial with the roots of its derivative.

Exercises

1. Let $A \in \mathrm{M}_n(\mathbb{C})$ be a normal matrix with diagonal elements $a_{11}, \ldots, a_{nn}$ and eigenvalues $\lambda_1, \ldots, \lambda_n$. Prove that there exists a doubly stochastic matrix $S \in \mathrm{M}_n(\mathbb{R})$ such that $\mathbf{a} = S\lambda$, where $\mathbf{a} = (a_{kk})_{k=1}^n$ and $\lambda = (\lambda_k)_{k=1}^n$.

   [*Hint:* Adapt the argument used in the proof of Schur's lemma 4.2.2.]

2. Suppose that $(\lambda_k)_{k=1}^n$ and $(\mu_k)_{k=1}^{n-1}$ are respectively the roots of a polynomial P and of its derivative P' and $n \geq 2$. Assuming that they are all positive, prove that

$$\left(\prod\nolimits_{k=1}^{n} \lambda_k\right)^{1/n} \leq \left(\prod\nolimits_{k=1}^{n-1} \mu_k\right)^{1/n-1}.$$

   This result can be easily generalized within the class of elementary symmetric functions, taking into account that $e_k^{1/k}(x_1, x_2, \ldots, x_n)$ is a concave function. See Section C3, Appendix C.

3. The Legendre polynomial of order $n \geq 0$ is defined by the formula $P_n(x) = \frac{1}{n!2^n} \frac{\mathrm{d}^n}{\mathrm{d}x^n} \left(x^2 - 1\right)^n$. Prove that its roots $x_1 < \cdots < x_n$ verify the formula

$$\frac{1}{n} \sum\nolimits_{k=1}^{n} f(x_k) \leq \frac{f(-1) + f(1)}{2}$$

   for all convex functions defined on $[-1, 1]$ and all $n \geq 2$.

4. Consider two interlace sequences

$$\lambda_1 \leq \mu_1 \leq \lambda_2 \leq \cdots \leq \mu_{n-1} \leq \lambda_n.$$

For example, the roots of successive Legendre polynomials interlace (and the same works in the case of Hermite and Laguerre polynomials). See G. Andrews et al. [18].

(a) Prove the Markov–Krein identity,

$$\sum_{k=1}^{n} \frac{p_k}{z-\lambda_k} = \frac{(z-\mu_1)\cdots(z-\mu_{n-1})}{(z-\lambda_1)\cdots(z-\lambda_n)},$$

where $p_1, ..., p_n > 0$ are suitable weights with $\sum_{k=1}^{n} p_k = 1$.

(b) Let $\alpha, \beta \in \mathbb{C}$, $\alpha \neq 0$. Prove that

$$\sum_{k=1}^{n-1} f(\alpha\mu_k + \beta) \leq \sum_{k=1}^{n} (1-p_k) f(\alpha\lambda_k + \beta),$$

for every convex function $f : \mathbb{C} \to \mathbb{R}$.

4.8 Comments

In their celebrated book on *Inequalities*, G. H. Hardy, J. E. Littlewood, and G. Pólya [209] proved a number of important characterizations of convex functions based on a reflexive and transitive relation defined on $\mathbb{R}^N$, called by them *majorization*. Their approach nicely incorporated previous work done by R. F. Muirhead, H. Dalton, I. Schur, and J. Karamata. Around the 1950s, motivated by the subject of comparison of experiments, researchers like D. Blackwell, S. Sherman, C. Stein, V. Strassen, P. Cartier, J. M. G. Fell, and P. A. Meyer added significant results that connected the majorization theory with statistical decision theory. See the treatise of E. Torgersen [472]. Choquet's theory [106], [398], on representation of points of a convex compact set as barycenters of probability measures supported by their extreme points enlarged the fields of applications to quantum statistical mechanics. Nowadays, the theory of majorization became an important tool in numerous other applications to economics, combinatorics, signal processing, etc.

The authoritative monograph on majorization theory is that by A. W. Marshall, I. Olkin and B. C. Arnold [305]. The reader will find there almost all significative results published before 2011.

The integral version of the Hardy–Littlewood–Pólya inequality of majorization is based on the concept of decreasing rearrangement.

Given an integrable function $f\colon [0,1] \to \mathbb{R}$, its *complementary distribution function* is defined as

$$\lambda_f(y) = \mathcal{L}(\{x : f(x) > y\}) \quad \text{for } y \in \mathbb{R},$$

where $\mathcal{L}$ denotes the 1-dimensional Lebesgue measure. The *decreasing rearrangement* of f is defined by the formula

$$f^{\downarrow}(x) = \inf\{y : \lambda_f(y) \leq x\} \quad \text{for } x \in [0,1].$$

Equivalently, the decreasing rearrangement of f is the (unique) right continuous and decreasing function $f^{\downarrow}$ such that f and $f^{\downarrow}$ are *equimeasurable* in the sense that

$$\mathcal{L}(\{x : f(x) > t\}) = \mathcal{L}\left(\{x : f^{\downarrow}(x) > t\}\right) \quad \text{for all } t > 0.$$

4.8.1 Theorem (Hardy, Littlewood, and Pólya [208]) *Suppose that* $f, g \in L^1([0,1])$. *Then*

$$\int_0^1 \varphi(f(x))\,\mathrm{d}x \leq \int_0^1 \varphi(g(x))\,\mathrm{d}x$$

for every continuous convex function φ *if and only if* $f \prec_{HLP} g$, *that is,*

$$\int_0^x f^{\downarrow}(t)\,\mathrm{d}t \leq \int_0^x g^{\downarrow}(t)\,\mathrm{d}t \quad \textit{for } 0 \leq x < 1$$

and

$$\int_0^1 f^{\downarrow}(t)\,\mathrm{d}t = \int_0^1 g^{\downarrow}(t)\,\mathrm{d}t.$$

For more general results see K.-M. Chong [104]. Applications to mathematical risk analysis can be found in the book of L. Rüschendorf [431].

The book of E. H. Lieb and M. Loss [282] offers a thorough presentation of the theory of symmetric decreasing rearrangements and its applications.

Majorization is sometimes related to inequality measurement, but in a series of problems like tax progression and income redistribution, Condorcet jury theorems, "fair representation" in parliaments, etc. the use of *Lorenz curves* seems to be more appropriate. See B. C. Arnold [19]. Suppose that X is a positive random variable with $0 < E(X) < \infty$. The Lorenz curve associated to it is parametrized by the formula

$$L_X(u) = \frac{1}{E(X)} \int_0^u F_X^{-1}(t)\mathrm{d}t \quad \text{for } u \in [0,1],$$

where F_X^{-1} represents the generalized inverse of the cumulative distribution function F_X. Clearly, the function L_X is continuous, increasing and convex and moreover, $L_X(0) = 0$ and $L_X(1) = 1$. In risk analysis it is customary to define

$$Y \geq_L X \text{ if and only if } L_Y \leq L_X,$$

meaning that Y is more unequal (or more spread out, or more variable) than X in the Lorenz sense.

The results in Section 4.6 concerning the connection between the roots of a polynomial and the roots of its derivative represent the starting point for a sequence of much better inequalities based on other variants of majorization.

4.8.2 Definition *Let $\mathbf{x} = \{x_1, ..., x_m\}$ and $\mathbf{y} = \{y_1, ..., y_n\}$ be two strings of vectors in $\mathbb{R}^N$, with $m \leq n$. The doubly stochastic majorization,*

$$\mathbf{x} \prec_{ds} \mathbf{y},$$

means the existence of elements $x_{m+1}, ..., x_n$ and of a doubly stochastic matrix $S \in \mathrm{M}_n(\mathbb{R})$ such that $\widetilde{\mathbf{x}} = \{x_1, ..., x_n\} = S\mathbf{y}$. The special case when S can be chosen of the form $S = (|u_{ij}|^2)_{i,j}$, for a suitable unitary matrix $U = (u_{ij})_{i,j}$ will be denoted

$$\mathbf{x} \prec_{uds} \mathbf{y}.$$

Given a vector $(\lambda_j)_{j=1}^n$ in $\mathbb{C}^n$ and a number $k \in \{1, ..., n\}$ one can attach to them the following element of $\mathbb{C}^{\binom{n}{k}}$:

$$C_k\left((\lambda_j)_{j=1}^n\right) = \{\lambda_{i_1} \cdots \lambda_{i_k} : 1 \leq i_1 < \cdots < i_k \leq n\}.$$

4.8.3 Theorem (S. M. Malamud [297]) *Let $\{\lambda_1, ..., \lambda_n\}$ and $\{\mu_1, ..., \mu_{n-1}\}$ be two families of complex numbers. Then, for the existence of a normal matrix A such that $\sigma(A) = (\lambda_j)_{j=1}^n$ and $\sigma(A_{n-1}) = (\mu_k)_{k=1}^{n-1}$ (where A_{n-1} denotes the principal submatrix of A obtained by deleting the last row and last column) it is necessary that the condition*

$$C_k\left((\mu_j - w)_{j=1}^{n-1}\right) \prec_{uds} C_k\left((\lambda_j - w)_{j=1}^n\right)$$

is fulfilled for any complex number w and any $k \in \{1, ..., n-1\}$ and it is sufficient to be fulfilled for $k = n-1$ and all $w \in \{\lambda_1, ..., \lambda_n\}$.

This leads to the following improvement of Theorem 4.7.4, also due to S. M. Malamud [297]:

4.8.4 Theorem *Let $f : \mathbb{C} \to \mathbb{R}$ be a convex function and α a complex number. Then for any polynomial P of degree $n \geq 2$ and any integer $k \in \{1, ..., n-1\}$ we have*

$$\frac{1}{\binom{n-1}{k}} \sum_{1 \leq i_1 < \cdots < i_k \leq n} f(\prod\nolimits_{j=1}^k \mu_{i_j} - \alpha)$$
$$\leq \frac{1}{\binom{n}{k}} \sum_{1 \leq i_1 < \cdots < i_k \leq n} f(\prod\nolimits_{j=1}^k \lambda_{i_j} - \alpha),$$

where $(\lambda_k)_{k=1}^n$ and $(\mu_k)_{k=1}^{n-1}$ are respectively the roots of P and P'.

Similar results were obtained by J. Borcea [69], who discussed the distribution of the equilibrium points of logarithmic potentials of the form

$$U(z) = \sum\nolimits_k a_k \log\left|1 - \frac{z}{z_k}\right|.$$

Chapter 5

Convexity in Spaces of Matrices

In this chapter we investigate three subjects concerning the convexity of functions defined on a space of matrices (or just on a convex subset of it). The first one is devoted to the convex spectral functions, that is, to the convex functions $F : \mathrm{Sym}(n, \mathbb{R}) \to \mathbb{R}$ whose values $F(A)$ depend only on the spectrum of A. The main result concerns their description as superpositions $f \circ \Lambda$ between convex functions $f : \mathbb{R}^n \to \mathbb{R}$ invariant under permutations, and the eigenvalues map Λ.

The second subject, the matrix convexity, is aimed to clarify the conditions under which the functional calculus associates to a convex function $f : \mathbb{R} \to \mathbb{R}$ a convex function on $\mathrm{Sym}(n, \mathbb{R})$.

The third subject concerns the hyperbolic geometry of $\mathrm{Sym}^{++}(n, \mathbb{R})$ and the geodesic convexity structure associated to it. Being at the frontier between differential geometry, linear algebra, and spectral theory, a presentation of this subject with full details would be quite involving, so our option was to offer the reader just a glimpse of what is going on in spaces with a curved geometry.

5.1 Convex Spectral Functions

The structure of convex functions under the presence of symmetry can be surprisingly simple. For example, this is the case of functions f defined on the space $\mathrm{Sym}(n, \mathbb{R})$ and invariant under the action of the *orthogonal group*

$$\mathrm{O}(n) = \{U \in \mathrm{M}_n(\mathbb{R}) : U^*U = UU^* = \mathrm{I}\}.$$

The elements of $\mathrm{O}(n)$ are called *orthogonal matrices* because their columns (as well as their rows) form orthonormal families of vectors. Indeed,

$$U \in \mathrm{O}(n) \text{ implies } \langle U\mathbf{x}, U\mathbf{y}\rangle = \langle \mathbf{x}, \mathbf{y}\rangle \text{ for all } \mathbf{x}, \mathbf{y} \in \mathbb{R}^n.$$

C. P. Niculescu and L.-E. Persson, *Convex Functions and Their Applications*, CMS Books in Mathematics, https://doi.org/10.1007/978-3-319-78337-6_5

The group $\mathrm{O}(n)$ acts on $\mathrm{Sym}(n, \mathbb{R})$ via the map

$$\Phi : O(n) \times \mathrm{Sym}(n, \mathbb{R}) \to \mathrm{Sym}(n, \mathbb{R}), \quad \Phi(U, A) = U^* AU.$$

A subset $\mathcal{K}$ of $\mathrm{Sym}(n, \mathbb{R})$ is called $\mathrm{O}(n)$-*invariant* if

$$A \in \mathcal{K} \text{ implies } U^* AU \in \mathcal{K} \text{ for all } U \in \mathrm{O}(n),$$

while a function F defined on an $\mathrm{O}(n)$-invariant subset of $\mathrm{Sym}(n, \mathbb{R})$ is called $\mathrm{O}(n)$-*invariant* if

$$F(A) = F(U^* AU) \text{ for all } U \in \mathrm{O}(n).$$

Some simple examples of $\mathrm{O}(n)$-invariant convex functions defined on $\mathrm{Sym}(n, \mathbb{R})$ are trace and the largest eigenvalue. The same is true for their restrictions to $\mathrm{Sym}^+(n, \mathbb{R})$, $\mathrm{Sym}^{++}(n, \mathbb{R})$, $\{A \in \mathrm{Sym}(n, \mathbb{R}) : \operatorname{trace} A = 0\}$ (as well as to any other $\mathrm{O}(n)$-invariant convex subset of $\mathrm{Sym}(n, \mathbb{R})$).

According to the spectral decomposition theorem, the orbit of any symmetric matrix A,

$$\mathcal{O}(A) = \{U^* AU : U \in \mathrm{O}(n)\},$$

contains all diagonal matrices carrying the various permutations of eigenvalues of A with their multiplicities. As a consequence, the values $F(A)$ of every $\mathrm{O}(n)$-invariant function F depend only on the spectrum of the matrix A. That's why the $\mathrm{O}(n)$-invariant functions are also called *spectral functions.*

The spectral functions can be identified with the functions of n real variables, invariant under the permutation of coordinates. If $F : \mathcal{K} \to \mathbb{R}$ is a spectral function, then $\mathcal{C} = \{\mathbf{x} \in \mathbb{R}^n : \mathrm{Diag}(\mathbf{x}) \in \mathcal{K}\}$ is a subset of $\mathbb{R}^n$ invariant under permutations and the function

$$f : \mathcal{C} \to \mathbb{R}, \quad f(\mathbf{x}) = F(\mathrm{Diag}(\mathbf{x})),$$

called *the symmetric function associated to* F, is invariant under permutations.

Conversely, every function $f : \mathcal{C} \to \mathbb{R}$, invariant under permutations, represents the symmetric function associated to a spectral function, more precisely to the function

$$F : \mathcal{K} \to \mathbb{R}, \quad F = f \circ \Lambda,$$

where $\mathcal{K} = \{A \in \mathrm{Sym}(n, \mathbb{R}) : \Lambda(A) \in \mathcal{C}\}$ and $\Lambda : \mathrm{Sym}(n, \mathbb{R}) \to \mathbb{R}^n$ is the eigenvalues map that associates to each symmetric matrix A the string $\lambda^{\downarrow}(A)$, of its eigenvalues with multiplicities, listed downward. It is a simple exercise that $\mathcal{C}$ is a convex set when $\mathcal{K}$ is convex (and vice-versa).

As was shown in Example 3.1.5, the convexity of the spectral function $-\log\det(A)$ on $\mathrm{Sym}^{++}(n, \mathbb{R})$ can be derived from the convexity of its associated symmetric function, $-\log(x_1 \cdots x_n)$ on $\mathbb{R}^n_{++}$. Remarkably, this fact works for any spectral function.

5.1.1 Theorem (C. Davis [124]) *A spectral function* $F : \mathcal{K} \to \mathbb{R}$ *is convex if and only if its associated symmetric function* $f : \mathcal{C} \to \mathbb{R}$ *is convex.*

This result remains valid in the framework of log-convex functions, concave functions, log-concave functions, quasiconvex functions, etc.

Proof. If F is convex, then for every $\mathbf{x}, \mathbf{y} \in \mathcal{C}$ and $\lambda \in [0, 1]$ we have

$$\begin{aligned} f((1-\lambda)\mathbf{x} + \lambda\mathbf{y}) &= F(\operatorname{Diag}((1-\lambda)\mathbf{x} + \lambda\mathbf{y})) \\ &= F((1-\lambda)\operatorname{Diag}(\mathbf{x}) + \lambda\operatorname{Diag}(\mathbf{y})) \\ &\leq (1-\lambda)F(\operatorname{Diag}(\mathbf{x})) + \lambda F(\operatorname{Diag}(\mathbf{y})) \\ &= (1-\lambda)f(\mathbf{x}) + \lambda f(\mathbf{y}), \end{aligned}$$

so that f is convex too.

Suppose now that f is convex and fix arbitrarily A, B in $\mathcal{K}$ and λ in $[0, 1]$. Since F is $\mathrm{O}(n)$-invariant, we may assume that $C = (1-\lambda)A + \lambda B$ is diagonal. Denote by A_C and B_C the matrices having the same main diagonal as A and B respectively, but whose off diagonal elements are zero. Then $C = (1-\lambda)A_C + \lambda B_C$ and the fact that f is convex implies

$$F(C) \leq (1-\lambda)F(A_C) + \lambda F(B_C).$$

We next show that $F(A_C) \leq F(A)$ and $F(B_C) \leq F(B)$, a fact that will complete the proof.

Let $\mathbf{e}_1, ..., \mathbf{e}_n$ be the natural basis of $\mathbb{R}^n$ and let $\mathbf{u}_1, ..., \mathbf{u}_n$ be an orthonormal basis of it consisting of eigenvectors of A. The eigenvalues α_i of A can be counted so that $A\mathbf{u}_i = \alpha_i\mathbf{u}_i$ for every i. Then, the diagonal elements of A and A_C are given by the formula

$$a_{ii} = \langle A\mathbf{e}_i, \mathbf{e}_i\rangle = \langle\sum_{j=1}^{n} \alpha_j\langle\mathbf{e}_i, \mathbf{u}_j\rangle\mathbf{u}_j, \mathbf{e}_i\rangle = \sum_{j=1}^{n} \alpha_j \left|\langle\mathbf{e}_i, \mathbf{u}_j\rangle\right|^2.$$

According to the Pythagorean theorem, the matrix $D = (D_{ij})_{i,j}$ with coefficients $D_{ij} = |\langle\mathbf{e}_i, \mathbf{u}_j\rangle|^2$ is doubly stochastic. Taking into account Birkhoff's Theorem 2.3.6, we infer that D is a convex combination $D = \sum_{\pi\in\Pi(n)} d_\pi P^{(\pi)}$ of the $n!$ permutation matrices $P^{(\pi)}$ with coefficients

$$P_{ij}^{(\pi)} = \begin{cases} 1 & \text{if } j = \pi(i) \\ 0 & \text{otherwise.} \end{cases}$$

Notice that a permutation matrix is a doubly stochastic matrix whose elements are either 0 or 1. Then

$$\begin{aligned} F(A_C) &= f((a_{11}, ..., a_{nn})) \\ &= f\left(\sum_{j=1}^{n} D_{1,j}\alpha_j, ..., \sum_{j-1}^{n} D_{n,j}\alpha_j\right) \\ &= f\left(\sum_{\pi} d_\pi \sum_{j=1}^{n} P_{1,j}^{(\pi)}\alpha_j, ..., \sum_{\pi} d_\pi \sum_{j=1}^{n} P_{n,j}^{(\pi)}\alpha_j\right) \end{aligned}$$

which is bounded from above by

$$\sum_{\pi} d_{\pi} f\left(\sum_{j=1}^{n} P_{1,j}^{(\pi)} \alpha_j, ..., \sum_{j=1}^{n} P_{n,j}^{(\pi)} \alpha_j\right)$$
$$= \sum_{\pi} d_{\pi} f\left(\alpha_{\pi(1)}, ..., \alpha_{\pi(n)}\right) = f\left(\alpha_1, ..., \alpha_n\right) = F(A).$$

In the same manner one can show that $F(B_C) \leq F(B)$, which completes the proof. ■

5.1.2 Corollary *The function* $F(A) = \log \operatorname{trace}\left(\mathrm{e}^{A}\right)$ *is convex on the space* $\operatorname{Sym}(n, \mathbb{R})$.

Proof. Indeed, if $U \in \mathrm{O}(n)$, then

$$\begin{aligned} \mathrm{e}^{U^{*}AU} &= I + \frac{1}{1!}\left(U^{*}AU\right) + \frac{1}{2!}\left(U^{*}AU\right)^{2} + \cdots \\ &= U^{*}\left(I + \frac{1}{1!}A + \frac{1}{2!}A^{2} + \cdots\right)U = U^{*}\mathrm{e}^{A}U, \end{aligned}$$

so F is a spectral function. The symmetric function associated to F is the convex function $f(x) = \log\left(\sum_{k=1}^{n} e^{x_k}\right)$, and Theorem 5.1.1 applies. ■

5.1.3 Corollary *The set* $\mathbf{S}_n = \left\{A \in \operatorname{Sym}^{+}(n, \mathbb{R}) : \operatorname{trace}(A) = 1\right\}$ *of the so-called density matrices is an* $O(n)$*-invariant convex subset of* $\operatorname{Sym}(n, \mathbb{R})$ *and the map*

$$S(\rho) = -\operatorname{trace}\left(\rho \log \rho\right)$$

is strictly concave on $\mathbf{S}_n$.

In quantum statistical mechanics, $S(\rho)$ represents the *von Neumann entropy* of ρ. See Exercise 7 for additional information. Since $S(\rho)$ is a differentiable strictly concave function, we have the inequality

$$\begin{aligned} S(\rho) &< S(\sigma) + \mathrm{d}S(\sigma)(\rho - \sigma) \\ &= S(\sigma) + \langle -\log \sigma - \mathrm{I}, \rho - \sigma\rangle_{HS}, \end{aligned}$$

that is,

$$\operatorname{trace}\left[(\rho \log \rho) - \rho \log \sigma - (\rho - \sigma)\right] \geq 0, \tag{5.1}$$

and equality holds only if $\rho = \sigma$; the last inequality expresses the fact that the *quantum relative entropy* is positive.

If we extend the von Neumann entropy to $\operatorname{Sym}(n, \mathbb{R})$ via the formula

$$S(A) = \begin{cases} -\operatorname{trace}\left(A \log A\right) & \text{if } A \in \mathbf{S}_n \\ -\infty & \text{otherwise,} \end{cases}$$

then one can show that the functions $\log \operatorname{trace}(\mathrm{e}^A)$ and $-S(A)$ are conjugate to each other, that is,

$$-S(A) = \sup\left\{\operatorname{trace}(AH) - \log \operatorname{trace}(\mathrm{e}^H) : H \in \operatorname{Sym}(n, \mathbb{R})\right\}$$

for all $A \in \operatorname{Sym}(n, \mathbb{R})$. This yields

$$\log \operatorname{trace}(\mathrm{e}^A) = \sup\left\{\operatorname{trace}(AH) + S(\mathrm{e}^H) : H \in \operatorname{Sym}(n, \mathbb{R})\right\}$$

for all $A \in \operatorname{Sym}(n, \mathbb{R})$. A very short argument is available in the paper of E. A. Carlen [92], Theorem 2.13.

5.1.4 Corollary (S. Lehmich, P. Neff, and J. Lankeit [276]) *Let h be a twice continuously differentiable real function defined on the interval $(0, \infty)$. Then the function $H(A) = h(\det A)$ is convex on $\operatorname{Sym}^{++}(n, \mathbb{R})$ for $n \geq 2$, if and only if*

$$nth''(t) + (n-1)h'(t) \geq 0 \text{ and } h'(t) \leq 0 \text{ for all } t > 0. \tag{5.2}$$

Proof. (M. Šilhavý [447]) According to Theorem 5.1.1, the function $H(A)$ is convex if and only if the function $\varphi(\lambda) = h(\lambda_1 \cdots \lambda_n)$ is convex on $(0, \infty)^n$. Since φ is twice Fréchet differentiable, the second derivative test of convexity applies. Letting $t = \lambda_1 \cdots \lambda_n$, a simple computation shows that for all $\lambda = (\lambda_1, ..., \lambda_n) \in (0, \infty)^n$ and $\mathbf{x} = (x_1, ..., x_n) \in \mathbb{R}^n$ we have

$$\begin{aligned}
\mathrm{d}^2\varphi(\lambda)(\mathbf{x}, \mathbf{x}) &= \sum_{i,j=1}^{n} \frac{\partial^2 \varphi}{\partial \lambda_i \partial \lambda_j}(\lambda_1, ..., \lambda_n) x_i x_j \\
&= \left(t^2 h''(t) + th'(t)\right) \left(\sum_{i=1}^{n} \frac{x_i}{\lambda_i}\right)^2 - th'(t) \sum_{i=1}^{n} \left(\frac{x_i}{\lambda_i}\right)^2.
\end{aligned}$$

The function φ is convex if and only if $\mathrm{d}^2\varphi(\lambda)(\mathbf{x}, \mathbf{x}) \geq 0$ for all λ and $\mathbf{x}$. The necessity of the conditions (5.2) follows by considering separately the following two particular cases

a) $x_i/\lambda_i = 1$ for $i = 1, ..., n$;

b) $x_1/\lambda_1 = 1,\ x_2/\lambda_2 = -1$ and $x_i/\lambda_i = 0$ for $i \geq 3$.

As concerns their sufficiency, one has to remark that due to the Cauchy–Bunyakovsky–Schwarz inequality,

$$\mathrm{d}^2\varphi(\lambda)(\mathbf{x}, \mathbf{x}) \geq \frac{t}{n}\left(nth''(t) + (n-1)h'(t)\right) \left(\sum_{i=1}^{n} \frac{x_i}{\lambda_i}\right)^2 \geq 0$$

for all λ and $\mathbf{x}$. ■

As in the case of convex functions, it is sometimes convenient to extend a spectral function to the whole space ($\operatorname{Sym}(n, \mathbb{R})$ or $\mathbb{R}^n$) by assigning ∞ at each point outside the invariant set representing the effective domain of definition.

In particular, this is the case when dealing with their subdifferentiability. See Exercises 4 and 5.

The following result, due to H. H. Bauschke, O. Güler, A. S. Lewis, and H. S. Sendov [35], extends Theorem 5.1.1 to the context of hyperbolic polynomials.

5.1.5 Theorem *If $f : \mathbb{R}^m \to \mathbb{R} \cup \{\infty\}$ is convex and symmetric and Λ is the eigenvalues map of a hyperbolic polynomial of degree m defined on a finite-dimensional vector space V, then $f \circ \Lambda$ is also convex on V.*

Proof. If $\mathbf{x}$ and $\mathbf{y}$ are vectors in V and $\alpha \in [0,1]$, then $\mathbf{u} = (1-\alpha)\Lambda(\mathbf{x}) + \alpha\Lambda(\mathbf{y})$ and $\mathbf{v} = \Lambda\left((1-\alpha)\mathbf{x} + \alpha\mathbf{y}\right)$ both belong to the monotone cone $\mathbb{R}^m_{\geq}$. According to Exercise 5, Section 4.6, if $\mathbf{w}$ is an arbitrarily fixed vector in $\bar{\mathbb{R}}^m_{\geq}$, then the function $\mathbf{x} \to \langle \mathbf{w}, \Lambda(\mathbf{x})\rangle$ is convex on V. Hence

$$\langle \mathbf{w}, \Lambda((1-\alpha)\mathbf{x} + \alpha\mathbf{y})\rangle \leq (1-\alpha)\langle \mathbf{w}, \Lambda(\mathbf{x})\rangle + \alpha\langle \mathbf{w}, \Lambda(\mathbf{y})\rangle,$$

equivalently,

$$\langle \mathbf{w}, \mathbf{u} - \mathbf{v}\rangle \geq 0.$$

As $\mathbf{w}$ was arbitrarily fixed in $\mathbb{R}^m_{\geq}$, the last inequality shows that $\mathbf{u} - \mathbf{v} \in \left(\mathbb{R}^m_{\geq}\right)^*$. Under these circumstances we infer from Remark 4.3.5 that $\mathbf{v} \prec_{HLP} \mathbf{u}$. Then $f(\mathbf{u}) \geq f(\mathbf{v})$, by Theorem 4.1.8, which means that

$$f\left(\Lambda\left((1-\alpha)\mathbf{x} + \alpha\mathbf{y}\right)\right) \leq f\left((1-\alpha)\Lambda(\mathbf{x}) + \alpha\Lambda(\mathbf{y})\right).$$

The proof ends by noticing that $f\left((1-\alpha)\Lambda(\mathbf{x}) + \alpha\Lambda(\mathbf{y})\right) \leq (1-\alpha)\Lambda(\mathbf{x}) + \alpha\Lambda(\mathbf{y})$, due to the property of convexity of f. ∎

Applying Theorem 5.1.5 to the function $f(x_1, ..., x_m) = \max\{x_k : 1 \leq k \leq m\}$, one retrieves the fact that the largest eigenvalue of a hyperbolic polynomial is a convex function on V. See Theorem 4.6.6 (f).

Exercises

1. Prove the convexity of the following functions:

   (a) $F(A) = A^2$, for $A \in \mathrm{Sym}(n, \mathbb{R})$;

   (b) the partial sums

   $$A \to \sum_{k=1}^{m} \lambda_k^{\downarrow}(A) = \sup_{J \subset \{1,...,n\},\ |J| = m} \sum_{k \in J} \lambda_k,$$

   where $A \in \mathrm{Sym}(n, \mathbb{R})$ and $1 \leq m \leq n$;

   (c) $-\operatorname{trace}\left(A^{1/2}\right)$, for $A \in \mathrm{Sym}^+(n, \mathbb{R})$;

   (d) $\operatorname{trace}\left(A^{-1}\right)$, for $A \in \mathrm{Sym}^{++}(n, \mathbb{R})$.

2. Prove that $|\operatorname{trace} A| \leq \operatorname{trace} |A|$ for all $A \in \mathrm{Sym}(n, \mathbb{R})$.

3. Let $F : \mathrm{Sym}(n, \mathbb{R}) \to \mathbb{R} \cup \{\infty\}$ be an $\mathrm{O}(n)$-invariant function. Prove that f is a lower semicontinuous proper convex function if, and only if, $F \circ \mathrm{Diag}$ is likewise on $\mathbb{R}^n$.

4. Prove that the differentiability of a convex spectral function F is characterized by that of $f = F \circ \mathrm{Diag}$. More precisely, if $F : \mathrm{Sym}(n, \mathbb{R}) \to \mathbb{R} \cup \{\infty\}$ is an $\mathrm{O}(n)$-invariant lower semicontinuous proper convex function, then F is Gâteaux differentiable at A if and only if f is Gâteaux differentiable at $\Lambda(A)$.

5. Suppose that $F : \mathrm{Sym}(n, \mathbb{R}) \to \mathbb{R} \cup \{\infty\}$ is an $\mathrm{O}(n)$-invariant lower semicontinuous proper convex function. Prove that $B \in \partial F(A)$ if and only if A and B admit simultaneous spectral decomposition,
$$U^*(\mathrm{Diag}(x))U = A \text{ and } U^*(\mathrm{Diag}(y))U = B,$$
for some U with $U^*U = \mathrm{I}$ and some $y \in \partial f(x)$.

6. Since the function $\det A$ is concave and strictly positive on $\mathrm{Sym}^{++}(n, \mathbb{R})$, its reciprocal $1/\det A$ is convex, so $h(t) = 1/t$ is an example of function for which Corollary 5.1.4 applies. Indicate other examples.

7. Prove that the von Neumann entropy (as defined in Corollary 5.1.3) verifies $0 \leq S(\rho) \leq \log n$ for all $\rho \in \mathbf{S}_n$. There is equality on the left if and only if $\rho = \langle \cdot, \mathbf{v} \rangle \mathbf{v}$ with $\|\mathbf{v}\| = 1$ (and there is equality on the right if and only if $\rho = (1/n)\mathrm{I}$).

   Remark. The operators $\rho = \langle \cdot, \mathbf{v} \rangle \mathbf{v}$ with $\mathbf{v} \in \mathbb{R}^n$, $\|\mathbf{v}\| = 1$, are known as the *pure states* of the algebra $\mathrm{M}_n(\mathbb{R})$. They represent the extreme points of the set $\mathbf{S}_n$ of density matrices. See S. Strătilă and L. Zsidó [465].

8. (P. M. Alberti and A. Uhlmann) Given two symmetric matrices $A, B \in \mathrm{Sym}(n, \mathbb{R})$, we say that A is majorized by B with respect to spectral order (abbreviated, $A \prec_{sp} B$) if $\Lambda(A) \prec_{HLP} \Lambda(B)$. Prove that the following assertions are equivalent:

   (a) $A \prec_{sp} B$;

   (b) there exist positive numbers λ_k and orthogonal matrices U_k such that
$$\sum_{k=1}^{n} \lambda_k = 1 \text{ and } A = \sum_{k=1}^{n} \lambda_k U_k^* B U_k;$$

   (c) there exists a doubly stochastic map $\Phi : \mathrm{M}_n(\mathbb{R}) \to \mathrm{M}_n(\mathbb{R})$ such that $A = \Phi(B)$, that is, a linear map that verifies the following three properties:
$$\Phi(\mathrm{I}) = \mathrm{I}, \ \Phi(\mathrm{Sym}^+(n, \mathbb{R})) \subset \mathrm{Sym}^+(n, \mathbb{R})$$
$$\text{and trace } \Phi(A) = \text{trace } A \text{ for all } A.$$

9. Suppose that $F : \mathrm{Sym}(n, \mathbb{R}) \to \mathbb{R}$ is an $\mathrm{O}(n)$-invariant convex function. Prove that $F(A) \leq F(B)$ whenever $A \prec_{sp} B$.

10. Prove that the rank function is quasiconcave on $\mathrm{Sym}^+(n, \mathbb{R})$ and its convex hull is the trace function.

5.2 Matrix Convexity

The concept of convex function can be extended verbatim to the framework of vector-valued functions, provided that the range is endowed with an order structure.

5.2.1 Definition *A function f, defined on a convex subset U of a normed linear space E and taking values in a regularly ordered Banach space F, is called* convex *if for all points $\mathbf{x}, \mathbf{y} \in U$ and all $\lambda \in [0, 1]$ we have*

$$f((1-\lambda)\mathbf{x} + \lambda\mathbf{y}) \leq (1-\lambda)f(\mathbf{x}) + \lambda f(\mathbf{y}),$$

equivalently,

$$(1-\lambda)f(\mathbf{x}) + \lambda f(\mathbf{y}) - f((1-\lambda)\mathbf{x} + \lambda\mathbf{y}) \in F_+.$$

A function $f : U \to F$ is called concave *if $-f$ is convex.*

A function $f : U \to F$ is convex if and only if $x^* \circ f$ has this property for all positive linear functionals $x^* \in F^*$; see Theorem 2.5.3. Therefore, for $F = \mathbb{R}^n$ (endowed with the coordinate ordering), the convexity of f is equivalent to the convexity of all its components. An immediate consequence is the fact that many important results such as Proposition 3.1.2 (the equivalence of convexity with convexity along linear segments), Corollary 3.1.3 (the extension of Jensen's criterion of convexity) and Proposition 3.2.2 (the discrete case of Jensen's inequality) can be extended to the vector-valued context.

Typical examples are the so-called operator convex functions.

According to the continuous functional calculus (see the Comments at the end of Chapter 2), one can associate to each self-adjoint operator $A \in \mathcal{A}(H)$ (acting on the Hilbert space H) an application $\pi : C(\sigma(A)) \to \mathcal{A}(H)$ that transforms every real polynomial $P(t) = t^n + c_1 t^{n-1} + \cdots + c_n$ into the self-adjoint operator $P(A) = A^n + c_1 A^{n-1} + \cdots + c_n \mathrm{I}$ and, in general, every continuous real function $f \in C(\sigma(A))$ into an operator $f(A) \in \mathcal{A}(H)$ such that

$$\begin{cases} (f+g)(A) = f(A) + g(A) \\ (fg)(A) = f(A)g(A) \\ f \geq 0 \text{ implies } f(A) \geq 0 \\ \|f(A)\| = \sup\{|f(\lambda)| : \lambda \in \sigma(A)\}, \end{cases}$$

for all $f, g \in C(\sigma(A))$. Reversing the roles, and starting with a real-valued continuous function f defined on an interval I, one obtains the set of self-adjoint bounded operators $f(A)$ with $\sigma(A) \subset I$.

It is worth noticing that when $\sigma(A) \subset [m, M]$ and $\sigma(B) \subset [m, M]$, then $m\mathrm{I} \leq A \leq M\mathrm{I}$ and $m\mathrm{I} \leq B \leq M\mathrm{I}$, so that $m\mathrm{I} \leq (1-\lambda)A + \lambda B \leq M\mathrm{I}$. See Theorem 2.6.4. Therefore the set $\mathcal{A}_I(H)$ of all self-adjoint bounded operators A with $\sigma(A) \subset I$ constitutes a convex set in $\mathcal{A}(H)$.

5.2.2 Definition *A continuous function $f : I \to \mathbb{R}$ is said to be n-matrix convex if*

$$f((1-\lambda)A + \lambda B) \leq (1-\lambda)f(A) + \lambda f(B) \tag{5.3}$$

for every $\lambda \in [0,1]$ and every pair of matrices A, $B \in \mathrm{Sym}(n,\mathbb{R})$ with spectra in I. If the above inequality holds for every Hilbert space H and every pair of self-adjoint bounded operators A, $B \in \mathcal{A}_I(H)$, then f is called operator convex.

The concepts of strict n-matrix convexity and strict operator convexity refer to the case where the inequality (5.3) is strict when $\lambda \in (0,1)$ and $A \neq B$.

The matrix/operator companions to concave functions are defined by reversing the inequality sign in (5.3).

5.2.3 Example *The function $f(t) = 1/t$ is strictly operator convex on $(0,\infty)$: for each Hilbert space H, the induced function*

$$f : \mathcal{A}(H)_{++} \to \mathcal{A}(H), \quad f(A) = A^{-1},$$

is strictly convex, that is,

$$((1-\lambda)A + \lambda B)^{-1} < (1-\lambda)A^{-1} + \lambda B^{-1}$$

for all $A, B \in \mathcal{A}(H)_{++}$ with $A \neq B$, and $\lambda \in (0,1)$. For $B = \mathrm{I}$ and $A \neq \mathrm{I}$, this reduces to the inequality

$$((1-\lambda)A + \lambda \mathrm{I})^{-1} < (1-\lambda)A^{-1} + \lambda \mathrm{I},$$

which easily follows by simple algebraic manipulations. To obtain from here the general case we have to replace A by $A^{-1/2}BA^{-1/2}$.

5.2.4 Example *The preceding example can be used to prove the strict operator convexity of the function $-\log$ on $(0,\infty)$. Indeed, using the representation formula*

$$-\log x = \int_0^\infty \left(\frac{1}{x+t} - \frac{1}{1+t}\right) \mathrm{d}t,$$

we infer that for every $Z \in \mathcal{A}(H)_{++}$,

$$-\log Z = \int_0^\infty \left[(Z + t\mathrm{I})^{-1} - (\mathrm{I} + t\mathrm{I})^{-1}\right] \mathrm{d}t,$$

whence

$$\begin{aligned}
-\log((1-\lambda)A + \lambda B) &= \int_0^\infty \left[((1-\lambda)A + \lambda B + t\mathrm{I})^{-1} - (\mathrm{I} + t\mathrm{I})^{-1}\right] \mathrm{d}t \\
&= \int_0^\infty \left[((1-\lambda)(A + t\mathrm{I}) + \lambda(B + t\mathrm{I}))^{-1} - (\mathrm{I} + t\mathrm{I})^{-1}\right] \mathrm{d}t \\
&< (1-\lambda)\int_0^\infty \left[(A + t\mathrm{I})^{-1} - (I + tI)^{-1}\right] \mathrm{d}t \\
&\quad + \lambda \int_0^\infty \left[(B + tI)^{-1} - (I + tI)^{-1}\right] \mathrm{d}t \\
&= -(1-\lambda)\log A - \lambda \log B,
\end{aligned}$$

for all $A, B \in \mathcal{A}(H)_{++}$, $A \neq B$, and $\lambda \in (0,1)$.

It was shown by J. Bendat and S. Sherman [40] that a continuous real function f is operator convex on the interval $[0,\infty)$ if and only if it admits a representation of the form

$$f(t) = c_0 + c_1 t + c_2 t^2 + \int_0^\infty \frac{st^2}{s+t} \mathrm{d}\mu(s),$$

where $c_0, c_1 \in \mathbb{R}$, $c_2 \geq 0$ and μ is a finite measure on $[0,\infty)$. This implies that f must be analytic. Very good accounts of Bendat–Sherman theory can be found in the books of R. Bhatia [49] and B. Simon [450]. Both relate the study of operator convex/concave functions to the class of operator monotone functions.

5.2.5 Definition *A function $f : I \to \mathbb{R}$ is called n-matrix monotone if, for every pair of matrices A, $B \in \operatorname{Sym}(n,\mathbb{R})$ with spectra in I, the following implication holds:*

$$A \leq B \text{ implies } f(A) \leq f(B).$$

The function f is called operator monotone if the above inequality holds for every Hilbert space H and every pair of self-adjoint operators A, $B \in \mathcal{A}_I(H)$.

The function $-1/t$ is operator concave and operator monotone on $(0,\infty)$. For monotonicity, notice that if $A, B \in \mathcal{A}(H)_{++}$, and $A \leq B$, then

$$A^{-1} - B^{-1} = A^{-1/2}[\mathrm{I} - (\mathrm{I} + C)^{-1}]A^{-1/2} \geq 0,$$

since $C = A^{-1/2}(B-A)A^{-1/2}$. Analogously, the log function is operator concave and operator monotone on $(0,\infty)$; the monotonicity follows from the Minkowski determinant inequality.

The following two results due to F. Hansen and G. K. Pedersen [201] describe the relationship between matrix monotone functions and matrix concave functions defined on the interval $(0,\infty)$.

5.2.6 Theorem *Every matrix $2n$-monotone function $f : (0,\infty) \to \mathbb{R}$ is also matrix n-concave.*

Therefore (see Exercise 1), every operator monotone function defined on $(0,\infty)$ is also operator concave.

See F. Hansen [199] for a simple direct argument, based on handling block matrices.

5.2.7 Theorem *Let $f : (0,\infty) \to [0,\infty)$ be a matrix n-concave function. Then f is also n-monotone.*

Therefore (see Exercise 1), every positive operator concave function defined on $(0,\infty)$ is also operator monotone.

Proof. Let A and B be strictly positive $n \times n$ matrices with $A < B$ and fix arbitrarily λ in $(0,1)$. Since λB can be represented as a convex combination of two strictly positive matrices,

$$\lambda B = \lambda A + (1-\lambda)\left(\lambda(1-\lambda)^{-1}(B-A)\right),$$

and f is n-concave, we obtain

$$f(\lambda B) \geq \lambda f(A) + (1-\lambda) f\left(\lambda(1-\lambda)^{-1}(B-A)\right) \geq \lambda f(A).$$

Since f is continuous we obtain $f(B) \geq f(A)$ by passing to the limit as $\lambda \to 1$. ■

Some other interesting examples are given by the following result, whose proof can be found in the surveys of E. A. Carlen [92] and F. Hansen [199]. See also R. Bhatia [49], Chapter V.

5.2.8 Theorem (The Löwner–Heinz theorem) *For $-1 \leq p \leq 0$, the function $f(t) = -t^p$ is operator monotone and operator concave on $(0,\infty)$. For $0 \leq p \leq 1$, the function $f(t) = t^p$ is operator monotone and operator concave on $[0,\infty)$. For $1 \leq p \leq 2$, the function $f(t) = t^p$ is operator convex on $[0,\infty)$ and so is the function $t \log t$.*

5.2.9 Corollary *Since*

$$A \log A = \lim_{p \to 1-} \frac{A^p - A}{p-1}$$

and the property of convexity is preserved under taking limits, it follows that the function $t \log t$ is operator convex on $[0,\infty)$.

The analogue of Jensen's inequality in the framework of operator convex functions makes necessary the consideration of non-commutative convex combinations of n-tuples of self-adjoint operators. This remark is due to F. Hansen and G. K. Pedersen [202], who proved the following important result:

5.2.10 Theorem (Jensen's operator inequality) *For a continuous function f defined on an interval I the following two conditions are equivalent:*

(a) *f is operator convex;*

(b) *For each integer number $n \geq 1$ we have the inequality*

$$f\left(\sum_{k=1}^{n} C_k^* A_k C_k\right) \leq \sum_{k=1}^{n} C_k^* f(A_k) C_k \tag{5.4}$$

for every n-tuple $(A_1, ..., A_n)$ of bounded, self-adjoint operators on an arbitrary Hilbert space H, with spectra contained in I, and every n-tuple $(C_1, ..., C_n)$ of bounded operators on H with $\sum_{k=1}^{n} C_k^* C_k = \mathrm{I}$.

Moreover, this theorem admits the following contractive version: Let f be a continuous function defined on an interval I and suppose that $0 \in I$. Then f is operator convex and $f(0) \leq 0$ if and only if for every natural number $n \geq 1$, the inequality (5.4) is valid for every n-tuple $(A_1, ..., A_n)$ of bounded self-adjoint operators on an arbitrary Hilbert space H, with spectra contained in I, and every n-tuple $(C_1, ..., C_n)$ of bounded operators on H with $\sum_{k=1}^{n} C_k^* C_k \leq \mathrm{I}$.

A nice application of Theorem 5.2.10 is offered by the noncommutative analogue of the notion of generalized perspective function, introduced first by E. G. Effros [142] (in the commutative case) and then by A. Ebadian, I. Nikoufar, and M. E. Gordji [140] (in the general case). We will consider here only a simplified version of their results, which refers to the perspective function associated to a continuous function $f : [0, \infty) \to \mathbb{R}$:

$$\tilde{f} : \mathcal{A}(H)_+ \times \mathcal{A}(H)_{++} \to \mathbb{R}, \quad \tilde{f}(A, B) = B^{1/2} f\left(B^{-1/2} A B^{-1/2}\right) B^{1/2}.$$

As was noticed by E. G. Effros and F. Hansen [143], the function $\tilde{f}$ is the unique extension of the corresponding commutative perspective function associated to f that preserves homogeneity and convexity.

5.2.11 Theorem *The function f is operator convex/concave if, and only if, the function $\tilde{f}$ is jointly convex/concave.*

Proof. Suppose first that f is operator convex and consider $A_1, A_2 \in \mathcal{A}(H)_+$, $B_1, B_2 \in \mathcal{A}(H)_{++}$ and $\lambda \in [0, 1]$. Put $A = (1 - \lambda)A_1 + \lambda A_2$ and $B = C_1 = (1 - \lambda)B_1 + \lambda B_2$. Then the operators $C_1 = ((1 - \lambda)B_1)^{1/2} B^{-1/2}$ and $C_2 = (\lambda B_2)^{1/2} B^{-1/2}$ verify $C_1^* C_1 + C_2 C_2 = \mathrm{I}$. According to Theorem 5.2.10,

$$\begin{aligned}
\tilde{f}(A, B) &= B^{1/2} f\left(B^{-1/2} A B^{-1/2}\right) B^{1/2} \\
&= B^{1/2} f(C_1^* B_1^{-1/2} A_1 B_1^{-1/2} C_1 + C_2^* B_2^{-1/2} A_2 B_2^{-1/2} C_2) B^{1/2} \\
&\leq B^{1/2} \left[C_1^* f(B_1^{-1/2} A_1 B_1^{-1/2}) C_1 + C_2^* f(B_2^{-1/2} A_2 B_2^{-1/2}) C_2\right] B^{1/2} \\
&= (1 - \lambda) B_1^{1/2} f(B_1^{-1/2} A_1 B_1^{-1/2}) B_1^{1/2} + \lambda B_2^{1/2} f(B_2^{-1/2} A_2 B_2^{-1/2}) B_2^{1/2} \\
&= (1 - \lambda)\tilde{f}(A_1, B_1) + \lambda \tilde{f}(A_2, B_2),
\end{aligned}$$

and the proof of joint convexity of $\tilde{f}$ is done.

The converse implication is straightforward since $f(A) = \tilde{f}(A, \mathrm{I})$. ■

The aforementioned paper of E. G. Effros includes a number of deep results that can be derived from Theorem 5.2.11. An example is offered by the case of *Kullback–Leibler divergence* of ρ from σ,

$$D_{KL}(\rho||\sigma) = \operatorname{trace}(\rho \log \rho - \rho \log \sigma),$$

where $\rho, \sigma \in \left\{A \in \mathrm{Sym}^{++}(n, \mathbb{R}) : \operatorname{trace}(A) = 1\right\}$.

5.2.12 Corollary (G. Lindblad [285]) *The Kullback–Leibler divergence is a jointly convex function.*

Proof. We apply Theorem 5.2.11 for the real Hilbert space $H = \mathrm{M}_n(\mathbb{R})$ (endowed with the Hilbert–Schmidt norm) and the continuous function $f(t) = t \log t$; according to Corollary 5.2.9, this function is operator convex on $\mathcal{A}(\mathcal{H})_+$.

The operators $A, B : \mathrm{M}_n(\mathbb{R}) \to \mathrm{M}_n(\mathbb{R})$ given by $A(X) = \rho X$ and $B(X) = X\sigma$ are positive and commute; for example,

$$\langle A(X), X\rangle_{HS} = \operatorname{trace}(X^*\rho X) = \operatorname{trace}\left(\left(\rho^{1/2}X\right)^*\left(\rho^{1/2}X\right)\right) \geq 0$$

for all $X \in \mathrm{M}_n(\mathbb{R})$, hence A is positive. Therefore,

$$\begin{aligned}\langle \tilde{f}(A,B)(\mathrm{I}), \mathrm{I}\rangle_{HS} &= \langle (A\log A - A\log B)(\mathrm{I}), \mathrm{I}\rangle_{HS} \\ &= \operatorname{trace}(\rho\log\rho - \rho\log\sigma) = D_{KL}(\rho\|\sigma)\end{aligned}$$

is jointly convex. ■

Lindblad's original argument for Corollary 5.2.12 is sketched in Exercise 9.

Exercises

1. The property of operator convexity of a continuous function $f : (a, b) \to \mathbb{R}$ is equivalent to the property of matrix convexity of all orders. Prove that f is operator convex if, and only if, the matrix functions $A \to f(A)$ are convex on $\{A \in \mathrm{Sym}(n,\mathbb{R}) : \sigma(A) \subset (a, b)\}$, whenever $n \geq 1$.

   A similar statement holds in the case of operator monotone functions.

   [*Hint*: In case H is a real Hilbert space having a countably infinite orthonormal basis $(\mathbf{u}_n)_n$, one denotes by P_n the orthogonal projection on $\mathrm{span}\{\mathbf{u}_1, ..., \mathbf{u}_n\}$. Then for every $A \in \mathcal{A}(H)$ we have $A_n = P_nAP_n + ((a+b)/2)(\mathrm{I} - P_n) \to A$ pointwise, whence $f(A_n) \to f(A)$ pointwise. On the other hand, by the functional calculus, $f(A_n) = f(P_nAP_n) + f((a+b)/2)(\mathrm{I} - P_n)$.]

2. (F. A. Berezin) Let P be an orthogonal projection in $\mathbb{R}^n$ and $A \in \mathrm{Sym}(n,\mathbb{R})$. Prove that

   $$\operatorname{trace}(Pf(PAP)P) \leq \operatorname{trace}(Pf(A)P),$$

   for every convex function $f : \mathbb{R} \to \mathbb{R}$.

   [*Hint:* Use Lemma 1.9.1 and Lemma 1.9.2 to reduce the proof to the case where f is the absolute value function.]

3. (Peierls' inequality) Suppose that $A \in \mathrm{Sym}(n,\mathbb{R})$ is a self-adjoint matrix whose spectrum is contained in the interval $[m, M]$. Infer from the spectral decomposition of self-adjoint matrices that for every continuous convex function $f : [m, M] \to \mathbb{R}$ and every $\mathbf{x} \in \mathbb{R}^n$ with $\|\mathbf{x}\| = 1$ we have

   $$f(\langle A\mathbf{x}, \mathbf{x}\rangle) \leq \langle f(A)\mathbf{x}, \mathbf{x}\rangle.$$

   An immediate consequence is the fact that

   $$\sum_{k=1}^{n} f(\langle A\mathbf{u}_k, \mathbf{u}_k\rangle) \leq \operatorname{trace} f(A),$$

   for every orthonormal base $(\mathbf{u}_k)_k$ of $\mathbb{R}^n$.

4. Infer from Peierls' inequality that if $f : \mathbb{R} \to \mathbb{R}$ is a convex function, then the function
$$g(A) = \text{trace } f(A)$$
is convex on $\text{Sym}(n, \mathbb{R})$.

5. (The Peierls–Bogoliubov inequality). Suppose that $A, B \in \text{Sym}(n, \mathbb{R})$ and $f : \mathbb{R} \to \mathbb{R}$ is a convex function. Prove that
$$f\left(\frac{\text{trace }\left(A\mathrm{e}^B\right)}{\text{trace } \mathrm{e}^B}\right) \leq \frac{\text{trace }\left(f(A)\mathrm{e}^B\right)}{\text{trace } \mathrm{e}^B}.$$
[*Hint:* Replacing B by $B+\lambda I$ (if necessary) we may assume that trace $\left(\mathrm{e}^B\right) = 1$. Then, if $\mathrm{e}^B = \sum_k \mathrm{e}^{\lambda_k} \langle \cdot, \mathbf{v}_k \rangle \mathbf{v}_k$ is the spectral decomposition of e^B, we have $\sum_k \mathrm{e}^{\lambda_k} = 1$ and
$$\text{trace }\left(f(A)\mathrm{e}^B\right) = \sum_k \mathrm{e}^{\lambda_k} \langle f(A)\mathbf{v}_k, \mathbf{v}_k \rangle.$$
Continue by using the result of Exercise 3.]

6. (Hadamard determinant inequality) Infer from Peierls' inequality the Ha-damard determinant inequality, that is,
$$\det(A) \leq \prod\nolimits_{k=1}^{n} a_{kk}$$
for every $A \in \text{Sym}^{++}(n, \mathbb{R})$, $A = (a_{ij})_{i,j}$.
[*Hint:* Apply the result of Exercise 3 for $f(x) = -\log x$, the operator associated to A, and the natural basis of $\mathbb{R}^n$. Notice that $\mathrm{e}^{\text{trace}\,(\log A)} = \det(A)$.]

7. (An operator form of the Hermite-Hadamard inequality due to B. Mond and J. Pečarić [333]) Suppose that H is a real Hilbert space and $A \in \mathcal{A}(H)$ is a self-adjoint bounded operator whose spectrum is contained in the interval $[m, M]$. Prove that for every continuous convex function $f : [m, M] \to \mathbb{R}$ and every vector $\mathbf{x} \in H$ with $\|\mathbf{x}\| = 1$, we have
$$f(\langle A\mathbf{x}, \mathbf{x} \rangle) \leq \langle f(A)\mathbf{x}, \mathbf{x} \rangle \leq \frac{M - \langle A\mathbf{x}, \mathbf{x} \rangle}{M - m} f(m) + \frac{\langle A\mathbf{x}, \mathbf{x} \rangle - m}{M - m} f(M).$$
[*Hint:* For the left-hand side inequality, apply Theorem 1.6.5 to deduce for each $\varepsilon > 0$ the existence of an affine function $mt + n$ such that $f(t) \geq mt + n$ for all $t \in [m, M]$, and $m\langle A\mathbf{x}, \mathbf{x} \rangle + n \geq f\left(\langle A\mathbf{x}, \mathbf{x} \rangle\right) - \varepsilon$. For the right-hand side inequality, notice that $\frac{M-t}{M-m} f(m) + \frac{t-m}{M-m} f(M) - f(t) \geq 0$, so, according to the continuous functional calculus it follows that
$$\frac{M - A}{M - m} f(m) + \frac{A - m}{M - m} f(M) - f(A) \geq 0.]$$

8. Infer from Theorem 5.2.11 the following special case of an inequality due to E. H. Lieb: If $0 < p < 1$, then the function $F(A, B) = \text{trace}\left(A^p B^{1-p}\right)$ is jointly concave on the strictly positive $n \times n$ dimensional matrices A, B.

   [*Hint:* Adapt the argument of Corollary 5.2.12, by taking into account the fact that the function $-t^p$ is operator convex.]

   Remark. The *Lieb concavity inequality* asserts that for every $n \times n$ dimensional matrix X and every $0 < t < 1$, the function
$$f(A, B) = \text{trace}\,(X^* A^t X B^{1-t}),$$
   defined on pairs of positive matrices, is jointly concave. See B. Simon [449] Theorem 8.10, p. 73, for the proof.

9. According to Exercise 8, the map $(A, B) \to \frac{1}{p-1}\,\text{trace}\left(A^p B^{1-p} - A\right)$ is jointly convex. Prove that
$$\lim_{p \to 1} \frac{\text{trace}\left(A^p B^{1-p} - A\right)}{p - 1} = \text{trace}\,(A \log A - A \log B) = D_{KL}(A||B),$$
   and conclude that $D_{KL}(A||B)$ is jointly convex (since convexity is preserved by passing to the limit).

10. Infer from Theorem 5.2.11 that the function $(A, B) \to AB^{-1}A$ is jointly convex on $\text{Sym}(n, \mathbb{R}) \times \text{Sym}^{++}(n, \mathbb{R})$.

11. (Convexity inequality of O. Klein). Let A, B be self-adjoint bounded operators with spectrum in the domain of definition of the continuous convex function f. Prove that
$$\text{trace}\,[f(A) - f(B) - (A - B)\,f'(B)] \geq 0.$$

12. (A convex, but not an operator convex function) Prove that the function
$$f : \text{M}_{m \times n}(\mathbb{R}) \to \text{Sym}(n, \mathbb{R}), \quad f(A) = A^* A,$$
   is convex (though it is not operator convex).

5.3 The Trace Metric of $\text{Sym}^{++}(n, \mathbb{R})$

In every Hilbert space H, the middle point of a linear segment $[x, y]$ can be characterized as the unique point z such that
$$d(x, y)^2 + 4d(z, w)^2 = 2d(x, w)^2 + 2d(y, w)^2 \quad \text{for all } w \in H;$$
see the parallelogram law. Clearly, this remark also works when H is replaced by a convex subset of it.

$\mathrm{Sym}^{++}(n,\mathbb{R})$ (which is an open convex subset of the Hilbert space $\mathrm{Sym}(n,\mathbb{R})$) admits another metric structure that allows the presence of a valuable substitute of the parallelogram law. We refer here to the *trace metric*, defined on $\mathrm{Sym}^{++}(n,\mathbb{R})$ by the formula

$$\delta(A,B) = \left(\sum_{k=1}^{n} \log^2 \lambda_k\right)^{1/2}, \tag{5.5}$$

where $\lambda_1,\ldots,\lambda_n$ are the eigenvalues of AB^{-1}. The fact that this metric is well-defined is based on the property of invariance of the spectrum relative to similarity, that is,

$$\sigma(C) = \sigma(X^{-1}CX)$$

for every $C \in \mathrm{M}_n(\mathbb{R})$ and every X in $\mathrm{GL}(n,\mathbb{R})$ (the group of all nonsingular matrices belonging to $\mathrm{M}_n(\mathbb{R})$). Indeed, the matrix AB^{-1} is similar with

$$A^{-1/2}(AB^{-1})A^{1/2} = A^{1/2}B^{-1/2}(A^{1/2}B^{-1/2})^{\star} > 0,$$

and this fact assures the strict positivity of the eigenvalues of AB^{-1}. As a consequence, the trace metric is invariant under the natural action of the group $\mathrm{GL}(n,\mathbb{R})$ on $\mathrm{Sym}^{++}(n,\mathbb{R})$ via the bijective maps

$$\Gamma_C(A) = C^*AC, \quad A \in \mathrm{Sym}^{++}(n,\mathbb{R}),$$

associated to each $C \in \mathrm{GL}(n,\mathbb{R})$. Indeed,

$$\delta(\Gamma_C(A),\Gamma_C(B)) = \delta\,(A,B)\,.$$

In dimension 1, $\mathrm{Sym}^{++}(n,\mathbb{R})$ coincides with the interval $(0,\infty)$ and the trace metric is given by the formula

$$\delta(a,b) = |\log a - \log b|\,.$$

Then, for every two points $a,b \in (0,\infty)$, there is a unique point $c \in (0,\infty)$ such that

$$\delta(a,c) = \delta(c,b) = \frac{1}{2}\delta(a,b).$$

This unique point c is nothing but the geometric mean $\sqrt{ab}$ of a and b. Remarkably, all these facts extend to all dimensions.

The *geometric mean* $A\#_{1/2}B$ of two strictly positive matrices A and B was introduced by W. Pusz and S. L. Woronowicz [410] as the unique strictly positive solution

$$X = A^{1/2}(A^{-1/2}BA^{-1/2})^{1/2}A^{1/2} \tag{5.6}$$

of the so-called *Riccati equation,*

$$XA^{-1}X = B.$$

That $A\#_{1/2}B$ is such a solution is trivial. For uniqueness, suppose that X and Y are two strictly positive matrices such that $XA^{-1}X = B$ and $YA^{-1}Y = B$. Then

$$\left(A^{-1/2}XA^{-1/2}\right)^2 = A^{-1/2}\left(XA^{-1}X\right)A^{-1/2}$$
$$= A^{-1/2}\left(YA^{-1}Y\right)A^{-1/2} = \left(A^{-1/2}YA^{-1/2}\right)^2.$$

Because a strictly positive matrix has a unique strictly positive square root it follows that $A^{-1/2}XA^{-1/2} = A^{-1/2}YA^{-1/2}$. This easily implies that $X = Y$.

Some simple but useful properties are listed below:

$$A\#_{1/2}B = (AB)^{1/2} \quad \text{if } A \text{ and } B \text{ commute;}$$
$$A\#_{1/2}B = B\#_{1/2}A;$$
$$(C^*AC)\#_{1/2}(C^*BC) = C^*(A\#_{1/2}B)C \quad \text{for all } C \in \mathrm{GL}(n,\mathbb{R}).$$

See Exercises 1–3.

The *geometric–arithmetic mean inequality* for strictly positive matrices,

$$A\#_{1/2}B \leq \frac{A+B}{2},$$

can be easily derived from the following result on simultaneous diagonalization of noncommuting matrices:

5.3.1 Lemma *If $A, B \in \mathrm{Sym}^{++}(n,\mathbb{R})$, then there exists a nonsingular matrix C such that $\Gamma_C(A) = \mathrm{I}$ and $\Gamma_C(B)$ is diagonal.*

Proof. According to the spectral decomposition, one can find an orthogonal matrix C_1 such that $C_1^*AC_1$ is diagonal, and a strictly positive diagonal matrix D such that $D(C_1^*AC_1) = \mathrm{I}$. There exists also an orthogonal matrix C_2 such that $C_2^*(D^{1/2}C_1^*BC_1D^{1/2})C_2$ is diagonal. Put $C = C_1D^{1/2}C_2$. Then $\Gamma_C(A) = \mathrm{I}$ and $\Gamma_C(B)$ is diagonal. ■

Some other inequalities that can be deduced via Lemma 5.3.1 make the objective of Exercises 4 and 5.

5.3.2 Theorem *Given two matrices A and B in $\mathrm{Sym}^{++}(n,\mathbb{R})$, their geometric mean $A\#_{1/2}B$ is the unique matrix X in $\mathrm{Sym}^{++}(n,\mathbb{R})$ such that*

$$\delta(A,X) = \delta(B,X) = \frac{1}{2}\,\delta(A,B).$$

The proof follows from the following result:

5.3.3 Lemma *Suppose that A and B are two strictly positive matrices. Then, the map*

$$G_{(A,B)} : \mathrm{Sym}^{++}(n,\mathbb{R}) \to \mathrm{Sym}^{++}(n,\mathbb{R}),$$

defined by the formula $G_{(A,B)}X = (A\#_{1/2}B)X^{-1}(A\#_{1/2}B)$, *has the following properties:*

(a) $G_{(A,B)}$ *is a bijective isometry relative to the trace metric;*

(b) $G_{(A,B)}(A) = B$ *and* $G_{(A,B)}(B) = A$;

(c) $A\#_{1/2}B$ *is the unique fixed point of* $G_{(A,B)}$ *and*

$$\delta(G_{(A,B)}(X), X) = 2\delta(A\#_{1/2}B, X).$$

Proof. The proof of the assertions (a) and (b) is straightforward.

As concerns the first part of the assertion (c), if X is a fixed point of $G_{(A,B)}$, then

$$(A\#_{1/2}B)X^{-1}(A\#_{1/2}B) = X,$$

which yields $\left(X^{-1/2}(A\#_{1/2}B)X^{-1/2}\right)^2 = \mathrm{I}$ and $X^{-1/2}(A\#_{1/2}B)X^{-1/2} = \mathrm{I}$, by the uniqueness of the square root. Thus $X = A\#_{1/2}B$.

The second part of (c) is equivalent to the fact that

$$\delta((A\#_{1/2}B)X^{-1}(A\#_{1/2}B), X) = 2\delta(X, A\#_{1/2}B)$$

for every $X \in \mathrm{Sym}^{++}(n,\mathbb{R})$. This follows from the definition of the trace metric, by noticing that $\sigma(C^2) = \{\lambda^2 : \lambda \in \sigma(C)\}$ for all C in $\mathrm{Sym}^{++}(n,\mathbb{R})$. ■

The Hilbert space $\mathrm{Sym}(n,\mathbb{R})$ (endowed with the Hilbert–Schmidt norm) and the metric space $\mathrm{Sym}^{++}(n,\mathbb{R})$ can be related via the exponential map $\exp : \mathrm{Sym}(n,\mathbb{R}) \to \mathrm{Sym}^{++}(n,\mathbb{R})$, which is given by the formula

$$\exp(A) = \sum\nolimits_{k=1}^{r} \mathrm{e}^{\lambda_k} P_k,$$

provided that A has the spectral decomposition $A = \sum_{k=1}^{r} \lambda_k P_k$. Clearly this map is a continuous bijection; its continuity follows from Weyl's perturbation Theorem 4.4.7. If we denote its inverse by log, then

$$\delta(A,B) = \left\| \log(A^{-1/2}BA^{-1/2}) \right\|_{HS}.$$

It is worth mentioning that the above definition of the exponential map agrees with that used in functional analysis,

$$\exp(A) = \sum\nolimits_{k=1}^{\infty} \frac{1}{k!} A^k.$$

5.3.4 Lemma *The exponential is a continuous map from* $\mathrm{Sym}(n,\mathbb{R})$ *(endowed with the Hilbert–Schmidt norm) onto the space* $\mathrm{Sym}^{++}(n,\mathbb{R})$ *(endowed with the trace metric).*

Proof. Indeed, if $A_m \to A$ in $\mathrm{Sym}(n,\mathbb{R})$, then $\mathrm{e}^{-A_m}\mathrm{e}^{A}$ converges to I in the Hilbert–Schmidt norm, which implies that $\lambda_k(\mathrm{e}^{-A_m}\mathrm{e}^{A}) \to 1$ for each $k \in \{1,...n\}$. Taking into account the definition of the trace metric, we conclude that $\delta\left(\mathrm{e}^{A_m}, \mathrm{e}^{A}\right) \to 0$. ■

In order to outline deeper properties of the metric space $\mathrm{Sym}^{++}(n,\mathbb{R})$ we shall need the following technical result:

5.3.5 Lemma (The exponential metric increasing property) *We have*

$$\|\log A - \log B\|_{HS} \leq \delta(A, B) \quad \text{for all } A, B \in \mathrm{Sym}^{++}(n, \mathbb{R}), \tag{5.7}$$

The equality occurs when A and B commute with each other.

For details see R. Bhatia [52], Theorem 6.1.4 and Proposition 6.1.5. A proof of the inequality (5.7) (not including the equality part) makes the objective of Exercise 6.

5.3.6 Corollary *The metric space $\mathrm{Sym}^{++}(n, \mathbb{R})$ is complete relative to the trace metric.*

Proof. Let $(A_m)_m$ be a Cauchy sequence in $\mathrm{Sym}^{++}(n, \mathbb{R})$. By the exponential metric increasing property, the operators $B_m = \log A_m$ form a Cauchy sequence in $\mathrm{Sym}(n, \mathbb{R})$, so $B_m \to B$ for some B in $\mathrm{Sym}(n, \mathbb{R})$. By Lemma 5.3.4, the sequence $(A_m)_m$ converges to $A = \mathrm{e}^B$ in the metric space $\mathrm{Sym}^{++}(n, \mathbb{R})$. ■

Another basic feature of the geometry of $\mathrm{Sym}^{++}(n, \mathbb{R})$ is as follows.

5.3.7 Theorem (The semiparallelogram law) *Let A and B be two matrices in $\mathrm{Sym}^{++}(n, \mathbb{R})$ and let $M = A\#_{1/2}B$. Then, for every matrix $C \in \mathrm{Sym}^{++}(n, \mathbb{R})$, we have*

$$\delta^2(M, C) \leq \frac{\delta^2(A, C) + \delta^2(B, C)}{2} - \frac{1}{4}\delta^2(A, B).$$

Proof. By applying the isometry $\Gamma_{M^{-1/2}}$ to all matrices involved we may assume that $M = \mathrm{I}$. Then

$$\delta(M, C) = \|\log M - \log C\|_{HS} \quad \text{and} \quad \delta(A, B) = \|\log A - \log B\|_{HS},$$

according to the exponential metric increasing property. By the parallelogram law in $\mathrm{Sym}(n, \mathbb{R})$,

$$\begin{aligned}
\delta^2(M, C) &= \|\log M - \log C\|_{HS}^2 \\
&= \frac{\|\log A - \log C\|_{HS}^2 + \|\log B - \log C\|_{HS}^2}{2} - \frac{\|\log A - \log B\|_{HS}^2}{4} \\
&\qquad = \frac{\|\log A - \log C\|_{HS}^2 + \|\log B - \log C\|_{HS}^2}{2} - \frac{1}{4}\delta^2(A, B).
\end{aligned}$$

The proof ends by taking again into account the exponential metric increasing property. ■

The discussion above can be summarized saying that $\mathrm{Sym}^{++}(n, \mathbb{R})$ endowed with the trace metric is a global NPC space in the sense of the following definition:

5.3.8 Definition *A metric space $M = (M, d)$ is called a global NPC space (that is, a space with a global nonpositive curvature) if it is complete and for* *each pair of points $x_0, x_1 \in M$ there exists a point $y \in M$ with the property that for all $z \in M$ we have*

$$d^2(z,y) \le \frac{1}{2}d^2(z,x_0) + \frac{1}{2}d^2(z,x_1) - \frac{1}{4}d^2(x_0,x_1). \tag{5.8}$$

The point y that appears in Definition 5.3.8 is called the *midpoint* of x_0 and x_1; it is the unique point $x_{1/2}$ in M such that

$$d(x_0, x_{1/2}) = d(x_{1/2}, x_1) = \frac{1}{2}d(x_0, x_1).$$

The existence of geodesics in the space $\mathrm{Sym}^{++}(n,\mathbb{R})$ follows from the following result, whose proof is borrowed from K. T. Sturm [466].

5.3.9 Theorem *If $M = (M,d)$ is a global NPC space, then each pair of points $x_0, x_1 \in M$ can be connected by a geodesic, that is, by a rectifiable and continuous map $\gamma : [0,1] \to M$ such that $\gamma(0) = x_0$, $\gamma(1) = x_1$ and the length of $\gamma|_{[s,t]}$ is $d(\gamma(s), \gamma(t))$ for all $0 \le s \le t \le 1$. Moreover, this geodesic is unique and the inequality* (5.8) *can be extended as*

$$d^2(z,\gamma(t)) \le (1-t)d^2(z,x_0) + td^2(z,x_1) - t(1-t)d^2(x_0,x_1), \tag{5.9}$$

for every $z \in M$ and every $t \in [0,1]$.

Proof. Given $x_0, x_1 \in M$, we infer from Definition 5.3.8 the existence of their midpoint $x_{1/2}$. Then the points $x_{1/4}$ and $x_{3/4}$ are obtained as midpoints of x_0 and $x_{1/2}$ and $x_{1/2}$ and x_1, respectively. In this way, we obtain the points x_t for all dyadic $t \in [0,1]$ and, clearly, $d(x_s, x_t) = d(x_s, x_r) + d(x_r, x_t)$ for all dyadic $0 \le s < r < t \le 1$. By completeness of M, it yields the existence of $x_t \in M$ for all $t \in [0,1]$ such that $\gamma : [0,1] \to M$, $\gamma(t) = x_t$, is a geodesic connecting x_0 and x_1. The uniqueness of the geodesic follows from the uniqueness of the midpoints.

As concerns the inequality (5.9), it suffices to prove it for dyadic $t \in [0,1]$. It obviously holds for $t = 0$ and $t = 1$. Assuming that it holds for all $t = k2^{-n}$ with $k = 0, 1, ..., 2^n$, we will show that (5.9) also holds for all $t = k2^{-n-1}$ with $k = 0, 1, ..., 2^{n+1}$. The case of even k is covered by our assumption, so it remains the consider the case of points $t = k2^{-n-1}$ with k an odd number. Put $\Delta = 2^{-n-1}$. Then, according to the definition of global NPC spaces, for all $z \in M$ we have

$$d^2(z,x_t) \le \frac{1}{2}d^2(z,x_{t-\Delta}) + \frac{1}{2}d^2(z,x_{t+\Delta}) - \frac{1}{4}d^2(x_{t-\Delta},x_{t+\Delta}),$$

and, by (5.9), for multiples of 2^{-n}, we have

$$\begin{aligned} d^2(z,x_{t\pm\Delta}) \le (1-t\mp\Delta)d^2(z,x_0) &+ (t\pm\Delta)d^2(z,x_1) \\ &- (t\pm\Delta)(1-t\mp\Delta)d^2(x_0,x_1), \end{aligned}$$

for all $z \in M$. Combining them, we conclude that

$$d^2(z, x_t) \le (1-t)d^2(z, x_0) + td^2(z, x_1) - t(1-t)d^2(x_0, x_1).$$

■

Among the Banach spaces, only Hilbert spaces are global NPC spaces. In their case, the geodesics are the line segments.

In the case of $\mathrm{Sym}^{++}(n, \mathbb{R})$, the geodesic connecting the points A and B is parametrized by the function

$$\gamma(t) = A\#_t B, \quad t \in [0,1],$$

where $A\#_t B = A^{1/2}(A^{-1/2}BA^{-1/2})^t A^{1/2}$.

Exercises

1. Use spectral decomposition to prove that $A\#_{1/2}B = (AB)^{1/2}$ if the strictly positive matrices A and B commute.

2. Infer from the Riccati equation that

$$\left(A\#_{1/2}B\right)^{-1} = \left(A^{-1}\right)\#_{1/2}\left(B^{-1}\right) \text{ and } \left(A\#_{1/2}B\right)^{-1} = \left(B^{-1}\right)\#_{1/2}\left(A^{-1}\right)$$

   for all $A, B \in \mathrm{Sym}^{++}(n, \mathbb{R})$. This implies the commutativity of the geometric mean,

$$A\#_{1/2}B = B\#_{1/2}A.$$

3. Prove that the transformations $\Gamma_C(A) = C^*AC$ preserve the geometric mean, that is,

$$(C^*AC)\#_{1/2}(C^*BC) = C^*(A\#_{1/2}B)C$$

   for all $A, B \in \mathrm{Sym}^{++}(n, \mathbb{R})$ and $C \in \mathrm{GL}(n, \mathbb{R})$.

   [*Hint:* Use the fact that the geometric mean of two strictly positive matrices A and B is the unique strictly positive solution of the equation $XA^{-1}X = B$.]

4. Suppose that $A, B \in \mathrm{Sym}^{++}(n, \mathbb{R})$ and consider the following two iterative sequences defined by $A_0 = A$, $B_0 = B$ and

$$A_{n+1} = 2(A_n^{-1} + B_n^{-1})^{-1}, \; B_{n+1} = (A_n + B_n)/2 \quad \text{for } n \ge 0.$$

   Prove that

   (a) $A_n \le A_{n+1} \le A\#_{1/2}B \le B_{n+1} \le B_n \le$ for all $n \ge 1$;

   (b) Both sequences $(A_n)_n$ and $(B_n)_n$ converge to $A\#_{1/2}B$.

   [*Hint:* Apply Lemma 5.3.1 to reduce ourselves to the case where $A = I$ and B is a diagonal matrix.]

5. Infer from the preceding exercise the monotonicity property of the geometric mean: if $A \leq A'$ and $B \leq B'$ in $\mathrm{Sym}^{++}(n, \mathbb{R})$, then

$$A\#_{1/2}B \leq A'\#_{1/2}B'.$$

An immediate consequence is the following special case of the Löwner–Heinz Theorem: if $A, B \in \mathrm{Sym}^{+}(n, \mathbb{R})$ and $A \leq B$, then $A^{1/2} \leq B^{1/2}$.

6. (R. Bhatia [52]) The aim of this exercise is to sketch a proof of the exponential metric increasing property based on the formula (5.10) of J.-B. Duhamel, F. J. Dyson, R. P. Feynman, and J. Schwinger, concerning the differential of the exponential.

(a) Let $A \in \mathrm{Sym}(n, \mathbb{R})$. The differential of the exponential map is defined by the formula $\mathrm{de}^{A}(B) = \frac{\mathrm{d}}{\mathrm{d}t}\mathrm{e}^{A+tB}\big|_{t=0}$. Prove that

$$\mathrm{de}^{A}(B) = \int_0^1 \mathrm{e}^{tA} B \mathrm{e}^{(1-t)A} \mathrm{d}t. \tag{5.10}$$

(b) Suppose that $X = (x_{ij})_{i,j} \in \mathrm{Sym}(n, \mathbb{R})$ and $A \in \mathrm{Sym}^{++}(n, \mathbb{R})$ is a diagonal matrix with eigenvalues $\lambda_1, ..., \lambda_n$. Prove that the entries of the matrix $\int_0^1 A^t X A^{1-t} \mathrm{d}t$ are of the form

$$\left(\int_0^1 \lambda_i^t \lambda_j^{1-t} \mathrm{d}t \right) x_i x_j = L(\lambda_i, \lambda_j)\, x_i x_j,$$

where $L(\lambda_i, \lambda_j)$ denotes the logarithmic mean of λ_i and λ_j. Infer that in this case

$$\left\| A^{1/2} X A^{1/2} \right\|_{HS} \leq \left\| \int_0^1 A^t X A^{1-t} \mathrm{d}t \right\|_{HS}.$$

(c) Use the assertion (b) to prove that for all $A, B \in \mathrm{Sym}(n, \mathbb{R})$ we have

$$\|B\|_{HS} \leq \left\| \mathrm{e}^{-A/2} \mathrm{de}^{A}(B) \mathrm{e}^{-A/2} \right\|_{HS}.$$

[*Hint:* (b) The logarithmic mean is greater than or equal with the geometric mean. (c) Apply (b) for $X = \mathrm{e}^{-A/2} B \mathrm{e}^{-A/2}$.]

5.4 Geodesic Convexity in Global NPC Spaces

In any global NPC space $M = (M, d)$, one can introduce the concepts of convex set and convex functions by using the geodesics.

5.4.1 Definition *A set $C \subset M$ is called convex if for every pair of points x_0 and x_1 in C, the image of every geodesic $\gamma : [0, 1] \to M$ joining $x_0 = \gamma(0)$ and $x_1 = \gamma(1)$ is contained in C.*

A function $\varphi : C \to \mathbb{R}$ is called convex if the function $\varphi \circ \gamma : [0,1] \to \mathbb{R}$ is convex for each geodesic $\gamma : [0,1] \to C$, $\gamma(t) = \gamma_t$, that is,

$$\varphi(\gamma_t) \leq (1-t)\varphi(\gamma_0) + t\varphi(\gamma_1)$$

for all $t \in [0,1]$.

The related notions of log-convex function, strictly convex function, strongly convex function, etc. can be introduced in the same manner, imitating the case of normed linear spaces.

According to Theorem 5.3.9, for z arbitrarily fixed in M, the function

$$f(x) = d^2(x,z)$$

is strongly convex on M. As a consequence, the balls are convex sets.

One can even prove that the distance function d is convex on $M \times M$ as a function of both variables. For details, see K. T. Sturm [466], Corollary 2.5. The global NPC product structure on $M \times M$ is provided by the product metric

$$\omega((x_0,x_1),(y_0,y_1)) = \left(d^2(x_0,y_0) + d^2(x_1,y_1)\right)^{1/2}.$$

Every closed convex subset of a global NPC space is a space of the same nature.

The determinant function is convex on $\mathrm{Sym}^{++}(n,\mathbb{R})$ due to the identity

$$\det A^{1/2}\left(A^{-1/2}BA^{-1/2}\right)^t A^{1/2} = (\det A)^{1-t}(\det B)^t$$

and the *AM–GM* inequality.

Every positive linear functional on $\mathrm{M}_n(\mathbb{R})$ induces a log-convex function on $\mathrm{Sym}^{++}(n,\mathbb{R})$ in the sense of Definition 5.4.1. In particular, the trace is a log-convex function. See Exercise 3.

5.4.2 Remark *More examples of convex functions defined on the global NPC space $\mathrm{Sym}^{++}(n,\mathbb{R})$ can be obtained by taking into account the following facts:*

If $f : \mathrm{Sym}^{++}(n,\mathbb{R}) \to \mathbb{R}$ is convex and $g : \mathbb{R} \to \mathbb{R}$ is increasing and convex, then $g \circ f$ is convex.

If $\Psi : \mathrm{M}_n(\mathbb{R}) \to \mathrm{M}_n(\mathbb{R})$ is a strictly positive linear map and f is increasing and convex, then $f \circ \Psi|_{\mathrm{Sym}^{++}(n,\mathbb{R})}$ is convex.

Many results mentioned in Chapter 4 to work in the framework of linear normed spaces remain valid also in the context of NPC spaces.

For example, any continuous midpoint convex function $f : M \to \mathbb{R}$ is convex.

A function $f : M \to \mathbb{R}$ is convex if, and only if, its epigraph is a convex set. All level sets of a convex function are convex sets.

A function $f : M \to \mathbb{R}$ is lower semicontinuous if, and only if, its level sets are convex (if, and only if, its epigraph is closed).

One serious difference between usual convexity and convexity in global NPC spaces concerns the concept of convex combination. The key remark indicating the right approach of this problem is offered by the following alternative definition of the notion of midpoint.

5.4.3 Lemma *Let x_0 and x_1 be a pair of points in the global NPC space M. Then their midpoint $x_{1/2}$ is the unique minimizer of the strongly convex function*

$$V(x) = \frac{1}{2}d^2(x, x_0) + \frac{1}{2}d^2(x, x_1).$$

Proof. According to the definition of a global NPC space, we have

$$d^2(x, x_{1/2}) \leq V(x) - \frac{1}{4}d^2(x_0, x_1) = V(x) - V(x_{1/2}),$$

whence we infer that $x_{1/2}$ is a minimizer of V. Since V is strictly convex, we conclude that $x_{1/2}$ is in fact the unique minimizer of V. ■

This remark suggests to define the NPC analog $\lambda_1 x_1 \boxplus \cdots \boxplus \lambda_n x_n$ of a convex combination $\sum_{k=1}^n \lambda_k x_k$ as the unique point minimizing the strongly convex function

$$W(x) = \sum_{k=1}^{n} \lambda_k d^2(x, x_k).$$

It turns out that this point is nothing but the barycenter $\operatorname{bar}(\mu)$ of the discrete Borel probability measure $\mu = \sum_{k=1}^n \lambda_k \delta_{x_k}$.

The discrete case of Jensen's inequality in the context of global NPC spaces can be settled by adapting the argument of Proposition 3.2.2.

5.4.4 Proposition *If f is a convex function on the global NPC space M, then for every family of points $x_1, ..., x_n$ in M and every family of positive weights $\lambda_1, ..., \lambda_n$ with $\sum_{k=1}^n \lambda_k = 1$ we have*

$$f(\lambda_1 x_1 \boxplus \cdots \boxplus \lambda_n x_n) \leq \sum_{k=1}^{n} \lambda_k f(x_k).$$

The existence of barycenters in more general situations needs the following result.

5.4.5 Lemma *If M is a global NPC space, then every lower semicontinuous strongly convex function $f : M \to \mathbb{R}$ which is bounded below admits a unique minimizer $z \in M$.*

Proof. Let $(x_n)_n$ be a sequence of points such that

$$\lim_{n \to \infty} f(x_n) = \inf_{x \in M} f(x) = \alpha.$$

Assuming that f is strongly convex with parameter $C > 0$, we have

$$\alpha \leq f\left(\frac{1}{2}x_m \boxplus \frac{1}{2}x_n\right) \leq \frac{f(x_m) + f(x_n)}{2} - \frac{C}{4}d^2(x_m, x_n),$$

whence we infer that $(x_n)_n$ is a Cauchy sequence. Due to the completeness of M there exists a point x such that $x_n \to x$ as $n \to \infty$. Then $f(x) = \alpha$ because f is lower semicontinuous and its lower bound is α.

The uniqueness of the minimizer is a consequence of the strict convexity of f. ■

Under the circumstances of Lemma 5.4.5 we denote

$$z = \arg\min_{x \in M} f(x).$$

The notion of barycenter for an arbitrary Borel probability measure μ defined on a global NPC space M was introduced by E. Cartan [95], inspired by Gauss' least squares method.

We denote by $\mathcal{P}^2(M)$ the set of all Borel probability measures μ on M under which all functions $d^2(\cdot, z)$ are integrable.

5.4.6 Definition *The barycenter* $\mu \in \mathcal{P}^2(M)$ *is defined by the formula*

$$\mathrm{bar}(\mu) = \arg\min_{z \in M} \frac{1}{2} \int_M d^2(z, x) \mathrm{d}\mu(x).$$

The fact that the notion of barycenter is well defined follows from Lemma 5.4.5, by noticing that strong convexity of the function

$$I(x) = \int_M d^2(z, x) \mathrm{d}\mu(z).$$

The generalization of Jensen's inequality to the context of global NPC spaces was done by J. Jost [237]. We state it here in the formulation of J. Eells and B. Fuglede [141], Proposition 12.3, p. 242:

5.4.7 Theorem (Jensen's Inequality). *For any lower semicontinuous convex function* $f : M \to \mathbb{R}$ *and any Borel probability measure* $\mu \in \mathcal{P}^2(M)$ *we have the inequality*

$$f(\mathrm{bar}(\mu)) \leq \int_M f(x) \mathrm{d}\mu(x),$$

provided the right-hand side is well defined.

The proof of Eells and Fuglede is based on the following result concerning barycenters:

5.4.8 Lemma *If a Borel probability measure* μ *is supported by a convex closed set* K, *then its barycenter* $\mathrm{bar}(\mu)$ *lies in* K.

The proof of Lemma 5.4.8 is an immediate consequence of the basic separation Theorem 2.3.1.

Exercises

1. Prove that for every three matrices A, B, C in $\mathrm{Sym}^{++}(n, \mathbb{R})$ we have

$$\delta(A\#_{1/2}B, A\#_{1/2}C) \leq \frac{\delta(A, B)}{2}.$$

2. Prove that the geometric mean $A\#_{1/2}B$ is continuous as a function of pairs of strictly positive matrices.

3. Suppose that $A, B, A', B' \in \mathrm{Sym}^{++}(n, \mathbb{R})$. Prove that the function

$$f(t) = \delta(A\#_t B, A'\#_t B')$$

is convex on [0,1].

[*Hint*: It suffices to show that f is midpoint convex.]

4. Prove that every positive linear functional on $\mathrm{M}_n(\mathbb{R})$ induces a log-convex function on $\mathrm{Sym}^{++}(n, \mathbb{R})$ in the sense of Definition 5.4.1.

[*Hint*: Reduce the problem to the case of pure states, that is, to the case of functionals $\omega : A \to \langle A\mathbf{u}, \mathbf{u}\rangle$ associated to the unit vectors $\mathbf{u} \in \mathbb{R}^n$.]

5. (The variance inequality) Suppose that M is a global NPC space and $\mu \in \mathcal{P}^2(M)$. Prove that for all $z \in M$,

$$\int_M d^2(z, x)\mathrm{d}\mu(x) \geq d^2(z, \mathrm{bar}(\mu)) + \int_M d^2(\mathrm{bar}(\mu), x)\mathrm{d}\mu(x).$$

Compare to Lagrange's barycentric identity. It is remarkable that the above inequality characterizes the global NPC spaces among the complete metric spaces. See K. T. Sturm [466], Theorem 4.9.

5.5 Comments

The theory of convex spectral functions goes back to Minkowski's determinant inequality and to von Neumann's paper from 1937, relating the concepts of unitarily invariant matrix norm and symmetric gauge function. See [349]. A norm on $\mathrm{M}_N(\mathbb{C})$ is *unitarily invariant* if $\|UAV\| = \|A\|$ for all unitary U and V. Examples are the operator norm and the Hilbert–Schmidt norm. More generally, if $f : \mathbb{R}^N \to \mathbb{R}$ is a *symmetric gauge function*, that is, a norm invariant under permutations on $\mathbb{R}^N$ such that $f(\mathbf{x}) = f(|\mathbf{x}|)$ for every $\mathbf{x}$, then

$$\|A\|_f = f(s_1(A)..., s_N(A))$$

is a unitarily invariant matrix norm. Von Neumann has proved that every unitarily invariant matrix norm can be written in this form. This was generalized to the infinite-dimensional setting by I. C. Gohberg and M. G. Krein. See the book of B. Simon [449] for details. More recently, J. M. Borwein, A. S. Lewis, Q. J. Zhu et al. have added significant contributions concerning differentiability of convex spectral functions. See [73] and references therein.

The notion of spectral function in the framework of C^*-algebras was studied by F. Hansen [200] in connection with Jensen-type inequalities for convex functions of several operator variables and the construction of multivariate geometric means.

The subject of matrix (and operator) convexity is important in quantum statistical mechanics, quantum information theory, free convexity, etc. Our presentation is based on the expository papers of E. A. Carlen [92] and F. Hansen [199].

More advanced material is made available by the books of R. Bhatia [49] and B. Simon [449], [450].

The extension of trace inequalities from the case of matrices to that of operators is possible for the operators of trace class. Recall that a bounded linear operator A (acting on the separable Hilbert space H) is said to be of *trace class* (or *nuclear*) if

$$\|A\|_1 = \sum_n \langle |A| \mathbf{u}_n, \mathbf{u}_n \rangle < \infty \tag{5.11}$$

for some (and hence all) orthonormal bases $(\mathbf{u}_n)_n$ of H. In this case, the *trace* of A is given by the formula

$$\text{trace } A = \sum_n \langle A\mathbf{u}_n, \mathbf{u}_n \rangle,$$

and is independent of the choice of the orthonormal basis. A result due to V. B. Lidskii asserts that

$$\text{trace } A = \sum_n \lambda_n(A),$$

where $\lambda_1(A), \lambda_2(A), \lambda_3(A), ...$ are the eigenvalues of A, enumerated with their algebraic multiplicities. Therefore, a positive self-adjoint operator is trace class if, and only if, trace $A < \infty$.

Notice that every trace class operator is compact (but not every compact operator is of trace class).

The trace class operators $A : H \to H$ constitute a Banach space relative to the *nuclear norm* $\|\cdot\|_1$ defined by the formula (5.11). This space, denoted $N(H,H)$, is the dual of the Banach space $K(H,H)$, of all compact operators from H into itself, endowed with the operator norm. The dual of $N(H,H)$ is the space $L(H,H)$, of all continuous linear operators.

A nice survey of the trace class operators (including details of all results mentioned above) is offered by the book of B. Simon [449].

The world of matrix inequalities is full of many exotic examples, which seems to appear from nowhere. So is the case of *Golden–Thompson inequality,* which asserts that if $A, B \in \mathrm{Sym}(N, \mathbb{R})$, then

$$\mathrm{trace}\left(\mathrm{e}^{A+B}\right) \leq \mathrm{trace}\left(\mathrm{e}^{A}\mathrm{e}^{B}\right).$$

For the proof and generalizations see the aforementioned books of R. Bhatia and B. Simon. Interestingly, the det companion of this inequality is an identity,

$$\det\left(\mathrm{e}^{A+B}\right) = \det\left(\mathrm{e}^{A}\mathrm{e}^{B}\right).$$

The theory of global NPC spaces was initiated by A. D. Alexandrov and Yu. G. Reshetnyak and was strongly influenced by the work of M. Gromov. A very readable introductory paper is that by J. D. Lawson and Y. Lim [274]. Some more advanced material clarifying many geometric aspects is available in the books of R. Bhatia [52] and J. Jost [237]. A probabilistic view of the theory of global NPC spaces is offered by the excellent survey of K. T. Sturm [466].

During the last decade much activity was done to extend the concept of mean of several variables to the framework of global NPC spaces. See the paper of J. D. Lawson and Y. Lim [275] and references therein.

For an extension of the majorization inequality of Hardy–Littlewood–Pólya, see C. P. Niculescu and I. Roventa [373].

Chapter 6

Duality and Convex Optimization

Convex optimization is one of the main applications of the theory of convexity and Legendre–Fenchel duality is a basic tool, making more flexible the approach of many concrete problems. The diet problem, the transportation problem, and the optimal assignment problem are among the many problems that during the Second World War and immediately after led L. Kantorovich, T. C. Koopmans, F. L. Hitchcock, and G. B. Danzig to develop the mathematical theory of linear programming. Soon it was realized that most results extend to the framework of convex functions, which marked the birth of convex programming. Later on, W. Fenchel, R. T. Rockafellar, and J. J. Moreau laid the foundations of convex analysis.

Since even the briefest presentation of the aims and scope of convex optimization requires more than a single chapter, our discussion here will be limited to its basic aspects. The following nice books provide a considerably more systematic treatment of the subject and much more significant applications: V. Barbu and T. Precupanu [29], J. M. Borwein and A. S. Lewis [72], J. M. Borwein and J. Vanderwerff [73], F. H. Clarke [110], I. Ekeland and R. Temam [145], R. T. Rockafellar and R. J.-B. Wets [422].

6.1 Legendre–Fenchel Duality

An immediate consequence of the separation theorems presented in Section 2.3 is the fact that the closed convex sets can be described either as a union of points or as an intersection of closed half-spaces. This dual description has important consequences when applied to epigraphs.

Given a real Banach space E, the *conjugate* (or the *Legendre–Fenchel transform*) of a proper function $f\colon E \to \mathbb{R} \cup \{\infty\}$ is the function $f^*\colon E^* \to \mathbb{R} \cup \{\infty\}$

C. P. Niculescu and L.-E. Persson, *Convex Functions and Their Applications*,
CMS Books in Mathematics, https://doi.org/10.1007/978-3-319-78337-6_6

defined by the formula

$$f^*(x^*) = \sup\{\langle \mathbf{x}, x^*\rangle - f(\mathbf{x}) : \mathbf{x} \in E\}. \tag{6.1}$$

Here $\langle \mathbf{x}, x^*\rangle = x^*(\mathbf{x})$ represents the duality map between E and E^*. It is clear that the formula (6.1) can be restated as

$$f^*(x^*) = \sup\{\langle \mathbf{x}, x^*\rangle - f(\mathbf{x}) : \mathbf{x} \in \mathrm{dom}(f)\},$$

a fact which shows that the function f^* is convex and lower semicontinuous since it is the pointwise supremum of the family of continuous affine functions $x^* \to \langle \mathbf{x}, x^*\rangle - f(\mathbf{x})$. Moreover, f^* takes values in $\mathbb{R} \cup \{\infty\}$. Notice that these properties of f^* do not assume that f is convex.

Geometrically, the Legendre–Fenchel conjugate solves the problem of finding for a given $x^* \in E^*$, the greatest affine function $y = \langle \mathbf{x}, x^*\rangle - \alpha$ majorized by f and having the gradient x^*. The solution is $\mathbf{y} = \langle \mathbf{x}, x^*\rangle - f^*(x^*)$.

The following result extends Young's inequality (1.5). See Example 6.1.6.

6.1.1 Proposition (The Fenchel–Young inequality)*If $f : E \to \mathbb{R} \cup \{\infty\}$ is a lower semicontinuous proper convex function, then so is f^* and*

$$f(\mathbf{x}) + f^*(x^*) \geq x^*(\mathbf{x}) \quad \textit{for all } \mathbf{x} \in E, x^* \in E^*. \tag{6.2}$$

Equality holds if and only if $x^* \in \partial f(\mathbf{x})$ (and also if and only if $\mathbf{x} \in \partial f^*(x^*)$, see Theorem 6.1.2 below).

Proof. According to Theorem 3.3.2, f admits a continuous affine minorant h. Since this minorant is of the form $h(\mathbf{x}) = z^*(\mathbf{x}) + \alpha$ for suitable $z^* \in E^*$ and $\alpha \in \mathbb{R}$, we have $f^*(z^*) \leq \sup\{z^*(\mathbf{x}) - h(\mathbf{x}) : \mathbf{x} \in E\} \leq -\alpha$, so f^* is a proper function. The inequality (6.2) is now clear. As concerns the equality case, notice that

$$\begin{aligned}
x^*(\mathbf{x}) = f(\mathbf{x}) + f^*(x^*) &\Leftrightarrow x^*(\mathbf{x}) \geq f(\mathbf{x}) + f^*(x^*) \\
&\Leftrightarrow x^*(\mathbf{x}) \geq f(\mathbf{x}) + x^*(\mathbf{y}) - f(\mathbf{y}) \text{ for all } \mathbf{y} \in E \\
&\Leftrightarrow f(\mathbf{y}) \geq f(\mathbf{x}) + x^*(\mathbf{y} - \mathbf{x}) \text{ for all } \mathbf{y} \in E \\
&\Leftrightarrow x^* \in \partial f(\mathbf{x}).
\end{aligned}$$

■

By definition, the *biconjugate* f^{**} of f is given by the formula

$$f^{**}(\mathbf{x}) := \sup\{x^*(\mathbf{x}) - f^*(x^*) : x^* \in E^*\}, \quad \mathbf{x} \in E.$$

In general,

$$f \leq g \text{ implies } f^* \geq g^*$$

and thus $f^{**} \leq g^{**}$.

The circumstances under which $f = f^{**}$ make the objective of the following result.

6.1.2 Theorem (The Fenchel–Hörmander–Moreau theorem on biconjugation) *Suppose that $f\colon E \to \mathbb{R} \cup \{\infty\}$ is a proper function. Then the following assertions are equivalent:*

(a) *f is lower semicontinuous and convex;*
(b) *$f = g^*$ for some proper function g;*
(c) *$f = f^{**}$.*

Proof. It is obvious that (c) implies (b), which in turn implies (a). As concerns the implication (a) $\Rightarrow$ (c), notice that every continuous affine minorant h of f verifies $h = h^{**} \leq f^{**} \leq f$. According to Theorem 3.3.2, f is the pointwise supremum of the family of its continuous and affine minorants and thus $f = f^{**}$.

Alternatively, the proof of the implication (a) $\Rightarrow$ (c) is a simple consequence of the Fenchel–Young inequality when $\operatorname{dom} f = E$. Indeed, in this case $\partial f(\mathbf{x}) \neq \emptyset$ for every $\mathbf{x}$, so by choosing $x^* \in \partial f(\mathbf{x})$ we have

$$f(\mathbf{x}) = x^*(\mathbf{x}) - f^*(x^*) \leq \sup_{\mathbf{y}^*} [y^*(\mathbf{x}) - f^*(y^*)] = f^{**}(\mathbf{x}),$$

whence we infer that $f = f^{**}$. The general case can be fixed via an approximation argument. We will come back to this fact in Section 6.5, Exercise 6. ■

6.1.3 Corollary *If E is a real Banach space and $f\colon E \to \mathbb{R} \cup \{\infty\}$ is a proper convex function, then $f^{**} = \operatorname{cl} f$.*

Combining Proposition 6.1.1 and Theorem 6.1.2, one obtains the following inversion rule for subgradients: Under the assumptions of Corollary 6.1.3, we have

$$\partial f^* = (\partial f)^{-1} \text{ and } \partial f = (\partial f^*)^{-1};$$

see Section 3.3 for the inverse of a set-valued map.

Since any real Hilbert space H is self-dual, we infer from Theorem 6.1.2, that the Legendre–Fenchel transform $f \to f^*$ is an involution of the set $\Gamma(H)$ (of all lower semicontinuous proper convex functions $f\colon H \to \mathbb{R} \cup \{\infty\}$). It acts according to the formulas

$$\begin{aligned} f^*(\mathbf{v}) &= \sup\{\langle \mathbf{x}, \mathbf{v}\rangle - f(\mathbf{x}) : \mathbf{x} \in H\} \\ f(\mathbf{x}) = f^{**}(\mathbf{x}) &= \sup\{\langle \mathbf{x}, \mathbf{v}\rangle - f^*(\mathbf{v}) : \mathbf{v} \in H\}. \end{aligned}$$

6.1.4 Example *For any nonempty convex subset C of a real Banach space, the conjugate of the indicator function ι_C is the support function σ_C. According to Theorem 6.1.2, if C is also closed, then the conjugate of σ_C is ι_C.*

6.1.5 Example *If E is a real Banach space, then the conjugate of the norm function $\mathrm{N}(\mathbf{x}) = \|\mathbf{x}\|$ is the indicator function of the closed unit ball of E^*, that is,*

$$\mathrm{N}^*(x^*) = \sup\{x^*(\mathbf{x}) - \|\mathbf{x}\| : \mathbf{x} \in E\} = \begin{cases} 0 & \text{if } \|x^*\| \leq 1 \\ \infty & \text{if } \|x^*\| > 1. \end{cases}$$

Indeed, if $\|x^*\| \leq 1$, *then* $x^*(\mathbf{x}) - \|\mathbf{x}\| \leq \|\mathbf{x}\|(\|x^*\| - 1) \leq 0$ *and the supremum is attained at* $\mathbf{x} = 0$. *If* $\|x^*\| > 1$, *then there exists an element* $\mathbf{x} \in E$ *such that* $\|\mathbf{x}\| = 1$ *and* $\varepsilon = x^*(\mathbf{x}) - 1 > 0$. *Then* $x^*(n\mathbf{x}) - \|n\mathbf{x}\| = n\varepsilon$ *for every* $n \in \mathbb{N}$, *whence* $\mathrm{N}^*(x^*) = \infty$.

6.1.6 Example *Suppose that* $\varphi : \mathbb{R} \to \mathbb{R}$ *is an even function and* $\varphi \in \Gamma(\mathbb{R})$. *Then, for every real Banach space* E, *the functions*

$$F(\mathbf{x}) = \varphi(\|\mathbf{x}\|), \quad \mathbf{x} \in E,$$

and

$$G(x^*) = \varphi^*(\|x^*\|), \quad x^* \in E^*,$$

are conjugate to each other. Indeed, both functions are lower semicontinuous, convex and proper, so it suffices to show that $F^* = G$. *This is done as follows:*

$$\begin{aligned} F^*(x^*) &= \sup_{\mathbf{x} \in E} (\langle \mathbf{x}, x^* \rangle - F(\mathbf{x})) \\ &= \sup_{t > 0} \sup_{\mathbf{x} \in E, \|\mathbf{x}\| = t} (\langle \mathbf{x}, x^* \rangle - \varphi(\|\mathbf{x}\|)) \\ &= \sup_{t > 0} (t\|x^*\| - \varphi(t)) \\ &= \sup_{t \in \mathbb{R}} (t\|x^*\| - \varphi(t)) = \varphi^*(\|x^*\|) = G(x^*). \end{aligned}$$

If $p \in (1, \infty)$, *then, as was noticed in Section 1.6, the real function* $\varphi(t) = |t|^p / p$ *is an example of even continuous convex function whose conjugate is* $\varphi^*(t) = |t|^q / q$, *where* $1/p + 1/q = 1$. *Therefore,*

$$F(\mathbf{x}) = \frac{1}{p}\|\mathbf{x}\|^p \quad \text{and} \quad G(x^*) = \frac{1}{q}\|x^*\|^q$$

are conjugate to each other in every real Banach space.

In the case of continuously differentiable convex functions, one can easily compute f^* by using the differential of f. A simple example is as follows:

6.1.7 Example *Consider the strictly convex function* $f(\mathbf{x}) = \frac{1}{2}\langle A\mathbf{x}, \mathbf{x}\rangle$, *associated to a matrix* $A \in \mathrm{Sym}^{++}(N, \mathbb{R})$. *For each* $\mathbf{p} \in \mathbb{R}^N$, *the function* $f_{\mathbf{p}} : \mathbf{x} \to \langle \mathbf{x}, \mathbf{p}\rangle - \frac{1}{2}\langle A\mathbf{x}, \mathbf{x}\rangle$ *is strictly concave and differentiable, so it attains its maximum at the point where* $\nabla f_{\mathbf{p}}(\mathbf{x}) = \mathbf{p} - A\mathbf{x} = 0$, *that is, at* $\mathbf{x} = A^{-1}\mathbf{p}$. *Therefore*

$$f^*(\mathbf{p}) = \frac{1}{2}\langle A^{-1}\mathbf{p}, \mathbf{p}\rangle \quad \text{for } \mathbf{p} \in \mathbb{R}^N.$$

For $A = \mathrm{I}$, *we infer that the function* $\varphi(\mathbf{x}) = \frac{1}{2}\|\mathbf{x}\|^2$ *is self-conjugate (that is, equal to its Legendre–Fenchel transform). Moreover,* φ *is the only function defined on a finite dimensional real Hilbert space* H *with this property. Indeed, from the Fenchel–Young inequality it follows that* $\varphi(\mathbf{x}) \geq \frac{1}{2}\|\mathbf{x}\|^2$, *whence, by conjugation we deduce the other inequality,* $\varphi(\mathbf{x}) = \varphi^*(\mathbf{x}) \leq \frac{1}{2}\|\mathbf{x}\|^2$.

Example 6.1.7 illustrates the classical Legendre transform. This applies only to continuously differentiable strictly convex functions and it was W. Fenchel who extended the applicability of this transform using the formula (6.1).

6.1.8 Theorem (The classical Legendre transform) *Let $f : \mathbb{R}^N \to \mathbb{R}$ be a function of class C^2, whose Hessian matrix is strictly positive at every point $\mathbf{x} \in \mathbb{R}^N$. Then the conjugate of f is a function $f^* : \mathbb{R}^N \to \mathbb{R}$ of the same nature. More precisely, the following properties hold true:*

(a) *the gradient map $\mathbf{x} \to \nabla f(\mathbf{x})$ is a homeomorphism from $\mathbb{R}^N$ onto itself;*
(b) $f^\star(\mathbf{p}) = \langle \mathbf{p}, (\nabla f)^{-1} \mathbf{p}\rangle - f((\nabla f)^{-1} \mathbf{p})$ *for every* $\mathbf{p} \in \mathbb{R}^N$;
(c) $f^\star$ *is a C^1 function and* $\nabla f^\star = (\nabla f)^{-1}$;
(d) *the matrices* $\text{Hess}_{\mathbf{x}} f$ *and* $\text{Hess}_{\mathbf{p}} f^*$ *are inverse to each other when* $\mathbf{p} = \nabla f(\mathbf{x})$ *(or, equivalently, when* $\mathbf{x} = \nabla f^\star(\mathbf{p})$*).*

Therefore, the classical Legendre transform can be interpreted as an encoding of the function's epigraph in terms of its supporting hyperplanes.

Proof. (a) By our hypotheses, f is strongly convex and supercoercive; see Exercise 5, Section 3.8, and Proposition 3.10.8. Hence for every $\mathbf{x}, \mathbf{y} \in \mathbb{R}^N$ with $\mathbf{x} \neq \mathbf{y}$ and $\mathbf{y} \neq 0$, the function $g(t) = f(\mathbf{x} + t\mathbf{y})$ is strictly convex on $\mathbb{R}$, whence $g'(1) - g'(0) = \langle \nabla f(\mathbf{x} + \mathbf{y}) - \nabla f(\mathbf{y}), \mathbf{y}\rangle > 0$. This shows that ∇f is one-to-one. Let $\mathbf{p} \in \mathbb{R}^N$. Since $g(\mathbf{x}) = f(\mathbf{x}) - \langle \mathbf{x}, \mathbf{p}\rangle$ is convex, coercive, and of class C^1, it attains a global minimum at a point $\mathbf{x}$ for which $\nabla g(\mathbf{x}) = \nabla f(\mathbf{x}) - \mathbf{p} = 0$. Hence the gradient map $\nabla f(\mathbf{x})$ is also onto. The subgradient inequality $f(\mathbf{0}) + \langle \nabla f(\mathbf{x}), \mathbf{x}\rangle \geq f(\mathbf{x})$ and the property of supercoercivity of f yield $\|\nabla f(\mathbf{x})\| \to \infty$ as $\|\mathbf{x}\| \to \infty$. Therefore the inverse image under ∇f of every compact set is compact too, a fact which assures the continuity of $(\nabla f)^{-1}$. For (b), notice that the reasoning above demonstrates that the point $\mathbf{x}$ at which the map $\langle \mathbf{x}, \mathbf{p}\rangle - f(\mathbf{x})$ attains its supremum verifies $\nabla f(\mathbf{x}) - \mathbf{p} = 0$, whence $f^\star(\mathbf{p}) = \langle \mathbf{p}, (\nabla f)^{-1} \mathbf{p}\rangle - f((\nabla f)^{-1} \mathbf{p})$. The assertions (c) and (d) can be deduced by differentiation, using the inverse mapping theorem. ■

The classical Legendre transform plays an important role in the theory of Hamilton–Jacobi equations. See Section 6.6.

An example of conjugacy involving the Hilbert space of symmetric matrices is offered by the following result.

6.1.9 Theorem (A. S. Lewis [278]) *Suppose that $f : \mathbb{R}^N \to \mathbb{R} \cup \{\infty\}$ is a symmetric function and $\Lambda : \text{Sym}\,(N, \mathbb{R}) \to \mathbb{R}$ is the eigenvalues map that associates to each symmetric matrix A the string $\lambda^{\downarrow}(A)$, of its eigenvalues with multiplicities, listed downward. Then*

$$(f \circ \Lambda)^* = f^* \circ \Lambda.$$

Proof. According to Corollary 4.4.11, for every matrix $Y \in \text{Sym}(N, \mathbb{R})$ we have

$$\begin{aligned}(f \circ \Lambda)^* (Y) &= \sup \{\text{trace}\, XY - f(\Lambda(X)) : X \in \text{Sym}\,(N, \mathbb{R})\} \\ &\leq \sup \{\langle \Lambda(X), \Lambda(Y)\rangle - f(\Lambda(X)) : X \in \text{Sym}\,(N, \mathbb{R})\} \\ &\leq \sup \{\langle \mathbf{x}, \Lambda(Y)\rangle - f(\mathbf{x}) : \mathbf{x} \in \mathbb{R}^N\} = f^*(\Lambda(Y)).\end{aligned}$$

For the reverse inequality, notice that Y admits a factorization of the form $Y = U\Lambda(Y)U^*$, with $U \in \mathrm{O}(N)$, which implies that

$$\begin{aligned} f^*(\Lambda(Y)) &= \sup\left\{\langle \mathbf{x}, \Lambda(Y)\rangle - f(\mathbf{x}) : \mathbf{x} \in \mathbb{R}^N\right\} \\ &= \sup\left\{\operatorname{trace}(\operatorname{Diag}(\mathbf{x})U^*YU) - f(\mathbf{x}) : \mathbf{x} \in \mathbb{R}^N\right\} \\ &= \sup\left\{\operatorname{trace}(U\operatorname{Diag}(\mathbf{x})U^*Y) - f(\Lambda(U\operatorname{Diag}(\mathbf{x})U^*)) : \mathbf{x} \in \mathbb{R}^N\right\} \\ &\leq \sup\left\{\operatorname{trace}(XY) - f(\Lambda(X)) : X \in \operatorname{Sym}(N,\mathbb{R})\right\} = (f\circ\Lambda)^*(Y). \end{aligned}$$

■

6.1.10 Corollary *Suppose that $f : \mathbb{R}^N \to \mathbb{R} \cup \{\infty\}$ is a lower semicontinuous proper convex function. If f is also symmetric, then the function $f \circ \Lambda$ is lower semicontinuous and convex.*

Proof. Indeed, $f = f^{**}$ by Theorem 6.1.2 and since f^* is symmetric we have $f \circ \Lambda = (f^*)^* \circ \Lambda = (f^* \circ \Lambda)^*$ by Theorem 6.1.9. Therefore $f \circ \Lambda$ is a conjugate function and this fact implies its convexity and also its lower semicontinuity. ■

6.1.11 Corollary *Suppose that $f : \mathbb{R}^N \to \mathbb{R} \cup \{\infty\}$ is a symmetric convex function and $X, Y \in \operatorname{Sym}(N,\mathbb{R})$. Then $Y \in \partial(f\circ\Lambda)(X)$ if, and only if, $\Lambda(Y) \in \partial f(\Lambda(X))$ and there exists an orthogonal matrix V such that $V^*XV = \operatorname{Diag}(\Lambda(X))$ and $V^*YV = \operatorname{Diag}(\Lambda(Y))$.*

Proof. According to Proposition 6.1.1 and Theorem 6.1.9, $Y \in \partial(f\circ\Lambda)(X)$ if, and only if,

$$\operatorname{trace} XY = (f\circ\Lambda)(X) + (f\circ\Lambda)^*(Y) = f(\Lambda(X)) + f^*(\Lambda(Y)).$$

On the other hand, $f(\Lambda(X)) + f^*(\Lambda(Y)) \geq \langle\Lambda(X), \Lambda(Y)\rangle$ and $\langle\Lambda(X), \Lambda(Y)\rangle \geq \operatorname{trace} XY$, by Corollary 4.4.11. This shows that $Y \in \partial(f\circ\Lambda)(X)$ if, and only if, $\Lambda(Y) \in \partial f(\Lambda(X))$. The second part of the statement of Corollary 6.1.11 follows from Corollary 4.4.11. ■

Since the conjugate of the lower semicontinuous proper convex function

$$f : \mathbb{R}^N \to \mathbb{R} \cup \{\infty\}, \quad f(\mathbf{x}) = \begin{cases} -\sum_{k=1}^N \log x_k & \text{if } \mathbf{x} \in \mathbb{R}^N_{++} \\ \infty & \text{otherwise,} \end{cases}$$

is the function

$$f^* : \mathbb{R}^N \to \mathbb{R} \cup \{\infty\}, \quad f^*(\mathbf{x}) = \begin{cases} f(-\mathbf{x}) - N & \text{if } \mathbf{x} \in -\mathbb{R}^N_{++} \\ \infty & \text{otherwise,} \end{cases}$$

it follows from Theorem 6.1.9 that the conjugate of its spectral function

$$F : \operatorname{Sym}(N,\mathbb{R}) \to \mathbb{R} \cup \{\infty\}, \quad F(A) = \begin{cases} -\log\det A & \text{if } A \in \operatorname{Sym}^{++}(N,\mathbb{R}) \\ \infty & \text{otherwise,} \end{cases}$$

verifies $F^*(A) = -\log\det(-A) - N$ for all $A \in -\operatorname{Sym}^{++}(N,\mathbb{R})$.

Exercises

1. Prove that every continuous affine function $h : \mathbb{R}^N \to \mathbb{R}$ verifies

$$h = h^{**}.$$

2. Suppose that H is a Hilbert space and $f \colon H \to \mathbb{R} \cup \{\infty\}$ is a lower semicontinuous proper convex function. Prove that:

   (a) the Legendre–Fenchel transform of $f(\mathbf{x}) + \alpha$ is $f^*(\mathbf{p}) - \alpha$;

   (b) if $\lambda > 0$, then the Legendre–Fenchel transform of $\lambda f(\mathbf{x})$ is $\lambda f^*(\mathbf{p}/\lambda)$;

   (c) if $\lambda > 0$, then the Legendre–Fenchel transform of $\lambda f(\mathbf{x}/\lambda)$ is $\lambda f^*(\mathbf{p})$;

   (d) the Legendre–Fenchel transform of $f(\mathbf{x}) + \langle \mathbf{x}, q \rangle$ is $f^*(\mathbf{p} - \mathbf{q})$.

3. Let E be a real Banach space. Prove that for every family of functions $f_\alpha : E \to \mathbb{R} \cup \{\infty\}$ $(\alpha \in A)$ we have

$$(\inf_{\alpha \in A} f_\alpha)^* = \sup_{\alpha \in A} f_\alpha^* \quad \text{and} \quad (\sup_{\alpha \in A} f_\alpha)^* \leq \inf_{\alpha \in A} f_\alpha^*.$$

4. Prove that the Legendre–Fenchel transform of the log-sum-exp function $f(\mathbf{x}) = \log\left(\sum_{k=1}^N \mathrm{e}^{x_k}\right)$ is the negative entropy function restricted to the probability simplex, that is,

$$f^*(\mathbf{y}) = \begin{cases} \sum_{k=1}^N y_k \log y_k & \text{if } \mathbf{y} \geq 0, \ \sum_{k=1}^N y_k = 1 \\ \infty & \text{otherwise.} \end{cases}$$

5. The *Kullback–Leibler divergence* from Q to P is defined by the formula

$$D_{KL}(P||Q) = \sum_{k=1}^n p_k \log \frac{p_k}{q_k},$$

where $P = (p_1, ..., p_n)$ and $Q = (q_1, ..., q_n)$ belong to the compact convex set Prob $(n) = \{(\lambda_1, ..., \lambda_n) \in (0, \infty)^n : \sum_{k=1}^n \lambda_k = 1\}$. According to Corollary 3.5.2, this is a jointly convex function. Prove that the conjugate of $D_{KL}(P||Q)$ with respect to the variable P is the function

$$f^*(\mathbf{x}) = \log\left(\sum_{k=1}^n q_k \mathrm{e}^{x_k}\right).$$

6. (Support function) The notion of a support function can be attached to any nonempty convex set C in $\mathbb{R}^N$, by defining it as the conjugate of the indicator function of C, that is, $\sigma_C = \iota_C^*$. Prove that the support function of $C = \{(x, y) \in \mathbb{R}^2 : x + y^2/2 \leq 0\}$ is $\sigma_C(x, y) = y^2/2x$ if $x > 0$, $\sigma_C(0, 0) = 0$ and $\sigma_C(x, y) = \infty$ otherwise. Infer that σ_C is a lower semicontinuous proper convex function.

7. Calculate the support function for the convex set

$$C = \{A \in \text{Sym}^{++}(N, \mathbb{R}) : \text{trace}(A) = 1\}.$$

8. Suppose that $g, h : \mathbb{R}^N \to \mathbb{R} \cup \{\infty\}$ are two proper functions such that h is lower semicontinuous and convex. Prove that the conjugate of the function

$$f(\mathbf{x}) = \begin{cases} g(\mathbf{x}) - h(\mathbf{x}) & \text{if } \mathbf{x} \in \operatorname{dom} g \\ \infty & \text{if } \mathbf{x} \notin \operatorname{dom} g \end{cases}$$

is given by the formula

$$f^*(\mathbf{y}) = \sup_{\mathbf{x} \in \operatorname{dom} g} \left(\langle \mathbf{x}, \mathbf{y} \rangle - g(\mathbf{x}) + h(\mathbf{x}) \right) = \sup_{\mathbf{z} \in \operatorname{dom} h^*} \left(g^*(\mathbf{y} + \mathbf{z}) - h^*(\mathbf{z}) \right).$$

[*Hint:* Notice that $h = h^{**}$.]

9. (The Tolland–Singer formula) Suppose that $g, h : \mathbb{R}^N \to \mathbb{R} \cup \{\infty\}$ are two lower semicontinuous proper convex functions. Infer from the preceding exercise that

$$\inf_{\mathbf{x} \in \operatorname{dom} g} \left(g(\mathbf{x}) - h(\mathbf{x}) \right) = \sup_{\mathbf{z} \in \operatorname{dom} h^*} \left(g^*(\mathbf{z}) - h^*(\mathbf{z}) \right).$$

10. (F. Barthe [31]) Let $\mathbf{v}_1, \ldots, \mathbf{v}_m$ be vectors in $\mathbb{R}^N$ ($m \geq N$), and $c_1, \ldots, c_m$ be positive numbers such that $\sum_{k=1}^m c_k = N$. For $\lambda = (\lambda_1, \ldots, \lambda_m)$ in $(0, \infty)^m$, consider the following two norms on $\mathbb{R}^N$:

$$M_\lambda(\mathbf{x}) = \inf \left\{ \left(\sum_{k=1}^m c_k \theta_k^2 / \lambda_k \right)^{1/2} : \mathbf{x} = \sum_{k=1}^m c_k \theta_k v_k,\ \theta_k \in \mathbb{R} \right\}$$

and

$$N_\lambda(\mathbf{x}) = \left(\sum_{k=1}^m c_k \lambda_k \langle \mathbf{x}, \mathbf{v}_k \rangle^2 \right)^{1/2}.$$

Prove that $\mathcal{E}_\lambda = \{\mathbf{x} \in \mathbb{R}^N : M_\lambda(\mathbf{x}) \leq 1\}$ is the polar of the ellipsoid $\mathcal{F}_\lambda = \{\mathbf{x} \in \mathbb{R}^N : N_\lambda(\mathbf{x}) \leq 1\}$ and

$$\operatorname{Vol}_N(\mathcal{E}_\lambda) \operatorname{Vol}_N(\mathcal{F}_\lambda) = \operatorname{Vol}_N(B)^2,$$

where B denotes the closed unit ball of the Euclidean space $\mathbb{R}^N$.

[*Hint:* Notice that the support function of the polar of $\mathcal{F}_\lambda$ is

$$\sigma_{\mathcal{F}_\lambda^\circ}(\mathbf{x}) = \left(\sum_{k=1}^m c_k \lambda_k \langle \mathbf{x}, \mathbf{v}_k \rangle^2 \right)^{1/2},$$

which equals the support function of $\mathcal{E}_\lambda$,

$$\sigma_{\mathcal{E}_\lambda}(\mathbf{x}) = \sup \left\{ \sum_{k=1}^m c_k \theta_k \langle \mathbf{x}, \mathbf{v}_k \rangle : \sum_{k=1}^m c_k \theta_k^2 / \lambda_k \leq 1, \theta_k \in \mathbb{R} \right\}.]$$

11. (D. Kramkov and W. Schachermayer) Suppose that $U : (0, \infty) \to \mathbb{R}$ is an increasing, strictly concave, and continuously differentiable function that satisfies the Inada conditions, that is,

$$U'(0) = \lim_{x \to 0} U'(x) = \infty \text{ and } U'(\infty) = \lim_{x \to \infty} U'(x) = 0.$$

Prove that

$$V(y) = \sup_{x>0} \{U(x) - xy\}, \quad y > 0$$

is a continuously differentiable, decreasing and strictly convex function satisfying $V'(0) = -\infty$, $V'(\infty) = 0$, $V(0) = U(\infty)$ and $V(\infty) = U(0)$. Moreover,

$$U(x) = \inf_{y>0} (V(y) - xy), \quad x > 0.$$

6.2 The Correspondence of Properties under Duality

Conjugacy offers a convenient way to get information about the behavior of a function through its conjugate. For example, it allows us to recognize properties like coercivity and convexity, or to compute the optimal value of a lower semicontinuous proper convex function.

Throughout this section, H denotes a real Hilbert space.

6.2.1 Theorem (J. J. Moreau and R. T. Rockafellar) *A lower semicontinuous proper convex function $f\colon H \to \mathbb{R} \cup \{\infty\}$ has bounded level sets if and only if its conjugate is continuous at the origin.*

Proof. By Lemma 3.10.3, the function f has bounded level sets if and only if it verifies an inequality of the form $f(\mathbf{x}) \geq \alpha \|\mathbf{x}\| + b$ with $a, b \in \mathbb{R}$ and $\alpha > 0$. Taking conjugates, this is equivalent to f^* being bounded above on a neighborhood of 0, which in turn is equivalent to f^* being continuous at 0. See Proposition 3.1.11. ■

6.2.2 Corollary (Hörmander–Moreau–Rockafellar) *Under the hypotheses of Theorem 6.2.1, if $\mathbf{y} \in H$, then $f - \langle \cdot, \mathbf{y} \rangle$ is coercive if and only if f^* is continuous at $\mathbf{y}$.*

The duality of supercoercive convex functions makes the objective of Exercises 1–3.

6.2.3 Theorem (V. Soloviev) *Suppose that $g\colon H \to \mathbb{R} \cup \{\infty\}$ is a lower semicontinuous proper function. If the conjugate of g is differentiable, then g is necessarily convex.*

The details can be found in the paper of J.-B. Hiriart-Urruty [215].

There is a dual correspondence between strict convexity of a function and the smoothness of its conjugate.

6.2.4 Theorem (a) *If $f : H \to \mathbb{R}$ is a strictly convex and supercoercive function, then f^* is continuously differentiable on H.*

(b) *If $g : H \to \mathbb{R}$ is a supercoercive, differentiable and convex function, then g^* is strictly convex on H.*

Proof. (a) Since f is supercoercive, the effective domain of f^* is the whole space H. See Exercise 1. Then, with the help of Exercise 6, Section 3.3, we infer that $\partial f^*(\mathbf{x})$ is a singleton for every $\mathbf{x}$. According to Corollary 3.6.10, this implies that f is differentiable. The continuity of the differential follows from Exercise 2, Section 3.7.

For (b), if g^* fails to be strictly convex, then (by Exercise 6, Section 3.3) there exist distinct points $\mathbf{x}$ and $\mathbf{y}$ and $\mathbf{z} \in \partial g^*(\mathbf{x}) \cap \partial g^*(\mathbf{y})$. This implies that $\mathbf{x}, \mathbf{y} \in \partial g(\mathbf{z})$, a contradiction with differentiability of g. ■

A stronger version of Theorem 6.2.4 is offered by the Comments at the end of this chapter.

Let f be a lower semicontinuous proper convex function defined on the real Hilbert space H and let $\mathbf{z} \in \operatorname{dom} f$ arbitrarily fixed. The *recession function* of f is defined by the formula

$$(\operatorname{rec} f)(\mathbf{y}) = \sup_{\mathbf{x}\in H} [f(\mathbf{x}+\mathbf{y}) - f(\mathbf{x})] = \lim_{t\to\infty} \frac{f(\mathbf{z}+t\mathbf{y}) - f(\mathbf{z})}{t} \quad \text{for } \mathbf{y} \in H.$$

The second equality is motivated by the three chords inequality. Hence, the limit exists and is independent of $\mathbf{z}$. See Exercise 9.

The *recession* function is proper, lower semicontinuous, and sublinear. If f is bounded below, then $\operatorname{rec} f \geq 0$.

6.2.5 Proposition (P. J. Laurent [273]) *The recession function of f is the support function of* $\operatorname{dom} f^*$, *while rec (f^*) is the support function of* $\operatorname{dom} f$.

Proof. According to Theorem 6.1.2, it suffices to prove the second assertion. Let $\mathbf{u} \in \operatorname{dom} f^*$ be arbitrarily fixed. Then for every $\mathbf{v} \in H$ and $t > 0$ we have

$$\begin{aligned} f^*(\mathbf{u}+t\mathbf{v}) &= \sup\{\langle \mathbf{y}, \mathbf{u}+t\mathbf{v}\rangle - f(\mathbf{y}) : \mathbf{y} \in \operatorname{dom} f\} \\ &\leq \sup\{\langle \mathbf{y}, \mathbf{u}\rangle - f(\mathbf{y}) : \mathbf{y} \in \operatorname{dom} f\} + t \sup_{\mathbf{y}\in\operatorname{dom} f} \langle \mathbf{y}, \mathbf{v}\rangle \\ &\leq f^*(\mathbf{u}) + t\sigma_{\operatorname{dom} f}(\mathbf{v}), \end{aligned}$$

whence rec $(f^*) \leq \sigma_{\operatorname{dom} f}$. On the other hand, for $\mathbf{v} \in H$ and $\alpha \in \mathbb{R}$ with $(\operatorname{rec} f^*)(\mathbf{v}) < \alpha$ one has $f^*(\mathbf{u}+t\mathbf{v}) \leq f^*(\mathbf{u}) + t\alpha$ for all $t > 0$, hence for any $\mathbf{x} \in H$,

$$\begin{aligned} f(\mathbf{x}) &\geq \langle \mathbf{x}, \mathbf{u}+t\mathbf{v}\rangle - f^*(\mathbf{u}+t\mathbf{v}) \\ &\geq \langle \mathbf{x}, \mathbf{u}\rangle - f^*(\mathbf{u}) + t(\langle \mathbf{x}, \mathbf{v}\rangle - \alpha). \end{aligned}$$

This yields $\langle \mathbf{x}, \mathbf{v}\rangle \leq \alpha$ for all $x \in \operatorname{dom} f$, whence $\sigma_{\operatorname{dom} f} \leq \operatorname{rec}\,(f^*)$. Therefore rec $(f^*) = \sigma_{\operatorname{dom} f}$. ■

We next discuss the dualization of optimization problems. We start with an easy (but important) lemma.

6.2.6 Lemma (*Dual properties in minimization*) *Suppose that* $f : H \to \mathbb{R} \cup \{\infty\}$ *is a lower semicontinuous proper convex function. Then:*

(a) $\inf f = -f^*(\mathbf{0})$ *and* $\arg\min f = \partial f^*(\mathbf{0})$.

(b) $\arg\min f = \{\bar{\mathbf{x}}\}$ *if and only if* f^* *is differentiable at* $\mathbf{0}$ *with* $\nabla f^*(\mathbf{0}) = \bar{\mathbf{x}}$.

Proof. The assertion (a) is clear. The assertion (b) follows from Theorem 3.6.11. ■

The above result can be embedded into a very general duality scheme concerning optimality under convex perturbations.

Let E and F be two real Hilbert spaces and consider the product space $E \times F$ endowed with the Hilbertian norm associated to the scalar product

$$\langle(\mathbf{x},\mathbf{y}),(\mathbf{u},\mathbf{v})\rangle = \langle\mathbf{x},\mathbf{u}\rangle + \langle\mathbf{y},\mathbf{v}\rangle \quad \text{for } (\mathbf{x},\mathbf{y}),(\mathbf{u},\mathbf{v}) \in E \times F.$$

Given a function $\Phi : E \times F \to \overline{\mathbb{R}}$, $\Phi = \Phi(\mathbf{x},\mathbf{y})$, we associate to it the functions

$$\varphi(\mathbf{x}) = \Phi(\mathbf{x},\mathbf{0}) \quad \text{and} \quad \psi(\mathbf{v}) = -\Phi^*(\mathbf{0},\mathbf{v}),$$

and the following two optimization problems: the *primal* problem, which is the minimization problem

$$(\mathcal{P}) \qquad \inf_{\mathbf{x}\in E} \varphi(\mathbf{x}),$$

and the *dual* problem, which is the maximization problem

$$(\mathcal{P}^*) \qquad \sup_{\mathbf{v}\in F} \psi(\mathbf{v}).$$

Here

$$\Phi^*(\mathbf{u},\mathbf{v}) = \sup\left\{\langle\mathbf{x},\mathbf{u}\rangle + \langle\mathbf{y},\mathbf{v}\rangle - \Phi(\mathbf{x},\mathbf{y}) : (\mathbf{x},\mathbf{y}) \in E \times F\right\}$$

and hence

$$\Phi^*(\mathbf{0},\mathbf{v}) = \sup\left\{\langle\mathbf{y},\mathbf{v}\rangle - \Phi(\mathbf{x},\mathbf{y}) : (\mathbf{x},\mathbf{y}) \in E \times F\right\}.$$

Since $\Phi^{***} = \Phi^*$, the duals of higher order of $(\mathcal{P})$ coincide either with $(\mathcal{P})$ or with $(\mathcal{P}^*)$ (depending on parity). When Φ is a lower semicontinuous proper convex function, then we have a complete duality between $(\mathcal{P})$ and $(\mathcal{P}^*)$, because they are dual to each other.

The function Φ appears as the source of *perturbations* for a given function f.

The Fenchel–Moreau duality theorem deals with the perturbations of the form

$$\Phi(\mathbf{x},\mathbf{y}) = f(\mathbf{x}) - g(A\mathbf{x} - \mathbf{y}),$$

where

(FM1) $f : E \to \mathbb{R} \cup \{\infty\}$ is a lower semicontinuous proper convex function,

(FM2) $g : F \to \mathbb{R} \cup \{\infty\}$ is an upper semicontinuous proper concave function, and

(FM3) $A : E \to F$ is a continuous linear operator.

In this case,

$$\begin{aligned}\Phi^*(\mathbf{u},\mathbf{v}) &= \sup_{(\mathbf{x},\mathbf{y})\in E\times F} \{\langle \mathbf{x},\mathbf{u}\rangle + \langle \mathbf{y},\mathbf{v}\rangle - f(\mathbf{x}) + g(A\mathbf{x}-\mathbf{y})\} \\ &= \sup_{\mathbf{x}\in E}\sup_{\mathbf{z}\in F} \{\langle \mathbf{x},\mathbf{u}\rangle + \langle A\mathbf{x},\mathbf{v}\rangle - f(\mathbf{x}) + g(\mathbf{z}) - \langle \mathbf{z},\mathbf{v}\rangle\} \\ &= \sup_{\mathbf{x}\in E} \{\langle \mathbf{x},\mathbf{u}\rangle + \langle \mathbf{x},A^*\mathbf{v}\rangle - f(\mathbf{x})\} - \inf_{\mathbf{z}\in F}\{\langle \mathbf{z},\mathbf{v}\rangle - g(\mathbf{z})\} \\ &= f^*\left(A^*\mathbf{v}+\mathbf{u}\right) - g_*(\mathbf{v}),\end{aligned}$$

where A^* denotes the adjoint operator (see R. Bhatia [51], p. 111, for details) and

$$g_*(\mathbf{v}) = \inf_{\mathbf{z}\in F}\{\langle \mathbf{z},\mathbf{v}\rangle - g(\mathbf{z})\}$$

represents the *concave conjugate* of g. Therefore,

$$\Phi^*(\mathbf{0},\mathbf{v}) = f^*\left(A^*\mathbf{v}\right) - g_*(\mathbf{v}).$$

Now, consider the minimization problem

$$p = \inf\{f(\mathbf{x}) - g(A\mathbf{x}) : \mathbf{x}\in X\} \tag{6.3}$$

and its dual maximization problem

$$d = \sup\{-f^*(A^*\mathbf{y}) + g_*(\mathbf{y}) : \mathbf{y}\in Y\}, \tag{6.4}$$

under the hypotheses

$$\mathbf{0}\in A(\text{dom } f) - \text{dom } g \quad \text{and} \quad \mathbf{0}\in A^*(\text{dom } g^*) - \text{dom } f^*,$$

which assure that p and d are finite; indeed, according to the Fenchel–Young inequality we have

$$f(\mathbf{x}) - g(A\mathbf{x}) + f^*(A^*\mathbf{y}) - g_*(\mathbf{y}) \geq \langle \mathbf{x}, A^*\mathbf{y}\rangle - \langle A\mathbf{x},\mathbf{y}\rangle \geq 0,$$

which yields

$$-\infty < d \leq p < \infty.$$

The following result provides sufficient conditions under which $p = d$.

6.2.7 Theorem (Fenchel–Moreau duality theorem) *Suppose that f, g, and A verify the conditions* (FM1)–(FM3) *above.*

If there is a point $\mathbf{x}_0 \in \text{dom}\, f$ such that g is continuous at $A\mathbf{x}_0$, then

$$\inf_{\mathbf{x}\in E}\{f(\mathbf{x}) - g(A\mathbf{x})\} = \sup_{\mathbf{y}\in F}\{-f^*(A^*\mathbf{y}) + g_*(\mathbf{y})\}$$

and the supremum in the dual maximization problem is attained.

Proof. Consider the auxiliary function $h : F \to \overline{\mathbb{R}}$ defined by

$$h(\mathbf{y}) = \inf_{\mathbf{x}\in E} \Phi(\mathbf{x},\mathbf{y}) = \inf_{\mathbf{x}\in E} \{f(\mathbf{x}) - g(A\mathbf{x} - \mathbf{y})\}.$$

This is a lower semicontinuous proper convex function with domain dom $h = A(\text{dom } f) - \text{dom } g$. Since

$$\begin{aligned} h^*(\mathbf{y}) &= \sup_{\mathbf{v}\in F} \{\langle \mathbf{v},\mathbf{y}\rangle - h(\mathbf{v})\} = \sup_{\mathbf{v}\in F}\{\langle \mathbf{v},\mathbf{y}\rangle - \inf_{\mathbf{x}\in E} \Phi(\mathbf{x},\mathbf{v})\} \\ &= \sup_{(\mathbf{x},\mathbf{v})\in E\times F} \{\langle \mathbf{v},\mathbf{y}\rangle - \Phi(\mathbf{x},\mathbf{v})\} = \Phi^*(\mathbf{0},\mathbf{y}), \end{aligned}$$

it follows that

$$d = \sup_{\mathbf{y}\in F} - \Phi^*(\mathbf{0},\mathbf{y}) = \sup_{\mathbf{y}\in F} \{\langle \mathbf{0},\mathbf{y}\rangle - h^*(\mathbf{y})\} = h^{**}(\mathbf{0}).$$

Since

$$p = h(\mathbf{0}),$$

the equality $p = d$ reduces to the fact that $h(\mathbf{0}) = h^{**}(\mathbf{0})$.

We next show that h is bounded above on a suitable neighborhood of the origin. Indeed, since g is concave and continuous at $A\mathbf{x}_0$, it follows that the function $\mathbf{y} \to g(A\mathbf{x}_0 - \mathbf{y})$ is bounded below on a neighborhood V of the origin. Taking into account that $h(\mathbf{y}) \leq f(\mathbf{x}_0) - g(A\mathbf{x}_0 - \mathbf{y})$, we conclude that the function h is bounded above on V (and thus is continuous there by Proposition 3.1.11).

By Corollary 6.1.3, $h^{**} = \text{cl } h$, while from the continuity of h at the origin we infer that

$$\text{cl } h(\mathbf{0}) = \liminf_{\mathbf{x}\to\mathbf{0}} h(\mathbf{x}) = h(\mathbf{0}).$$

Therefore $h(\mathbf{0}) = h^{**}(\mathbf{0})$ and the proof of the equality $p = d$ is done.

An element $\mathbf{y}_0 \in F$ is a solution to the dual problem if, and only if,

$$-\Phi^*(\mathbf{0},\mathbf{y}_0) \geq -\Phi^*(\mathbf{0},\mathbf{y}) \quad \text{for all } \mathbf{y} \in F,$$

equivalently, $h^*(\mathbf{y}) \geq h^*(\mathbf{y}_0)$ for all $\mathbf{y} \in F$. This means that y_0 is a point of minimum for h^*, a fact that can be restated as $\mathbf{0} \in \partial h^*(\mathbf{y}_0)$. Because h^* is lower semicontinuous and convex, we infer from Proposition 6.1.1 that this latter condition is equivalent to $\mathbf{y}_0 \in \partial h^{**}(\mathbf{0})$. Therefore the supremum in the dual maximization problem is attained. ∎

For $E = F$ and A the identity of E, we obtain:

6.2.8 Corollary (W. Fenchel) *Let f and $-g$ be lower semicontinuous proper convex functions on E such that* dom $f \cap$ dom g *contains a point where f and g are continuous. Then*

$$\inf\{f(\mathbf{x}) - g(\mathbf{x}) : \mathbf{x} \in E\} = \max\{-f^*(\mathbf{y}) + g_*(\mathbf{y}) : \mathbf{y} \in E\}.$$

The infimum is attained, for example, if there is a point $\bar{\mathbf{x}}$ such that $\partial f(\bar{\mathbf{x}}) \cap \partial g(\bar{\mathbf{x}}) \neq \emptyset$.

Last, but not least, it is worth noticing that the above argument for Fenchel–Moreau duality Theorem 6.2.7 and its Corollary 6.2.8 can be easily adapted to obtain their extension to the framework of Banach spaces. The details are left to the reader and make the objective of Exercise 7.

The aforementioned extension plays an important role in the theory of optimal mass transportation, nicely presented by C. Villani in his books [477] and [478]. A very brief presentation of the key points of this connection is as follows.

Consider two Borel probability measures μ and ν on $\mathbb{R}^N$ and a lower semicontinuous positive function $c : \mathbb{R}^N \times \mathbb{R}^N \to \mathbb{R}$ (called *cost* function). Kantorovich's mass transportation problem consists in the minimization of the function

$$I(\pi) = \int_{\mathbb{R}^N \times \mathbb{R}^N} c(\mathbf{x}, \mathbf{y}) \mathrm{d}\pi(\mathbf{x}, \mathbf{y}),$$

over the nonempty convex set $\Pi(\mu, \nu)$ of all Borel probability measures π on $\mathbb{R}^N \times \mathbb{R}^N$, with marginals μ and ν. This means that

$$\pi(A, \mathbb{R}^N) = \mu(A) \quad \text{for all } A \in \mathcal{B}(\mathbb{R}^N)$$

and

$$\pi(\mathbb{R}^N, B) = \nu(B) \quad \text{for all } B \in \mathcal{B}(\mathbb{R}^N).$$

It is easy to show that $\pi \in \Pi(\mu, \nu)$ if, and only if, it is a positive Borel measure on $\mathbb{R}^N \times \mathbb{R}^N$ such that for all pairs $(\varphi, \psi) \in L^\infty(\mu) \times L^\infty(\nu)$ we have

$$\int_{\mathbb{R}^N \times \mathbb{R}^N} (\varphi(\mathbf{x}) + \psi(\mathbf{y}))\, \mathrm{d}\pi(\mathbf{x}, \mathbf{y}) = \int_{\mathbb{R}^N} \varphi(\mathbf{x}) \mathrm{d}\mu(\mathbf{x}) + \int_{\mathbb{R}^N} \psi(\mathbf{y}) \mathrm{d}\nu(\mathbf{y}).$$

The measures π should be viewed as *transportation plans* and the values $I(\pi)$ as *costs of transportation*. The existence of an optimal transportation plan is assured by the following result:

6.2.9 Lemma *The function $I(\pi)$ admits a minimizer.*

Proof. Since $I(\pi) \geq 0$, for all π, there is a sequence $(\pi_n)_n$ such that $I(\pi_n) \to \min I(\pi)$. The set $\Pi(\mu, \nu)$ is a weak-star compact subset of the set Prob $(\mathbb{R}^N \times \mathbb{R}^N)$, of all Borel probabilities on $\mathbb{R}^N \times \mathbb{R}^N$. Use the Banach–Alaoglu theorem *B.1.6.* Therefore, we may assume (by passing to a subsequence if necessary) that $(\pi_n)_n$ is weak-star convergent to some π_∞ in $\Pi(\mu, \nu)$. The next step is to prove that $I(\pi)$ is sequentially weak-star lower semicontinuous.

Choose $\rho_R \in C_c(\mathbb{R}^N \times \mathbb{R}^N)$ such that $0 \leq \rho_R \leq 1$ and $\rho_R = 1$ on the ball of center 0 and radius $R > 0$. Then

$$\begin{aligned}\int_{\mathbb{R}^N\times\mathbb{R}^N} \rho_R(\mathbf{x},\mathbf{y})c(\mathbf{x},\mathbf{y})\mathrm{d}\pi_\infty(\mathbf{x},\mathbf{y}) &= \lim_{n\to\infty}\int_{\mathbb{R}^N\times\mathbb{R}^N} \rho_R(\mathbf{x},\mathbf{y})c(\mathbf{x},\mathbf{y})\mathrm{d}\pi_n(\mathbf{x},\mathbf{y}) \\ &\leq \lim_{n\to\infty}\int_{\mathbb{R}^N\times\mathbb{R}^N} c(\mathbf{x},\mathbf{y})\mathrm{d}\pi_n(\mathbf{x},\mathbf{y}) \\ &= \inf_{\pi\in\Pi(\mu,\nu)}\int_{\mathbb{R}^N\times\mathbb{R}^N} c(\mathbf{x},\mathbf{y})\mathrm{d}\pi(\mathbf{x},\mathbf{y}).\end{aligned}$$

Letting $R \to \infty$, we obtain that

$$\int_{\mathbb{R}^N\times\mathbb{R}^N} c(\mathbf{x},\mathbf{y})\mathrm{d}\pi_\infty(\mathbf{x},\mathbf{y}) \leq \inf_{\pi\in\Pi(\mu,\nu)}\int_{\mathbb{R}^N\times\mathbb{R}^N} c(\mathbf{x},\mathbf{y})\mathrm{d}\pi(\mathbf{x},\mathbf{y}),$$

whence we conclude that π_∞ is a minimizer for $I(\pi)$. ■

As was already noticed in this section, a convex minimization problem like that concerning the function $I(\pi)$ admits a dual formulation involving the function

$$J(\varphi,\psi) = \int_{\mathbb{R}^N} \varphi(\mathbf{x})\mathrm{d}\mu(\mathbf{x}) + \int_{\mathbb{R}^N} \psi(\mathbf{y})\mathrm{d}\nu(\mathbf{y}),$$

which is defined on the nonempty convex set $\mathcal{F}_c$, of all functions $(\varphi,\psi) \in L^1(\mu)\times L^1(\nu)$ such that

$$\varphi(\mathbf{x}) + \psi(\mathbf{y}) \leq c(\mathbf{x},\mathbf{y})$$

for μ-almost all $\mathbf{x} \in X$ and ν-almost all $\mathbf{y} \in Y$.

6.2.10 Theorem (Kantorovich duality) *We have*

$$\inf_{\pi\in\Pi(\mu,\nu)} I(\pi) = \sup_{(\varphi,\psi)\in\mathcal{F}_c} J(\varphi,\psi).$$

C. Villani [477] uses the Fenchel–Moreau duality theorem to settle the particular case where both μ and ν have compact supports. The general case follows via an approximation argument.

In the case of quadratic costs $c(\mathbf{x},\mathbf{y}) = \|\mathbf{x}-\mathbf{y}\|^2$, it is shown that $\pi \in \Pi(\mu,\nu)$ is an optimal transportation plan if, and only if, there exists a lower semicontinuous proper convex function φ such that for π-almost all $(\mathbf{x},\mathbf{y})$ we have

$$\mathbf{y} \in \partial\varphi(\mathbf{x}).$$

See [477], Theorem 2.12, p. 66 and W. Gangbo and R. J. McCann [174]. An important ingredient is Rockafellar's characterization of convexity via cyclic monotonicity. See Theorem 3.3.14.

Does there exist a map $T : \mathbb{R}^N \to \mathbb{R}^N$ such that the optimal plan π is of the form $\pi = (\mathrm{I} \times T)\#\mu$? The answer is provided by Brenier's theorem, which asserts that if μ is an absolutely continuous probability measure and $c(\mathbf{x},\mathbf{y}) = \|\mathbf{x}-\mathbf{y}\|^2$, then there exists a convex function $\varphi : \mathbb{R}^N \to \mathbb{R}\cup\{\infty\}$ whose gradient $T = \nabla\varphi$ pushes μ forward to ν. Apart from changes on a set of μ-measure zero,

T is the only map to arise in this way. The details can be found either in Y. Brenier's paper [77] or in the aforementioned books of C. Villani.

Exercises

1. Suppose that f is a lower semicontinuous proper convex function defined on $\mathbb{R}^N$. Prove that dom $f^* = \mathbb{R}^N$ if f is supercoercive.

   [*Hint*: According to Theorem 3.3.2, f admits an affine minorant.]

2. Prove that the conjugate of a supercoercive convex function $f : \mathbb{R}^N \to \mathbb{R}$ is also supercoercive.

   [*Hint*: According to Exercise 1, f^* is real valued and defined on $\mathbb{R}^N$. Then, infer from Fenchel–Young inequality that $f^*(\mathbf{y})/\|\mathbf{y}\| \geq \alpha \|\mathbf{y}\| - f(\alpha\mathbf{y}/\|\mathbf{y}\|)$, whenever $\alpha > 0$ and $\mathbf{y} \in \mathbb{R}^N \setminus \{\mathbf{0}\}$.]

3. Suppose that $f : \mathbb{R}^N \to \mathbb{R}$ is a lower semicontinuous proper convex function. Prove that f (respectively f^*) is supercoercive if, and only if, f^* (respectively f) is bounded on bounded sets.

   Remark. Combining Corollary 6.2.2 with Exercise 3, one can easily prove that a function $h : \mathbb{R}^N \to \mathbb{R} \cup \{\infty\}$ is supercoercive if, and only if, $h(\cdot) - \langle \cdot, x^* \rangle$ has bounded level sets whenever $x^* \in \mathbb{R}^N$.

4. Suppose $f : \mathbb{R}^N \to \mathbb{R} \cup \{\infty\}$ is a lower semicontinuous proper convex function. Prove that f is Lipschitz with Lipschitz constant Lip $(f) = K > 0$ if, and only if, dom f^* is included in the closed ball $\bar{B}_K(\mathbf{0})$.

5. (A proof of the Bunt-Motzkin Theorem via Asplund's function) Given a nonempty closed subset S of $\mathbb{R}^n$ one can associate to it the function

$$\varphi_S(\mathbf{x}) = \frac{1}{2}\left(\|\mathbf{x}\|^2 - d_S^2(\mathbf{x})\right),$$

   where $d_S(\mathbf{x}) = \inf\{\|\mathbf{x} - \mathbf{s}\| : \mathbf{s} \in S\}$ represents the distance from $\mathbf{x}$ to S. Prove that:

   (a) $\varphi_S(\mathbf{x}) = \sup\{\langle \mathbf{x}, \mathbf{s}\rangle - \frac{1}{2}\|\mathbf{s}\|^2 : \mathbf{s} \in S\}$, which implies that φ_S is a convex function.

   (b) φ_S is the conjugate of the function $f_S(\mathbf{x}) = \|\mathbf{x}\|^2/2$ if $\mathbf{x} \in S$ and $f_S(\mathbf{x}) = \infty$ otherwise.

   Infer from Theorem 6.2.3 the Bunt-Motzkin theorem 2.2.2.

6. (A second proof of the Bunt-Motzkin Theorem) Let C be a nonempty closed subset of $\mathbb{R}^N$ and let

$$d_C : \mathbb{R}^N \to \mathbb{R}, \quad d_C(\mathbf{x}) = \inf\{\|\mathbf{x} - \mathbf{y}\| : \mathbf{y} \in C\}$$

   be the distance function from C.

(a) Prove that $\varphi = d_C^2$ verifies the relation

$$\varphi(\mathbf{x}+\mathbf{y}) = \varphi(\mathbf{x}) + \varphi'(\mathbf{x};\mathbf{y}) + \varepsilon(\mathbf{y})\|\mathbf{y}\|,$$

where $\varphi'(\mathbf{x};\mathbf{y}) = \min\{\langle 2\mathbf{y}, \mathbf{x}-\mathbf{z}\rangle : \mathbf{z} \in \mathcal{P}_C(\mathbf{x})\}$ and $\lim_{\mathbf{y}\to\mathbf{0}} \varepsilon(\mathbf{y}) = \varepsilon(\mathbf{0}) = 0$. In particular, φ is Gâteaux differentiable everywhere. Here $\mathcal{P}_C(\mathbf{x})$ denotes the set of best approximation to $\mathbf{x}$ from C.

(b) Suppose that C is also convex. Infer the formula

$$\nabla \frac{d_C^2}{2}(\mathbf{x}) = \mathbf{x} - P_C(\mathbf{x}) \quad \text{for all } \mathbf{x} \in \mathbb{R}^N.$$

(c) Prove that d_C is differentiable at a point $\mathbf{x} \in \mathbb{R}^N \backslash C$ if, and only if, $\mathcal{P}_C(\mathbf{x})$ is a singleton.

(d) Consider the function $f_C(\mathbf{x}) = \|\mathbf{x}\|^2/2$ if $\mathbf{x} \in C$, and $f_C(\mathbf{x}) = \infty$ if $\mathbf{x} \in \mathbb{R}^n \backslash C$ (where C is a nonempty closed subset of $\mathbb{R}^N$). Notice that $f_C^*(\mathbf{y}) = [\|\mathbf{y}\|^2 - d_C^2(\mathbf{y})]/2$ and infer from Theorem 6.2.3 the conclusion of Bunt-Motzkin Theorem 2.2.2.

7. Adapt the argument of Fenchel–Moreau duality Theorem 6.2.7 and of its Corollary 6.2.8, to extend them to the framework of Banach spaces.

8. Let C be a nonempty closed convex set in a Hilbert space H. The *recession cone* of C is the cone rec C, consisting of all vectors $\mathbf{x} \in H$ such that $\mathbf{x}_0 + \lambda\mathbf{x} \in C$ for all $\mathbf{x}_0 \in C$ and $\lambda > 0$. Prove that:

   (a) rec $C = \{\mathbf{x} \in H : \mathbf{x} + C \subset C\}$;

   (b) rec $C = \{\mathbf{x} : \mathbf{x}_0 + \lambda\mathbf{x} \in C \text{ for all } \lambda > 0\} = \bigcap_{\lambda>0} \lambda(C - \mathbf{x}_0)$, where $\mathbf{x}_0 \in C$ is arbitrarily fixed.

9. The recession function attached to a lower semicontinuous, convex and proper function $f : H \to \mathbb{R} \cup \{\infty\}$ can be defined as the function rec f whose epigraph is the recession cone of epi f. Prove that

$$(\text{rec } f)(\mathbf{y}) = \sup_{t>0} \frac{f(\mathbf{x}_0 + t\mathbf{y}) - f(\mathbf{x}_0)}{t},$$

   where $\mathbf{x}_0 \in$ dom f is arbitrarily fixed.

   [*Hint:* A point $(\mathbf{y}, \lambda) \in H \times \mathbb{R}$ belongs to epi (rec f) if, and only if, it belongs to rec (epi f), that is, for $(\mathbf{x}_0, f(\mathbf{x}_0)) \in$ epi f arbitrarily fixed, we have $(\mathbf{x}_0, f(\mathbf{x}_0)) + t(\mathbf{y}, \lambda) \in$ epi f for all $t > 0$. Put $\varphi(t) = (f(\mathbf{x}_0 + t\mathbf{y}) - f(\mathbf{x}_0))/t$. Since f is convex, the three chords inequality implies that the function φ is increasing and $\alpha = \sup_{t>0} \varphi(t) = \lim_{t\to\infty} \varphi(t) \leq \infty$. Thus the condition $(\mathbf{y}, \lambda) \in H \times \mathbb{R}$ is equivalent to $\lambda \in [a, \infty)$ and $\mathbf{y} \in$ dom (rec f) if, and only if, $a \in \mathbb{R}$ (in which case (rec f) $(\mathbf{y}) = a$.]

10. (The closure of the perspective function) Assuming C is a convex cone in a Hilbert space H, the perspective of a convex function $f : C \to \mathbb{R}$ was defined in Section 3.5 as $\widetilde{f} : C \times (0, \infty) \to \mathbb{R}$, $\widetilde{f}(\mathbf{x}, t) = tf(\mathbf{x}/t)$. This does not necessarily produce a lower semicontinuous function. Prove that the lower semicontinuous envelope of the perspective function of f is

$$\widetilde{f} : H \times \mathbb{R} \to \mathbb{R}, \quad \widetilde{f}(\mathbf{x}, t) = \begin{cases} tf(\mathbf{x}/t) & \text{if } t > 0 \\ (\operatorname{rec} f)(\mathbf{0}) & \text{if } t = 0 \\ \infty & \text{otherwise.} \end{cases}$$

Notice that $\tilde{f}$ is proper, convex and positively homogeneous.

6.3 The Convex Programming Problem

The aim of this section is to discuss the problem of minimizing a convex function over a convex set defined by a system of convex inequalities. The main result is the equivalence of this problem to the so-called saddle-point problem. Assuming the differentiability of the functions concerned, the solution of the saddle-point problem is characterized by the Karush–Kuhn–Tucker conditions, which will be made explicit in Theorem 6.3.2 below. In what follows $f, g_1, \ldots, g_m$ will denote convex functions on $\mathbb{R}^N$. The *convex programming problem* for these data is to minimize $f(x)$ over the convex set

$$X = \{\mathbf{x} \in \mathbb{R}^N : \mathbf{x} \geq \mathbf{0},\ g_1(\mathbf{x}) \leq 0, \ldots, g_m(\mathbf{x}) \leq 0\}.$$

In optimization theory, f represents a *cost*, which is minimized over the *feasible set* X. Among the constraints $g_k \leq 0$ we may include linear equalities $A\mathbf{x} = \mathbf{b}$ with A a real $m \times N$-dimensional matrix and $\mathbf{b} \in \mathbb{R}^m$ (since an equality of this form is equivalent to two linear inequalities $A\mathbf{x} - \mathbf{b} \leq \mathbf{0}$ and $-A\mathbf{x} + \mathbf{b} \leq \mathbf{0}$).

A particular case is the canonical form of the *linear programming problem*. In this problem we seek to maximize a linear function

$$L(\mathbf{x}) = -\langle \mathbf{x}, \mathbf{c} \rangle = -\sum_{k=1}^{N} c_k x_k$$

subject to the constraints

$$\mathbf{x} \geq \mathbf{0} \quad \text{and} \quad A\mathbf{x} \leq \mathbf{b}.$$

Notice that this problem can be easily converted into a minimization problem, by replacing L by $-L$. According to Theorem 3.10.11, L attains its global maximum at an extreme point of the convex set $\{\mathbf{x} : \mathbf{x} \geq \mathbf{0},\ A\mathbf{x} \leq \mathbf{b}\}$. This point can be found by using numerical algorithms.

Linear programming has many applications in planning, scheduling, resource allocation, and design in transportation, communication, banking, energy, health, agriculture, etc. This motivates a very active area of research for finding more efficient algorithms. See J. Gondzio [184] for a survey on the recent methods in convex optimization.

The following example is due to E. Stiefel. Consider a matrix $A = (a_{ij})_{i,j} \in \mathrm{M}_{m\times N}(\mathbb{R})$ and a vector $\mathbf{b} \in \mathbb{R}^m$ such that the system $A\mathbf{x} = \mathbf{b}$ has no solution. Typically this occurs when we have more equations than unknowns. The error in the equation of rank i is a function of the form

$$e_i(\mathbf{x}) = \sum_{j=1}^{N} a_{ij}x_j - b_i.$$

The problem of Chebyshev approximation is to minimize the maximum absolute error

$$Z = \max\{|e_i(\mathbf{x})| : i = 1, \ldots, m\}.$$

Letting Z be a new unknown, this problem can be read as

$$\text{minimize } Z$$

subject to the inequalities

$$-Z \leq \sum_{j=1}^{N} a_{ij}x_j - b_i \leq Z \quad (i = 1, \ldots, m),$$

which can be easily converted into a canonical linear programming problem.

We pass now to the convex programming problem. As in the case of any constrained extremal problem, one can apply the method of Lagrange multipliers in order to eliminate the constraints (at the cost of increasing the number of variables). An excellent exposition of the theoretical aspects of this method is offered by the paper of B. H. Pourciau [409].

The *Lagrangian function* associated with the convex programming problem is the function

$$F(\mathbf{x}, \mathbf{y}) = f(\mathbf{x}) + y_1 g_1(\mathbf{x}) + \cdots + y_m g_m(\mathbf{x})$$

of $N+m$ real variables $x_1, \ldots, x_N, y_1, \ldots, y_m$ (the components of $\mathbf{x}$ and respectively of $\mathbf{y}$). A *saddle point* of F is any point $(\mathbf{x}^0, \mathbf{y}^0)$ of $\mathbb{R}^N \times \mathbb{R}^m$ such that

$$\mathbf{x}^0 \geq \mathbf{0}, \quad \mathbf{y}^0 \geq \mathbf{0}$$

and

$$F(\mathbf{x}^0, \mathbf{y}) \leq F(\mathbf{x}^0, \mathbf{y}^0) \leq F(\mathbf{x}, \mathbf{y}^0)$$

for all $\mathbf{x} \geq \mathbf{0}$, $\mathbf{y} \geq \mathbf{0}$. The saddle points of F will provide solutions to the convex programming problem that generates F:

6.3.1 Theorem *Let $(\mathbf{x}^0, \mathbf{y}^0)$ be a saddle point of the Lagrangian function F. Then $\mathbf{x}^0$ is a solution to the convex programming problem and*

$$f(\mathbf{x}^0) = F(\mathbf{x}^0, \mathbf{y}^0).$$

Proof. The condition $F(\mathbf{x}^0, \mathbf{y}) \leq F(\mathbf{x}^0, \mathbf{y}^0)$ yields

$$y_1 g_1(\mathbf{x}^0) + \cdots + y_m g_m(\mathbf{x}^0) \leq y_1^0 g_1(\mathbf{x}^0) + \cdots + y_m^0 g_m(\mathbf{x}^0).$$

By keeping $y_2, \ldots, y_m$ fixed and taking the limit as $y_1 \to \infty$ we infer that $g_1(\mathbf{x}^0) \leq 0$. Similarly, $g_2(\mathbf{x}^0) \leq 0, \ldots, g_m(\mathbf{x}^0) \leq 0$. Thus $\mathbf{x}^0$ belongs to the feasible set X. From $F(\mathbf{x}^0, \mathbf{0}) \leq F(\mathbf{x}^0, \mathbf{y}^0)$ and from the definition of X we infer

$$0 \leq y_1^0 g_1(\mathbf{x}^0) + \cdots + y_m^0 g_m(\mathbf{x}^0) \leq 0,$$

that is, $y_1^0 g_1(\mathbf{x}^0) + \cdots + y_m^0 g_m(\mathbf{x}^0) = 0$. Then $f(\mathbf{x}^0) = F(\mathbf{x}^0, \mathbf{y}^0)$. Since $F(\mathbf{x}^0, \mathbf{y}^0) \leq F(\mathbf{x}, \mathbf{y}^0)$ for all $x \geq 0$, we have

$$f(\mathbf{x}^0) \leq f(\mathbf{x}) + y_1^0 g_1(\mathbf{x}) + \cdots + y_m^0 g_m(\mathbf{x}) \leq f(\mathbf{x})$$

for all $\mathbf{x}$ in X, which shows that $\mathbf{x}^0$ is a solution to the convex programming problem. ■

6.3.2 Theorem (The Karush–Kuhn–Tucker conditions) *Suppose that the convex functions $f, g_1, \ldots, g_m$ are differentiable on $\mathbb{R}^N$. Then $(\mathbf{x}^0, \mathbf{y}^0)$ is a saddle point of the Lagrangian function F if, and only if,*

$$\mathbf{x}^0 \geq 0, \tag{6.5}$$

$$\frac{\partial F}{\partial x_k}(\mathbf{x}^0, \mathbf{y}^0) \geq 0, \quad \textit{for } k = 1, \ldots, N, \tag{6.6}$$

$$\frac{\partial F}{\partial x_k}(\mathbf{x}^0, \mathbf{y}^0) = 0 \quad \textit{whenever } x_k^0 > 0, \tag{6.7}$$

and

$$\mathbf{y}^0 \geq 0, \tag{6.8}$$

$$\frac{\partial F}{\partial y_j}(\mathbf{x}^0, \mathbf{y}^0) = g_j(\mathbf{x}^0) \leq 0, \quad \textit{for } j = 1, \ldots, m, \tag{6.9}$$

$$\frac{\partial F}{\partial y_j}(\mathbf{x}^0, \mathbf{y}^0) = 0 \quad \textit{whenever } y_j^0 > 0. \tag{6.10}$$

Proof. If $(\mathbf{x}^0, \mathbf{y}^0)$ is a saddle point of F, then (6.5) and (6.8) are clearly fulfilled. Also,

$$F(\mathbf{x}^0 + t\mathbf{e}_k, \mathbf{y}^0) \geq F(\mathbf{x}^0, \mathbf{y}^0) \quad \text{for all } t \geq -x_k^0.$$

If $x_k^0 = 0$, then

$$\frac{\partial F}{\partial x_k}(\mathbf{x}^0, \mathbf{y}^0) = \lim_{t \to 0+} \frac{F(\mathbf{x}^0 + t\mathbf{e}_k, \mathbf{y}^0) - F(\mathbf{x}^0, \mathbf{y}^0)}{t} \geq 0.$$

If $x_k^0 > 0$, then $\frac{\partial F}{\partial x_k}(\mathbf{x}^0, \mathbf{y}^0) = 0$ by Fermat's theorem on interior extrema. In a similar way one can prove (6.9) and (6.10). Suppose now that the conditions (6.5)–(6.10) are satisfied. As $F(\mathbf{x}, \mathbf{y}^0)$ is a differentiable convex function of $\mathbf{x}$

(being a linear combination, with positive coefficients, of such functions), it verifies the assumptions of Theorem 3.8.1. Taking into account the conditions (6.5)–(6.7), we are led to

$$\begin{aligned} F(\mathbf{x},\mathbf{y}^0) &\geq F(\mathbf{x}^0,\mathbf{y}^0) + \langle \mathbf{x}-\mathbf{x}^0, \nabla_x F(\mathbf{x}^0,\mathbf{y}^0)\rangle \\ &= F(\mathbf{x}^0,\mathbf{y}^0) + \sum_{k=1}^{N} (x_k - x_k^0)\,\frac{\partial F}{\partial x_k}(\mathbf{x}^0,\mathbf{y}^0) \\ &= F(\mathbf{x}^0,\mathbf{y}^0) + \sum_{k=1}^{n} x_k\,\frac{\partial F}{\partial x_k}(\mathbf{x}^0,\mathbf{y}^0) \geq F(\mathbf{x}^0,\mathbf{y}^0), \end{aligned}$$

for all $\mathbf{x} \geq \mathbf{0}$. On the other hand, by (6.9)–(6.10), for $\mathbf{y} \geq \mathbf{0}$, we have

$$\begin{aligned} F(\mathbf{x}^0,\mathbf{y}) &= F(\mathbf{x}^0,\mathbf{y}^0) + \sum_{j=1}^{m} (y_j - y_j^0)\, g_j(\mathbf{x}^0) \\ &= F(\mathbf{x}^0,\mathbf{y}^0) + \sum_{j=1}^{m} y_j g_j(\mathbf{x}^0) \\ &\leq F(\mathbf{x}^0,\mathbf{y}^0). \end{aligned}$$

Consequently, $(\mathbf{x}^0,\mathbf{y}^0)$ is a saddle point of F. ∎

We shall illustrate Theorem 6.3.2 by the following example:

$$\text{minimize } (x_1-2)^2 + (x_2+1)^2 \quad \text{subject to } 0 \leq x_1 \leq 1 \text{ and } 0 \leq x_2 \leq 2.$$

Here $f(x_1,x_2) = (x_1-2)^2+(x_2+1)^2$, $g_1(x_1,x_2) = x_1-1$ and $g_2(x_1,x_2) = x_2-2$. The Lagrangian function attached to this problem is

$$F(x_1,x_2,y_1,y_2) = (x_1-2)^2 + (x_2+1)^2 + y_1(x_1-1) + y_2(x_2-2)$$

and the Karush–Kuhn–Tucker conditions give us the equations

$$\begin{cases} x_1(2x_1 - 4 + y_1) = 0, \\ x_2(2x_2 + 2 + y_2) = 0, \\ y_1(x_1 - 1) = 0, \\ y_2(x_2 - 2) = 0, \end{cases} \tag{6.11}$$

and the inequalities

$$\begin{cases} 2x_1 - 4 + y_1 \geq 0, \\ 2x_2 + 2 + y_2 \geq 0, \\ 0 \leq x_1 \leq 1 \text{ and } 0 \leq x_2 \leq 2, \\ y_1, y_2 \geq 0. \end{cases} \tag{6.12}$$

The system of equations (6.11) admits 9 solutions,

$$\begin{gathered} (1,0,2,0),\ (1,2,2,-6),\ (1,-1,2,0),\ (0,0,0,0),\ (2,0,0,0), \\ (0,-1,0,0),\ (2,-1,0,0),\ (0,0,0,-1),\ (2,0,0,-1), \end{gathered}$$

of which only $(1, 0, 2, 0)$ verifies also the inequalities (6.12). Consequently,

$$\inf_{\substack{0 \le x_1 \le 1 \\ 0 \le x_2 \le 2}} f(x_1, x_2) = f(1, 0) = 2.$$

We next indicate a fairly general situation under which the convex programming problem is equivalent to the saddle-point problem. For this we shall need the following convex analogue of Farkas' Theorem of the alternative (see Exercise 3, Section 2.3):

6.3.3 Lemma *Let $f_1, \ldots, f_m$ be convex functions defined on a nonempty convex set Y in $\mathbb{R}^N$. Then either there exists $\mathbf{y}$ in Y such that $f_1(\mathbf{y}) < 0, \ldots, f_m(\mathbf{y}) < 0$, or there exist positive numbers $a_1, \ldots, a_m$, not all zero, such that*

$$a_1 f_1(\mathbf{y}) + \cdots + a_m f_m(\mathbf{y}) \ge 0 \quad \text{for all } \mathbf{y} \in Y.$$

Proof. Assume that the first alternative does not work and consider the set

$$C = \Big\{(t_1, \ldots, t_m) \in \mathbb{R}^m : \text{there is } \mathbf{y} \in Y \text{ with } f_k(\mathbf{y}) < t_k \text{ for all } k = 1, \ldots, m\Big\}.$$

Then C is an open convex set that does not contain the origin of $\mathbb{R}^m$. According to Theorem 2.3.2, the set C and the origin can be separated by a closed hyperplane, that is, there exist scalars $a_1, \ldots, a_m$ not all zero, such that for all $\mathbf{y} \in Y$ and all $\varepsilon_1, \ldots, \varepsilon_m > 0$,

$$a_1(f_1(\mathbf{y}) + \varepsilon_1) + \cdots + a_m(f_m(\mathbf{y}) + \varepsilon_m) \ge 0. \tag{6.13}$$

Keeping $\varepsilon_2, \ldots, \varepsilon_m$ fixed and letting $\varepsilon_1 \to \infty$, we infer that $a_1 \ge 0$. Similarly, $a_2 \ge 0, \ldots, a_m \ge 0$. Letting $\varepsilon_1 \to 0, \ldots, \varepsilon_m \to 0$ in (6.13) we conclude that $a_1 f_1(\mathbf{y}) + \cdots + a_m f_m(\mathbf{y}) \ge 0$ for all $\mathbf{y}$ in Y. ∎

6.3.4 Theorem (Slater's condition) *Suppose that $\mathbf{x}^0$ is a solution of the convex programming problem. If there exists $\bar{\mathbf{x}} \ge 0$ such that $g_1(\bar{\mathbf{x}}) < 0, \ldots, g_m(\bar{\mathbf{x}}) < 0$, then one can find an $\mathbf{y}^0$ in $\mathbb{R}^m$ for which $(\mathbf{x}^0, \mathbf{y}^0)$ is a saddle point of the associated Lagrangian function F.*

Proof. By Lemma 6.3.3, applied to the functions $g_1, \ldots, g_m, f - f(\mathbf{x}^0)$ and the set $Y = \mathbb{R}^N_+$, we can find $a_1, \ldots, a_m, a_0 \ge 0$, not all zero, such that

$$a_1 g_1(\mathbf{x}) + \cdots + a_m g_m(\mathbf{x}) + a_0(f(\mathbf{x}) - f(\mathbf{x}^0)) \ge 0 \tag{6.14}$$

for all $\mathbf{x} \ge \mathbf{0}$. A moment's reflection shows that $a_0 > 0$. Put $y_j^0 = a_j / a_0$ and $\mathbf{y}^0 = (y_1^0, \ldots, y_m^0)$. By (6.14) we infer that $f(\mathbf{x}^0) \le f(\mathbf{x}) + \sum_{j=1}^m y_j^0 g_j(\mathbf{x}) = F(\mathbf{x}, \mathbf{y}^0)$ for all $\mathbf{x} \ge \mathbf{0}$. Particularly, for $\mathbf{x} = \mathbf{x}^0$, this yields

$$f(\mathbf{x}^0) \le f(\mathbf{x}^0) + \sum_{j=1}^{m} y_j^0 g_j(\mathbf{x}^0) \le f(\mathbf{x}^0)$$

that is, $\sum_{j=1}^{m} y_j^0 g_j(\mathbf{x}^0) = 0$, whence $F(\mathbf{x}^0, \mathbf{y}^0) = f(\mathbf{x}^0) \leq F(\mathbf{x}, \mathbf{y}^0)$ for all $\mathbf{x} \geq \mathbf{0}$. On the other hand, for $\mathbf{y} \geq \mathbf{0}$ we have

$$F(\mathbf{x}^0, \mathbf{y}^0) = f(\mathbf{x}^0) \geq f(\mathbf{x}^0) + \sum_{j=1}^{m} y_j g_j(\mathbf{x}^0) = F(\mathbf{x}^0, \mathbf{y}),$$

so that $(\mathbf{x}^0, \mathbf{y}^0)$ is a saddle point. ■

A nice geometric application of convex programming (more precisely, of quadratic programming), was noted by J. Franklin [169], in his beautiful introduction to mathematical methods of economics. It is about a problem of J. Sylvester, requiring the least circle which contains a given set of points in the plane. Suppose the given points are $\mathbf{a}_1, \ldots, \mathbf{a}_m$. They lie inside the circle of center $\mathbf{x}$ and radius r if

$$\|\mathbf{a}_k - \mathbf{x}\|^2 \leq r^2 \quad \text{for } k = 1, \ldots, m. \tag{6.15}$$

We want to find $\mathbf{x}$ and r so as to minimize r. Letting

$$x_0 = \frac{1}{2}(r^2 - \|x\|^2),$$

we can replace the quadratic constraints (6.15) by linear ones,

$$x_0 + \langle \mathbf{a}_k, \mathbf{x} \rangle \geq b_k \quad \text{for } k = 1, \ldots, m;$$

here $b_k = \|\mathbf{a}_k\|^2/2$. In this way, Sylvester's problem becomes a problem of quadratic programming,

$$\text{minimize } 2x_0 + x_1^2 + x_2^2,$$

subject to the m linear inequalities

$$x_0 + a_{k1}x_1 + a_{k2}x_2 \geq b_k \quad (k = 1, \ldots, m).$$

We end this section with a short discussion about the duality in linear programming. Calling the problem

$$\mathcal{P} : \text{minimize } f(\mathbf{x}) = \langle \mathbf{c}, \mathbf{x} \rangle \text{ subject to } A\mathbf{x} \leq \mathbf{b}, \mathbf{x} \geq \mathbf{0}, \tag{6.16}$$

the *primal* problem, the *dual* problem is

$$\mathcal{P}^*: \text{maximize } g(\mathbf{y}) = -\langle \mathbf{b}, \mathbf{y} \rangle \text{ subject to } A^*\mathbf{y} \geq -\mathbf{c}, \mathbf{y} \geq \mathbf{0}. \tag{6.17}$$

It is easy to verify that the dual of the dual problem is the primal problem.

The following classical result can be found in any textbook on linear programming. See, for example, Ch. L. Byrne [87].

6.3.5 Theorem (*The duality theorem of linear programming*) *Consider the primal and dual problems $\mathcal{P}$ and $\mathcal{P}^*$ with feasible sets respectively $\mathcal{F}$ and $\mathcal{F}^*$. Then:*

(a) *If* $\mathbf{x} \in \mathcal{F}$ *and* $\mathbf{y} \in \mathcal{F}^*$, *then* $f(\mathbf{x}) \leq g(\mathbf{y})$.

(b) *If for some* $\bar{\mathbf{x}} \in \mathcal{F}$ *and some* $\bar{\mathbf{y}} \in \mathcal{F}^*$ *we have* $f(\bar{\mathbf{x}}) = g(\bar{\mathbf{y}})$, *then* $\bar{\mathbf{x}}$ *is an optimal solution to* $\mathcal{P}$ *and* $\bar{\mathbf{y}}$ *is an optimal solution to* $\mathcal{P}^*$.

(c) *Conversely, if one of the problems* $\mathcal{P}$ *or* $\mathcal{P}^*$ *has an optimal solution* $\bar{\mathbf{x}}$ *or* $\bar{\mathbf{y}}$, *then so does the other and* $f(\bar{\mathbf{x}}) = g(\bar{\mathbf{y}})$.

(d) *If* $\mathcal{F} \neq \emptyset$ *and* f *is bounded above on* $\mathcal{F}$, *or* $\mathcal{F}^* \neq \emptyset$ *and* g *is bounded below on* $\mathcal{F}^*$, *then both problems* $\mathcal{P}$ *and* $\mathcal{P}^*$ *have optimal solutions.*

Exercises

1. Minimize $x^2 + y^2 - 6x - 4y$, subject to $x \geq 0$, $y \geq 0$ and $x^2 + y^2 \leq 1$.

2. Prove the assertions (a) and (b) of Theorem 6.3.5.

3. Any textbook on linear programming includes a proof of the fact that every primal problem (6.16) can be put in standard form,

$$\mathcal{P}_s : \text{minimize } \tilde{f}(\tilde{\mathbf{x}}) = \langle \tilde{\mathbf{c}}, \tilde{\mathbf{x}} \rangle \quad \text{subject to } \tilde{A}\tilde{\mathbf{x}} = \tilde{\mathbf{b}}, \tilde{\mathbf{x}} \geq \mathbf{0},$$

with $\tilde{\mathbf{b}} \geq \mathbf{0}$. See, for example, [87]. Use the theorems of the alternative to provide conditions under which this problem admits feasible solutions.

4. (Orthogonal projection) Consider a point $\mathbf{x}_0$ in the Euclidean space $\mathbb{R}^N$ such that $\|\mathbf{x}_0\| > 1$. Solve the minimization problem

$$\min_{\|\mathbf{x}\| \leq 1} \|\mathbf{x} - \mathbf{x}_0\|^2$$

by considering the dual problem

$$\max_{\lambda > 0} \min_{\|\mathbf{x}\| \leq 1} \left\{ \|\mathbf{x} - \mathbf{x}_0\|^2 + \lambda \left(\|\mathbf{x}\|^2 - 1 \right) \right\}.$$

5. (The convex multiplier rule of S. Karlin and H. Uzawa) Suppose that $f, g_1, \ldots, g_m$ are convex functions on $\mathbb{R}^N$ and that there is a point $\mathbf{x}_0 \in \mathbb{R}^N$ such that $g_i(\mathbf{x}_0) < 0$ for $i = 1, \ldots, m$. Then $\mathbf{a} \in \mathbb{R}^N$ is a solution of the convex programming problem for these data if, and only if, there is a vector $\mathbf{w} = (w_1, \ldots, w_m) \in \mathbb{R}_+^m$ such that

$$\mathbf{0} \in \partial f(\mathbf{a}) + w_1 \partial g_1(\mathbf{a}) + \cdots + w_m \partial g_m(\mathbf{a})$$

and

$$g_i(\mathbf{a}) \leq 0, \quad w_i g_i(\mathbf{a}) = 0 \quad \text{for } i = 1, \ldots, m.$$

6.4 Ky Fan Minimax Inequality

Brouwer's fixed point theorem asserts that every continuous map of a nonempty compact convex subset of $\mathbb{R}^N$ into itself has a fixed point. The book of J. Franklin [168] includes several elegant proofs of this result (including that by A. Garsia, based on Green's theorem).

The following consequence of Brouwer's fixed point theorem is known as the *KKM theorem*:

6.4.1 Theorem (The Knaster–Kuratowski–Mazurkiewicz theorem) *Suppose that X is a nonempty subset of $\mathbb{R}^N$ and M is a function which associates to each $\mathbf{x} \in X$ a closed nonempty subset $M(\mathbf{x})$ of X. If*

$$\operatorname{conv}(F) \subset \bigcup_{\mathbf{x}\in F} M(\mathbf{x})$$

for every finite subset $F \subset X$, then $\bigcap_{\mathbf{x}\in F} M(\mathbf{x}) \neq \emptyset$ for every finite subset $F \subset X$. Moreover, $\bigcap_{\mathbf{x}\in X} M(\mathbf{x}) \neq \emptyset$ if X is compact.

Proof. If $\bigcap_{\mathbf{x}\in F} M(\mathbf{x})$ is empty for some finite subset F, then consider the map

$$\mathbf{y} \in \operatorname{conv}(F) \to \Big[\sum_{\mathbf{x}\in F} d_{M(\mathbf{x})}(\mathbf{y})\mathbf{x}\Big] \Big/ \Big[\sum_{\mathbf{x}\in F} d_{M(\mathbf{x})}(\mathbf{y})\Big].$$

By Brouwer's fixed point theorem, this map admits a fixed point $\mathbf{z}$. Letting $G = \{\mathbf{x} \in F : \mathbf{z} \notin M(\mathbf{x})\}$, then $\mathbf{z}$ should be in $\operatorname{conv}(G)$, and this leads to a contradiction. ∎

In turn, the KKM theorem yields the finite dimensional case of the *Ky Fan minimax inequality* [156]:

6.4.2 Theorem *Suppose that C is a nonempty, compact, and convex subset of $\mathbb{R}^N$. If $f\colon C \times C \to \mathbb{R}$, $f = f(\mathbf{x}, \mathbf{y})$, is quasiconcave in the first variable and lower semicontinuous in the second variable, then there exists a point $\mathbf{y}$ in C such that*

$$\sup_{\mathbf{x}\in C} f(\mathbf{x}, \mathbf{y}) \leq \sup_{\mathbf{x}\in C} f(\mathbf{x}, \mathbf{x}).$$

Proof. Put $\alpha = \sup_{\mathbf{x}\in C} f(\mathbf{x}, \mathbf{x})$ and consider the family of sets $M(\mathbf{x}) = \{\mathbf{y} \in C : f(\mathbf{x}, \mathbf{y}) \leq \alpha\}$ for $\mathbf{x} \in C$. Clearly, $\mathbf{x} \in M(\mathbf{x})$ for every $\mathbf{x}$ and the sets $M(\mathbf{x})$ are closed because the function $f(\mathbf{x}, \mathbf{y})$ is lower semicontinuous in the second variable. For every convex combination $\sum_{k=1}^n \lambda_k \mathbf{x}_k$ of points in C we have

$$\min_k f(\mathbf{x}_k, \mathbf{y}) \leq f\left(\sum_{k=1}^n \lambda_k \mathbf{x}_k, \mathbf{y}\right),$$

whence every point in $M(\sum_{k=1}^n \lambda_k \mathbf{x}_k)$ (in particular, $\sum_{k=1}^n \lambda_k \mathbf{x}_k$) must be in one of the sets $M(\mathbf{x}_k)$. Hence the KKM theorem applies and this implies that $\bigcap_{\mathbf{x}\in C} M(\mathbf{x}) \neq \emptyset$. The conclusion of Theorem 6.4.2 is now obvious. ∎

It is worth noticing that all three results stated above (Brouwer's fixed point theorem, KKM theorem, and the Ky Fan minimax inequality) can be deduced from each other. See Exercise 1.

The Ky Fan minimax inequality has many important applications to mathematical economics and game theory, from which we mention here the following one:

6.4.3 Theorem (The minimax theorem of von Neumann) *Let X and Y be compact convex subsets (of suitable Euclidean spaces) and let $f : X \times Y \to \mathbb{R}$ be a function which verifies the following two conditions:*

(a) $\mathbf{x} \to f(\mathbf{x}, \mathbf{y})$ *is upper semicontinuous and quasiconcave on X for each fixed $\mathbf{y} \in Y$;*

(b) $\mathbf{y} \to f(\mathbf{x}, \mathbf{y})$ *is lower semicontinuous and quasiconvex on Y for each fixed $\mathbf{x} \in X$.*

Then

$$\max_{\mathbf{x}\in X} \min_{\mathbf{y}\in Y} f(\mathbf{x}, \mathbf{y}) = \min_{\mathbf{y}\in Y} \max_{\mathbf{x}\in X} f(\mathbf{x}, \mathbf{y}).$$

Proof. The inequality $\leq$ is straightforward. As concerns the other inequality, notice that the function

$$\varphi : (X \times Y) \times (X \times Y) \to \mathbb{R}, \quad \varphi\left((\mathbf{u}, \mathbf{v}), (\mathbf{x}, \mathbf{y})\right) = f(\mathbf{u}, \mathbf{y}) - f(\mathbf{x}, \mathbf{v}),$$

is quasiconcave with respect to the variable $(\mathbf{u}, \mathbf{v})$ and lower semicontinuous with respect to the variable $(\mathbf{x}, \mathbf{y})$. According to the Ky Fan minimax inequality, there exists a point $(\bar{\mathbf{x}}, \bar{\mathbf{y}})$ such that

$$\varphi\left((\mathbf{u}, \mathbf{v}), (\bar{\mathbf{x}}, \bar{\mathbf{y}})\right) \leq \sup_{(\mathbf{u},\mathbf{v})\in X\times Y} \varphi\left((\mathbf{u}, \mathbf{v}), (\mathbf{u}, \mathbf{v})\right) = 0.$$

Therefore $f(\mathbf{u}, \bar{\mathbf{y}}) \leq f(\bar{\mathbf{x}}, \mathbf{v})$ for all $\mathbf{u} \in X$, $\mathbf{v} \in Y$, which yields

$$f(\mathbf{x}, \bar{\mathbf{y}}) \leq f(\bar{\mathbf{x}}, \bar{\mathbf{y}}) \leq f(\bar{\mathbf{x}}, \mathbf{y}) \quad \text{for all } \mathbf{x} \in X, \ \mathbf{y} \in Y.$$

As a consequence,

$$\min_{\mathbf{y}\in Y} \max_{\mathbf{x}\in X} f(\mathbf{x}, \mathbf{y}) \leq \max_{\mathbf{x}\in X} f(\mathbf{x}, \bar{\mathbf{y}}) = f(\bar{\mathbf{x}}, \bar{\mathbf{y}}) = \min_{\mathbf{y}\in Y} f(\bar{\mathbf{x}}, \mathbf{y}) \leq \max_{\mathbf{x}\in X} \min_{\mathbf{y}\in Y} f(\mathbf{x}, \mathbf{y}),$$

and the proof of the minimax equality is done. ■

In order to understand the meaning of Theorem 6.4.3 from the point of game theory, let us consider the class of games called two-person zero-sum games, that is, games with only two players in which one player wins what the other player loses. X is the set of strategies of Player I and Y is the set of strategies of Player II. Player I chooses $x \in X$ and Player II chooses $y \in Y$, each unaware of the choice of the other. Then their choices are made known and I wins the amount $f(x, y)$ from II. According to Theorem 6.4.3, the maximum value of the minimum expected gain for one player is equal to the minimum value of the maximum expected loss for the other; moreover each player has a mixed strategy which realizes this equality.

The next application of the Ky Fan minimax inequality refers to the existence of Nash equilibrium. A group of players are in *Nash equilibrium* if each one is making the best decision possible, taking into account the decisions of the others in the game as long as the other parties' decisions remain unchanged.

6.4.4 Theorem (Nash equilibrium) *Consider the product set $C = C_1 \times \cdots \times C_n$, where each set C_k is a nonempty, compact, and convex subset of $\mathbb{R}^N$. Consider also continuous functions $f_1, \ldots, f_n \colon C \to \mathbb{R}$ such that, for each k, the function*

$$x_k \in C_k \to f_k(y_1, \ldots, y_{k-1}, x_k, y_k, \ldots, y_n)$$

is convex on C_k for all $y_i \in C_i$, $i \neq k$. Then there exists an element $c = (c_1, \ldots, c_n) \in C$ such that

$$f_k(c) \leq f_k(c_1, \ldots, x_k, \ldots, c_n) \quad \text{for all } x_k \in C_k,\ k \in \{1, \ldots, n\}.$$

Proof. Apply the Ky Fan minimax inequality to the function

$$f(\mathbf{x}, \mathbf{y}) = \sum_{k=1}^{n} [f_k(\mathbf{y}) - f_k(y_1, \ldots, x_k, \ldots, y_n)].$$

■

A very attractive introduction to game theory is provided by the book of P. D. Straffin [464].

We end with a variant of the Ky Fan minimax inequality, that proves useful for establishing delicate inequalities. See J. Diestel, H. Jarchow, A. Tonge [132], p. 191.

6.4.5 Theorem (Ky Fan Lemma [156]) *Let E be a separated locally convex space, and let C be a compact convex subset of E. Let M be a set of functions on C with values in $\mathbb{R} \cup \{\infty\}$, having the following three properties:*

(a) *Each $f \in M$ is convex and lower semicontinuous.*

(b) *If $g \in \operatorname{conv}(M)$, then there is $f \in M$ with $g(\mathbf{x}) \leq f(\mathbf{x})$ for all $\mathbf{x} \in C$.*

(c) *There is an $r \in \mathbb{R}$ such that each $f \in M$ has a value less than or equal to r.*

Then there is a point $\bar{\mathbf{x}} \in C$ such that $f(\bar{\mathbf{x}}) \leq r$ for all $f \in M$.

Proof. For $f \in M$ and $\varepsilon > 0$ denote

$$S(f, \varepsilon) := \{\mathbf{x} \in C : f(\mathbf{x}) \leq r + \varepsilon\}.$$

By (c), each set $S(f, \varepsilon)$ is nonempty; it is also closed since f is lower semicontinuous. We have to prove that the intersection of the whole family of sets $S(f, \varepsilon)$ is nonempty. Since C is a compact set, it suffices to prove that the intersection of every finite subfamily $S(f_1, \varepsilon_1), \ldots, S(f_n, \varepsilon_n)$ is nonempty.

For this, consider the following two auxiliary subsets of $\mathbb{R}^n$: the open and convex set $A = \prod_{k=1}^{n}(-\infty, r + \varepsilon_k)$ and the convex hull B of

$$\{(f_k(\mathbf{x}))_{k=1}^{n} : \mathbf{x} \in C\} \bigcap \mathbb{R}^n;$$

the properties (b) and (c) ensure that B is nonempty. We shall prove (by contradiction) that $A \cap B \neq \emptyset$.

Assume that $A \cap B = \emptyset$. Then, by the Hahn–Banach Separation Theorem, there are $\mu = (\mu_1, ..., \mu_n) \in \mathbb{R}^n$ and $\lambda \in \mathbb{R}$ such that $\langle \mu, \mathbf{a} \rangle < \lambda \leq \langle \mu, \mathbf{b} \rangle$ for all $\mathbf{a} \in A$ and $\mathbf{b} \in B$. The strict inequality forces $\mu \neq \mathbf{0}$, and so we may, without loss of generality, normalize μ so that $\sum_{k=1}^n |\mu_k| = 1$. Necessarily, $\mu_k \geq 0$ for each k. To check this, note that, no matter how we choose $\rho \leq r$, the point $\mathbf{a} = \sum_{j \neq k} r\mathbf{e}_j + \rho \mathbf{e}_k$ belongs to A and so $\langle \mu, \mathbf{a} \rangle = r \sum_{j \neq k} \mu_j + \rho \mu_k < \lambda$. Letting $\rho \to -\infty$, we see that the case $\mu_k < 0$ contradicts the existence of the upper bound λ. Therefore, $\mu_k \geq 0$.

This allows us to use hypothesis (b) to find an $f \in M$ such that for $\mathbf{x} \in C$, either $f(\mathbf{x}) = \infty$ or $f(x) < \infty$ and

$$\begin{aligned} f(\mathbf{x}) &\geq \textstyle\sum_{k=1}^n \mu_k f_k(\mathbf{x}) = \langle \mu, (f_k(\mathbf{x}))_{k=1}^n \rangle \\ &\geq \lambda > \langle \mu, (r + \varepsilon_k/2)_{k=1}^n \rangle = \textstyle\sum_{k=1}^n \mu_k (r + \varepsilon_k/2) > r. \end{aligned}$$

This contradicts hypothesis (c). Therefore $A \cap B \neq \emptyset$.

To end the proof, choose $s = (s_1, ..., s_n)$ in $A \cap B$. By the definition of B, we can write $s_k = \sum_{j=1}^n \alpha_j f_k(\mathbf{x}_j)$ $(1 \leq k \leq n)$, where the $\mathbf{x}_j$'s are in C and the α_j's are positive and sum to 1. Since C is convex, the point $\mathbf{x} = \sum_{j=1}^n \alpha_j \mathbf{x}_j$ belongs to C. We have $f_k(\mathbf{x}) \leq \sum_{j=1}^n \alpha_j f_k(\mathbf{x}_j) = s_k$ $(1 \leq k \leq n)$, due to the convexity of the functions f_k. Finally, since $s \in A$, we conclude that $\mathbf{x} \in S(f_k, \varepsilon_k)$ whenever $k \in \{1, ..., n\}$. ∎

Exercises

1. Prove that the Brouwer fixed point theorem, the KKM theorem, and the Ky Fan minimax inequality are equivalent in the sense that each one implies the other ones.

   [*Hint*: For example, in order to show that the Ky Fan minimax inequality implies Brouwer's fixed point theorem, consider a continuous self-map f of a compact convex set $C \subset \mathbb{R}^N$ and attach to it the function $\varphi : C \times C \to \mathbb{R}$ given by $\varphi(\mathbf{x}, \mathbf{y}) = \langle \mathbf{x} - f(\mathbf{x}), \mathbf{x} - \mathbf{y} \rangle$.]

2. (The minimax flavor of the Fenchel–Moreau duality theorem) Consider the perturbation function $\Phi(\mathbf{x}, \mathbf{y}) = f(\mathbf{x}) - g(A\mathbf{x} - \mathbf{y})$, as in the statement of Theorem 6.2.7, and attached to it the *Hamiltonian* $\mathcal{H} : E \times F \to \mathbb{R}$, defined by

   $$\mathcal{H}(\mathbf{x}, \mathbf{v}) = \sup \{ \langle \mathbf{y}, \mathbf{v} \rangle - \Phi(\mathbf{x}, \mathbf{y}) : \mathbf{y} \in F \}.$$

   Notice that $\mathcal{H}(\mathbf{x}, \mathbf{v})$ is lower semicontinuous and convex with respect to the second variable (being a conjugate with respect to this variable) and $\mathcal{H}(\mathbf{x}, \mathbf{v})$ is lower semicontinuous and concave with respect to the first variable. Prove that the following two assertions are equivalent for a pair $(\bar{\mathbf{x}}, \bar{\mathbf{v}}) \in E \times F$:

   (a) $\mathcal{H}(\mathbf{x}, \bar{\mathbf{v}}) \leq \mathcal{H}(\bar{\mathbf{x}}, \bar{\mathbf{v}}) \leq \mathcal{H}(\bar{\mathbf{x}}, \mathbf{v})$ for all $(\mathbf{x}, \mathbf{v}) \in E \times F$ (that is, $(\bar{\mathbf{x}}, \bar{\mathbf{v}})$ is a saddle point for the function $\mathcal{H}(\mathbf{x}, \mathbf{v})$);

(b) $\bar{\mathbf{x}}$ is an optimal solution to the primal problem (6.3), $\bar{\mathbf{v}}$ is an optimal solution to the dual problem (6.4) and the extreme values p and d are equal.

3. Infer from the von Neumann minimax theorem the equality $f = f^{**}$ stated in the Fenchel–Moreau theorem on biconjugation.

6.5 Moreau–Yosida Approximation

A convex analogue of mollification is provided by the *infimal convolution*, which for two proper convex functions f, g defined on a real Hilbert space H is defined by the formula

$$(f\Box g)(\mathbf{x}) = \inf\{f(\mathbf{x}-\mathbf{y}) + g(\mathbf{y}) : \mathbf{y} \in H\};$$

the value $-\infty$ is allowed. Clearly, $f\Box g = g\Box f$. The fact that $f\Box g$ is a convex function is left as Exercise 1.

If $(f\Box g)(\mathbf{x}) > -\infty$ for all $\mathbf{x}$, then $f\Box g$ is a proper convex function. For example, this happens when both functions f and g are positive or, more generally, when there exists an affine function $h\colon H \to \mathbb{R}$ such that $f \geq h$ and $g \geq h$.

Notice that the distance function from a convex set C is an example of infimal convolution:

$$d_C(\mathbf{x}) = \inf_{\mathbf{y}\in C} \|\mathbf{x}-\mathbf{y}\| = (\|\cdot\|\Box\iota_C)(x).$$

6.5.1 Lemma *If f and g are two proper functions defined on H, then*

$$(f\Box g)^* = f^* + g^*.$$

Proof. Indeed, for $\mathbf{z}$ arbitrarily fixed in H we have

$$\begin{aligned}
(f\Box g)^*(\mathbf{z}) &= \sup_{\mathbf{x}\in H} \left[\langle \mathbf{x},\mathbf{z}\rangle - (f\Box g)(\mathbf{x})\right] \\
&= \sup_{\mathbf{x}\in H} \left[\langle \mathbf{x},\mathbf{z}\rangle - \inf_{\mathbf{y}\in H} (f(\mathbf{y}) + g(\mathbf{x}-\mathbf{y}))\right] \\
&= \sup_{\mathbf{x}\in H}\sup_{\mathbf{y}\in H} \left[\langle \mathbf{y},\mathbf{z}\rangle - f(\mathbf{y}) + \langle \mathbf{x}-\mathbf{y},\mathbf{z}\rangle - g(\mathbf{x}-\mathbf{y})\right] \\
&= \sup_{\mathbf{y}\in H}\sup_{\mathbf{v}\in H} \left([\langle \mathbf{y},\mathbf{z}\rangle - f(\mathbf{y})] + [\langle \mathbf{v},\mathbf{z}\rangle - g(\mathbf{v})]\right) \\
&= f^*(\mathbf{z}) + g^*(\mathbf{z}).
\end{aligned}$$

■

See Exercise 2 for a dual property.

A useful way to approximate from below a lower semicontinuous proper convex function $f\colon H \to \mathbb{R}\cup\{\infty\}$ is provided by the *Moreau–Yosida approximation.*

This is done via the functions

$$\begin{aligned} f_\lambda(\mathbf{x}) &= \left(f \square \frac{1}{2\lambda} \|\cdot\|^2 \right)(\mathbf{x}) \\ &= \inf_{\mathbf{y}\in H} \left(f(\mathbf{y}) + \frac{1}{2\lambda}\|\mathbf{x}-\mathbf{y}\|^2 \right) \end{aligned}$$

for $\mathbf{x} \in H$ and $\lambda > 0$. The functions f_λ are well defined and *finite* for all $\mathbf{x} \in H$ due to the existence of an affine minorant $\langle \cdot, \mathbf{z} \rangle + \alpha$ for f (assured by Theorem 3.3.1). Indeed,

$$\begin{aligned} f(\mathbf{y}) + \frac{1}{2\lambda}\|\mathbf{x}-\mathbf{y}\|^2 &\geq \langle \mathbf{y}, \mathbf{z} \rangle + \alpha + \frac{1}{2\lambda}\|\mathbf{x}-\mathbf{y}\|^2 \\ &\geq \langle \mathbf{y}-\mathbf{x}, \mathbf{z} \rangle + \langle \mathbf{x}, \mathbf{z} \rangle + \alpha + \frac{1}{2\lambda}\|\mathbf{x}-\mathbf{y}\|^2 \\ &\geq \langle \mathbf{x}, \mathbf{z} \rangle + \alpha - \left\| \frac{\mathbf{y}-\mathbf{x}}{\sqrt{\lambda}} \right\| \left\| \mathbf{z}\sqrt{\lambda} \right\| + \frac{1}{2\lambda}\|\mathbf{x}-\mathbf{y}\|^2 \\ &\geq \langle \mathbf{x}, \mathbf{z} \rangle + \alpha - \frac{\lambda}{2}\|\mathbf{z}\|^2, \end{aligned}$$

for all $\mathbf{y} \in H$.

Notice that the infimum in the definition of $f_\lambda(\mathbf{x})$ is actually a minimum, that is, there is $\mathbf{x}_\lambda \in H$ such that

$$f_\lambda(\mathbf{x}) = f(\mathbf{x}_\lambda) + \frac{1}{2\lambda}\|\mathbf{x}-\mathbf{x}_\lambda\|^2 \quad \text{for } \mathbf{x} \in H. \tag{6.18}$$

To prove this, consider a minimizing sequence $\mathbf{x}_n \in H$ such that

$$f(\mathbf{x}_n) + \frac{1}{2\lambda}\|\mathbf{x}-\mathbf{x}_n\|^2 \leq f_\lambda(\mathbf{x}) + \frac{1}{n} \tag{6.19}$$

for all indices $n \geq 1$. According to the parallelogram law,

$$\|\mathbf{x}_m - \mathbf{x}_n\|^2 = 2\|\mathbf{x}_m - \mathbf{x}\|^2 + 2\|\mathbf{x}_n - \mathbf{x}\|^2 - 4\left\|\mathbf{x} - \frac{\mathbf{x}_m + \mathbf{x}_n}{2}\right\|^2,$$

and combing this fact with the inequality (6.19) and the convexity of f we infer that

$$\begin{aligned} \|\mathbf{x}_m - \mathbf{x}_n\|^2 &\leq 4\lambda \left(\frac{1}{m} + \frac{1}{n} + 2f_\lambda(\mathbf{x}) - f(\mathbf{x}_m) - f(\mathbf{x}_n) \right) \\ &\quad + 8\lambda \left(f\left(\frac{\mathbf{x}_m + \mathbf{x}_n}{2} \right) - f_\lambda(\mathbf{x}) \right) \\ &= 4\lambda \left(\frac{1}{m} + \frac{1}{n} + 2f\left(\frac{\mathbf{x}_m + \mathbf{x}_n}{2} \right) - f(\mathbf{x}_m) - f(\mathbf{x}_n) \right) \\ &\leq 4\lambda \left(\frac{1}{m} + \frac{1}{n} \right). \end{aligned}$$

Thus $(\mathbf{x}_n)_n$ is convergent to an element $\mathbf{x}_\lambda$ of H since H is complete. Taking into account that f is lower semicontinuous we obtain

$$f(\mathbf{x}_\lambda) + \frac{1}{2\lambda}\|\mathbf{x} - \mathbf{x}_\lambda\|^2 \leq \liminf_{n\to\infty}\left(f(\mathbf{x}_n) + \frac{1}{2\lambda}\|\mathbf{x} - \mathbf{x}_n\|^2\right) \leq f_\lambda(\mathbf{x}),$$

whence $f_\lambda(\mathbf{x}) = f(\mathbf{x}_\lambda) + \frac{1}{2\lambda}\|\mathbf{x} - \mathbf{x}_\lambda\|^2$.

6.5.2 Lemma *The solutions* $\mathbf{x}_\lambda$ *of the equation* (6.18) *are the same with the solutions of the variational inequality*

$$\frac{1}{\lambda}\langle \mathbf{x}_\lambda - \mathbf{x}, \mathbf{x}_\lambda - \mathbf{y}\rangle + f(\mathbf{x}_\lambda) - f(\mathbf{y}) \leq 0 \quad \text{for all } \mathbf{y} \in H. \tag{6.20}$$

Proof. Indeed, if $\mathbf{x}_\lambda$ is a solution of the equation (6.18) and $\mathbf{y} \in H$, then for every $\theta \in (0,1)$ we have $\mathbf{x}_\lambda + \theta(\mathbf{y} - \mathbf{x}_\lambda) = (1-\theta)\mathbf{x}_\lambda + \theta\mathbf{y}$, so that

$$\begin{aligned} f(\mathbf{x}_\lambda) + \frac{1}{2\lambda}\|\mathbf{x} - \mathbf{x}_\lambda\|^2 &\leq f(\mathbf{x}_\lambda + \theta(\mathbf{y} - \mathbf{x}_\lambda)) + \frac{1}{2\lambda}\|\mathbf{x} - \mathbf{x}_\lambda - \theta(\mathbf{y} - \mathbf{x}_\lambda)\|^2 \\ &\leq (1-\theta)f(\mathbf{x}_\lambda) + \theta f(y) + \frac{1}{2\lambda}\|\mathbf{x} - \mathbf{x}_\lambda\|^2 \\ &\quad - \frac{\theta}{\lambda}\langle \mathbf{x} - \mathbf{x}_\lambda, \mathbf{y} - \mathbf{x}_\lambda\rangle + \frac{\theta^2}{2\lambda}\|y - \mathbf{x}_\lambda\|^2, \end{aligned}$$

whence

$$f(\mathbf{x}_\lambda) - f(\mathbf{y}) + \frac{1}{\lambda}\langle \mathbf{x}_\lambda - \mathbf{x}, \mathbf{x}_\lambda - \mathbf{y}\rangle \leq \frac{\theta}{2\lambda}\|\mathbf{y} - \mathbf{x}_\lambda\|^2.$$

Passing to the limit as $\theta \to 0$, we infer that $\mathbf{x}_\lambda$ is a solution of the variational inequality (6.20). Conversely, if $\mathbf{x}_\lambda$ is a solution of the variational inequality (6.20), then

$$\begin{aligned} 0 &\geq f(\mathbf{x}_\lambda) - f(\mathbf{y}) + \frac{1}{\lambda}\langle \mathbf{x}_\lambda - \mathbf{x}, \mathbf{x}_\lambda - \mathbf{y}\rangle \\ &\geq f(\mathbf{x}_\lambda) + \frac{1}{2\lambda}\|\mathbf{x} - \mathbf{x}_\lambda\|^2 - f(\mathbf{y}) - \frac{1}{2\lambda}\|\mathbf{x} - \mathbf{y}\|^2, \end{aligned}$$

for all $\mathbf{y} \in H$ and the proof is done. ∎

6.5.3 Corollary *The equation* (6.18) *has a unique solution.*

Proof. The existence part was already noticed. As concerns the uniqueness, suppose that $\mathbf{x}'$ is another solution of the equation (6.18). According to Lemma 6.5.2, it follows that

$$\frac{1}{\lambda}\langle \mathbf{x}_\lambda - \mathbf{x}, \mathbf{x}_\lambda - \mathbf{x}'\rangle + f(\mathbf{x}_\lambda) - f(\mathbf{x}') \leq 0$$

and

$$\frac{1}{\lambda}\langle \mathbf{x}' - \mathbf{x}, \mathbf{x}' - \mathbf{x}_\lambda\rangle + f(\mathbf{x}') - f(\mathbf{x}_\lambda) \leq 0.$$

Summing up one obtains $\|\mathbf{x}_\lambda - \mathbf{x}'\| \leq 0$, that is, $\mathbf{x}' = \mathbf{x}_\lambda$. ■

In what follows we will deal with the maps

$$J_\lambda : H \to \operatorname{dom} f, \quad J_\lambda(\mathbf{x}) = \mathbf{x}_\lambda,$$

and

$$A_\lambda : H \to H, \quad A_\lambda(\mathbf{x}) = \frac{1}{\lambda}\left(\mathbf{x} - J_\lambda(\mathbf{x})\right).$$

According to Lemma 6.5.2,

$$\frac{1}{\lambda}\langle J_\lambda(\mathbf{x}) - \mathbf{x}, J_\lambda(\mathbf{x}) - \mathbf{y}\rangle + f\left(J_\lambda(\mathbf{x})\right) - f(\mathbf{y}) \leq 0 \quad \text{for all } \mathbf{y} \in H, \tag{6.21}$$

which implies that

$$A_\lambda(\mathbf{x}) \in \partial f(J_\lambda(\mathbf{x})) \quad \text{for all } \mathbf{x} \in H \tag{6.22}$$

and

$$\frac{1}{\lambda}\langle J_\lambda(\mathbf{x}) - \mathbf{x}, J_\lambda(\mathbf{x}) - J_\lambda(\mathbf{y})\rangle + f\left(J_\lambda(\mathbf{x})\right) - f(J_\lambda(\mathbf{y})) \leq 0.$$

Adding the last inequality with its companion obtained by interchanging $\mathbf{x}$ and $\mathbf{y}$, we obtain

$$\langle J_\lambda(\mathbf{x}) - J_\lambda(\mathbf{y}), \mathbf{x} - \mathbf{y}\rangle \geq \|\mathbf{x} - \mathbf{y}\|^2. \tag{6.23}$$

Since

$$\begin{aligned}\|\mathbf{x} - \mathbf{y}\|^2 &= \|(\mathbf{x} - J_\lambda(\mathbf{x})) - (\mathbf{y} - J_\lambda(\mathbf{y})) + J_\lambda(\mathbf{x}) - J_\lambda(\mathbf{y})\|^2 \\ &= \|(\mathbf{x} - J_\lambda(\mathbf{x})) - (\mathbf{y} - J_\lambda(\mathbf{y}))\|^2 + \|J_\lambda(\mathbf{x}) - J_\lambda(\mathbf{y})\|^2 \\ &\quad + 2\langle(\mathbf{x} - J_\lambda(\mathbf{x})) - (\mathbf{y} - J_\lambda(\mathbf{y}))\,, J_\lambda(\mathbf{x}) - J_\lambda(\mathbf{y})\rangle,\end{aligned}$$

we infer (based on inequality (6.23)) that

$$\|\mathbf{x} - \mathbf{y}\|^2 \geq \|(\mathbf{x} - J_\lambda(\mathbf{x})) - (\mathbf{y} - J_\lambda(\mathbf{y}))\|^2 + \|J_\lambda(\mathbf{x}) - J_\lambda(\mathbf{y})\|^2. \tag{6.24}$$

This yields the following result:

6.5.4 Lemma *The maps J_λ and $\mathrm{I} - J_\lambda$ are both Lipschitz, with Lipschitz constant less than* 1.

We are now in a position to present the key features of the Moreau–Yosida approximants:

6.5.5 Theorem *Suppose that $f\colon H \to \mathbb{R} \cup \{\infty\}$ is a lower semicontinuous proper convex function. Then the Moreau–Yosida approximants f_λ are Fréchet differentiable convex functions on H and*

$$\nabla f_\lambda(\mathbf{x}) = \frac{1}{\lambda}(\mathbf{x} - J_\lambda(\mathbf{x})) \in \partial f(J_\lambda(\mathbf{x})).$$

Moreover, $f_\lambda \to f$ as $\lambda \to 0$.

Proof. We first prove that

$$A_\lambda(\mathbf{x}) = \frac{1}{\lambda}(\mathbf{x} - J_\lambda(\mathbf{x})) \in \partial f_\lambda(\mathbf{x}) \quad \text{for all } \mathbf{x} \in H. \tag{6.25}$$

Indeed, since $A_\lambda(\mathbf{x}) \in \partial f(J_\lambda(\mathbf{x}))$, we have

$$\begin{aligned} f_\lambda(\mathbf{x}) - f_\lambda(\mathbf{y}) &= f(J_\lambda(\mathbf{x})) - f(J_\lambda(\mathbf{y})) + \frac{\lambda}{2}\|A_\lambda(\mathbf{x})\|^2 - \frac{\lambda}{2}\|A_\lambda(\mathbf{y})\|^2 \\ &\leq \langle A_\lambda(\mathbf{x}), J_\lambda(\mathbf{x}) - J_\lambda(\mathbf{y})\rangle + \frac{\lambda}{2}\|A_\lambda(\mathbf{x})\|^2 - \frac{\lambda}{2}\|A_\lambda(\mathbf{y})\|^2 \\ &\overset{J_\lambda = \mathrm{I} - \lambda A_\lambda}{=} \langle A_\lambda(\mathbf{x}), \mathbf{x} - \mathbf{y}\rangle - \lambda\langle A_\lambda(\mathbf{x}), A_\lambda(\mathbf{x}) - A_\lambda(\mathbf{y})\rangle \\ &\qquad + \frac{\lambda}{2}\|A_\lambda(\mathbf{x})\|^2 - \frac{\lambda}{2}\|A_\lambda(\mathbf{y})\|^2 \\ &= \langle A_\lambda(\mathbf{x}), \mathbf{x} - \mathbf{y}\rangle - \frac{\lambda}{2}\|A_\lambda(\mathbf{x}) - A_\lambda(\mathbf{y})\|^2 \leq \langle A_\lambda(\mathbf{x}), \mathbf{x} - \mathbf{y}\rangle, \end{aligned}$$

and this implies the relation (6.25). Therefore

$$\begin{aligned} f_\lambda(\mathbf{x}) - f_\lambda(\mathbf{y}) &\geq \langle A_\lambda(\mathbf{y}), \mathbf{x} - \mathbf{y}\rangle \\ &= \langle A_\lambda(\mathbf{x}), \mathbf{x} - \mathbf{y}\rangle - \langle A_\lambda(\mathbf{x}) - A_\lambda(\mathbf{y}), \mathbf{x} - \mathbf{y}\rangle \\ &\geq \langle A_\lambda(\mathbf{x}), \mathbf{x} - \mathbf{y}\rangle - \|A_\lambda(\mathbf{x}) - A_\lambda(\mathbf{y})\|\,\|\mathbf{x} - \mathbf{y}\| \\ &\geq \langle A_\lambda(\mathbf{x}), \mathbf{x} - \mathbf{y}\rangle - \frac{1}{\lambda}\|\mathbf{x} - \mathbf{y}\|^2, \end{aligned}$$

according to Lemma 6.5.4. Thus

$$0 \geq f_\lambda(\mathbf{x}) - f_\lambda(\mathbf{y}) - \langle A_\lambda(\mathbf{x}), \mathbf{x} - \mathbf{y}\rangle \geq -\frac{1}{\lambda}\|\mathbf{x} - \mathbf{y}\|^2,$$

equivalently,

$$\frac{|f_\lambda(\mathbf{x}) - f_\lambda(\mathbf{y}) - \langle A_\lambda(\mathbf{x}), \mathbf{x} - \mathbf{y}\rangle|}{\|\mathbf{x} - \mathbf{y}\|} \leq \frac{1}{\lambda}\|\mathbf{x} - \mathbf{y}\|,$$

which shows that f_λ is Fréchet differentiable and $\nabla f_\lambda(\mathbf{x}) = A_\lambda(\mathbf{x})$. We pass now to the problem of convergence of Moreau–Yosida approximants. Let $\mathbf{x} \in \operatorname{dom}(f)$ and $\mathbf{x}^* \in \operatorname{dom} f^*$. Then

$$f(J_\lambda(\mathbf{x})) + \frac{1}{2\lambda}\|\mathbf{x} - J_\lambda(\mathbf{x})\|^2 = f_\lambda(\mathbf{x}) \leq f(\mathbf{x})$$

and

$$-f(J_\lambda(\mathbf{x})) \leq f^*(\mathbf{x}^*) - \langle J_\lambda(\mathbf{x}), \mathbf{x}^*\rangle,$$

whence

$$\begin{aligned} \frac{1}{2\lambda}\|\mathbf{x} - J_\lambda(\mathbf{x})\|^2 &\leq f(\mathbf{x}) + f^*(\mathbf{x}^*) - \langle \mathbf{x}, \mathbf{x}^*\rangle - \langle J_\lambda(\mathbf{x}) - \mathbf{x}, \mathbf{x}^*\rangle \\ &\leq \frac{1}{4\lambda}\|\mathbf{x} - J_\lambda(\mathbf{x})\|^2 + f(\mathbf{x}) + f^*(\mathbf{x}^*) - \langle \mathbf{x}, \mathbf{x}^*\rangle + \lambda\|\mathbf{x}^*\|^2, \end{aligned}$$

since

$$-\langle J_\lambda(\mathbf{x}) - \mathbf{x}, \mathbf{x}^*\rangle \leq \|J_\lambda(\mathbf{x}) - \mathbf{x}\| \, \|\mathbf{x}^*\| \leq \frac{1}{4\lambda}\|\mathbf{x} - J_\lambda(\mathbf{x})\|^2 + \lambda \|\mathbf{x}^*\|^2 .$$

Therefore

$$\|\mathbf{x} - J_\lambda(\mathbf{x})\|^2 \leq 4\lambda \left(f(\mathbf{x}) + f^*(\mathbf{x}^*) - \langle \mathbf{x}, \mathbf{x}^*\rangle + \lambda \|\mathbf{x}^*\|^2 \right),$$

which implies that $\lim_{\lambda\to 0} J_\lambda(\mathbf{x}) \to \mathbf{x}$. Since f is lower semicontinuous and $f(J_\lambda(\mathbf{x})) \leq f_\lambda(\mathbf{x}) \leq f(\mathbf{x})$ we infer that

$$f(\mathbf{x}) \leq \liminf_{\lambda\to 0} f(J_\lambda(\mathbf{x})) \leq \liminf_{\lambda\to 0} f_\lambda(\mathbf{x}) \leq \limsup_{\lambda\to 0} f_\lambda(\mathbf{x}) \leq f(\mathbf{x}),$$

that is, $\lim_{\lambda\to 0} f_\lambda(\mathbf{x}) = f(\mathbf{x})$. To end the proof, we have to prove that $f_\lambda(\mathbf{x})$ tends to $f(\mathbf{x})$ also in the case where $f(\mathbf{x}) = \infty$. Suppose, by reductio ad absurdum, that there exists a sequence λ_n converging to 0 such that $f_{\lambda_n}(\mathbf{x}) \leq C < \infty$ for all n. Then $f\left(J_{\lambda_n}(\mathbf{x})\right) \leq f_{\lambda_n}(\mathbf{x}) \leq C$ for all n. Since $J_{\lambda_n}(\mathbf{x}) \to \mathbf{x}$ and f is lower semicontinuous, this would imply that $f(\mathbf{x}) \leq C$, a contradiction. ■

The lower semicontinuous proper convex functions f are determined by their Moreau–Yosida approximants. Indeed, according to Lemma 6.5.1 and Example 6.1.6,

$$f_\lambda^* = f^* + \frac{\lambda}{2} \|\cdot\|^2 , \tag{6.26}$$

so if two lower semicontinuous proper convex functions $f, g\colon H \to \mathbb{R} \cup \{\infty\}$ verify the condition $f_\lambda = g_\lambda$ for some $\lambda > 0$, then necessarily they must be equal.

The convergence of the functions f_λ to f is much better when f is Lipschitz. To show this we need the following auxiliary result:

6.5.6 Lemma *Suppose that $f : H \to \mathbb{R}$ is a Lipschitz convex function and $(f_n)_n$ is a sequence of convex functions on H such that $f_n \leq f$ for all n. If $f_n^* \to f^*$ uniformly, then $f_n \to f$ uniformly on bounded subsets of H.*

Proof. Let A be a bounded subset of H and let $\varepsilon > 0$. Since f is bounded on bounded sets, it follows that $\partial f(A)$ is bounded. See Exercise 5, Section 3.3. Since $f_n^* \to f^*$ uniformly, there exists a natural number N such that $f_n^*(\mathbf{y}) - f^*(\mathbf{y}) < \varepsilon$ for all $n \geq N$ and $\mathbf{y} \in \partial f(A)$. Then, for all $\mathbf{x} \in A$ and $\mathbf{y} \in \partial f(\mathbf{x})$, we have

$$f^*(\mathbf{y}) = \langle \mathbf{x}, \mathbf{y}\rangle - f(\mathbf{x}) \geq f_n^*(\mathbf{y}) - \varepsilon \geq \langle \mathbf{x}, \mathbf{y}\rangle - f_n(\mathbf{x}) - \varepsilon,$$

whenever $n \geq N$. Therefore $f(\mathbf{x}) - \varepsilon \leq f_n(\mathbf{x}) \leq f(\mathbf{x})$ for all $\mathbf{x} \in A$ and $n \geq N$. ■

6.5.7 Corollary *If $f : H \to \mathbb{R}$ is a Lipschitz convex function, then the Moreau–Yosida approximants f_λ converge uniformly to f on bounded sets as $\lambda \to 0$.*

The infimal convolution and the Moreau–Yosida approximation are useful in the theory of Hamilton–Jacobi equations.

Exercises

1. (Huber function). Prove that the Moreau–Yosida approximants of the absolute value function, $f(x) = |x|$ for $x \in \mathbb{R}$, are given by the formulas

$$f_\lambda(x) = \begin{cases} x^2/(2\lambda) & \text{for } |x| \leq \lambda \\ |x| - \lambda/2 & \text{for } |x| \geq \lambda. \end{cases}$$

2. Prove that the infimal convolution $f \square g$ of two proper convex functions $f, g \colon H \to \mathbb{R} \cup \{\infty\}$ is a convex function.

3. Suppose that $f, g \colon H \to \mathbb{R} \cup \{\infty\}$ are two proper convex functions. Prove that
$$(f+g)^* = \text{cl } (f^* \square g^*).$$
Here the closure operation is superfluous when $0 \in \text{int}\,(\text{dom } f - \text{dom } g)$ (in particular, when $\text{int}\,(\text{dom } f) \cap \text{dom } g \neq \emptyset$ or $\text{dom } f \cap \text{int}\,(\text{dom } g) \neq \emptyset$). See Corollary 6.2.8.

4. Suppose that H is a Hilbert space and $f : H \to \mathbb{R} \cup \{\infty\}$ is a lower semicontinuous proper convex function. Prove that the Moreau–Yosida approximants $f_\lambda(\mathbf{x})$ tend to $-f^*(\mathbf{0}) = \inf \{f(\mathbf{y}) : \mathbf{y} \in H\}$ as $\lambda \to \infty$.

5. If f is as in Exercise 4, prove that for all $\lambda > 0$,
$$\inf_{\mathbf{x} \in H} \left(f(\mathbf{x}) + \frac{1}{2\lambda} \|\mathbf{x}\|^2 \right) + \inf_{\mathbf{y} \in H} \left(f^*(\mathbf{y}) + \frac{1}{2\lambda} \|\mathbf{y}\|^2 \right) = 0.$$

6. Use the Moreau–Yosida approximation to complete the proof of the implication (a) $\Rightarrow$ (c) in Theorem 6.1.2, when $E = \mathbb{R}^N$.
[*Hint*: Notice that $(f_\lambda)^{**} = f_\lambda$ (since $\text{dom } f_\lambda = \mathbb{R}^N$) and
$$f^{**}(\mathbf{x}) \geq \liminf_{\lambda \to 0} f_\lambda^{**}(\mathbf{x}) = \liminf_{\lambda \to 0} f_\lambda(\mathbf{x}) = f(\mathbf{x}).]$$

7. (J. J. Moreau [338]) Given a lower semicontinuous proper convex function $f \colon H \to \mathbb{R} \cup \{\infty\}$, the *proximity* operator of f is
$$\text{prox}_f : H \to H, \quad \text{prox}_f(\mathbf{x}) = \arg\min_{\mathbf{y} \in H} \left(f(\mathbf{y}) + \frac{1}{2} \|\mathbf{x} - \mathbf{y}\|^2 \right).$$
Prove that:
(a) For every $\mathbf{x}, \mathbf{p} \in H$, we have $\mathbf{p} = \text{prox}_f(\mathbf{x})$ if, and only if, $\mathbf{x} - \mathbf{p} \in \partial f(\mathbf{p})$;
(b) If C is a nonempty closed convex subset of H and $f = \iota_C$, then $\text{prox}_f = P_C$ (the orthogonal projection on C);
(c) If $\alpha > 0$ and $\mathbf{x} \in H$, then $\mathbf{x} = \text{prox}_{\alpha f}\, \mathbf{x} + \alpha\, \text{prox}_{f^*/\alpha}(\mathbf{x}/\alpha)$.

6.6 The Hopf–Lax Formula

Suppose that $\mathcal{L} : \mathbb{R}^N \times \mathbb{R}^N \to \mathbb{R}$, $\mathcal{L} = \mathcal{L}(\mathbf{x}, \mathbf{q})$, is a given smooth function, referred to as the *Lagrangian*; its variables represent usually the *generalized position* $\mathbf{x}$ and *generalized velocity* $\mathbf{q}$. Consider the variational problem of minimizing the *action* functional,

$$I(\mathbf{w}) = \int_0^t \mathcal{L}(\mathbf{w}, \dot{\mathbf{w}})ds,$$

over the class of admissible functions,

$$\mathcal{A} = \left\{\mathbf{w} \in C^2\left([0,t], \mathbb{R}^N\right) : \mathbf{w}(0) = \mathbf{x}_0 \text{ and } \mathbf{w}(t) = \mathbf{x}\right\}.$$

Here $\mathbf{x}_0$ and $\mathbf{x}$ are two given points in $\mathbb{R}^N$ and $\dot{\mathbf{w}}$ denotes the derivative of $\mathbf{w}$ with respect to the variable s (classical notation in mechanics introduced by Newton). As well known (for example, see L. C. Evans [151], pp. 117–118), the minimizer $\mathbf{u}$, if it exists, must verify the Euler–Lagrange vector equation,

$$-\frac{d}{ds}\left(\nabla_{\mathbf{q}}\mathcal{L}(\mathbf{u}(s), \dot{\mathbf{u}}(s))\right) + \nabla_{\mathbf{x}}\mathcal{L}(\mathbf{u}(s), \dot{\mathbf{u}}(s)) = 0 \quad \text{for } s \in [0,t], \tag{6.27}$$

and the boundary conditions $u(0) = \mathbf{x}_0$ and $u(t) = \mathbf{x}$. Here

$$\nabla_{\mathbf{q}} = \left(\frac{\partial}{\partial q_1}, ..., \frac{\partial}{\partial q_N}\right)$$

and

$$\nabla_{\mathbf{x}} = \left(\frac{\partial}{\partial x_1}, ..., \frac{\partial}{\partial x_N}\right).$$

Thus (6.27) is equivalent to a system of N scalar equations of second order:

$$-\frac{d}{ds}\left(\frac{\partial \mathcal{L}}{\partial q_i}(\mathbf{u}(s), \dot{\mathbf{u}}(s))\right) + \frac{\partial \mathcal{L}}{\partial x_i}(\mathbf{u}(s), \dot{\mathbf{u}}(s)) = 0 \quad \text{for } s \in [0,t] \text{ and } i = 1, ..., N.$$

It can be converted into a system of $2N$ scalar equations of first order by *replacing* the generalized velocity with *generalized momentum*,

$$\mathbf{p}(s) = D_{\mathbf{q}}\mathcal{L}(\mathbf{u}(s), \dot{\mathbf{u}}(s)), \tag{6.28}$$

and *assuming* that the equation (6.28) can be inverted to yield a unique smooth function

$$\mathbf{q} = \mathbf{q}(\mathbf{x}, \mathbf{p}).$$

Under these circumstances, we attach to the Lagrangian $\mathcal{L}$ the *Hamiltonian* function,

$$\mathcal{H}(\mathbf{x}, \mathbf{p}) = \langle \mathbf{p}, \mathbf{q}(\mathbf{x}, \mathbf{p})\rangle - \mathcal{L}(\mathbf{x}, \mathbf{q}(\mathbf{x}, \mathbf{p})), \tag{6.29}$$

whose meaning in mechanics is that of *total energy.*

A straightforward computation yields the following fact:

6.6.1 Lemma *The Euler–Lagrange equation* (6.27) *is equivalent to Hamilton's equations*

$$\begin{cases} \dot{\mathbf{x}}(s) = \nabla_p \mathcal{H}(\mathbf{x}(s), \mathbf{p}(s)) \\ \dot{\mathbf{p}}(s) = -\nabla_x \mathcal{H}(\mathbf{x}(s), \mathbf{p}(s)) \end{cases}$$

for $s \in [0, t]$.

Notice also that the function $s \to \mathcal{H}(\mathbf{x}(s), \mathbf{p}(s))$ is constant. Indeed,

$$\begin{aligned} \frac{d}{ds}\mathcal{H}(\mathbf{x}(s), \mathbf{p}(s)) &= \sum_{i=1}^{N} \left(\frac{\partial \mathcal{H}}{\partial p_i} \dot{p}_i + \frac{\partial \mathcal{H}}{\partial x_i} \dot{x}_i \right) \\ &= \sum_{i=1}^{N} \frac{\partial \mathcal{H}}{\partial p_i} \left(-\frac{\partial \mathcal{H}}{\partial x_i} \right) + \sum_{i=1}^{N} \frac{\partial \mathcal{H}}{\partial x_i} \left(\frac{\partial \mathcal{H}}{\partial p_i} \right) = 0. \end{aligned}$$

Comparing the formula (6.29) with the classical Legendre transform (see Theorem 6.1.8), we see that the true meaning of the Hamiltonian is that of Legendre–Fenchel conjugate of the Lagrangian with respect to the second variable.

Therefore, the natural framework that assures not only the existence of a unique global minimizer for the action functional but also the transformation of the Euler–Lagrange equation into Hamilton's equations can be briefly described as follows:

(H) The Lagrangian $\mathcal{L} : \mathbb{R}^N \times \mathbb{R}^N \to \mathbb{R}$, $\mathcal{L} = \mathcal{L}(\mathbf{x}, \mathbf{q})$, is a function of class C^2, strictly convex and supercoercive with respect to the second variable.

The existence of a unique global minimizer of the action functional follows from Theorem 3.10.1 and Corollary 3.10.4.

According to Theorem 6.1.8, one can introduce the Hamiltonian associated to $\mathcal{L}$ via the formula

$$\mathcal{H}(\mathbf{x}, \mathbf{p}) = \sup \left\{ \langle \mathbf{q}, \mathbf{p} \rangle - \mathcal{L}(\mathbf{x}, \mathbf{q}) : \mathbf{q} \in \mathbb{R}^N \right\}.$$

This is a continuously differentiable strictly convex function. See Theorem 6.2.4. Also, the map

$$(\mathbf{x}, \mathbf{v}) \to (\mathbf{x}, \nabla_{\mathbf{v}} \mathcal{L}(\mathbf{x}, \mathbf{q}))$$

is bijective and of class C^1. Hence, under the hypothesis (H), the Euler–Lagrange equation can be indeed restated in terms of Hamilton's equations.

6.6.2 Remark *The motion* $\mathbf{x} = \mathbf{x}(t)$ *of a mass-*m *particle in the potential* $V(\mathbf{x})$ *outlines the following conjugate pair,*

$$\mathcal{L}(\mathbf{x}, \mathbf{q}) = \frac{1}{2} m \|\mathbf{q}\|^2 - V(\mathbf{x}) \text{ and } \mathcal{H}(\mathbf{x}, \mathbf{p}) = \frac{1}{2m} \|\mathbf{p}\|^2 + V(\mathbf{x}).$$

In this case, the Euler–Lagrange equation coincide with Newton's law of motion $m \frac{\mathrm{d}^2 \mathbf{x}}{\mathrm{d}t^2} = -\nabla V$.

Following L. C. Evans [151], we next discuss the Hopf–Lax formula, which provides a solution for the Hamilton–Jacobi equation with given initial data, in the particular case of Hamiltonians that depend only on the variable $\mathbf{p}$.

6.6.3 Theorem *Assume that $\mathcal{H} : \mathbb{R}^N \to \mathbb{R}$, $\mathcal{H} = \mathcal{H}(\mathbf{p})$, is a supercoercive convex function of class C^2 with conjugate $\mathcal{L} = \mathcal{L}(\mathbf{q})$, and $u_0 : \mathbb{R}^N \to \mathbb{R}$ is a Lipschitz function. Then the function $u = u(t, \mathbf{x})$ given by the Hopf–Lax formula,*

$$u(t,\mathbf{x}) = \begin{cases} \min\limits_{\mathbf{y}\in\mathbb{R}^N} \left(t\mathcal{L}\left(\frac{\mathbf{x}-\mathbf{y}}{t}\right) + u_0(\mathbf{y})\right) & \text{for } (t,\mathbf{x}) \in (0,\infty)\times\mathbb{R}^N \\ u_0(\mathbf{x}) & \text{for } t = 0 \text{ and } \mathbf{x}\in\mathbb{R}^N \end{cases} \tag{6.30}$$

has the following properties:

(a) *u is a Lipschitz function, differentiable almost everywhere in $(0,\infty)\times\mathbb{R}^N$;*

(b) *At every point $(t, \mathbf{x}) \in (0,\infty)\times\mathbb{R}^N$ where u is differentiable it verifies the Hamilton–Jacobi equation*

$$u_t(t,\mathbf{x}) + \mathcal{H}(\nabla u(t,\mathbf{x})) = 0; \tag{6.31}$$

(c) *u verifies the initial condition*

$$\lim_{t\to 0} u(t,\mathbf{x}) = u_0(\mathbf{x}) \quad \text{for } \mathbf{x}\in\mathbb{R}^N. \tag{6.32}$$

The source of the Hopf–Lax formula is provided by the variational approach of the Hamilton–Jacobi equation (6.31), when accompanied by the initial condition (6.32):

6.6.4 Lemma (The Bellman principle) *The Hopf–Lax formula (6.30) represents the solution of the following minimization problem:*

$$u(t,\mathbf{x}) = \inf_{w\in\mathcal{A}(t,\mathbf{x})} \left\{ \int_0^t \mathcal{L}(\dot{\mathbf{w}}(s))\mathrm{d}s + u_0(\mathbf{w}(\mathbf{0})) \right\},$$

where $\mathcal{A}(t,\mathbf{x}) = \left\{\mathbf{w} : \mathbf{w} \in C^1\left([0,t], \mathbb{R}^N\right) \text{ and } \mathbf{w}(t) = \mathbf{x}\right\}.$

Proof. Indeed, since the function $\mathbf{w}(s) = \mathbf{y} + s(\mathbf{x}-\mathbf{y})/t$ belongs to $\mathcal{A}(t,\mathbf{x}))$, we have

$$u(t,\mathbf{x}) \le \int_0^t \mathcal{L}(\dot{\mathbf{w}}(s))\mathrm{d}s + u_0(\mathbf{y}) = t\mathcal{L}\left(\frac{\mathbf{x}-\mathbf{y}}{t}\right) + u_0(\mathbf{y}),$$

whence $u(t,\mathbf{x}) \le \inf_{\mathbf{y}\in\mathbb{R}^N}\left(t\mathcal{L}\left(\frac{\mathbf{x}-\mathbf{y}}{t}\right) + u_0(\mathbf{y})\right)$. On the other hand, for every smooth function $\mathbf{w}$ such that $w(0) = \mathbf{y}$ and $\mathbf{w}(t) = \mathbf{x}$, we infer from Jensen's inequality that

$$\mathcal{L}\left(\frac{1}{t}\int_0^t \dot{\mathbf{w}}(s)\mathrm{d}s\right) \le \frac{1}{t}\int_0^t \mathcal{L}(\dot{\mathbf{w}}(s))\mathrm{d}s.$$

Thus

$$t\mathcal{L}\left(\frac{\mathbf{x}-\mathbf{y}}{t}\right)+u_0(\mathbf{y})\leq\int_0^t \mathcal{L}(\dot{\mathbf{w}}(s))\mathrm{d}s+u_0(\mathbf{y}),$$

which yields $u(t,\mathbf{x})\geq\inf_{\mathbf{y}\in\mathbb{R}^N}\left(t\mathcal{L}\left(\frac{\mathbf{x}-\mathbf{y}}{t}\right)+u_0(\mathbf{y})\right)$, and thus the equality

$$u(t,\mathbf{x})=\inf_{\mathbf{y}\in\mathbb{R}^N}\left(t\mathcal{L}\left(\frac{\mathbf{x}-\mathbf{y}}{t}\right)+u_0(\mathbf{y})\right).$$

The fact that this infimum is attained follows from Corollary 3.10.4. ∎

We pass now to the proof of Theorem 6.6.3, that makes the objective of Lemma 6.6.5, and Lemma 6.6.7.

6.6.5 Lemma *The function $u=u(t,\mathbf{x})$ given by the Hopf–Lax formula verifies the initial condition (6.32) and is Lipschitz with respect to both variables.*

Proof. Notice first that $u(t,\mathbf{x})\leq t\mathcal{L}(\mathbf{0})+u_0(\mathbf{x})$, which yields

$$u(t,\mathbf{x})-u_0(\mathbf{x})\leq t\mathcal{L}(\mathbf{0}).$$

On the other hand,

$$\begin{aligned}
u(t,\mathbf{x}) &= \min_{y\in\mathbb{R}^N}\left(t\mathcal{L}\left(\frac{\mathbf{x}-\mathbf{y}}{t}\right)+u_0(\mathbf{y})\right)\\
&= u_0(\mathbf{x})+\min_{y\in\mathbb{R}^N}\left(t\mathcal{L}\left(\frac{\mathbf{x}-\mathbf{y}}{t}\right)+u_0(\mathbf{y})-u_0(\mathbf{x})\right)\\
&\geq u_0(\mathbf{x})-\max_{y\in\mathbb{R}^N}\left(\|\mathbf{y}-\mathbf{x}\|\operatorname{Lip}(u_0)-t\mathcal{L}\left(\frac{\mathbf{x}-\mathbf{y}}{t}\right)\right)\\
&= u_0(\mathbf{x})-t\max_{\mathbf{z}\in\mathbb{R}^N}\left(\|\mathbf{z}\|\operatorname{Lip}(u_0)-\mathcal{L}(\mathbf{z})\right)\\
&= u_0(\mathbf{x})-t\max_{\mathbf{v}\in\overline{B}(\mathbf{0},\operatorname{Lip}(u_0))}\left(\max_{\mathbf{z}}\left(\langle\mathbf{v},\mathbf{z}\rangle-\mathcal{L}(\mathbf{z})\right)\right)\\
&= u_0(\mathbf{x})-t\max_{\mathbf{v}\in\overline{B}(\mathbf{0},\operatorname{Lip}(u_0))}\mathcal{H}(v),
\end{aligned}$$

where $\overline{B}(\mathbf{0},\operatorname{Lip}(u_0))$ denotes the closed ball centered at the origin and having the radius $\operatorname{Lip}(u_0)$. Therefore

$$|u(t,\mathbf{x})-u_0(\mathbf{x})|\leq Ct\quad\text{for all }(t,\mathbf{x})\in(0,\infty)\times\mathbb{R}^N,$$

where

$$C=\max\left\{|\mathcal{L}(0)|,\max_{\mathbf{v}\in\overline{B}(\mathbf{0},\operatorname{Lip}(u_0))}|\mathcal{H}(\mathbf{v})|\right\}.$$

The inside maximum is motivated by the continuity of $\mathcal{H}$.

As concerns the Lipschitzianity of u, let's fix arbitrarily $t,\bar{t}>0$ and points $\mathbf{x},\bar{\mathbf{x}}$ in $\mathbb{R}^N$ and choose $\mathbf{y}$ such that

$$u(t,\mathbf{x})=t\mathcal{L}\left(\frac{\mathbf{x}-\mathbf{y}}{t}\right)+u_0(\mathbf{y}).$$

Then

$$\mathcal{L}\left(\frac{\mathbf{x}-\mathbf{y}}{t}\right) = \frac{u(t,\mathbf{x})-u_0(\mathbf{x})}{t} + \frac{u_0(\mathbf{x})-u_0(\mathbf{y})}{t}$$
$$\leq \mathcal{L}(\mathbf{0}) + \mathrm{Lip}(u_0)\frac{\|\mathbf{x}-\mathbf{y}\|}{t}. \tag{6.33}$$

Since $\mathcal{L}$ is supercoercive, there is a constant $C_1 > 0$ (that depends on $\mathcal{L}$ and Lip (u_0)) such that

$$\mathcal{L}(\mathbf{q}) \geq \mathcal{L}(\mathbf{0}) + \text{ Lip } (u_0)\|\mathbf{q}\| \quad \text{for all } \|\mathbf{q}\| > C_1.$$

Combining this fact with the inequality (6.33) we infer that $\|\mathbf{x}-\mathbf{y}\|/t \leq C_1$. Letting $\bar{\mathbf{y}} = \bar{\mathbf{x}} - (\bar{t}/t)\,(\mathbf{x}-\mathbf{y})$, we have

$$\frac{\mathbf{x}-\mathbf{y}}{t} = \frac{\bar{\mathbf{x}}-\bar{\mathbf{y}}}{\bar{t}} \text{ and } \bar{\mathbf{y}}-\mathbf{y} = \bar{\mathbf{x}}-\mathbf{x}+(t-\bar{t})\,\frac{\mathbf{x}-\mathbf{y}}{t}.$$

Hence

$$u(\bar{t},\bar{\mathbf{x}}) - u(t,\mathbf{x}) \leq \bar{t}\mathcal{L}\left(\frac{\bar{\mathbf{x}}-\bar{\mathbf{y}}}{\bar{t}}\right) + u_0(\bar{\mathbf{y}}) - t\mathcal{L}\left(\frac{\mathbf{x}-\mathbf{y}}{t}\right) - u_0(\mathbf{y})$$
$$= (\bar{t}-t)\,\mathcal{L}\left(\frac{\mathbf{x}-\mathbf{y}}{t}\right) + u_0(\bar{\mathbf{y}}) - u_0(\mathbf{y})$$
$$\leq |\bar{t}-t| \max_{\|\mathbf{q}\|\leq C_1} \mathcal{L}(\mathbf{q}) + \text{ Lip } (u_0)\|\bar{\mathbf{y}}-\mathbf{y}\|$$
$$\leq \left(C_1 \text{ Lip } (u_0) + \max_{\|\mathbf{q}\|\leq C_1} \mathcal{L}(\mathbf{q})\right)|\bar{t}-t| + \text{ Lip}(u_0)\|\bar{\mathbf{y}}-\mathbf{y}\|.$$

Interchanging the roles of $(t,\mathbf{x})$ and $(\bar{t},\bar{\mathbf{x}})$, we conclude that u is a Lipschitz function. ∎

The proof of the assertion (b) of Theorem 6.6.3 needs the following technical result, describing a semigroup property:

6.6.6 Lemma *For each $\mathbf{x} \in \mathbb{R}^N$ and $s \in [0,t)$ we have*

$$u(t,\mathbf{x}) = \min_{\mathbf{y}\in\mathbb{R}^N} \left\{(t-s)\mathcal{L}\left(\frac{\mathbf{x}-\mathbf{y}}{t-s}\right) + u(s,\mathbf{y})\right\}.$$

Proof. (L. C. Evans [151], p. 125) Choose $\mathbf{z}$ in $\mathbb{R}^N$ such that

$$u(t,\mathbf{x}) = t\mathcal{L}\left(\frac{\mathbf{x}-\mathbf{z}}{t}\right) + u_0(\mathbf{z}).$$

For $\mathbf{y} = \frac{s}{t}\mathbf{x} + (1-\frac{s}{t})\mathbf{z}$ we have $(\mathbf{x}-\mathbf{y})/(t-s) = (\mathbf{x}-\mathbf{z})/t = (\mathbf{y}-\mathbf{z})/s$ and thus

$$(t-s)\mathcal{L}\left(\frac{\mathbf{x}-\mathbf{y}}{t-s}\right) + u(s,\mathbf{y}) \leq (t-s)\,\mathcal{L}\left(\frac{\mathbf{x}-\mathbf{z}}{t}\right) + s\mathcal{L}\left(\frac{\mathbf{y}-\mathbf{z}}{s}\right) + u_0(\mathbf{z})$$
$$\leq t\mathcal{L}\left(\frac{\mathbf{x}-\mathbf{z}}{t}\right) + u_0(\mathbf{z}) = u(t,\mathbf{x}).$$

This shows that $\min_{\mathbf{y}\in\mathbb{R}^N}\left\{(t-s)\mathcal{L}\left(\frac{\mathbf{x}-\mathbf{y}}{t-s}\right)+u(s,\mathbf{y})\right\}\le u(t,\mathbf{x})$. For the reverse inequality, let's fix arbitrarily $\mathbf{y}\in\mathbb{R}^N$ and $s\in(0,t)$. Choose $\mathbf{w}\in\mathbb{R}^N$ such that

$$u(s,\mathbf{y})=s\mathcal{L}\left(\frac{\mathbf{y}-\mathbf{w}}{s}\right)+u_0(\mathbf{w}).$$

Since $\mathcal{L}$ is a convex function and

$$\frac{\mathbf{x}-\mathbf{w}}{t}=(1-\frac{s}{t})\frac{\mathbf{x}-\mathbf{y}}{t-s}+\frac{s}{t}\frac{\mathbf{y}-\mathbf{w}}{s},$$

we have

$$\mathcal{L}\left(\frac{\mathbf{x}-\mathbf{w}}{t}\right)\le(1-\frac{s}{t})\mathcal{L}\left(\frac{\mathbf{x}-\mathbf{y}}{t-s}\right)+\frac{s}{t}\mathcal{L}\left(\frac{\mathbf{y}-\mathbf{w}}{s}\right),$$

which implies that

$$\begin{aligned}u(t,\mathbf{x})&\le t\mathcal{L}\left(\frac{\mathbf{x}-\mathbf{w}}{t}\right)+u_0(\mathbf{w})\le(t-s)\mathcal{L}\left(\frac{\mathbf{x}-\mathbf{y}}{t-s}\right)+s\mathcal{L}\left(\frac{\mathbf{y}-\mathbf{w}}{s}\right)+u_0(\mathbf{w})\\&=(t-s)\mathcal{L}\left(\frac{\mathbf{x}-\mathbf{y}}{t-s}\right)+u(s,\mathbf{y}).\end{aligned}$$

The conclusion is now obvious. ■

Combining Lemma 6.6.5 with Rademacher's theorem (see Appendix D, Theorem D.1.1) we infer that the Hopf–Lax formula yields a function differentiable almost everywhere.

6.6.7 Lemma *At every point $(t,\mathbf{x})\in(0,\infty)\times\mathbb{R}^N$ where the function $u=u(t,\mathbf{x})$ defined by the Hopf–Lax formula is differentiable, it verifies the Hamilton–Jacobi equation*

$$u_t(t,\mathbf{x})+\mathcal{H}(\nabla u(t,\mathbf{x}))=0.$$

Proof. Let $(t,\mathbf{x})\in(0,\infty)\times\mathbb{R}^N$ be a point at which the function $u=u(t,\mathbf{x})$ is differentiable. Fix $\mathbf{q}\in\mathbb{R}^N$ and $h>0$. According to Lemma 6.6.6,

$$\begin{aligned}u(t+h,\mathbf{x}+h\mathbf{q})&=\min_{\mathbf{y}\in\mathbb{R}^N}\left\{h\mathcal{L}\left(\frac{\mathbf{x}+h\mathbf{q}-\mathbf{y}}{h}\right)+u(t,\mathbf{y})\right\}\\&\le h\mathcal{L}(\mathbf{q})+u(t,\mathbf{x}),\end{aligned}$$

which implies that

$$\frac{u(t+h,\mathbf{x}+h\mathbf{q})-u(t,\mathbf{x})}{h}\le\mathcal{L}(\mathbf{q}).$$

Passing to the limit as $h\to0$, we obtain $\langle\nabla u(t,\mathbf{x}),\mathbf{q}\rangle+u_t(t,\mathbf{x})\le\mathcal{L}(\mathbf{q})$ (for all $\mathbf{q}\in\mathbb{R}^N$). Therefore

$$u_t(t,\mathbf{x})+\mathcal{H}(\nabla u(t,\mathbf{x}))=u_t(t,\mathbf{x})+\max_{\mathbf{q}\in\mathbb{R}^N}\{\langle\nabla u(t,\mathbf{x}),\mathbf{q}\rangle-\mathcal{L}(\mathbf{q})\}\le0.$$

In order to prove the other inequality, choose $\mathbf{z}$ such that $u(t,\mathbf{x}) = t\mathcal{L}\left(\frac{\mathbf{x}-\mathbf{z}}{t}\right) + u_0(\mathbf{z})$. Fix $h > 0$ and put $s = t-h$ and $\mathbf{y} = (s/t)\mathbf{x}+(1-s/t)\,\mathbf{z}$. Then $(\mathbf{x}-\mathbf{z})/t = (\mathbf{y}-\mathbf{z})/s$, which yields

$$\begin{aligned} u(t,\mathbf{x}) - u(s,\mathbf{y}) &\geq \left[t\mathcal{L}\left(\frac{\mathbf{x}-\mathbf{z}}{t}\right) + u_0(\mathbf{z})\right] - \left[s\mathcal{L}\left(\frac{\mathbf{y}-\mathbf{z}}{s}\right) + u_0(\mathbf{z})\right] \\ &= (t-s)\mathcal{L}\left(\frac{\mathbf{x}-\mathbf{z}}{t}\right), \end{aligned}$$

equivalently,

$$\frac{u(t,\mathbf{x}) - u\left(t-h, (1-h/t)\mathbf{x} + (h/t)\,\mathbf{z}\right)}{h} \geq \mathcal{L}\left(\frac{\mathbf{x}-\mathbf{z}}{t}\right).$$

Passing to the limit as $h \to 0$, we obtain that

$$\langle \nabla u(t,\mathbf{x}), \frac{\mathbf{x}-\mathbf{z}}{t}\rangle + u_t(t,\mathbf{x}) \geq \mathcal{L}\left(\frac{\mathbf{x}-\mathbf{z}}{t}\right).$$

Hence

$$\begin{aligned} u_t(t,\mathbf{x}) + \mathcal{H}(\nabla u(t,\mathbf{x})) &= u_t(t,\mathbf{x}) + \max_{\mathbf{q}\in\mathbb{R}^N} \left\{\langle \nabla u(t,\mathbf{x}), \mathbf{q}\rangle - \mathcal{L}(\mathbf{q})\right\} \\ &\geq u_t(t,\mathbf{x}) + \langle \nabla u(t,\mathbf{x}), \frac{\mathbf{x}-\mathbf{z}}{t}\rangle - \mathcal{L}\left(\frac{\mathbf{x}-\mathbf{z}}{t}\right) \geq 0, \end{aligned}$$

which ends the proof. ■

The proof of Theorem 6.6.3 is now an obvious consequence of Lemma 6.6.5, and Lemma 6.6.7.

As is shown in Exercise 2, the initial-value problem for the Hamilton–Jacobi equation does not have a unique *weak solution*, that is, a solution that verifies the assertions (b) and (c) of Theorem 6.6.3. This situation imposed the consideration of stronger concepts of solution, like *viscosity solution* in the sense of M. G. Crandall and P.-L. Lions. Full details and applications can be found in the books by P. Cannarsa and C. Sinestrari [89] and L. C. Evans [151] and also in the survey by M. G. Crandall, H. Ishii and P.-L. Lions [118].

Exercises

1. Prove that the Hopf–Lax formula can be formulated via the inf-convolution operation as follows:

$$u(t,\mathbf{x}) = \left((t\mathcal{H})^* \,\Box u_0\right)(\mathbf{x}).$$

2. Consider the initial-value problem for the Hamilton–Jacobi equation,

$$\begin{aligned} &u_t + |u_x|^2 = 0 \text{ for a.e. } (t,x) \in (0,\infty)\times\mathbb{R} \\ &\lim_{t\to 0} u(t,x) = 0 \text{ for } x \in \mathbb{R}. \end{aligned}$$

Prove that this problem admits an infinity of solutions.

[*Hint:* For $\alpha \geq 0$, consider the function $u_\alpha(t,x) = 0$ for $|x| \geq at$ and $u_\alpha(t,x) = \alpha \left|x - \alpha^2 t\right|$ for $|x| \leq at$.]

3. Call a continuous function $f : \mathbb{R}^N \to \mathbb{R}$ *semiconcave* if there is a constant C such that

$$f(\mathbf{x}+\mathbf{z}) - 2f(\mathbf{x}) + f(\mathbf{x}-\mathbf{z}) \leq C \left\|\mathbf{z}\right\|^2,$$

for all $\mathbf{x}, \mathbf{z} \in \mathbb{R}^N$. Prove that a function f of class C^2 is semiconcave if, and only if, $\sup_{\mathbf{x}\in\mathbb{R}^N} \left\|d^2 f(\mathbf{x})\right\| < \infty$.

4. Under the hypothesis of Theorem 6.6.3, show that the Hopf–Lax formula yields a semiconcave solution if the initial data is semiconcave. One can prove that this is actually the *only* semiconcave weak solution. See L. C. Evans [151], Theorem 7, p. 132.

6.7 Comments

The notion of duality is one of the central concepts both in geometry and in analysis. As was noticed by S. Artstein-Avidan and V. Milman [21], in many examples of duality, the standard definitions arise, explicitly, from two very simple and natural properties: involution, and order reversion. In particular, this applies to the Legendre–Fenchel transform.

6.7.1 Theorem *Denote by $Cvx(\mathbb{R}^N)$ the set of lower semicontinuous convex functions $\varphi : \mathbb{R}^N \to \mathbb{R} \cup \{\infty\}$ and consider a map*

$$\mathcal{T} : Cvx(\mathbb{R}^N) \to Cvx(\mathbb{R}^N)$$

verifying the following two properties:
(a) $\mathcal{T}\mathcal{T}\varphi = \mathcal{T}\varphi$;
(b) $\varphi \leq \psi$ *implies* $\mathcal{T}\psi \geq \mathcal{T}\varphi$.
Then, $\mathcal{T}$ is essentially the classical Legendre–Fenchel transform, namely there exists a constant $C_0 \in \mathbb{R}$, a vector $\mathbf{v}_0 \in \mathbb{R}^N$, and an invertible symmetric linear transformation $B : \mathbb{R}^N \to \mathbb{R}^N$ such that

$$\left(\mathcal{T}\varphi\right)(\mathbf{x}) = \varphi^* \left(B\mathbf{x} + \mathbf{v}_0\right) + \langle \mathbf{x}, \mathbf{v}_0\rangle + C_0.$$

A nice introduction to the classical theory of Legendre transform from the point of view of physicists (as well as its importance to classical mechanics, statistical mechanics, and thermodynamics) can be found in the paper of R. K. P. Zia et al. [491]. A connection between the Legendre-Fenchel transform and the Laplace transform is outlined by the tropical analysis. For details, see the paper of G. L. Litvinov et al [289].

Theorem 6.1.2, on biconjugation, is due to W. Fenchel [161] (in the case of Euclidean spaces) and to J. J. Moreau [337] (in the general case). Notice that

L. Hörmander [226] proved the theorem of biconjugation in the case of support and indicator functions.

The dual correspondence between the strict convexity of a function and the smoothness of its conjugate, as was noticed in Theorem 6.2.4, admits a much more general version (at the cost of a considerably more involving argument):

6.7.2 Theorem (R. T. Rockafellar [421]) *The following properties are equivalent for a lower semicontinuous proper convex function* $f : \mathbb{R}^N \to \mathbb{R} \cup \{\infty\}$*:*

(a) f *is strictly convex on every convex subset of* $\operatorname{dom} \partial f$ *(in particular, on* $\operatorname{ri}(\operatorname{dom} f)$*);*

(b) f^* *is differentiable on the open convex set* $\operatorname{int}(\operatorname{dom} f^*)$ *which is nonempty, but* $\partial f^*(\mathbf{x}) = \emptyset$ *for all points* $\mathbf{x} \in (\operatorname{dom} f^*) \setminus \operatorname{int}(\operatorname{dom} f^*)$*, if any.*

Moreover, due to Theorem 6.1.2, the roles of f *and* f^* *can be interchanged.*

For details, see R. T. Rockafellar and R. J.-B. Wets [422], Theorem 11.13, pp. 483–484.

The Fenchel–Moreau duality theorem 6.2.7 was proved by W. Fenchel [161] in the case of Euclidean spaces and by J. J. Moreau [339] in the general case.

The development of the theory of optimal transportation led to a natural generalization of the concepts of convexity and duality, by ascribing them to arbitrary cost functions. Some of the historical developments of this problem and some of the basic results are described by W. Gangbo and R. J. McCann [174], L. Rüschendorf [430], S. T. Rachev and L. Rüschendorf [411], and C. Villani [477], [478].

Let X and Y be two nonempty sets and let $c : X \times Y \to \mathbb{R} \cup \{\infty\}$, $c = c(x, y)$, be a function (referred to as a *cost* function). For $c(x, y) = \langle x, y \rangle$ we recover the classical theory. A proper function $\varphi : X \to \mathbb{R} \cup \{\infty\}$ is said to be *c-convex* if it admits a representation of the form

$$\varphi(x) = \sup_{y \in Y} \{c(x, y) + \psi(y)\}$$

for some function ψ. The *c-conjugate* φ^c of φ is defined by the formula

$$\varphi^c(y) = \sup_{x \in X} \{c(x, y) - \varphi(x)\}.$$

Continuing with the *double conjugate*,

$$\varphi^{cc}(y) = \sup_{y \in X} \{c(x, y) - \varphi^c(y)\},$$

we see that φ^c and φ^{cc} are c-convex and, moreover, φ is c-convex if, and only if, $\varphi = \varphi^{cc}$. In this context, the Fenchel–Young inequality takes the form

$$\varphi(x) + \varphi^c(y) \geq c(x, y).$$

The concept of *c-subdifferential* (associated to a c-convex φ) is introduced by the formula

$$\partial_c \varphi(x) = \{y : \varphi(z) \geq \varphi(x) + c(z, y) - c(x, y) \text{ for all } z \in \operatorname{dom} \varphi\}.$$

As was noticed by W. Gangbo and R. J. McCann [174], if $c = c(x)$ is locally Lipschitz on $\mathbb{R}^N$, then the $c(x-y)$-convex functions are differentiable almost everywhere; for an extension see the book of C. Villani [478]. This book also includes a proof of the following description of c-concave functions:

6.7.3 Theorem *When $c = c(\mathbf{x})$ is a convex function defined on $\mathbb{R}^N$, then the $c(\mathbf{x}-\mathbf{y})$-concave functions are precisely the viscosity solutions of the Hamilton–Jacobi equation*

$$u_t + c^*(\nabla u) = 0$$

at time $t = 1$ (or, equivalently, at a time $t \geq 1$).

Last, but not least, let us mention that the theory of optimal transportation received a great impetus in the 90s when Y. Brenier used the Monge–Kantorovich formulation of optimal transportation to derive a far-reaching generalization of the polar factorization. Let Ω be an open bounded subset of $\mathbb{R}^N$. A mapping $\mathbf{u} : \Omega \to \mathbb{R}^N$ is said to be *nondegenerate* if the inverse image $\mathbf{u}^{-1}(N)$ of every zero measure set $N \subset \mathbb{R}^N$ has also zero measure.

6.7.4 Theorem (Y. Brenier [77]) *Every nondegenerate vector field*

$$\mathbf{u} \in L^\infty(\Omega, \mathbb{R}^N)$$

can be decomposed as $\mathbf{u}(x) = \nabla\varphi \circ S(x)$ almost everywhere in Ω, where $\varphi : \mathbb{R}^N \to \mathbb{R}$ is a convex function and $S : \bar{\Omega} \to \bar{\Omega}$ is a measure preserving transformation.

A map $T\colon \mathbb{R}^N \to \mathbb{R}^N$ (defined μ-almost everywhere) pushes μ forward to ν (equivalently, transports μ onto ν) if $\nu(B) = \mu(T^{-1}(B))$ for every Borel set B in $\mathbb{R}^N$. Y. Brenier [77] found a very special map pushing forward one probability to another. His result was reconsidered by R. J. McCann [314], [315], who noticed that the absolutely continuous Borel probability measures can be transported by maps of the form $T = \nabla\varphi$, where φ is convex. These maps are usually referred to as *Brenier maps.* The differentiability properties of T (motivated by the existence of the Alexandrov Hessian of φ) make an easy handling of T possible. For example, if

$$\mu(B) = \int_B f(\mathbf{x})\,\mathrm{d}\mathbf{x} \quad \text{and} \quad \nu(B) = \int_B g(\mathbf{x})\,\mathrm{d}\mathbf{x},$$

and $T = \nabla\varphi$ is the Brenier map (pushing μ forward to ν), then

$$\int_{\mathbb{R}^N} h(\mathbf{y})g(\mathbf{y})\,\mathrm{d}\mathbf{y} = \int_{\mathbb{R}^N} h(\nabla\varphi(\mathbf{x}))f(\mathbf{x})\,\mathrm{d}\mathbf{x}$$

for all bounded Borel functions $h\colon \mathbb{R} \to \mathbb{R}_+$. Assuming the change of variable $\mathbf{y} = \nabla\varphi(\mathbf{x})$ is working, the last formula leads to the so-called *Monge–Ampère equation*,

$$f(\mathbf{x}) = \det(\operatorname{Hess}_{\mathbf{x}} \varphi) \cdot g(\nabla\varphi(\mathbf{x})). \tag{6.34}$$

As was noted by R. J. McCann [315], this equation is valid in general, provided that $\operatorname{Hess}_x \varphi$ is replaced by the Alexandrov Hessian of φ.

When $N = 1$, a map $T = T(t)$ that transports μ to ν is obtained by defining $T(t)$ as the smallest number such that

$$\int_{-\infty}^{t} f(x)\,\mathrm{d}x = \int_{-\infty}^{T(t)} g(x)\,\mathrm{d}x.$$

This is the key parametrization trick used in the proof of Theorem 3.11.6 (first noticed in the above form by R. Henstock and A. M. Macbeath [212]).

Chapter 7

Special Topics in Majorization Theory

The primary aim of this chapter is to discuss the connection between the Hermite–Hadamard double inequality and Choquet's theory. Noticed first by C. P. Niculescu [356], [357] (during the conference *Inequalities* 2001, in Timişoara), this connection led him to a partial extension of the majorization theory beyond the classical case of probability measures, using the so-called Steffensen–Popoviciu measures. Their main feature is to offer a large framework under which the Jensen–Steffensen inequality remains available. As a consequence, one obtains the extension of the left-hand side of Hermite–Hadamard inequality to a context involving signed Borel measures on arbitrary compact convex sets. A similar extension of the right-hand side of this inequality is known only in dimension 1, the higher dimensional case being still open.

7.1 Steffensen–Popoviciu Measures

The Jensen–Steffensen inequality (1.30) reveals an important case when Jensen's inequality works beyond the framework of positive measures. The aim of this section is to provide more insight into the relation between signed measures and Jensen's inequality.

7.1.1 Definition *Let K be a compact convex subset of a real Banach space E. A Steffensen–Popoviciu measure on K is any real Borel measure μ on K such that $\mu(K) > 0$ and*

$$\int_K f(x)\,\mathrm{d}\mu(x) \geq 0,$$

for every positive continuous convex function $f : K \to \mathbb{R}$.

Clearly, every Borel probability measure is an example of Steffensen–Popoviciu measure.

C. P. Niculescu and L.-E. Persson, *Convex Functions and Their Applications*, CMS Books in Mathematics, https://doi.org/10.1007/978-3-319-78337-6_7

Since every continuous convex function $f : K \to \mathbb{R}$ is the pointwise supremum of the family of its continuous and affine minorants (see Theorem 3.3.2), it follows that a real Borel measure μ on K is Steffensen–Popoviciu if, and only if, $\mu(K) > 0$ and

$$\int_K (x^*(x) + \alpha)^+ \, \mathrm{d}\mu(x) \geq 0 \quad \text{for all } x^* \in E^* \text{ and } \alpha \in \mathbb{R}. \tag{7.1}$$

The following result (due independently to T. Popoviciu [406], and A. M. Fink [164]) gives us a complete characterization of this class of measures in the case where K is a compact interval.

7.1.2 Theorem *Let μ be a real Borel measure on an interval $[a,b]$ with $\mu([a,b]) > 0$. Then μ is a Steffensen–Popoviciu measure if, and only if, it verifies the following condition of endpoints positivity:*

$$\int_a^t (t - x) \, \mathrm{d}\mu(x) \geq 0 \text{ and } \int_t^b (x - t) \, \mathrm{d}\mu(x) \geq 0 \tag{7.2}$$

for every $t \in [a,b]$.

Proof. The necessity of (7.2) is clear. For the sufficiency part, notice that any positive, continuous, and convex function $f : [a,b] \to \mathbb{R}$ is the uniform limit of a sequence $(g_n)_n$ of positive, piecewise linear, and convex functions. See Lemma 1.9.1. Or, according to Lemma 1.9.2, every such function g can be represented as a finite combination with positive coefficients of functions of the form 1, $(x-t)^+$ and $(t-x)^+$. Thus

$$\int_a^b f(x) \, \mathrm{d}\mu(x) = \lim_{n\to\infty} \int_a^b g_n(x) \, \mathrm{d}\mu(x) \geq 0.$$

■

An alternative argument, based on the integral representation of convex functions on intervals, was noticed by A. M. Fink [164].

7.1.3 Corollary *Suppose that $x_1 \leq \cdots \leq x_n$ are real points and $p_1, \ldots, p_n$ are real weights. Then the discrete measure $\mu = \sum_{k=1}^n p_k \delta_{x_k}$ is a Steffensen–Popoviciu measure if, and only if,*

$$\sum_{k=1}^n p_k > 0, \quad \sum_{k=1}^m p_k (x_m - x_k) \geq 0 \quad \text{and} \quad \sum_{k=m}^n p_k (x_k - x_m) \geq 0 \tag{7.3}$$

for all $m \in \{1, \ldots, n\}$.

7.1.4 Corollary (J. F. Steffensen [458]) *Suppose that $x_1 \leq \cdots \leq x_n$ are real points and $p_1, ..., p_n$ are real weights. Then, the discrete measure $\mu = \sum_{k=1}^n p_k \, \delta_{x_k}$ is a Steffensen–Popoviciu measure if*

$$\sum_{k=1}^n p_k > 0 \quad \text{and} \quad 0 \leq \sum_{k=1}^m p_k \leq \sum_{k=1}^n p_k \quad \text{for every } m \in \{1, ..., n\}. \tag{7.4}$$

Proof. Indeed, according to Theorem 7.1.2, we have to prove that (7.4) $\Rightarrow$ (7.3). This follows from Abel's summation formula (the discrete analogue of integration by parts). ■

The following result shows that Steffensen's condition (7.4) is stronger than (7.3).

7.1.5 Theorem *Suppose that* $p : [a, b] \to \mathbb{R}$ *is an integrable function. Then*

$$\int_a^b f(x)p(x)\mathrm{d}x \geq 0$$

for all positive, absolutely continuous and quasiconvex functions $f : [a, b] \to \mathbb{R}$ *if, and only if,*

$$\int_a^x p(t)\mathrm{d}t \geq 0 \text{ and } \int_x^b p(t)\mathrm{d}t \geq 0 \quad \text{for every } x \in [a, b]. \tag{7.5}$$

The absolutely continuous measures $\mu = p(x)\mathrm{d}x$ that verify $\mu([a, b]) > 0$ and also the condition (7.5) will be referred to as *strong Steffensen–Popoviciu measures*.

Proof. *Sufficiency.* According to Remark 1.1.7, there exists a point $c \in [a, b]$ such that f is decreasing on $[a, c]$ and increasing on $[c, b]$. Then

$$\begin{aligned}
\int_a^b f(x)p(x)\mathrm{d}x &= \int_a^c f(x)p(x)\mathrm{d}x + \int_c^b f(x)p(x)\mathrm{d}x \\
&= \int_a^c f(x)\mathrm{d}\left(\int_a^x p(t)\mathrm{d}t\right) - \int_c^b f(x)\mathrm{d}\left(\int_x^b p(t)\mathrm{d}t\right) \\
&= \left[f(x)\int_a^x p(t)\mathrm{d}t\right]\Bigg|_a^c - \int_a^c f'(x)\left(\int_a^x p(t)\mathrm{d}t\right)\mathrm{d}x \\
&\quad - \left[f(x)\int_x^b p(t)\mathrm{d}t\right]\Bigg|_c^b + \int_c^b f'(x)\left(\int_x^b p(t)\mathrm{d}t\right)\mathrm{d}x \\
&= f(c)\int_a^c p(t)\mathrm{d}t + \int_a^c (-f'(x))\left(\int_a^x p(t)\mathrm{d}t\right)\mathrm{d}x \\
&\qquad + f(c)\int_c^b p(t)\mathrm{d}t + \int_c^b f'(x)\left(\int_x^b p(t)\mathrm{d}t\right)\mathrm{d}x \geq 0,
\end{aligned}$$

as a sum of positive numbers. The integration by parts for absolutely continuous functions is motivated by Theorem 18.19, p. 287, in the monograph of E. Hewitt and K. Stromberg [214].

Necessity. Assuming $x_0 \in (a, b)$ and $\varepsilon > 0$ sufficiently small, the function

$$L_\varepsilon(x) = \begin{cases} 1 & \text{if } x \in [a, x_0 - \varepsilon] \\ -(x - x_0)/\varepsilon & \text{if } x \in [x_0 - \varepsilon, x_0] \\ 0 & \text{if } x \in [x_0, b] \end{cases}$$

is positive, decreasing, and also absolutely continuous. Therefore

$$\int_a^{x_0} p(t)\mathrm{d}t = \lim_{\varepsilon \to 0} \int_a^b L_\varepsilon(x)p(x)\mathrm{d}x \geq 0.$$

In a similar way one can prove that $\int_{x_0}^b p(t)\mathrm{d}t \geq 0$. ■

Since every continuous convex function $f : [a, b] \to \mathbb{R}$ is absolutely continuous, Theorem 7.1.5 implies that every strong Steffensen–Popoviciu measure is also a Steffensen–Popoviciu measure. However, the inclusion is strict. Indeed, a straightforward computation shows that $\left(x^2 + \lambda\right) \mathrm{d}x$ is an example of a Steffensen–Popoviciu measure on the interval $[-1, 1]$ if, and only if, $\lambda > -1/3$, and it is a strong Steffensen–Popoviciu measure if, and only if, $\lambda \geq -1/4$.

Several other examples of strong Steffensen–Popoviciu measures on $[-1, 1]$ are

$$\left(x^2 - \frac{1}{6}\right) \mathrm{d}x, \quad \left(x^2 \pm \frac{x}{2}\right) \mathrm{d}x, \quad \left(x^4 - \frac{1}{25}\right) \mathrm{d}x, \quad \left(x^2 - \frac{1}{6}\right)^3 \mathrm{d}x.$$

See Exercise 1 for more examples.

Using the pushing-forward technique of constructing image measures, one can indicate examples of (strong) Steffensen–Popoviciu measures supported by an arbitrarily given compact interval $[a, b]$ (with $a < b$). Notice also that these measures can be glued. For example, if $p(x)\mathrm{d}x$ and $q(x)\mathrm{d}x$ are two Steffensen–Popoviciu measures on the intervals $[a, c]$ and $[c, b]$ respectively, then

$$\left(p(x)\chi_{[a,c]} + q(x)\chi_{[c,b]}\right) \mathrm{d}x$$

is a Steffensen–Popoviciu measure on $[a, b]$.

A simple example of Steffensen–Popoviciu measures in dimension 2 is indicated below.

7.1.6 Example *If $p(x)\mathrm{d}x$ and $q(y)\mathrm{d}y$ are Steffensen–Popoviciu measures on the intervals $[a, b]$ and $[c, d]$ respectively, then $(p(x) + q(y))\, \mathrm{d}x\mathrm{d}y$ is a measure of the same type on $[a, b] \times [c, d]$. In fact, if $f : [a, b] \times [c, d] \to \mathbb{R}$ is a positive convex function, then*

$$x \to \left(\int_c^d f(x, y)\mathrm{d}y\right) \text{ and } y \to \left(\int_a^b f(x, y)\mathrm{d}x\right)$$

are also positive convex functions and thus

$$\int_a^b \int_c^d f(x, y)\,(p(x) + q(y))\,\mathrm{d}y\mathrm{d}x = \int_a^b \left(\int_c^d f(x, y)dy\right) p(x)dx$$
$$+ \int_c^d \left(\int_a^b f(x, y)\mathrm{d}x\right) q(y)\mathrm{d}y \geq 0$$

as a sum of positive numbers.

Using appropriate affine changes of variables, one can construct Steffensen–Popoviciu measures on every quadrilateral convex domain. Here it is worth noticing that the superposition of a convex function and an affine function is also a convex function.

The following result is a 2-dimensional analogue of Theorem 7.1.5. It proves the existence of strong Steffensen–Popoviciu measures on compact 2-dimensional intervals.

7.1.7 Theorem *Suppose that $p : [a, b] \to \mathbb{R}$ and $q : [c, d] \to \mathbb{R}$ are continuous functions such that at least one of the following two conditions is fulfilled:*

(i) $p \geq 0$ and $0 \leq \int_c^y q(s)\mathrm{d}s \leq \int_c^d q(s)\mathrm{d}s$ for all $y \in [c, d]$;

(ii) $q \geq 0$ and $0 \leq \int_a^x p(t)\mathrm{d}t \leq \int_a^b p(t)\mathrm{d}t$ for all $x \in [a, b]$.

Then

$$\int_a^b \int_c^d f(x, y)p(x)q(y)\mathrm{d}y\mathrm{d}x \geq 0$$

for every positive, continuously differentiable, and quasiconvex function $f : [a, b] \times [c, d] \to \mathbb{R}$.

The condition of continuous differentiability on f can be replaced by that of absolute continuity in the sense of Carathéodory. See J. Šremr [452] for details.

Proof. Assuming (i), we start by choosing a continuous path $x \to c^*(x)$ such that

$$f(x, c^*(x)) = \min \{f(x, y) : c \leq y \leq d\} \text{ for each } x \in [a, b].$$

Then $y \to f(x, y)$ is decreasing on $[c, c^*(x)]$ and increasing on $[c^*(x), d]$, which implies

$$\frac{\partial f}{\partial y}(x, y) \leq 0 \text{ on } [c, c^*(x)] \text{ and } \frac{\partial f}{\partial y}(x, y) \geq 0 \text{ on } [c^*(x), d].$$

Therefore

$$\begin{aligned}
&\int_a^b \int_c^d f(x, y)p(x)q(y)\mathrm{d}y\mathrm{d}x \\
&\qquad = \int_a^b \left[\int_c^{c^*(x)} f(x, y)\mathrm{d}\left(\int_c^y q(t)\mathrm{d}t\right)\right] p(x)\mathrm{d}x \\
&\qquad - \int_a^b \left[\int_{c^*(x)}^d f(x, y)\mathrm{d}\left(\int_y^d q(t)\mathrm{d}t\right)\right] p(x)\mathrm{d}x
\end{aligned}$$

$$= \int_a^b \left[f(x,y) \int_c^y q(t)\mathrm{d}t \Big|_c^{c^*(x)} - \int_c^{c^*(x)} \left(\frac{\partial f}{\partial y}(x,y) \int_c^y q(t)\mathrm{d}t \right) \mathrm{d}y \right] p(x)\mathrm{d}x$$

$$- \int_a^b \left[f(x,y) \int_y^d q(t)\mathrm{d}t \Big|_{c^*(x)}^d - \int_{c^*(x)}^d \left(\frac{\partial f}{\partial y}(x,y) \int_y^d q(t)\mathrm{d}t \right) \mathrm{d}y \right] p(x)\mathrm{d}x$$

$$= \int_a^b f(x, c^*(x)) \left(\int_c^{c^*(x)} q(t)\mathrm{d}t \right) p(x)\mathrm{d}x$$

$$+ \int_a^b \left[\int_c^{c^*(x)} \left(\frac{-\partial f}{\partial y}(x,y) \int_c^y q(t)\mathrm{d}t \right) \mathrm{d}y \right] p(x)\mathrm{d}x$$

$$+ \int_a^b f(x, c^*(x)) \left(\int_{c^*(x)}^d q(t)\mathrm{d}t \right) p(x)\mathrm{d}x$$

$$+ \int_a^b \left[\int_{c^*(x)}^d \left(\frac{\partial f}{\partial y}(x,y) \int_y^d q(t)\mathrm{d}t \right) \mathrm{d}y \right] p(x)\mathrm{d}x$$

$$= \int_c^d q(t)\mathrm{d}t \int_a^b f(x, c^*(x)) p(x)\mathrm{d}x - \int_a^b \left[\int_c^{c^*(x)} \left(\frac{\partial f}{\partial y}(x,y) \int_c^y q(t)\mathrm{d}t \right) \mathrm{d}y \right] p(x)\mathrm{d}x$$

$$+ \int_a^b \int_{c^*(x)}^d \left[\left(\frac{\partial f}{\partial y}(x,y) \int_y^d q(t)dt \right) dy \right] p(x)dx \geq 0$$

as a sum of positive numbers. The proof is done. ∎

Theorem 7.1.7 works for example for

$$p(x) = 1 \text{ and } q(y) = \left(\frac{2y - c - d}{d - c} \right)^2 - \frac{1}{4},$$

as well as for

$$p(x) = \left(\frac{2x - a - b}{b - a} \right)^2 - \frac{1}{4} \text{ and } q(y) = 1.$$

Combining these two cases we infer that

$$\int_a^b \int_c^d f(x,y) \left(\left(\frac{2x - a - b}{b - a} \right)^2 + \left(\frac{2y - c - d}{d - c} \right)^2 - \frac{1}{2} \right) \mathrm{d}y\mathrm{d}x \geq 0$$

for every positive, continuously differentiable, and quasiconvex function $f : [a,b] \times [c,d] \to \mathbb{R}$.

Exercises

1. Verify that the following real measures are strong Steffensen–Popoviciu on the indicated interval:

$$(4x^3 - 3x)\,\mathrm{d}x \quad \text{on } \left[-\frac{1}{2}\sqrt{3}, 1.5\right];\ \sin x\mathrm{d}x \quad \text{on } [0, 3\pi];$$

$$\left(1 - \frac{1-x}{\pi} - \sin \pi x\right)\mathrm{d}x \quad \text{on } [0,1];$$

$$\left[\left(\frac{2x-a-b}{b-a}\right)^2 + \lambda\right]\mathrm{d}x \quad \text{on } [a,b] \text{ (if } \lambda \geq -1/4\text{)};$$

$$\left[\left(\frac{2x-a-b}{b-a}\right)^2 - \lambda\frac{2x-a-b}{b-a}\right]\mathrm{d}x \quad \text{on } [a,b] \text{ (if } |\lambda| \leq 2/3\text{)}.$$

2. Suppose that $p(x)\mathrm{d}x$ is a Steffensen–Popoviciu measure on the interval $[a,b]$ and $q(y)\mathrm{d}y$ is a positive measure on the interval $[c,d]$. Prove that $p(x)q(y)\mathrm{d}x\mathrm{d}y$ is a Steffensen–Popoviciu measure on $[a,b]\times[c,d]$. Infer that every compact N-dimensional interval $\prod_{k=1}^{N}[a_k,b_k]$ carries a Steffensen–Popoviciu measure, which is not a positive Borel measure.

3. Suppose that $p(x)\mathrm{d}x$ is a Steffensen–Popoviciu measure on the interval $[0,R]$. Prove that

$$\frac{p(\sqrt{x^2+y^2})}{\sqrt{x^2+y^2}}\mathrm{d}x\mathrm{d}y$$

is a Steffensen–Popoviciu measure on the compact disc

$$\overline{D}_R(0) = \left\{(x,y) : x^2 + y^2 \leq R\right\}.$$

4. (Steffensen's inequalities [457]) Let $g\colon [a,b] \to \mathbb{R}$ be an integrable function such that $\lambda = \int_a^b g(t)\,\mathrm{d}t \in (0, b-a]$. Then the following two conditions are equivalent:

   (a) $0 \leq \int_a^x g(t)\,\mathrm{d}t \leq x - a$ and $0 \leq \int_x^b g(t)\,\mathrm{d}t \leq b - x$, for all $x \in [a,b]$;

   (b) $\int_a^{a+\lambda} f(t)\,\mathrm{d}t \leq \int_a^b f(t)g(t)\,\mathrm{d}t \leq \int_{b-\lambda}^b f(t)\,\mathrm{d}t$, for all increasing functions $f\colon [a,b] \to \mathbb{R}$.

   Remark. Condition (a) asserts that $g(x)\mathrm{d}x$ and $(1-g(x))\mathrm{d}x$ are strong Steffensen–Popoviciu measures on $[a,b]$.

5. (A stronger version of Iyengar's inequality [368]) Consider a Riemann integrable function $f\colon [a,b] \to \mathbb{R}$ such that the slopes of the lines AC and

CB, joining the endpoints $A(a, f(a))$ and $B(b, f(b))$ to any other point $C(x, f(x))$ of the graph of f, vary between $-M$ and M. Then:

$$\left|\frac{1}{b-a}\int_a^b f(x)\,\mathrm{d}x - \frac{f(a)+f(b)}{2}\right| \leq \frac{M}{4}(b-a) - \frac{(f(b)-f(a))^2}{4M(b-a)}.$$

[*Hint*: Reduce the problem to the case where f is piecewise linear. Then notice that the function $g(t) = (f'(t) + M)/(2M)$ verifies the condition (a) in the preceding exercise.]

6. A real Borel measure μ on an interval $[a, b]$ is said to be a *dual Steffensen–Popoviciu measure* if $\mu([a, b]) > 0$ and

$$\int_a^b f(x)\,\mathrm{d}\mu(x) \geq 0$$

for every positive, continuous and concave function $f : [a, b] \to \mathbb{R}$. Prove that $\left(\left(\frac{2x-a-b}{b-a}\right)^2 - \frac{1}{6}\right)\mathrm{d}x$ is an example of such a measure. For applications, see C. P. Niculescu [363].

7. (R. Apéry) Let f be a decreasing function on $(0, \infty)$ and g be an integrable function on $[0, \infty)$ such that $0 \leq g \leq A$ for a suitable strictly positive constant A. Prove that

$$\int_0^\infty f(x)g(x)\,\mathrm{d}x \leq A\int_0^\lambda f(x)\,\mathrm{d}x,$$

where $\lambda = (\int_0^\infty g\,\mathrm{d}x)/A$.

7.2 The Barycenter of a Steffensen–Popoviciu Measure

Throughout this section K will denote a (nonempty) compact convex subset of a real locally convex Hausdorff space E, and $C(K)$ will denote the space of all real-valued continuous functions on K. We want to relate the geometry of K with the cone $\operatorname{Conv}(K)$, of all real-valued continuous convex functions defined on K.

The linear space $\operatorname{Conv}(K) - \operatorname{Conv}(K)$ generated by this cone is dense in $C(K)$. In fact, due to the formula

$$\sup\{f_1 - g_1, f_2 - g_2\} = \sup\{f_1 + g_2, f_2 + g_1\} - (g_1 + g_2),$$

the set $\operatorname{Conv}(K) - \operatorname{Conv}(K)$ is a linear sublattice of $C(K)$ which contains the unit and separates the points of K (since $\operatorname{Conv}(K)$ that contains all restrictions to K of the functionals $x' \in E'$). Thus, the Stone–Weierstrass theorem applies. See [214], Theorem 7.29, p. 94.

We shall also need the space

$$A(K) = \mathrm{Conv}(K) \bigcap -\mathrm{Conv}(K),$$

of all real-valued continuous affine functions on K. This is a rich space, as the following result shows:

7.2.1 Lemma *$A(K)$ contains*

$$E^*|_K + \mathbb{R} \cdot 1 = \{x^*|_K + \alpha : x^* \in E^* \text{ and } \alpha \in \mathbb{R}\}$$

as a dense subspace.

Proof. Let $f \in A(K)$ and $\varepsilon > 0$. The following two subsets of $E \times \mathbb{R}$,

$$J_1 = \{(\mathbf{x}, f(\mathbf{x})) : \mathbf{x} \in K\}$$

and

$$J_2 = \{(\mathbf{x}, f(\mathbf{x}) + \varepsilon) : \mathbf{x} \in K\},$$

are nonempty, compact, convex, and disjoint. By a geometric version of the Hahn–Banach theorem (see Theorem B.2.4), there exists a continuous linear functional L on $E \times \mathbb{R}$ and a number $\lambda \in \mathbb{R}$ such that

$$\sup L(J_1) < \lambda < \inf L(J_2).$$

Hence the equation $L(\mathbf{x}, g(\mathbf{x})) = \lambda$ defines an element $g \in E^*|_K + \mathbb{R} \cdot 1$ such that

$$f(\mathbf{x}) < g(\mathbf{x}) < f(\mathbf{x}) + \varepsilon \quad \text{for all } \mathbf{x} \in K.$$

In fact, $\lambda = L(\mathbf{x}, 0) + g(\mathbf{x})L(0, 1)$, and thus $g(\mathbf{x}) = (\lambda - L(\mathbf{x}, 0))/L(0, 1)$. This solves the approximation (within ε) of f by elements of $E^*|_K + \mathbb{R} \cdot 1$. ∎

The following example shows that the inclusion $E^*|_K + \mathbb{R} \cdot 1 \subset A(K)$ may be strict. For this, consider the set,

$$S = \{(a_n)_n : |a_n| \leq 1/n^2 \text{ for every } n\},$$

viewed as a subset of ℓ^2 endowed with the weak topology. Then S is compact and convex and the function $f((a_n)_n) = \sum_n a_n$ defines an element of $A(S)$. Moreover, $f(0) = 0$. However, there is no y in ℓ^2 such that $f(x) = \langle x, y \rangle$ for all $x \in S$.

Given a function f in $C(K)$, one can attach to it a *lower envelope*,

$$\check{f}(\mathbf{x}) = \sup\{h(\mathbf{x}) : h \in A(K) \text{ and } f \geq h\},$$

and an upper envelope,

$$\widehat{f}(\mathbf{x}) = \inf\{h(\mathbf{x}) : h \in A(K) \text{ and } h \geq f\}.$$

They are related by formulas of the form

$$\check{f} = -\widehat{(-f)},$$

so it suffices to investigate the properties of one type of envelope, say the upper one:

7.2.2 Lemma *The upper envelope $\widehat{f}$ of a function $f \in C(K)$ is concave, bounded, and upper semicontinuous. Moreover:*

(a) $f \leq \widehat{f}$ and $f = \widehat{f}$ if f is concave;

(b) if $f, g \in C(K)$, then $\widehat{f+g} \leq \widehat{f} + \hat{g}$ with equality if $g \in A(K)$; also, $\widehat{\alpha f} = \alpha \widehat{f}$ if $\alpha \geq 0$;

(c) the map $f \to \widehat{f}$ is nonexpansive in the sense that $|\widehat{f} - \widehat{g}| \leq \|f - g\|$.

Proof. Most of this lemma follows directly from the definitions. We shall concentrate here on the less obvious assertion, namely the second part of (a). It may be proved by reductio ad absurdum. Assume that $f(\mathbf{x}_0) < \widehat{f}(\mathbf{x}_0)$ for some $\mathbf{x}_0 \in K$. By Theorem B.2.4, there exists a closed hyperplane which strictly separates the convex sets $K_1 = \{(\mathbf{x}_0, \widehat{f}(\mathbf{x}_0))\}$ and $K_2 = \{(\mathbf{x}, r) : f(\mathbf{x}) \geq r\}$. This means the existence of a continuous linear functional L on $E \times \mathbb{R}$ and of a scalar λ such that

$$\sup_{(\mathbf{x},r)\in K_2} L(\mathbf{x}, r) < \lambda < L(\mathbf{x}_0, \widehat{f}(\mathbf{x}_0)). \tag{7.6}$$

Then $L(\mathbf{x}_0, \widehat{f}(\mathbf{x}_0)) > L(\mathbf{x}_0, f(\mathbf{x}_0))$, which yields $L(0, 1) > 0$. The function

$$h(\mathbf{x}) = \frac{\lambda - L(\mathbf{x}, 0)}{L(0, 1)}$$

belongs to $A(K)$ and $L(\mathbf{x}, h(\mathbf{x})) = \lambda$ for all $\mathbf{x}$. By (7.6), we infer that $h > f$ and $h(\mathbf{x}_0) < \widehat{f}(\mathbf{x}_0)$, a contradiction. ■

The connection between the points of a compact convex set K and the positive functionals on $C(K)$ is made visible through the concept of barycenter. In what follows we will show that this concept can be attached not only to probability Borel measures, but also to any Steffensen–Popoviciu measure.

As above K denotes a compact convex set in a real locally convex Hausdorff space E.

7.2.3 Lemma *For every Steffensen–Popoviciu measure μ on K there exists and is unique a point* $\operatorname{bar}(\mu)$ *in K such that*

$$x^*(\operatorname{bar}(\mu)) = \frac{1}{\mu(K)} \int_K x^*(\mathbf{x})\, \mathrm{d}\mu(\mathbf{x}) \tag{7.7}$$

for all continuous linear functionals x^* on E.

The point $\operatorname{bar}(\mu)$ *is called the barycenter of μ.*

The uniqueness of the point $\operatorname{bar}(\mu)$ is a consequence of the separability of the topology of E. See Corollary B.1.5. In the case of Euclidean spaces the norm and the weak convergence agree, so the formula (7.7) reduces to

$$\operatorname{bar}(\mu) = \frac{1}{\mu(K)} \int_K \mathbf{x}\, \mathrm{d}\mu(\mathbf{x}).$$

Due to Lemma 7.2.1, the equality (7.7) extends to all $f \in A(K)$.

Proof of Lemma 7.2.3. We have to prove that

$$\Big(\bigcap_{f\in E^*} H_f\Big) \cap K \neq \emptyset,$$

where $\mu(f)$ denotes the integral of f with respect to μ and H_f denotes the closed hyperplane $\{\mathbf{x} : f(\mathbf{x}) = \mu(f)/\mu(K)\}$ associated to $f \in E^*$. Since K is compact, it suffices to prove that

$$\Big(\bigcap_{k=1}^{n} H_{f_k}\Big) \cap K \neq \emptyset$$

for every finite family $f_1, \dots, f_n$ of functionals in E^*. Equivalently, attaching to any such family $f_1, \dots, f_n$ of functionals the operator

$$T\colon K \to \mathbb{R}^n, \quad T(\mathbf{x}) = (f_1(\mathbf{x}), \dots, f_n(\mathbf{x})),$$

we have to prove that $T(K)$ contains the point $\mathbf{p} = \frac{1}{\mu(K)}(\mu(f_1), \dots, \mu(f_n))$. In fact, if $\mathbf{p} \notin T(K)$, then a separation argument yields a point $\mathbf{a} = (a_1, \dots, a_n)$ in $\mathbb{R}^n$ such that

$$\langle \mathbf{p}, \mathbf{a}\rangle > \sup_{\mathbf{x}\in K} \langle T(\mathbf{x}), \mathbf{a}\rangle,$$

that is,

$$\frac{1}{\mu(K)} \sum_{k=1}^{n} a_k \mu(f_k) > \sup_{\mathbf{x}\in K} \sum_{k=1}^{n} a_k f_k(\mathbf{x}).$$

Then $g = \sum_{k=1}^{n} a_k f_k$ will provide an example of a continuous affine function on K for which $\mu(g)/\mu(K) > \sup_{\mathbf{x}\in K} g(\mathbf{x})$, a fact which contradicts the property of μ to be a Steffensen–Popoviciu measure. ∎

The extension of Jensen's inequality to the context of Steffensen–Popoviciu measure is straightforward.

7.2.4 Theorem (The generalized Jensen–Steffensen inequality) *Suppose that μ is a Steffensen–Popoviciu measure on the compact convex set K. Then*

$$f(\operatorname{bar}(\mu)) \leq \frac{1}{\mu(K)} \int_K f(\mathbf{x})\, \mathrm{d}\mu(\mathbf{x})$$

for all continuous convex functions $f\colon K \to \mathbb{R}$.

Proof. Indeed, according to Lemma 7.2.2,

$$\begin{aligned} f(\operatorname{bar}(\mu)) &= \sup\{h(\operatorname{bar}(\mu)) : h \in A(K),\ h \leq f\} \\ &= \sup\Big\{\frac{1}{\mu(K)} \int_K h\, \mathrm{d}\mu \,\Big|\, h \in A(K),\ h \leq f\Big\} \\ &\leq \frac{1}{\mu(K)} \int_K f\, \mathrm{d}\mu. \end{aligned}$$

∎

The classical Jensen–Steffensen inequality (see Theorem 1.9.4) represents the particular case where μ is a strong Steffensen–Popoviciu measure on a compact interval.

We end this section with a monotonicity result.

7.2.5 Proposition *Suppose that μ is a Borel probability measure on K and $f\colon K \to \mathbb{R}$ is a continuous convex function. Then the function*

$$M(t) = \int_K f(t\mathbf{x} + (1-t)\operatorname{bar}(\mu))\,\mathrm{d}\mu(\mathbf{x})$$

is convex and increasing on $[0,1]$.

When $E = \mathbb{R}^n$ and μ is the Lebesgue measure, the value of M at $t > 0$ equals the arithmetic mean of $f|_{K_t}$, that is,

$$M(t) = \frac{1}{\mu(K_t)} \int_{K_t} f(\mathbf{x})\,\mathrm{d}\mu(\mathbf{x});$$

here K_t denotes the image of K through the mapping $x \to tx + (1-t)\operatorname{bar}(\mu)$. Proposition 7.2.5 tells us that the arithmetic mean of $f|_{K_t}$ decreases to $f(\operatorname{bar}(\mu))$ when K_t shrinks to $\operatorname{bar}(\mu)$. The proof is based on the following approximation argument:

7.2.6 Lemma (Barycentric approximation) *Every Borel probability measure μ on K is the pointwise limit of a net of discrete Borel probability measures μ_α, each having the same barycenter as μ.*

Lemma 7.2.6 yields an immediate derivation of Jensen's integral inequality from Jensen's discrete inequality.

Proof. We have to prove that for each $\varepsilon > 0$ and each finite family $f_1, \dots, f_n$ of continuous real functions on K there exists a discrete Borel probability measure ν such that

$$\operatorname{bar}(\nu) = \operatorname{bar}(\mu) \quad \text{and} \quad \sup_{1 \le k \le n} |\nu(f_k) - \mu(f_k)| < \varepsilon.$$

As K is compact and convex and the functions f_k are continuous, there exists a finite covering $(D_\alpha)_\alpha$ of K by open convex sets such that the oscillation of each of the functions f_k on each set D_α is less than ε. Let $(\varphi_\alpha)_\alpha$ be a partition of unity, subordinated to the covering $(D_\alpha)_\alpha$ and put

$$\nu = \sum_\alpha \mu(\varphi_\alpha)\delta_{\mathbf{x}(\alpha)},$$

where $\mathbf{x}(\alpha)$ is the barycenter of the measure $f \to \mu(\varphi_\alpha f)/\mu(\varphi_\alpha)$. As D_α is convex and the support of φ_α is included in D_α, we have $\mathbf{x}(\alpha) \in \overline{D}_\alpha$. On the other hand,

$$\mu(h) = \sum_\alpha \mu(h\varphi_\alpha) = \sum_\alpha \frac{\mu(h\varphi_\alpha)}{\mu(\varphi_\alpha)}\,\mu(\varphi_\alpha) = \sum_\alpha h(\mathbf{x}(\alpha))\,\mu(\varphi_\alpha) = \nu(h)$$

for all continuous affine functions $h\colon K \to \mathbb{R}$. Consequently, μ and ν have the same barycenter. Finally, for each k,

$$\begin{aligned}|\nu(f_k) - \mu(f_k)| &= \Big|\sum_\alpha \mu(\varphi_\alpha) f_k(\mathbf{x}(\alpha)) - \sum_\alpha \mu(\varphi_\alpha f_k)\Big| \\ &= \Big|\sum_\alpha \mu(\varphi_\alpha)\Big[f_k(\mathbf{x}(\alpha)) - \frac{\mu(\varphi_\alpha f_k)}{\mu(\varphi_\alpha)}\Big]\Big| \\ &\le \varepsilon \cdot \sum_\alpha \mu(\varphi_\alpha) = \varepsilon.\end{aligned}$$

■

Proof of Proposition 7.2.5. A straightforward computation shows that $M(t)$ is convex and $M(t) \le M(1)$. Then, assuming the inequality $M(0) \le M(t)$, from the convexity of $M(t)$ we infer

$$\frac{M(t) - M(s)}{t - s} \ge \frac{M(s) - M(0)}{s} \ge 0$$

for all $0 \le s < t \le 1$ that is, $M(t)$ is increasing. To end the proof, it remains to show that $M(t) \ge M(0) = f(\mathrm{bar}(\mu))$. For this, choose a net $(\mu_\alpha)_\alpha$ of discrete Borel probability measures on K, as in Lemma 7.2.6 above. By Lemma 1.1.11,

$$f(\mathrm{bar}\,\mu) \le \int_K f(t\mathbf{x} + (1-t)\,\mathrm{bar}(\mu))\,\mathrm{d}\mu_\alpha(\mathbf{x}) \quad \text{for all } \alpha$$

and thus the desired conclusion follows by passing to the limit over α. ■

The question whether Lemma 7.2.6 remains true in the context of Steffensen–Popoviciu measures is open.

Exercises

1. Suppose that μ is a Steffensen–Popoviciu measure on a compact convex set K. Prove that
$$\|\mu|_{A(K)}\| = \mu(K).$$
Then, consider the particular case where $K = [-1,1]$ and $\mathrm{d}\mu = (x^2 + a)\mathrm{d}x$ to show that $\mu/\mu(K)$ as a functional on $C(K)$ can have a norm larger than 1.

2. Suppose that $\mathrm{d}\mu = p(x)\mathrm{d}x$ is an absolutely continuous Steffensen–Popoviciu measure on an interval $[a,b]$ and let $T : [a,b] \to [a,b]$ be a measurable transformation such that $\mu(T^{-1}(A)) = \mu(A)$ for every measurable set A; in other words, μ is T-invariant. Is the barycenter of μ necessarily a fixed point of T?

3. (Barycenter from symmetry) The measure $(x^2 - 1/6)\mathrm{d}x$ defined on $[-1,1]$ is symmetric with respect to the origin and thus this point is its barycenter. Apply a similar argument to find the barycenter of the Steffensen–Popoviciu measure $(x^2 - 1/6)(y + 1/y)\mathrm{d}x\mathrm{d}y$ defined on $[-1,1] \times [1/2, 2]$.

4. Suppose that μ is a strong Steffensen–Popoviciu measure on an interval $[a,b]$. Does the generalized Jensen–Steffensen inequality work for all continuous quasiconvex functions defined on $[a,b]$?

7.3 Majorization via Choquet Order

Measure majorization in the setting of Steffensen–Popoviciu measures extends the case of discrete Borel probability measures, already discussed in Section 4.6.

7.3.1 Definition *Given two Steffensen–Popoviciu measures μ and ν on a compact convex set K, we say that μ is majorized by ν in Choquet order (abbreviated, $\mu \prec_{Ch} \nu$) if*

$$\frac{1}{\mu(K)}\int_K f(\mathbf{x})\,\mathrm{d}\mu(\mathbf{x}) \leq \frac{1}{\nu(K)}\int_K f(\mathbf{x})\,\mathrm{d}\nu(\mathbf{x})$$

for all continuous convex functions $f\colon K \to \mathbb{R}$.

The relation $\prec_{Ch}$is indeed a partial order relation. Clearly, it is transitive and reflexive; the fact that $\mu \prec_{Ch} \nu$ and $\nu \prec_{Ch} \mu$ imply $\mu = \nu$ comes from the fact that $\mathrm{Conv}(K) - \mathrm{Conv}(K)$ is dense in $C(K)$.

If f is in $A(K)$, then both f and $-f$ are in $\mathrm{Conv}(K)$, so that $\mu/\mu(K)$ and $\nu/\nu(K)$ coincide on $A(K)$. As a consequence, they have the same barycenter.

As we will show in the next section, the Choquet order detects when one probability measure is more diffuse than another one in the sense that its support lies closer to the extreme boundary $\operatorname{ext} K$.

7.3.2 Lemma *Suppose that μ and ν are two Steffensen–Popoviciu measures such that $\mu(K) = \nu(K) = 1$ and*

$$\int_K f(\mathbf{x})\,\mathrm{d}\mu(\mathbf{x}) \leq \int_K f(\mathbf{x})\,\mathrm{d}\nu(\mathbf{x})$$

for all functions $f \in \mathrm{Conv}(K)$, $f \geq 0$. Then $\mu \prec_{Ch} \nu$.

In particular, $\delta_{\mathrm{bar}(\nu)} \prec_{Ch} \nu$ and $\delta_{\mathrm{bar}(\nu)}$ is the smallest Steffensen–Popoviciu measure of total mass 1 that is majorized by ν.

Proof. The trick is to observe that $\lambda_{\varepsilon,\mathbf{z}} = \nu - \mu + \varepsilon\delta_{\mathbf{z}}$ is a Steffensen–Popoviciu measure, whenever $\varepsilon > 0$ and $\mathbf{z} \in K$. According to the Jensen–Steffensen inequality, if $f \in \mathrm{Conv}(K)$, then

$$\int_K f(\mathbf{x})\,\mathrm{d}\nu(\mathbf{x}) - \int_K f(\mathbf{x})\,\mathrm{d}\mu(\mathbf{x}) + \varepsilon f(\mathbf{z}) \geq \varepsilon f(\mathrm{bar}(\lambda_{\varepsilon,\mathbf{z}})) \geq \varepsilon \min_{\mathbf{x}\in K} f(\mathbf{x}),$$

whence by letting $\varepsilon \to 0$ we infer that

$$\int_K f(\mathbf{x})\,\mathrm{d}\nu(\mathbf{x}) - \int_K f(\mathbf{x})\,\mathrm{d}\mu(\mathbf{x}) \geq 0.$$

■

The following variant of Lemma 7.3.2 in the case of Stieltjes measures, is due to V. I. Levin and S. B. Stečkin [277]:

7.3.3 Lemma *Let $F\colon [a,b] \to \mathbb{R}$ be a function with bounded variation such that $F(a) = 0$. Then*

$$\int_a^b f(x)\,\mathrm{d}F(x) \geq 0 \quad \text{for all } f \in \operatorname{Conv}(K)$$

if, and only if, the following three conditions are fulfilled:

$$F(b) = 0, \quad \int_a^b F(x)\,\mathrm{d}x = 0, \quad \text{and} \quad \int_a^x F(t)\,\mathrm{d}t \geq 0 \quad \text{for all } x \in (a,b).$$

Proof. Via an approximation argument we may restrict to the case where f is also piecewise linear. Then, by using twice integration by parts, we get

$$\int_a^b f(x)\,\mathrm{d}F(x) = -\int_a^b F(x)f'(x)\,\mathrm{d}x = \int_a^b \left(\int_a^x F(t)\,\mathrm{d}t\right) f''(x)\,\mathrm{d}x,$$

whence the sufficiency part is proved. For the necessity, notice that $\int_a^x F(t)\,\mathrm{d}t < 0$ for some $x \in (a,b)$ yields an interval I around x on which the integral is still negative. Choosing f such that $f'' = 0$ outside I, the above equalities lead to a contradiction. The necessity of the other two conditions follows by checking our statement for $f = 1, -1, x-a, a-x$ (in this order). ■

7.3.4 Corollary *Let $F, G\colon [a,b] \to \mathbb{R}$ be two functions with bounded variation such that $F(a) = G(a)$. Then, in order that*

$$\int_a^b f(x)\,\mathrm{d}F(x) \leq \int_a^b f(x)\,\mathrm{d}G(x)$$

for all continuous convex functions $f\colon [a,b] \to \mathbb{R}$, it is necessary and sufficient that F and G verify the following three conditions:

$$F(b) = G(b);$$
$$\int_a^x F(t)\,\mathrm{d}t \leq \int_a^x G(t)\,\mathrm{d}t \quad \text{for all } x \in (a,b); \text{ and}$$
$$\int_a^b F(t)\,\mathrm{d}t = \int_a^b G(t)\,\mathrm{d}t.$$

7.3.5 Corollary *Let $f, g \in L^1[a,b]$ be two functions. Then $f\mathrm{d}x \prec_{Ch} g\mathrm{d}x$ if and only if the following conditions are fulfilled:*

$$\int_a^b f(x)\,\mathrm{d}x = \int_a^b g(x)\,\mathrm{d}x; \quad \int_a^b xf(x)\,\mathrm{d}x = \int_a^b xg(x)\,\mathrm{d}x;$$
$$\int_a^x (x-t)f(t)\,\mathrm{d}t \leq \int_a^x (x-t)g(t)\,\mathrm{d}t, \quad \text{for all } x \in [a,b].$$

Exercises

1. A diffusion process from the middle point across the unit interval is driven by the following stochastic process:

$$P_t = (1-t)\delta_{1/2} + \frac{1}{2}\chi_{[1/2-t,1/2+t]} \text{ for } t \in [0,1].$$

Prove that $P_t \prec_{Ch} P_s$, whenever $0 \leq s \leq t \leq 1$.

2. Let μ and ν be two signed Borel measures on $[a,b]$ such that $\mu([a,b]) = \nu([a,b]) = 1$. Prove that $\mu \prec_{Ch} \nu$ if, and only if,

$$\int_a^t (t-x)\,\mathrm{d}\mu(x) \leq \int_a^t (t-x)\,\mathrm{d}\nu(x) \quad \text{and} \quad \int_t^b (x-t)\,\mathrm{d}\mu(x) \leq \int_t^b (x-t)\,\mathrm{d}\nu(x)$$

for all $t \in [a,b]$.

3. Let $p\colon [0,1] \to R$ be a continuous function which is increasing on $[0,1/2]$ and satisfies the condition $f(x) = f(1-x)$. Prove that

$$\int_0^1 f(x)p(x)\,\mathrm{d}x \leq \left(\int_0^1 f(x)\,\mathrm{d}x\right)\left(\int_0^1 p(x)\,\mathrm{d}x\right)$$

for all $f \in \operatorname{Conv}([0,1])$. Infer that

$$\int_0^1 x(1-x)f(x)\,\mathrm{d}x \leq \frac{1}{6}\int_0^1 f(x)\,\mathrm{d}x$$

and

$$\int_0^\pi f(x)\sin x\,\mathrm{d}x \leq \frac{2}{\pi}\int_0^\pi f(x)\,\mathrm{d}x,$$

provided that f is convex on appropriate intervals.

[*Hint*: It suffices to verify the conditions of Corollary 7.3.5 for $f = p$ and $g = \int_0^1 p(x)\,\mathrm{d}x$. The third condition in Corollary 7.3.5 reads as

$$\frac{x^2}{2}\int_0^1 p(t)\,\mathrm{d}t \geq \int_0^x \int_0^t p(s)\,\mathrm{d}s\mathrm{d}t \quad \text{for all } x \in [0,1].$$

For $x \in [0,1/2]$ we have to observe that $\int_0^x p(t)\,\mathrm{d}t$ is a convex function on $[0,1/2]$, which yields

$$\frac{1}{x}\int_0^x p(t)\,\mathrm{d}t \leq 2\int_0^{1/2} p(t)\,\mathrm{d}t = \int_0^1 p(t)\,\mathrm{d}t \quad \text{for all } x \in [0,1/2].]$$

7.4 Choquet's Theorem

The aim of this section is to present a full extension of the Hermite–Hadamard inequality (1.32) to the framework of continuous convex functions defined on arbitrary compact convex spaces (when the mean values are computed via Borel probability measures). This is done by combining the generalized Jensen–Steffensen inequality (see Theorem 7.2.4) with the majorization theorems of G. Choquet, E. Bishop, and K. de Leeuw. We start with the metrizable case, following the classical approach initiated by G. Choquet [105].

7.4.1 Theorem (Choquet's theorem: the metrizable case) *Let μ be a Borel probability measure on a metrizable compact convex subset K of a locally convex Hausdorff space E. Then there exists a Borel probability measure λ on K such that the following two conditions are verified:*

(a) *$\mu \prec_{Ch} \lambda$ and λ and μ have the same barycenter;*

(b) *the set $\operatorname{ext} K$, of all extreme points of K, is a Borel set and λ is concentrated on $\operatorname{ext} K$ (that is, $\lambda(K \setminus \operatorname{ext} K) = 0$).*

7.4.2 Corollary (Choquet's representation) *Given a point $x \in K$, there exists a Borel probability measure λ concentrated on $\operatorname{ext} K$ such that for all affine functions f defined on K we have*

$$f(\mathbf{x}) = \int_{\operatorname{ext} K} f(\mathbf{x}) \,\mathrm{d}\lambda(\mathbf{x}).$$

For the proof, apply Theorem 7.4.1 to $\mu = \delta_x$. Corollary 7.4.2 provides an infinite-dimensional extension of Minkowski's Theorem 2.3.5.

7.4.3 Corollary (The generalized Hermite–Hadamard inequality; the metrizable case) *Under the hypotheses of Theorem 7.4.1 we have*

$$f(\operatorname{bar}(\mu)) \leq \int_K f(\mathbf{x}) \,\mathrm{d}\mu(\mathbf{x}) \leq \int_{\operatorname{ext} K} f(\mathbf{x}) \,\mathrm{d}\lambda(\mathbf{x}) \tag{7.8}$$

for every continuous convex function $f \colon K \to \mathbb{R}$.

Notice that the right part of (7.8) reflects the maximum principle for convex functions.

In general, the measure λ is not unique, except for the case of simplices; see the book of R. R. Phelps [398, Section 10] for details.

As was noticed in Theorem 7.2.4, the left-hand side inequality in the formula (7.8) works in the more general framework of Steffensen–Popoviciu measures. This is no longer true for the right-hand side inequality. In fact, $\mu = \delta_{-1} - \delta_0 + \delta_1$ provides an example of Steffensen–Popoviciu measure on $K = [-1, 1]$ which is not majorized by any Steffensen–Popoviciu measure concentrated on $\operatorname{ext} K$ (that is, by any convex combination of δ_{-1} and δ_1). However, as will be shown in the next section, the Hermite–Hadamard inequality remains valid for a subclass of Steffensen–Popoviciu measures, still larger than the class of probability measures.

Proof of Theorem 7.4.1. This will be done in four steps.

Step 1. We start by proving that $\operatorname{ext} K$ is a countable intersection of open sets (and thus it is a Borel set). Here the assumption on metrizability is essential. Suppose that the topology of K is given by the metric d and for each integer $n \geq 1$ consider the set

$$K_n = \left\{x : \mathbf{x} = \frac{\mathbf{y}+\mathbf{z}}{2}, \text{ with } \mathbf{y}, \mathbf{z} \in K \text{ and } d(\mathbf{y}, \mathbf{z}) \geq 1/2^n\right\}.$$

Clearly, $\operatorname{ext} K = K \backslash \bigcup_n K_n$ and an easy compactness argument shows that each K_n is closed. Consequently, $\operatorname{ext} K = \bigcap_n \complement K_n$ is a countable intersection of open sets.

Step 2. We may choose a maximal Borel probability measure $\lambda \succ \nu$. To show that Zorn's lemma may be applied, consider a chain $C = (\lambda_\alpha)_\alpha$ in

$$\mathcal{P} = \{\lambda : \lambda \succ \nu, \lambda \text{ Borel probability measure on } K\}.$$

As $(\lambda_\alpha)_\alpha$ is contained in the weak-star compact set

$$\{\lambda : \lambda \in C(K), \ \lambda \geq 0, \ \lambda(1) = 1\},$$

by a compactness argument we may find a subnet $(\lambda_\beta)_\beta$ which converges to a measure $\tilde{\lambda}$ in the weak-star topology. A moment's reflection shows that $\tilde{\lambda}$ is an upper bound for C. Consequently, we may apply Zorn's lemma to choose a maximal Borel probability measure $\lambda \succ \nu$. It remains to prove that λ does the job.

Step 3. Since K is metrizable, it follows that $C(K)$ (and thus $A(K)$) is separable. This is a consequence of Urysohn's lemma in general topology. See, for example [214], Theorem 6.80, p. 75. Every sequence $(h_n)_n$ of affine functions with $\|h_n\| = 1$, which is dense in the unit sphere of $A(K)$, separates the points of K in the sense that for every $\mathbf{x} \neq \mathbf{y}$ in K there is an h_n such that $h_n(\mathbf{x}) \neq h_n(\mathbf{y})$. Consequently, the function

$$\varphi = \sum_{n=1}^{\infty} 2^{-n} h_n^2$$

is continuous and strictly convex, from which it follows that

$$\mathcal{E} = \{\mathbf{x} : \varphi(\mathbf{x}) = \widehat{\varphi}(\mathbf{x})\} \subset \operatorname{ext} K.$$

In fact, if $\mathbf{x} = (\mathbf{y}+\mathbf{z})/2$, where $\mathbf{y}$ and $\mathbf{z}$ are distinct points of K, then the strict convexity of φ implies that

$$\varphi(\mathbf{x}) < \frac{\varphi(\mathbf{y}) + \varphi(\mathbf{z})}{2} \leq \frac{\widehat{\varphi}(\mathbf{y}) + \widehat{\varphi}(\mathbf{z})}{2} \leq \widehat{\varphi}(\mathbf{x}).$$

Step 4. As a consequence of the maximality of λ, we shall show that

$$\lambda(\varphi) = \lambda(\widehat{\varphi}). \tag{7.9}$$

Then $\widehat{\varphi} - \varphi \geq 0$ and $\lambda(\widehat{\varphi} - \varphi) = 0$, which yields $\lambda(\{\mathbf{x} : \varphi(\mathbf{x}) = \widehat{\varphi}(\mathbf{x})\}) = 0$. Hence λ is concentrated on E. The proof of (7.9) is based on Lemma 7.2.2. Consider the sublinear functional $q\colon C(K) \to \mathbb{R}$, given by $q(f) = \lambda(\hat{f})$, and the linear functional L defined on $A(K) + \mathbb{R} \cdot \varphi$ by $L(h + \alpha\varphi) = \lambda(h) + \alpha\lambda(\widehat{\varphi})$. By Lemma 7.2.2, if $\alpha \geq 0$, then $L(h + \alpha\varphi) = q(h + \alpha\varphi)$, while if $\alpha < 0$, then

$$0 = \widehat{\alpha\varphi - \alpha\varphi} \leq \widehat{\alpha\varphi} + \widehat{(-\alpha\varphi)} = \widehat{\alpha\varphi} - \alpha\widehat{\varphi},$$

which yields

$$L(h + \alpha\varphi) = \lambda(h + \alpha\widehat{\varphi}) \leq \lambda(\widehat{h + \alpha\varphi}) = q(h + \alpha\varphi).$$

By the Hahn–Banach extension theorem, there exists a linear extension ω of L to $C(K)$ such that $\omega \leq q$. If $f \leq 0$, then $\hat{f} \leq 0$, so $\omega(f) \leq q(f) = \lambda(\hat{f}) \leq 0$. Therefore $\omega \geq 0$ and the Riesz–Kakutani representation theorem allows us to identify ω with a suitable Borel probability measure on K. If f is in $\operatorname{Conv}(K)$, then $-f$ is concave and Lemma 7.2.2 yields

$$\omega(-f) \leq q(-f) = \lambda(\widehat{-f}) = \lambda(-f)$$

that is, $\lambda \prec \omega$. Or, λ is maximal, which forces $\omega = \lambda$. Consequently,

$$\lambda(\varphi) = \omega(\varphi) = L(\varphi) = \lambda(\overline{\varphi}),$$

which ends the proof. ■

As E. Bishop and K. de Leeuw [57] noticed, if K is non-metrizable, then ext K need not to be a Borel set. However, they were able to prove a Choquet-type theorem. By combining their argument (as presented in [398], p. 17) with Theorem 7.4.1 above, one can prove the following more general result:

7.4.4 Theorem (The Choquet–Bishop–de Leeuw theorem [57]) *Let μ be a Borel measure on a compact convex subset K of a locally convex Hausdorff space E. Then there exists a Borel probability measure λ on K such that the following two conditions are fulfilled:*

(a) *$\lambda \succ \mu$ and λ and μ have the same barycenter;*

(b) *λ vanishes on every Baire subset of K which is disjoint from the set of extreme points of K.*

Choquet's theory has deep applications to many areas of mathematics such as function algebras, invariant measures, and potential theory. R. R. Phelps' book [308] contains a good account on this matter.

Three examples illustrating the generalized form of the Hermite–Hadamard inequality are presented below.

7.4.5 Example *Let f be a continuous convex function defined on an N-dimensional simplex $K = [\mathbf{a}_0, \ldots, \mathbf{a}_N]$ in $\mathbb{R}^N$ and let μ be a Borel measure on K. Then*

$$\begin{aligned} f(\operatorname{bar}(\mu)) &\leq \frac{1}{\mu(K)} \int_K f(\mathbf{x}) \, \mathrm{d}\mu \\ &\leq \frac{1}{\operatorname{Vol}_N(K)} \sum_{k=0}^{n} \operatorname{Vol}_n([\mathbf{a}_0, \ldots, \widehat{\mathbf{a}_k}, \ldots, \mathbf{a}_n] \cdot f(\mathbf{a}_k). \end{aligned}$$

Here $[\mathbf{a}_0, \ldots, \widehat{\mathbf{a}_k}, \ldots, \mathbf{a}_N]$ denotes the subsimplex obtained by replacing $\mathbf{a}_k$ by $\operatorname{bar}(\mu)$*; this is the subsimplex opposite to $\mathbf{a}_k$, when adding* $\operatorname{bar}(\mu)$ *as a new vertex.*

7.4.6 Example *In the case of closed balls $K = \overline{B}_R(\mathbf{a})$ in $\mathbb{R}^N$,* $\operatorname{ext} K$ *coincides with the sphere $S_R(\mathbf{a})$. According to Corollary 7.4.3, if $f \colon \overline{B}_R(\mathbf{a}) \to \mathbb{R}$ is a continuous convex function, then*

$$f(a) \leq \frac{1}{\operatorname{Vol}_N \overline{B}_R(\mathbf{a})} \iiint_{\overline{B}_R(\mathbf{a})} f(\mathbf{x}) \, \mathrm{d}V \leq \frac{1}{\operatorname{Area}_{N-1} S_R(\mathbf{a})} \iint_{S_R(\mathbf{a})} f(\mathbf{x}) \, \mathrm{d}S. \tag{7.10}$$

Results similar to Example 7.4.6 also work in the case of subharmonic functions. See the Comments at the end of this chapter. As noticed by P. Montel [335], in the context of C^2-functions on open convex sets in $\mathbb{R}^n$, the class of subharmonic functions is strictly larger than the class of convex function. For example, the function $2x^2 - y^2$ is subharmonic but not convex on $\mathbb{R}^2$.

Many inequalities relating weighted means represent averages over the probability simplex:

$$\Sigma(N) = \{\mathbf{u} = (u_1, \ldots, u_N) : u_1, \ldots, u_N \geq 0, \ u_1 + \cdots + u_N = 1\}.$$

This set is compact and convex and its extreme points are the "corners"

$$(1, 0, \ldots, 0), (0, 1, \ldots, 0) \ldots, (0, 0, \ldots, 1).$$

7.4.7 Example *If $f \colon [a, b] \to \mathbb{R}$ is a continuous convex function, then for every N-tuple $\mathbf{x} = (x_1, \ldots, x_N)$ of elements of $[a, b]$ and every Borel probability measure μ on $\Sigma(N)$ we have*

$$f\Big(\sum_{k=1}^{N} w_k x_k\Big) \leq \int_{\Sigma(N)} f(\langle \mathbf{x}, \mathbf{u} \rangle) \, \mathrm{d}\mu \leq \sum_{k=1}^{N} w_k f(x_k). \tag{7.11}$$

Here $(w_1, \ldots, w_N)$ denotes the barycenter of μ. The above inequalities work in the reverse way if f is a concave function.

Under the hypotheses of Example 7.4.7, the weighted identric mean $I(x, \mu)$ is defined by the formula

$$I(\mathbf{x}, \mu) = \exp \int_{\Sigma(N)} \log(\langle \mathbf{x}, \mathbf{u} \rangle) \, \mathrm{d}\mu(\mathbf{u}), \quad \mathbf{x} \in \mathbb{R}^N_{++},$$

while the weighted logarithmic mean $L(\mathbf{x},\mu)$ is defined by the formula

$$L(\mathbf{x},\mu)=\left(\int_{\Sigma(N)}\frac{1}{\langle\mathbf{x},\mathbf{u}\rangle}\,\mathrm{d}\mu(\mathbf{u})\right)^{-1},\quad \mathbf{x}\in\mathbb{R}^N_{++}.$$

The formula (7.11) easily implies that $L(\mathbf{x},\mu)\le I(\mathbf{x},\mu)$ and that both means lie between the weighted arithmetic mean $A(\mathbf{x},\mu)=\sum_{k=1}^N w_k x_k$ and the weighted geometric mean $G(\mathbf{x},\mu)=\prod_{k=1}^N x_k^{w_k}$, that is,

$$G(\mathbf{x},\mu)\le L(\mathbf{x},\mu)\le I(\mathbf{x},\mu)\le A(\mathbf{x},\mu),$$

a fact which constitutes the weighted geometric-logarithmic-identric-arithmetic mean inequality.

An important example of a Borel probability measure on $\Sigma(N)$ is the *Dirichlet measure* of parameters $p_1,\dots,p_N>0$,

$$\frac{\Gamma(p_1+\cdots+p_N)}{\Gamma(p_1)\cdots\Gamma(p_N)}x_1^{p_1-1}\cdots x_{N-1}^{p_{N-1}-1}(1-x_1-\cdots-x_{N-1})^{p_N-1}\,\mathrm{d}x_1\cdots\mathrm{d}x_{N-1}.$$

Its barycenter is the point $(\sum_{k=1}^N p_k)^{-1}\cdot(p_1,\dots,p_N)$.

Exercises

1. A higher dimensional analogue of the Hermite–Hadamard inequality [33]) Let f be a continuous concave function defined on a compact convex subset $K\subset\mathbb{R}^N$ of positive volume. Prove that

$$\frac{1}{N+1}\sup_{\mathbf{x}\in K}f(\mathbf{x})+\frac{N}{N+1}\inf_{\mathbf{x}\in\operatorname{ext}K}f(\mathbf{x})\le\frac{1}{\operatorname{Vol}_N(K)}\int_K f(\mathbf{x})\,\mathrm{d}\mathbf{x}\le f(\operatorname{bar}(\mu)),$$

   where x_μ is the barycenter of the probability measure $\mathrm{d}\mu=\mathrm{d}x/\operatorname{Vol}_N(K)$ on K.

2. (R. R. Phelps [398]) Prove that in every normed linear space E, a sequence $(\mathbf{x}_n)_n$ is weakly convergent to $\mathbf{x}$ if, and only if, the sequence is norm bounded and $\lim_{n\to\infty}x^*(\mathbf{x}_n)=x^*(\mathbf{x})$ for every extreme point x^* of the closed unit ball in E^*. As a consequence, $f_n\to f$ weakly in $C([0,1])$ if, and only if,

$$\sup\{|f_n(t)|:t\in[0,1],\ n\in\mathbb{N}\}<\infty$$

   and $f_n(t)\to f(t)$ for every $t\in[0,1]$.

   [*Hint*: Let K be the closed unit ball in E^*. Then K is convex and weak-star compact (see Theorem B.1.6). Each point $\mathbf{x}\in E$ gives rise to an affine mapping $A_x\colon K\to\mathbb{R}$, $A_{\mathbf{x}}(x^*)=x^*(\mathbf{x})$. Then apply Theorem 7.4.4.]

3. (R. Haydon) Let E be a real Banach space and let K be a weak-star compact convex subset of E^* such that $\operatorname{ext}K$ is norm separable. Prove that K is the norm closed convex hull of $\operatorname{ext}K$ (and hence is itself norm separable).

4. Let K be a nonempty compact convex set (in a locally convex Hausdorff space E) and let μ be a Borel probability measure on K. If $f \in C(K)$, one denotes by $\widehat{f}$ its upper envelope. Prove that:
   (a) $\widehat{f}(\mathbf{x}) = \inf\{g(\mathbf{x}) : g \in -\operatorname{Conv}(K) \text{ and } g \geq f\}$;
   (b) for each pair of functions $g_1, g_2 \in -\operatorname{Conv}(K)$ with $g_1, g_2 \geq f$, there is a function $g \in -\operatorname{Conv}(K)$ such that $g_1, g_2 \geq g \geq f$;
   (c) $\mu(\widehat{f}) = \inf\{\mu(g) : -g \in \operatorname{Conv}(K) \text{ and } g \geq f\}$.
5. (G. Mokobodzki) Infer from the preceding exercise that a Borel probability measure μ on K is maximal if, and only if, $\mu(f) = \mu(\widehat{f})$ for all continuous convex functions f on K (equivalently, for all functions $f \in C(K)$).

7.5 The Hermite–Hadamard Inequality for Signed Measures

In dimension 1, the Hermite–Hadamard inequality asserts that for every Borel probability measure μ on a compact interval $[a,b]$ and every continuous convex function $f : [a,b] \to \mathbb{R}$ we have

$$f(\operatorname{bar}(\mu)) \leq \int_a^b f(x)\,\mathrm{d}\mu(x) \leq \frac{b - \operatorname{bar}(\mu)}{b-a} \cdot f(a) + \frac{\operatorname{bar}(\mu) - a}{b-a} \cdot f(b). \tag{7.12}$$

See Section 1.10. The Jensen–Steffensen inequality shows that the left-hand side inequality works in the more general context of Steffensen–Popoviciu measures. Unfortunately, this does not apply to the right-hand side inequality. However, A. M. Fink [164] was able to indicate some few concrete examples of signed measures for which the double inequality (7.12) holds true. This made possible to speak on *Hermite–Hadamard measures*, that is, on those Steffensen–Popoviciu measures of total mass 1, for which the Hermite–Hadamard inequality works in its entirety. The complete characterization of these measures combines Theorem 7.1.2 with the following result due to A. Florea and C. P. Niculescu [167].

7.5.1 Theorem *Let μ be a real Borel measure on the interval $[a,b]$ such that $\mu([a,b]) = 1$. Then the inequality*

$$\frac{b - \operatorname{bar}(\mu)}{b-a} \cdot f(a) + \frac{\operatorname{bar}(\mu) - a}{b-a} \cdot f(b) \geq \int_a^b f(x)\,\mathrm{d}\mu(x) \tag{7.13}$$

works for all continuous convex functions $f : [a,b] \to \mathbb{R}$ if, and only if,

$$\frac{b-t}{b-a} \int_a^t (x-a)\mathrm{d}\mu(x) + \frac{t-a}{b-a} \int_t^b (b-x)\mathrm{d}\mu(x) \geq 0 \tag{7.14}$$

for all $t \in [a,b]$.

Proof. By an approximation argument we may reduce ourselves to the case of convex functions of class C^2. As well known, every such function f admits an integral representation of the form

$$f(x) = \frac{b-x}{b-a} \cdot f(a) + \frac{x-a}{b-a} \cdot f(b) + \int_a^b G(x,t) f''(t) \mathrm{d}t,$$

where

$$G(x,t) = \begin{cases} -(x-a)(b-t)/(b-a) & \text{if } a \leq x \leq t \leq b \\ -(t-a)(b-x)/(b-a) & \text{if } a \leq t \leq x \leq b \end{cases}$$

represents the Green function of the operator $\frac{\mathrm{d}^2}{\mathrm{d}t^2}$, with homogeneous boundary conditions $y(a) = y(b) = 0$.

Thus, for every convex functions $f \in C^2([a,b])$ we have

$$\begin{aligned}
\int_a^b f(x)\, \mathrm{d}\mu(x) &- \frac{b - \mathrm{bar}(\mu)}{b-a} \cdot f(a) + \frac{\mathrm{bar}(\mu) - a}{b-a} \cdot f(b) \\
&= \int_a^b \left[f(x) - \frac{b-x}{b-a} \cdot f(a) + \frac{x-a}{b-a} \cdot f(b) \right] \mathrm{d}\mu(x) \\
&= \int_a^b \left(\int_a^b G(x,t) f''(t) \mathrm{d}t \right) \mathrm{d}\mu(x) \\
&= \int_a^b f''(t) \left(\int_a^b G(x,t) \mathrm{d}\mu(x) \right) \mathrm{d}t = \int_a^b f''(t) y(t) \mathrm{d}t,
\end{aligned}$$

where $y(t) = \int_a^b G(x,t) \mathrm{d}\mu(x)$ is a continuous function.

The differential operator $\frac{\mathrm{d}^2}{\mathrm{d}t^2}$ maps $\mathrm{Conv}([a,b]) \cap C^2([a,b])$ onto the positive cone of $C([a,b])$, so the inequality (7.13) holds true if, and only if, $y(t) \leq 0$ for all $t \in [a,b]$. This ends the proof. ■

According to Theorem 7.1.2 and Theorem 7.5.1,

$$3\left(x^2 - \frac{1}{6}\right) \mathrm{d}x, \; \frac{3780}{499} \left(x^2 - \frac{1}{6}\right)^3 \mathrm{d}x \text{ and } \frac{25}{8} \left(x^4 - \frac{1}{25}\right) \mathrm{d}x$$

are examples of Hermite–Hadamard measure on the interval $[-1,1]$ (each of total mass 1 and barycenter 0).

Notice that $\left(x^2 - \frac{1}{4}\right) \mathrm{d}x$, which is a strong Steffensen–Popoviciu measure on $[-1,1]$, does not fulfill the condition (7.14) and thus it is not a Hermite–Hadamard measure. However, every dual Steffensen–Popoviciu measure verifies the condition (7.14) because it represents the integral of a positive concave function.

The extension of the Hermite–Hadamard in higher dimensions for signed measures remains an open problem. No general criterion like that given by Theorem 7.5.1 is known. A simple example proving the consistency of this problem is shown in Exercise 2.

Exercises

1. (A. M. Fink [164]) Consider the signed measure $(x^2-x)\,\mathrm{d}x$ on the interval $[-1,1]$. Its total mass is $2/3$ and its barycenter is the point -1. Prove that
$$\int_{-1}^{1} f(x)(x^2-x)\,\mathrm{d}x \le \frac{2}{3}\, f(-1)$$
for all continuous convex functions f in $C^2([-1,1])$.
Remark. One can show easily that the signed measure $(x^2-x)\,\mathrm{d}x$ is not a Steffensen–Popoviciu measure on the interval $[-1,1]$.

2. Consider the real Borel measure $\mathrm{d}\mu(x,y) = \frac{3}{2}\left(x^2 - \frac{1}{6}\right)\mathrm{d}x\mathrm{d}y$ on the square $[-1,1]\times[-1,1]$. Prove that:
(a) The total mass of this measure is 1 and its barycenter is the origin.
(b) For every continuous convex function $f:[-1,1]\times[-1,1]\to\mathbb{R}$,
$$\begin{aligned} f(0) &\le \frac{3}{2}\int_{-1}^{1}\int_{-1}^{1} f(x,y)\left(x^2-\frac{1}{4}\right)\mathrm{d}x\mathrm{d}y \\ &\le \frac{f(-1,-1)+f(-1,1)+f(1,-1)+f(1,1)}{4}. \end{aligned}$$

7.6 Comments

The theory of Steffensen–Popoviciu measures was initiated by C. P. Niculescu [356], [357] and developed by him and his collaborators in a series of papers: [167], [366], [365], [363], [377], [322], and [378]. Except the cases mentioned explicitly in the text, all results presented in this chapter represent the contribution of this research team.

Initially, it was thought that both inequalities composing the Hermite–Hadamard inequality work in the context of Steffensen–Popoviciu measures, but soon it became clear that the extension of the right-hand side inequality is more intricate. Thus, the claim of Theorems 4 and 5 in [356] concerning the existence of maximal Borel probability measures majorizing a given Steffensen–Popoviciu measure is false. The paper by A. Florea and C. P. Niculescu [167] solved the problem in dimension 1, but still we lack a result similar to that of Theorem 7.5.1 in higher dimensions.

An interesting example of a strong Steffensen–Popoviciu measure is offered by the case of Rayleigh measures, introduced by S. Kerov [248] in a paper dedicated to the relationship between probability distributions μ and certain bounded signed measures τ on the real line $\mathbb{R}$, satisfying the Markov–Krein identity,
$$\int_{-\infty}^{\infty}\frac{\mathrm{d}\mu(u)}{z-u} = \exp\int_{-\infty}^{\infty}\ln\frac{1}{z-u}\mathrm{d}\tau(u) \quad \text{for } z\in\mathbb{C},\ \operatorname{Im} z>0.$$

Precisely, the function $R(z)$ appearing in the right-hand side can be represented as a Cauchy–Stieltjes transform of a probability measure μ,

$$R(z) = \int_{-\infty}^{\infty} \frac{\mathrm{d}\mu(u)}{z-u},$$

whenever τ is a *Rayleigh measure*, that is, a bounded signed measure that verifies the following three properties:

$$(RM1)\ 0 \leq \tau\left(\{x : x < a\}\right) \leq 1 \quad \text{for every } a \in \mathbb{R};$$
$$(RM2)\ \tau\left((-\infty, \infty)\right) = 1;$$
$$(RM3)\ \int_{-\infty}^{\infty} \ln\left(1 + |x|\right) \mathrm{d}\left|\tau\right|(x) < \infty.$$

See [248], Corollary 2.4.1. An alternative approach of Rayleigh measures is offered by the notion of interlace measures. Two finite positive measures τ' and τ'' *interlace* if there exists a Rayleigh measure τ such that $\tau' = \tau^+$ and $\tau'' = \tau^-$. For example, if $\lambda_1 \leq \mu_1 \leq \lambda_2 \leq \cdots \leq \mu_{n-1} \leq \lambda_n$, then the discrete measure $\tau = \sum_{k=1}^{n} \delta_{\lambda_k} - \sum_{j=1}^{n-1} \delta_{\mu_j}$ is an example of Rayleigh measure and the corresponding Markov–Krein identity has the form

$$\sum_{k=1}^{n} \frac{p_k}{z - \lambda_k} = \frac{(z-\mu_1)\cdots(z-\mu_{n-1})}{(z-\lambda_1)\cdots(z-\lambda_n)}$$

for suitable $p_1, ..., p_n > 0$ with $\sum_{k=1}^{n} p_k = 1$.

Interesting examples of absolutely continuous Rayleigh measures can be found in the paper of D. Romik [425].

Choquet theory was born around 1956, when G. Choquet started to announce his results in a series of papers published in Comptes Rendus de l'Académie des Sciences (Paris). His paper [105] from 1960 together with his report [106] at the International Congress of Mathematicians at Stockholm (1962) provided a clear picture about the importance and scope of the Choquet theory. R. R. Phelps [398] contributed much to the popularization of this theory with a well written, readable and easily accessible book.

As noticed in Theorem 4.1.5, the relation of majorization $x \prec y$ (for vectors in $\mathbb{R}^N$) can be characterized by the existence of a doubly stochastic matrix P such that $x = Py$. Thinking of x and y as discrete probability measures, this fact can be rephrased as saying that y is a dilation of x. The book by R. R. Phelps [398] indicates the details of an extension (due to P. Cartier, J. M. G. Fell, and P. A. Meyer [96]) of this characterization to the general framework of Borel probability measures on compact convex sets (in a locally convex Hausdorff space).

Following our paper [366], we shall show that the Hermite–Hadamard inequality also works in the context of subharmonic functions.

Let Ω be a bounded open subset of $\mathbb{R}^N$ with smooth boundary. Then the Dirichlet problem

$$\begin{cases} \Delta\varphi = 1 & \text{on } \Omega \\ \varphi = 0 & \text{on } \partial\Omega \end{cases} \tag{7.15}$$

has a unique solution, which is negative on Ω, according to the maximum principle for elliptic problems. See [416]. By Green's formula, for every u in $C^2(\Omega) \cap C^1(\overline{\Omega})$ we have

$$\int_{\Omega} \begin{vmatrix} u & \varphi \\ \Delta u & \Delta\varphi \end{vmatrix} \, \mathrm{d}V = \int_{\partial\Omega} \begin{vmatrix} u & \varphi \\ \nabla u & \nabla\varphi \end{vmatrix} \cdot n \, \mathrm{d}S,$$

that is, using (7.15),

$$\begin{aligned} \int_{\Omega} u \, \mathrm{d}V &= \int_{\Omega} u\Delta\varphi \, \mathrm{d}V = \int_{\Omega} \varphi\Delta u \, \mathrm{d}V + \int_{\partial\Omega} u(\nabla\varphi \cdot \mathbf{n}) \, \mathrm{d}S - \int_{\partial\Omega} \varphi(\nabla u \cdot \mathbf{n}) \, \mathrm{d}S \\ &= \int_{\Omega} \varphi\Delta u \, \mathrm{d}V + \int_{\partial\Omega} u(\nabla\varphi \cdot \mathbf{n}) \, \mathrm{d}S \end{aligned}$$

for every $u \in C^2(\Omega) \cap C^1(\overline{\Omega})$. Here $\mathbf{n}$ denotes the outward-pointing unit normal vector on the boundary.

We are thus led to the following result:

7.6.1 Theorem (The Hermite–Hadamard inequality for subharmonic functions) *If $u \in C^2(\Omega) \cap C^1(\overline{\Omega})$ is subharmonic (that is, $\Delta u \geq 0$ on Ω) and φ satisfies* (7.15), *then*

$$\int_{\Omega} u \, \mathrm{d}V < \int_{\partial\Omega} u(\nabla\varphi \cdot \mathbf{n}) \, \mathrm{d}S$$

except for harmonic functions (when equality occurs).

The equality case needs the remark that $\int_{\Omega} \varphi\Delta u \, \mathrm{d}V = 0$ yields $\varphi\Delta u = 0$ on Ω, and thus $\Delta u = 0$ on Ω; notice that $\varphi\Delta u$ is continuous and $\varphi\Delta u \leq 0$ since $\varphi < 0$ on Ω.

In the case of balls $\Omega = B_R(\mathbf{a})$ in $\mathbb{R}^3$, the solution of the problem (7.15) is $\varphi(\mathbf{x}) = (\|\mathbf{x}\|^2 - R^2)/6$ and $\nabla\varphi \cdot \mathbf{n} = \langle \mathbf{x}/3, \mathbf{x}/\|\mathbf{x}\| \rangle = \|\mathbf{x}\|/3$, so that by combining the maximum principle for elliptic problems with the conclusion of the above theorem we obtain the following Hermite–Hadamard-type inequality for subharmonic functions:

$$u(\mathbf{a}) \leq \frac{1}{\operatorname{Vol} \overline{B}_R(\mathbf{a})} \iiint_{\overline{B}_R(\mathbf{a})} u(\mathbf{x}) \, \mathrm{d}V < \frac{1}{\operatorname{Area} S_R(\mathbf{a})} \iint_{S_R(\mathbf{a})} u(\mathbf{x}) \, \mathrm{d}S$$

for every $u \in C^2(B_R(\mathbf{a})) \cap C^1(\overline{B}_R(\mathbf{a}))$ with $\Delta u \geq 0$, which is not harmonic.

Theorem 7.6.1 still works when the Laplacian is replaced by a strictly elliptic self-adjoint linear differential operator of second order which admits a Green function. See [323].

A nice theory explaining the similarities between the convex functions and the subharmonic functions was developed by G. Choquet [106].

Appendix A

Generalized Convexity on Intervals

An important feature underlying the notion of convex function is the comparison of means under the action of a function. Indeed, by considering the *weighted arithmetic mean* of n variables,

$$A(x_1, \ldots, x_n; \lambda_1, \ldots, \lambda_n) = \sum_{k=1}^{n} \lambda_k x_k,$$

the convex sets are precisely the sets invariant under the action of these means, while the convex functions are the functions that verify estimates of the form

$$f\left(A(x_1, \ldots, x_n; \lambda_1, \ldots, \lambda_n)\right) \leq A(f(x_1), \ldots, f(x_n); \lambda_1, \ldots, \lambda_n). \tag{AA}$$

This motivates to call the usual concept of convexity also (A, A)-convexity.

It is natural to ask: What is happening if we consider other means?

The interval $(0, \infty)$ is closed under the action of *weighted geometric mean* of n variables,

$$G(x_1, \ldots, x_n; \lambda_1, \ldots, \lambda_n) = \prod_{k=1}^{n} x_k^{\lambda_k},$$

a fact that allowed us to consider in Sections 1.1 and 3.1 the log-convex functions, that is, the functions $f : U \to (0, \infty)$ (defined on convex sets) that verify the inequalities

$$f\left(A(x_1, \ldots, x_n; \lambda_1, \ldots, \lambda_n)\right) \leq G(f(x_1), \ldots, f(x_n); \lambda_1, \ldots, \lambda_n); \tag{AG}$$

this motivates the name of (A, G)-convex for the log-convex functions.

The quasiconvex functions also fit a similar description as they coincide with the (A, M_∞)-convex functions, where the *max mean* is defined by

$$M_\infty(x_1, \ldots, x_n; \lambda_1, \ldots, \lambda_n) = \max\{x_1, \ldots, x_n\}.$$

C. P. Niculescu and L.-E. Persson, *Convex Functions and Their Applications*,
CMS Books in Mathematics, https://doi.org/10.1007/978-3-319-78337-6

The aim of this appendix is to give a short account on the subject of convexity according to a pair of means (acting, respectively, on the domain and the codomain) and to bring into attention new classes of convex like functions.

A.1 Means

By a mean we understand a procedure M to associate to each discrete random variable $X : \{1, ..., n\} \to \mathbb{R}$ having the probability distribution

$$P(\{X(k) = x_k) = \lambda_k \quad \text{for } k = 1, ..., n,$$

a number $M(x_1, ..., x_n; \lambda_1, ..., \lambda_n)$ such that

$$M(x_1, ..., x_n; \lambda_1, ..., \lambda_n) \in [\min X, \max X].$$

We make the convention to omit the weights λ_k when they are equal to each other, that is, we put

$$M(x_1, ..., x_n) = M(x_1, ..., x_n; 1/n, ..., 1/n).$$

A mean is called:

symmetric if $M(x_1, ..., x_n) = M(x_{\sigma(1)}, ..., x_{\sigma(n)})$ for every finite family $x_1, ..., x_n$ of elements of I and every permutation σ of $\{1, ..., n\}$ with $n \geq 1$.

continuous when all functions $M(x_1, ..., x_n; \lambda_1, ..., \lambda_n)$ are globally continuous;

increasing/strictly increasing when all functions $M(x_1, ..., x_n; \lambda_1, ..., \lambda_n)$ with strictly positive weights are increasing/strictly increasing in each of the variables x_k (when the others are kept fixed).

When I is one of the intervals $(0, \infty)$, $[0, \infty)$ or $(-\infty, \infty)$, it is usual to consider *homogeneous* means, that is, means for which

$$M(\alpha x_1, ..., \alpha x_n; \lambda_1, ..., \lambda_n) = \alpha M(x_1, ..., x_n; \lambda_1, ..., \lambda_n)$$

whenever $\alpha > 0$.

From a probabilistic point of view, the means represent generalizations of the concept of expectation of a random variable.

The most known class of symmetric, continuous, and homogeneous means on $(0, \infty)$ is that of *Hölder's means* (also known as the *power means*):

$$M_p(x_1, ..., x_n; \lambda_1, ..., \lambda_n) = \begin{cases} \min\{x_1, ..., x_n\} & \text{if } p = -\infty \\ (\sum_{k=1}^n \lambda_k x_k^p,)^{1/p} & \text{if } p \in \mathbb{R}\backslash\{0\} \\ \prod_{k=1}^n x_k^{\lambda_k} & \text{if } p = 0 \\ \max\{x_1, ..., x_n\}, & \text{if } p = \infty. \end{cases} \tag{A.1}$$

This family brings together three of the most important means in mathematics: the *arithmetic mean* $A = M_1$, the *geometric mean* $G = M_0$ and the *harmonic mean* $H = M_{-1}$. Notice that all Hölder's means of real index and equal weights are strictly increasing and $M_0(s,t) = \lim_{p\to 0} M_p(s,t)$. Besides, all Hölder's means of odd index $p \geq 1$ makes sense for every family of real numbers.

Hölder's means also admit a natural extension to all positive random variables (discrete or continuous); see Exercise 1, Section 1.7.

A somewhat more general class of means is that of quasi-arithmetic means. In the discrete case, the *quasi-arithmetic means* are associated to strictly monotonic and continuous functions $\varphi : I \to \mathbb{R}$ via the formulas

$$M_\varphi(x_1, ..., x_n; \lambda_1, ..., \lambda_n) = \varphi^{-1}\left(\sum_{k=1}^{n} \lambda_k \varphi(x_k)\right);$$

in the case of an arbitrary random variable, they can be defined as Riemann–Stieltjes integrals,

$$M_\varphi(X) = \varphi^{-1}\left(\int \varphi(x)dF(x)\right),$$

where F represents the cumulative distribution function of X.

A. Kolmogorov [257] has developed the first axiomatic theory of means, providing the ubiquity of quasi-arithmetic means. A nice account on his contribution, as well as on some recent results concerning the probabilistic applications of quasi-arithmetic means, can be found in the paper of M. de Carvalho [129].

A.1.1 Lemma (K. Knopp and B. Jessen; see [209], p. 66) *Suppose that φ and ψ are two continuous functions defined in an interval I such that φ is strictly monotonic and ψ is strictly increasing. Then: $M_\varphi(x_1, \ldots, x_n; \lambda_1, \ldots, \lambda_n) = M_\psi(x_1, \ldots, x_n; \lambda_1, \ldots, \lambda_n)$ for every family $x_1, \ldots, x_n$ of elements of I and every family $\lambda_1, \ldots, \lambda_n$ of positive numbers with $\sum_{k=1}^{n} \lambda_k = 1$ $(n \in \mathbb{N}^\star)$ if and only if $\psi \circ \varphi^{-1}$ is affine, that is, $\psi = \alpha\varphi + \beta$ for some constants α and β, with $\alpha \neq 0$.*

An immediate consequence of Lemma A.1.1 is the fact that every power mean M_p of real index is a quasi-arithmetic mean M_φ, where $\varphi(x) = \log x$, if $p = 0$, and $\varphi(x) = (x^p - 1)/p$, if $p \neq 0$.

A.1.2 Theorem (M. Nagumo, B. de Finetti and B. Jessen; see [209], p. 68) *Let φ be a continuous increasing function on $(0, \infty)$ such that the quasi-arithmetic mean M_φ is positively homogeneous. Then M_φ is a power mean.*

Proof. By Lemma A.1.1, we can replace φ by $\varphi - \varphi(1)$, so we may assume that $\varphi(1) = 0$. The same argument yields two functions α and β such that $\varphi(cx) = \alpha(c)\varphi(x) + \beta(c)$ for all $x > 0$, $c > 0$. The condition $\varphi(1) = 0$ shows that $\beta = \varphi$ so, for reasons of symmetry,

$$\varphi(cx) = \alpha(c)\varphi(x) + \varphi(c) = \alpha(x)\varphi(c) + \varphi(x).$$

Letting $c \neq 1$ fixed, we obtain that α is of the form $\alpha(x) = 1 + k\varphi(x)$ for some constant k. Then φ verifies the functional equation

$$\varphi(xy) = k\varphi(x)\varphi(y) + \varphi(x) + \varphi(y)$$

for all $x > 0$, $y > 0$. When $k = 0$ we find that $\varphi(x) = C \log x$ for some constant C, so $M_\varphi = M_0$. When $k \neq 0$ we notice that $\chi = k\varphi + 1$ verifies $\chi(xy) = \chi(x)\chi(y)$ for all $x > 0$, $y > 0$. This leads to $\varphi(x) = (x^p - 1)/k$, for some $p \neq 0$, hence $M_\varphi = M_p$. ■

Some authors (such as P. S. Bullen, D. S. Mitrinović and P. M. Vasić [83]) consider a more general concept of mean, associated to functions $M : I \times I \to I$ with the single property that

$$\min\{x, y\} \leq M(x, y) \leq \max\{x, y\} \quad \text{for all } x, y \in I.$$

One can interpret $M(x, y)$ as the mean $M(x, y; 1/2, 1/2)$ of a random variable taking only two values, x and y, with probability 1/2. Unfortunately, the problem to indicate "natural" extensions of such means to the entire class of discrete random variables remains open. The difficulty of this problem is illustrated by the case of *logarithmic* and *identric* means, which are defined, respectively, by

$$L(x, y) = \frac{x - y}{\log x - \log y} \text{ and } I(x, y) = \frac{1}{e}\left(\frac{y^y}{x^x}\right)^{1/(y-x)}$$

for $x, y > 0$, $x \neq y$, and $L(x, x) = I(x, x) = x$ for $x > 0$. The inequalities noticed in Example 1.10.2 can be completed as follows:

$$G(x, y) < L(x, y) < M_{1/3}(x, y) < M_{2/3}(x, y) < I(x, y) < A(x, y)$$

for all $x, y > 0$, $x \neq y$. See T.-P. Lin [284] and K. Stolarsky [462] for the inner inequalities and also for the fact that the logarithmic mean and the identric mean are not power means. Since the logarithmic mean is positively homogeneous, Theorem A.1.2 allows us to conclude that this mean is not quasi-arithmetic.

A satisfactory extension of the logarithmic and the identric means for arbitrary random variables was obtained by E. Neuman [346], [348]. See also the comments by C. P. Niculescu [364].

A.2 Convexity According to a Pair of Means

According to G. Aumann [25], if M and N are means defined, respectively, on the intervals I and J, a function $f : I \to J$ is called (M, N)-*midpoint convex* if

$$f(M(x, y)) \leq N(f(x), f(y)) \text{ for all } x, y \in I;$$

it is called (M, N)-*midpoint concave* if the inequality works in the reverse way, and (M, N)-*midpoint affine* if the inequality sign is replaced by equality. The condition of midpoint affinity is essentially a functional equation and this explains why the theory of generalized convexity has much in common with the subject of functional equations.

In what follows we will be interested in a concept of convexity dealing with weighted means.

A.2.1 Definition *A function $f : I \to J$ is called* (M, N)-convex *if*

$$f(M(x, y; 1 - \lambda, \lambda)) \leq N(f(x), f(y); 1 - \lambda, \lambda)$$

*for all $x, y \in I$ and all $\lambda \in [0, 1]$. It is called (M, N)-strictly convex when the inequality is strict whenever x and y are distinct points and $\lambda \in (0, 1)$. If $-f$ is (M, N)-convex (respectively, strictly (M, N)-convex), then we say that f is (M, N)-*concave *(respectively, (M, N)-*strictly concave*).*

Jensen's criterion of convexity (Theorem 1.1.8 above) can be extended easily to the context of power means (and even to that of quasi-arithmetic means). Thus, in their case, every (M, N)-midpoint and continuous convex function $f : I \to J$ is also (M, N)-convex. Jensen's inequality works as well, providing the possibility to deal with rather general random variables. All these facts are consequences of the following lemma, which reduces the convexity with respect to a pair of quasi-arithmetic means to the usual convexity of a function derived via a change of variable and a change of function.

A.2.2 Lemma (Aczél correspondence principle [3]) *Suppose that φ and ψ are two continuous and strictly monotonic functions defined, respectively, on the intervals I and J. Then:*

(a) *if ψ is strictly increasing, a function $f\colon I \to J$ is (M_φ, M_ψ)-convex/concave if and only if $\psi \circ f \circ \varphi^{-1}$ is convex/concave on $\varphi(I)$;*

(b) *if ψ is strictly decreasing, a function $f\colon I \to J$ is (M_φ, M_ψ)-convex/concave if and only if $\psi \circ f \circ \varphi^{-1}$ is concave/convex on $\varphi(I)$.*

A.2.3 Corollary *Suppose that $\varphi, \psi\colon I \to \mathbb{R}$ are two strictly monotonic continuous functions. If ψ is strictly increasing, then*

$$M_\varphi \leq M_\psi$$

if and only if $\psi \circ \varphi^{-1}$ is convex.

Corollary A.2.3 has important consequences. For example, as was noticed by J. Lamperti [268], it yields Clarkson's inequalities as stated in Section 2.2. His basic remark is as follows:

A.2.4 Lemma *Suppose that $\Phi\colon [0, \infty) \to \mathbb{R}$ is a continuous increasing function with $\Phi(0) = 0$ and $\Phi(\sqrt{t})$ convex. Then*

$$\Phi(|z + w|) + \Phi(|z - w|) \geq 2\Phi(|z|) + 2\Phi(|w|), \tag{A.2}$$

for all $z, w \in C$, while if $\Phi(\sqrt{t})$ is concave, then the reverse inequality is true. Provided the convexity or concavity is strict, equality holds if and only if $zw = 0$.

Clarkson's inequalities follow for $\Phi(t) = t^p$; this function is strictly concave for $1 < p \leq 2$ and strictly convex for $2 \leq p < \infty$.

Proof. When $\Phi(\sqrt{t})$ is convex, we infer from Corollary A.2.3 and the parallelogram law that

$$\Phi^{-1}\left\{\frac{\Phi(|z+w|)+\Phi(|z-w|)}{2}\right\} \geq \left\{\frac{|z+w|^2+|z-w|^2}{2}\right\}^{1/2} = (|z|^2+|w|^2)^{1/2}. \tag{A.3}$$

On the other hand, taking into account the three chords inequality (as stated in Section 1.4), we infer from the convexity of $\Phi(\sqrt{t})$ and the fact that $\Phi(0)=0$ the increasing monotonicity of $\Phi(\sqrt{t})/t$; the monotonicity is strict provided that the convexity of $\Phi(\sqrt{t})$ is strict.

Then

$$|z|^2\frac{\Phi(|z|)}{|z|^2}+|w|^2\frac{\Phi(|w|)}{|w|^2} \leq (|z|^2+|w|^2)\frac{\Phi\left((|z|^2+|w|^2)^{1/2}\right)}{|z|^2+|w|^2},$$

and the inequality is strict when $\Phi(\sqrt{t})$ is strictly convex and $zw \neq 0$. Therefore $\Phi^{-1}\left(\Phi(|z|)+\Phi(|w|)\right) \leq (|z|^2+|w|^2)^{1/2}$ and this fact together with (A.3) ends the proof in the case where $\Phi(\sqrt{t})$ is convex. The case where $\Phi(\sqrt{t})$ is concave can be treated in a similar way. ■

According to Lemma A.2.2, a function $f:(0,\alpha)\to(0,\infty)$ is (H,G)-convex (concave) on $(0,\alpha)$ if and only if $\log f(1/x)$ is convex (concave) on $(1/\alpha,\infty)$. Clearly, the "strict" variant also works.

The following example of a (H,G)-strictly log-concave function is due to D. Borwein, J. Borwein, G. Fee and R. Girgensohn [67].

A.2.5 Example *Given $\alpha>1$, the function $V_\alpha(p)=2^\alpha\frac{\Gamma(1+1/p)^\alpha}{\Gamma(1+\alpha/p)}$ verifies the inequality*

$$V_\alpha^{1-\lambda}(p)V_\alpha^{\lambda}(q) < V_\alpha\left(\frac{1}{\frac{1-\lambda}{p}+\frac{\lambda}{q}}\right),$$

for all $p,q>0$ with $p\neq q$ and all $\lambda\in(0,1)$. This example has a geometric motivation. When $\alpha>1$ is an integer and $p\geq 1$ is a real number, $V_\alpha(p)$ represents the volume of the ellipsoid $\{x\in\mathbb{R}^\alpha : \|x\|_{L^p}\leq 1\}$.

In fact, according to Lemma A.2.2, it suffices to prove that the function

$$U_\alpha(x) = -\log(V_\alpha(1/x)/2^\alpha) = \log\mathfrak{d}(1+\alpha x) - \alpha\log\mathfrak{d}(1+x)$$

is strictly convex on $(0,\infty)$ for every $\alpha>1$. Using the psi function,

$$\mathrm{Psi}(x) = \frac{d}{dx}\log\mathfrak{d}(x),$$

we have

$$U_\alpha''(x) = \alpha^2\frac{d}{dx}\mathrm{Psi}(1+\alpha x) - \alpha\frac{d}{dx}\mathrm{Psi}(1+x).$$

The condition $U''_\alpha(x) > 0$ on $(0,\infty)$ is equivalent to $(x/\alpha)U''_\alpha(x) > 0$ on $(0,\infty)$, and the latter holds if the function $x \to x\frac{d}{dx}\operatorname{Psi}(1+x)$ is strictly increasing. Or, a classical formula in special function theory [18], [186] asserts that

$$\frac{d}{dx}\operatorname{Psi}(1+x) = \int_0^\infty \frac{ue^{ux}}{e^u - 1}\,du,$$

whence we infer that

$$\frac{d}{dx}\left(x\frac{d}{dx}\operatorname{Psi}(1+x)\right) = \int_0^\infty \frac{u[(u-1)e^u+1]e^{ux}}{(e^u-1)^2}\,du > 0,$$

and the statement in Example A.2.5 follows. Notice that the volume function $V_n(p)$ is neither convex nor concave for $n \geq 3$.

The following example is devoted to the presence of generalized convexity in the framework of hypergeometric functions.

A.2.6 Example *The Gaussian hypergeometric function (of parameters $a, b, c > 0$) is defined via the formula*

$$F(x) ={}_2F_1(x;a,b,c) = \sum_{n=0}^\infty \frac{(a,n)\,(b,n)}{(c,n)n!}x^n \quad \text{for } |x| < 1,$$

where $(a,n) = a(a+1)\cdots(a+n-1)$ if $n \geq 1$ and $(a,0) = 1$. G. D. Anderson, M. K. Vamanamurthy and M. Vuorinen [16] proved that if $a+b \geq c > 2ab$ and $c \geq a+b-1/2$, then the function $1/F(x)$ is concave on $(0,1)$. This implies

$$F\left(\frac{x+y}{2}\right) \leq \frac{1}{\frac{1}{2}\left(\frac{1}{F(x)} + \frac{1}{F(y)}\right)} \quad \text{for all } x, y \in (0,1),$$

whence it follows that the hypergeometric function $F(x)$ is (A,H)-convex.

Last but not least, Lemma A.2.2 offers a simple but powerful tool for deriving the entire theory of (M_r, M_s)-convex functions from that of usual convex functions. We will illustrate this fact in the next section, by discussing the multiplicative analogue of usual convexity.

A.3 A Case Study: Convexity According to the Geometric Mean

The following relative of usual convexity was first considered by P. Montel [335] in a beautiful paper discussing different analogues of convex functions in several variables.

A.3.1 Definition *A function $f: I \to J$, acting on subintervals of $(0,\infty)$, is called multiplicatively convex (equivalently, (G,G)-convex) if it verifies the condition*

$$f(x^{1-\lambda}y^{\lambda}) \leq f(x)^{1-\lambda} f(y)^{\lambda}, \tag{GG}$$

whenever $x, y \in I$ and $\lambda \in [0,1]$. The related notions of multiplicatively strictly convex function, multiplicatively concave function, multiplicatively strictly concave function and multiplicatively affine function can be introduced in a similar manner.

The basic tool in translating the results known for convex functions into results valid for multiplicatively convex functions (and vice versa) is the following particular case of Lemma A.2.2:

A.3.2 Lemma *Suppose that I is a subinterval of $(0,\infty)$ and $f: I \to (0,\infty)$ is a multiplicatively (strictly) convex function on I. Then*

$$F = \log \circ f \circ \exp\colon \log(I) \to \mathbb{R}$$

is a (strictly) convex function. Conversely, if J is an interval of $\mathbb{R}$ and $F: J \to \mathbb{R}$ is a (strictly) convex function, then

$$f = \exp \circ F \circ \log\colon \exp(J) \to (0,\infty)$$

is a multiplicatively (strictly) convex function.

The corresponding "concave" variant of this lemma also works.

This approach of multiplicative convexity follows C. P. Niculescu [351], [353].

According to Lemma A.3.2, every multiplicatively convex function f has finite lateral derivatives at each interior point of its domain (and the set of all points where f is not differentiable is at most countable). As a consequence, every multiplicatively convex function is continuous at the interior points of its domain. The following result represents the multiplicative analogue of Theorem 1.1.8 (Jensen's criterion of convexity).

A.3.3 Theorem *Suppose that I is a subinterval of $(0,\infty)$ and $f: I \to (0,\infty)$ is a function continuous on the interior of I. Then f is multiplicatively convex if, and only if,*

$$f(\sqrt{xy}) \leq \sqrt{f(x)f(y)} \text{ for all } x, y \in I.$$

A large class of strictly multiplicatively convex functions is indicated by the following result:

A.3.4 Proposition (G. H. Hardy, J. E. Littlewood and G. Pólya [209, Theorem 177, p. 125]) *Every polynomial $P(x)$ with positive coefficients is a multiplicatively convex function on $(0,\infty)$. More generally, every real analytic function $f(x) = \sum_{n=0}^{\infty} c_n x^n$ with positive coefficients is a multiplicatively convex function on $(0,R)$, where R denotes the radius of convergence.*

Notice that except for the case of functions Cx^n (with $C > 0$ and $n \in \mathbb{N}$), Proposition A.3.4 exhibits examples of strictly multiplicatively convex functions (which are also strictly increasing and strictly convex). So are:

exp, sinh, and cosh on $(0, \infty)$;
tan, sec, csc, and $\frac{1}{x} - \cot x$ on $(0, \pi/2)$;
arcsin on $(0, 1]$;
$-\log(1 - x)$ and $\frac{1+x}{1-x}$ on $(0, 1)$.
See the table of series in I. S. Gradshteyn and I. M. Ryzhik [186].

By continuity, it suffices to prove only the first assertion. Suppose that $P(x) = \sum_{n=0}^{N} c_n x^n$. According to Theorem A.3.3, we have to prove that

$$x, y > 0 \text{ implies } (P(\sqrt{xy}))^2 \leq P(x)P(y),$$

or, equivalently,

$$x, y > 0 \text{ implies } (P(xy))^2 \leq P(x^2)P(y^2).$$

The later implication is an easy consequence of Cauchy–Bunyakovsky–Schwarz inequality.

The following remark collects a series of useful facts concerning the multiplicative convexity of concrete functions:

A.3.5 Remark

(a) *If a function is log-convex and increasing, then it is multiplicatively convex.*

(b) *If a function f is multiplicatively convex, then the function $1/f$ is multiplicatively concave (and vice versa).*

(c) *If a function f is multiplicatively convex, increasing and one-to-one, then its inverse is multiplicatively concave (and vice versa).*

(d) *If a function f is multiplicatively convex, so is $x^\alpha[f(x)]^\beta$ (for all $\alpha \in \mathbb{R}$ and all $\beta > 0$).*

(e) *If f is continuous, and one of the functions $f(x)^x$ and $f(e^{1/\log x})$ is multiplicatively convex, then so is the other.*

(f) *The general form of multiplicatively affine functions is Cx^α with $C > 0$ and $\alpha \in \mathbb{R}$.*

We omit the easy proof.

The indefinite integral of a multiplicatively convex function has the same nature.

A.3.6 Proposition (P. Montel [335]) *Let $f\colon [0, a) \to [0, \infty)$ be a continuous function which is multiplicatively convex on $(0, a)$. Then*

$$F(x) = \int_0^x f(t)\, dt$$

is also continuous on $[0, a)$ and multiplicatively convex on $(0, a)$.

Proof. Due to the continuity of F, it suffices to show that

$$(F(\sqrt{xy}))^2 \leq F(x)F(y) \quad \text{for all } x, y \in [0, a),$$

which is a consequence of the corresponding inequality at the level of integral sums,

$$\Big[\frac{\sqrt{xy}}{n}\sum_{k=0}^{n-1} f\Big(k\frac{\sqrt{xy}}{n}\Big)\Big]^2 \leq \Big[\frac{x}{n}\sum_{k=0}^{n-1} f\Big(k\frac{x}{n}\Big)\Big]\Big[\frac{y}{n}\sum_{k=0}^{n-1} f\Big(k\frac{y}{n}\Big)\Big],$$

that is, of the inequality

$$\Big[\sum_{k=0}^{n-1} f\Big(k\frac{\sqrt{xy}}{n}\Big)\Big]^2 \leq \Big[\sum_{k=0}^{n-1} f\Big(k\frac{x}{n}\Big)\Big]\Big[\sum_{k=0}^{n-1} f\Big(k\frac{y}{n}\Big)\Big].$$

To see that the later inequality holds, first notice that

$$\Big[f\Big(k\frac{\sqrt{xy}}{n}\Big)\Big]^2 \leq \Big[f\Big(k\frac{x}{n}\Big)\Big]\Big[f\Big(k\frac{y}{n}\Big)\Big]$$

and then apply the Cauchy–Bunyakovsky–Schwarz inequality. ■

According to Proposition A.3.6, the *logarithmic integral*,

$$\operatorname{Li}(x) = \int_2^x \frac{dt}{\log t}, \quad x \geq 2,$$

is multiplicatively convex. This function is important in number theory. For example, if $\pi(x)$ counts the number of primes p such that $2 \leq p \leq x$, then an equivalent formulation of the Riemann hypothesis is the existence of a function $C\colon (0,\infty) \to (0,\infty)$ such that

$$|\pi(x) - \operatorname{Li}(x)| \leq C(\varepsilon)x^{1/2+\varepsilon} \quad \text{for all } x \geq 2 \text{ and all } \varepsilon > 0.$$

Since the function tan is continuous on $[0, \pi/2)$ and strictly multiplicatively convex on $(0, \pi/2)$, a repeated application of Proposition A.3.6 shows that the *Lobacevski's function*

$$\operatorname{L}(x) = -\int_0^x \log\cos t\, dt$$

is strictly multiplicatively convex on $(0, \pi/2)$.

Starting with $t/(\sin t)$ (which is strictly multiplicatively convex on $(0, \pi/2]$) and then switching to $(\sin t)/t$, a similar argument leads us to the fact that the *integral sine* function,

$$\operatorname{Si}(x) = \int_0^x \frac{\sin t}{t}\, dt,$$

is strictly multiplicatively concave on $(0, \pi/2]$.

A.3.7 Proposition *The gamma function is strictly multiplicatively convex on the interval* $[1, \infty)$.

Proof. In fact, $\log \Gamma(1+x)$ is strictly convex and increasing on $(1,\infty)$. Moreover, an increasing strictly convex function of a strictly convex function is strictly convex. Hence, $F(x) = \log \Gamma(1+e^x)$ is strictly convex on $(0,\infty)$ and thus $\Gamma(1+x) = \exp F(\log x)$ is strictly multiplicatively convex on $[1,\infty)$. As $\Gamma(1+x) = x\Gamma(x)$, we conclude that Γ itself is strictly multiplicatively convex on $[1,\infty)$. ■

As was noted by T. Trif [473], the result of Proposition A.3.7 can be improved: the gamma function is strictly multiplicatively concave on $(0,\alpha]$ and strictly multiplicatively convex on $[\alpha,\infty)$, where $\alpha = 0.21609...$ is the unique positive solution of the equation $\operatorname{Psi}(x) + x\frac{d}{dx}\operatorname{Psi}(x) = 0$.

D. Gronau and J. Matkowski [188] have proved the following multiplicative analogue of the Bohr–Mollerup theorem: *If $f\colon (0,\infty) \to (0,\infty)$ verifies the functional equation*

$$f(x+1) = xf(x),$$

the normalization condition $f(1) = 1$ and f is multiplicatively convex on an interval (a,∞), for some $a > 0$, then $f = \Gamma$.

Another application of Proposition A.3.7 is the strict multiplicative convexity of the function $\Gamma(2x+1)/\Gamma(x+1)$ on $[1,\infty)$. This can be seen by using the Gauss–Legendre duplication formula (see [107], Theorem 10.3.10, p. 356).

A.3.8 Remark *The following result due to J. Matkowski [309] connects convexity, multiplicative convexity, and (L,L)-convexity: every convex and multiplicatively convex function is (L,L)-convex. The "concave" variant of this result also works. As a consequence, the exponential function is an example of (L,L)-convex function on $(0,\infty)$, while the tangent function is an (L,L)-convex function on $(0,\pi/2)$.*

Exercises

1. Let $f\colon I \to (0,\infty)$ be a differentiable function defined on a subinterval I of $(0,\infty)$. Prove that the following assertions are equivalent:

   (a) f is multiplicatively convex;

   (b) the function $xf'(x)/f(x)$ is increasing;

   (c) f verifies the inequality

   $$\frac{f(x)}{f(y)} \geq \left(\frac{x}{y}\right)^{yf'(y)/f(y)} \quad \text{for all } x,y \in I.$$

   A similar statement works for the multiplicatively concave functions. Illustrate this fact by considering the restriction of $\sin(\cos x)$ to $(0,\pi/2)$.

2. Let $f\colon I \to (0,\infty)$ be a twice differentiable function defined on a subinterval I of $(0,\infty)$. Prove that f is multiplicatively convex if and only if it verifies the differential inequality

   $$x[f(x)f''(x) - f'^2(x)] + f(x)f'(x) \geq 0 \quad \text{for all } x > 0.$$

Infer that the integral sine function is multiplicatively concave.

3. (A multiplicative variant of Popoviciu's inequality) Prove that

$$\Gamma(x)\Gamma(y)\Gamma(z)\Gamma^3(\sqrt[3]{xyz}) \geq \Gamma^2(\sqrt{xy})\Gamma^2(\sqrt{yz})\Gamma^2(\sqrt{zx})$$

for all $x, y, z \geq 1$; equality occurs only for $x = y = z$.

4. (The multiplicative analogue of the Hermite–Hadamard inequality [364]). Assume that $f : [a, b] \to (0, \infty)$ is a multiplicatively convex function. Prove that

$$f(I(a,b)) \leq \frac{1}{b-a}\int_a^b f(x)dx \leq \frac{1}{b-a}\int_a^b f(a)^{\frac{\log b - \log x}{\log b - \log a}} f(b)^{\frac{\log x - \log a}{\log b - \log a}} dx$$
$$= \frac{L(af(a), bf(b))}{L(a,b)}.$$

[*Hint*: Apply the Hermite–Hadamard inequality to the function $g(t) = \log f(e^t)$.]

Appendix B

Background on Convex Sets

This appendix is devoted to Hahn–Banach theorem and its consequences to the theory of convex functions.

B.1 The Hahn–Banach Extension Theorem

Throughout, E will denote a *real* linear space.

A functional $p\colon E \to \mathbb{R}$ is *subadditive* if $p(\mathbf{x}+\mathbf{y}) \leq p(\mathbf{x}) + p(\mathbf{y})$ for all $\mathbf{x}, \mathbf{y} \in E$; p is *positively homogeneous* if $p(\lambda \mathbf{x}) = \lambda p(\mathbf{x})$ for each $\lambda \geq 0$ and each $\mathbf{x}$ in E; p is *sublinear* if it has both the above properties. A sublinear functional p is a seminorm if $p(\lambda \mathbf{x}) = |\lambda| p(\mathbf{x})$ for all λ. Finally, a seminorm p is a norm if

$$p(\mathbf{x}) = 0 \implies \mathbf{x} = 0.$$

If p is a sublinear functional, then $p(\mathbf{0}) = 0$ and $-p(-\mathbf{x}) \leq p(\mathbf{x})$. If p is a seminorm, then $p(\mathbf{x}) \geq 0$ for all $\mathbf{x}$ in E and $\{\mathbf{x} : p(\mathbf{x}) = 0\}$ is a linear subspace of E.

B.1.1 Theorem (The Hahn–Banach theorem) *Let p be a sublinear functional on E, let E_0 be a linear subspace of E, and let $f_0\colon E_0 \to \mathbb{R}$ be a linear functional dominated by p, that is, $f_0(\mathbf{x}) \leq p(\mathbf{x})$ for all $\mathbf{x} \in E_0$. Then f_0 has a linear extension f to E which is also dominated by p.*

Proof. We consider the set $\mathcal{P}$ of all pairs (h, H), where H is a linear subspace of E that contains E_0 and $h\colon H \to \mathbb{R}$ is a linear functional dominated by p that extends f_0. $\mathcal{P}$ is nonempty (as $(f_0, E_0) \in \mathcal{P}$). One can easily prove that $\mathcal{P}$ is inductively ordered with respect to the order relation

$$(h, H) \prec (h', H') \iff H \subset H' \text{ and } h'|_H = h,$$

so that by Zorn's lemma we infer that $\mathcal{P}$ contains a maximal element (g, C). It remains to prove that $G = E$. If $G \neq E$, then we can choose an element $\mathbf{z} \in E \backslash G$ and denote by G' the set of all elements of the form $\mathbf{x} + \lambda \mathbf{z}$, with

C. P. Niculescu and L.-E. Persson, *Convex Functions and Their Applications*, CMS Books in Mathematics, https://doi.org/10.1007/978-3-319-78337-6

$\mathbf{x} \in G$ and $\lambda \in \mathbb{R}$. Clearly, G' is a linear space that contains G strictly and the formula

$$g'(\mathbf{x} + \lambda \mathbf{z}) = g(\mathbf{x}) + \alpha\lambda$$

defines (for every $\alpha \in \mathbb{R}$) a linear functional on G' that extends g. We shall show that α can be chosen so that g' is dominated by p (a fact that contradicts the maximality of (g, G)). In fact, g' is dominated by p if

$$g(\mathbf{x}) + \alpha\lambda \leq p(\mathbf{x} + \lambda \mathbf{z})$$

for every $\mathbf{x} \in G$ and every $\lambda \in \mathbb{R}$. If $\lambda \geq 0$, this means

$$g(\mathbf{x}) + \alpha \leq p(\mathbf{x} + \mathbf{z}) \quad \text{for every } \mathbf{x} \in G.$$

If $\lambda < 0$, we get (after simplification by $-\lambda$),

$$g(\mathbf{x}) - \alpha \leq p(\mathbf{x} - \mathbf{z}) \quad \text{for every } \mathbf{x} \in G.$$

Therefore, we have to choose α such that

$$g(\mathbf{u}) - p(\mathbf{u} - \mathbf{z}) \leq \alpha \leq p(\mathbf{v} + \mathbf{z}) - g(\mathbf{v})$$

for every $\mathbf{u}, \mathbf{v} \in G$. This choice is possible because

$$g(\mathbf{u}) + g(\mathbf{v}) = g(\mathbf{u} + \mathbf{v}) \leq p(\mathbf{u} + \mathbf{v}) \leq p(\mathbf{u} - \mathbf{z}) + p(\mathbf{v} + \mathbf{z})$$

for all $\mathbf{u}, \mathbf{v} \in G$, which yields

$$\sup_{\mathbf{u} \in G} (g(\mathbf{u}) - p(\mathbf{u} - \mathbf{z})) \leq \inf_{\mathbf{v} \in G} (p(\mathbf{v} + \mathbf{z}) - g(\mathbf{v})).$$

The proof is now complete. ■

B.1.2 Corollary *If p is a sublinear functional on a real linear space E, then for every element $\mathbf{x}_0 \in E$ there exists a linear functional $f \colon E \to \mathbb{R}$ such that $f(\mathbf{x}_0) = p(\mathbf{x}_0)$ and $f(\mathbf{x}) \leq p(\mathbf{x})$ for all $\mathbf{x}$ in E.*

Proof. Take $E_0 = \{\lambda \mathbf{x}_0 : \lambda \in \mathbb{R}\}$ and $f_0(\lambda \mathbf{x}_0) = \lambda p(\mathbf{x}_0)$ in Theorem B.1.1. ■

The continuity of a linear functional on a topological linear space E means that it is bounded in a neighborhood of the origin. We shall denote by E^* the *dual space* of E that is, the space of all continuous linear functionals on E. The bilinear map

$$\langle \cdot, \cdot \rangle : E \times E^* \to \mathbb{R}, \quad \langle \mathbf{x}, x^* \rangle = x^*(\mathbf{x})$$

is called the *duality map* between E and E^*.

In the context of normed linear spaces, the remark above allows us to define the norm of a continuous linear functional $f \colon E \to \mathbb{R}$ by the formula

$$\|f\| = \sup_{\|\mathbf{x}\| \leq 1} |f(\mathbf{x})|.$$

With respect to this norm, the dual space of a normed linear space is always complete.

It is worth noting the following variant of Theorem B.1.1 in the context of real normed linear spaces:

B.1.3 Theorem (The Hahn–Banach theorem) *Let E_0 be a linear subspace of the normed linear space E, and let $f_0 \colon E_0 \to \mathbb{R}$ be a continuous linear functional. Then f_0 has a continuous linear extension f to E, with $\|f\| = \|f_0\|$.*

B.1.4 Corollary *If E is a normed linear space, then for each $\mathbf{x}_0 \in E$ with $x_0 \neq 0$ there exists a continuous linear functional $f \colon E \to \mathbb{R}$ such that $f(\mathbf{x}_0) = \|\mathbf{x}_0\|$ and $\|f\| = 1$.*

B.1.5 Corollary *If E is a normed linear space and $\mathbf{x}$ is an element of E such that $f(\mathbf{x}) = 0$ for all f in the dual space of E, then $\mathbf{x} = 0$.*

The *weak topology* on E is the locally convex topology associated to the family of seminorms

$$p_F(\mathbf{x}) = \sup\{|f(\mathbf{x})| : f \in F\},$$

where F runs over all nonempty finite subsets of E^*. A sequence $(\mathbf{x}_n)_n$ converges to $\mathbf{x}$ in the weak topology (abbreviated, $\mathbf{x}_n \xrightarrow{w} \mathbf{x}$) if and only if $f(\mathbf{x}_n) \to f(\mathbf{x})$ for every $f \in E^*$. When $E = \mathbb{R}^N$ this is the coordinatewise convergence and agrees with the norm convergence. In general, the norm function is only *weakly lower semicontinuous*, that is,

$$\mathbf{x}_n \xrightarrow{w} \mathbf{x} \quad \Longrightarrow \quad \|\mathbf{x}\| \leq \liminf_{n\to\infty} \|\mathbf{x}_n\|.$$

By Corollary B.1.5 it follows that E^* *separates* E in the sense that

$$\mathbf{x}, \mathbf{y} \in E \quad \text{and} \quad f(\mathbf{x}) = f(\mathbf{y}) \quad \text{for all } f \in E^* \Longrightarrow \mathbf{x} = \mathbf{y}.$$

As a consequence we infer that the weak topology is separated (equivalently, Hausdorff).

For E^* we can speak of the normed topology, of the weak topology (associated to $E^{**} = (E^*)^*$), and also of the *weak-star topology*, which is associated to the family of seminorms p_F defined as above, with the difference that F runs over all nonempty finite subsets of E. The weak-star topology on E^* is separated.

A net $(f_i)_{i \in I}$ (over some directed set I) converges to f in the weak-star topology (abbreviated, $f_i \xrightarrow{w^*} f$) if and only if $f_i(\mathbf{x}) \to f(\mathbf{x})$ for all $\mathbf{x} \in E$.

B.1.6 Theorem (The Banach–Alaoglu theorem) *If E is a normed linear space, then the closed unit ball of its dual space is compact in the weak-star topology. Consequently, each net of points of this ball has a converging subnet.*

See [128, p. 47] for details.

When E is a separable normed linear space, the closed unit ball of E^* is also a metrizable space in the weak-star topology (and in this case dealing with sequences suffices as well). We come to the separability situation very often, by replacing E with a subspace generated by a suitable sequence of elements.

B.1.7 Remark *According to the Banach–Alaoglu theorem, if E is a normed linear space, then each weak-star closed subset of the closed unit ball of the dual of E is weak-star compact. This is a big source of compact convex sets in mathematics. For example, so is the set* Prob(X), *of all Borel probability measures on a compact Hausdorff space X. These are the regular σ-additive measures μ on the Borel subsets of X with $\mu(X) = 1$. The Riesz–Kakutani representation theorem (see [214, p. 177]) allows us to identify* Prob(X) *with the following weak-star closed subset of norm-1 functionals of $C(X)^*$:*

$$K = \{L : L \in C(X)^*,\ L(1) = 1 = \|L\|\}.$$

Notice that K consists of positive functionals, *that is,*

$$f \in C(X), \quad f \geq 0 \text{ implies } L(f) \geq 0.$$

In fact, if the range of f is included in $[0, 2r]$, then $\|f - r\| \leq r$, so that $r \geq |L(f - r)| = |L(f) - r|$, that is, $L(f) \in [0, 2r]$.

Corollary B.1.4 yields an important canonical embedding of each normed linear space E into its *second dual* $E^{**} = (E^*)^*$:

$$J_E : E \to E^{**}, \quad J_E(\mathbf{x})(x^*) = x^*(\mathbf{x}).$$

One can easily show that J_E is a linear isometry.

A Banach space E is said to be *reflexive* if J_E is onto (that is, if E is isometric with its second dual through J_E). Besides the finite-dimensional Banach spaces, other examples of reflexive Banach spaces are Hilbert spaces and the spaces $L^p(\mu)$ for $1 < p < \infty$. One can easily prove the following permanence properties:

(R1) Every closed subspace of a reflexive space is reflexive.

(R2) The dual of a reflexive space is also a reflexive space.

(R3) Reflexivity preserves under renorming by an equivalent norm.

Property (R3) follows from the following characterization of reflexivity:

B.1.8 Theorem (The Eberlein–Šmulyan theorem) *A Banach space E is reflexive if and only if every bounded sequence of elements of E admits a weakly converging subsequence.*

The *necessity* part is a consequence of the Banach–Alaoglu theorem (Theorem B.1.6). In fact, we may restrict ourselves to the case where E is also separable. The *sufficiency* part follows from the remark that J_E maps the closed unit ball of E into a weak-star dense (and also weak-star closed) subset of the closed unit ball of E^{**}. Full details are available in books such as those by H. W. Alt [12], J. B. Conway [115] or M. M. Day [128].

B.1.9 Corollary *Every nonempty closed and convex subset of a reflexive space is proximinal.*

The proof follows easily from the Eberlein–Šmulyan theorem and the weak lower semicontinuity of the norm function.

B.2 Separation of Convex Sets

The notion of a hyperplane in a real linear space E was introduced in Section 3.3 as the translate of the kernel of a nonzero linear functional. It can be equally defined as a maximal proper affine subset. In fact, if $h\colon E \to \mathbb{R}$ is a nonzero linear functional, we may choose a $\mathbf{v} \in E$ with $h(\mathbf{v}) = 1$. Then all $\mathbf{x} \in E$ can be represented as

$$\mathbf{x} = (\mathbf{x} - h(\mathbf{x})\mathbf{v}) + h(\mathbf{x})\mathbf{v},$$

where $\mathbf{x} - h(\mathbf{x})\mathbf{v} \in \ker h$. This shows that $\ker h$ is a linear space of codimension 1, and thus all its translates are maximal proper affine subsets.

Conversely, if H is a maximal proper affine set in E and $\mathbf{x}_0 \in H$, then $-\mathbf{x}_0 + H$ is a linear subspace (necessarily of codimension 1). Hence there exists a vector $\mathbf{v} \neq \mathbf{0}$ such that E is the direct sum of $-\mathbf{x}_0 + H$ and $\mathbb{R}\mathbf{v}$, that is, all $\mathbf{x} \in E$ can be uniquely represented as

$$\mathbf{x} = (-\mathbf{x}_0 + \mathbf{y}) + \lambda\mathbf{v}$$

for suitable $\mathbf{y} \in H$ and $\lambda \in \mathbb{R}$. The formula $h(\mathbf{x}) = \lambda$ defines a linear functional h such that $h(\mathbf{v}) = 1$ and $h(\mathbf{x}) = 0$ if and only if $\mathbf{x} \in -\mathbf{x}_0 + H$. Consequently,

$$H = \{\mathbf{x} : h(\mathbf{x}) = h(\mathbf{x}_0)\}.$$

Suppose now that E is a Hausdorff linear topological space. Then the discussion above shows that the closed hyperplanes H in E coincide with the constancy sets of nonzero continuous and linear functionals. In fact, it suffices to consider the case where H is a closed subspace of codimension 1. In that case E/H is 1-dimensional and thus it is algebraically and topologically isomorphic to $\mathbb{R}$. By composing such an isomorphism with the canonical projection from E onto E/H we obtain a continuous linear functional h for which $H = \ker h$.

To each hyperplane $\{\mathbf{x} : h(\mathbf{x}) = \lambda\}$ we can attach two *half-spaces*,

$$\{\mathbf{x} : h(\mathbf{x}) \leq \lambda\} \quad \text{and} \quad \{\mathbf{x} : h(\mathbf{x}) \geq \lambda\}.$$

We say that two sets A and B are *separated* by a hyperplane H if they are contained in different half-spaces. The separation is *strict* if at least one of the two sets does not intersect H.

A basic result concerning the separability by hyperplanes is as follows:

B.2.1 Theorem (Mazur's theorem) *Let K be a convex set with nonempty interior in a real linear topological Hausdorff space E and let A be an affine subset which contains no interior point of K. Then there exists a closed hyperplane H such that $H \supset A$ and $H \cap K = \emptyset$.*

In other words, there exists a continuous linear functional $h\colon E \to \mathbb{R}$ and a number $\alpha \in \mathbb{R}$ such that $h(\mathbf{x}) = \alpha$ if $\mathbf{x} \in A$ and $h(\mathbf{x}) < \alpha$ if $\mathbf{x} \in K$.

Proof. We may assume that K is a convex neighborhood of the origin since otherwise we choose an interior point $\mathbf{x}_0$ in K and replace K and A by $K - \mathbf{x}_0$ and $A - \mathbf{x}_0$, respectively. Notice that translations are isomorphisms, so they preserve the nature of K and A. Denote by E_0 the linear span of A. Then A is a hyperplane in E_0, which yields a linear functional $f_0\colon E_0 \to \mathbb{R}$ such that

$$A = \{\mathbf{x} \in E_0 : f_0(\mathbf{x}) = 1\}.$$

The Minkowski functional of K,

$$p_K(\mathbf{x}) = \inf\{\lambda > 0 : \mathbf{x} \in \lambda K\},$$

is sublinear and $\{\mathbf{x} : p_k(\mathbf{x}) < 1\}$ coincides with the interior of K. In fact, if x is an interior point of K, then $\mathbf{x} + V \subset K$ for a convex neighborhood V of the origin. Due to the continuity of the map $\lambda \to \lambda\mathbf{x}$, there must exist a $\lambda \in (0,1)$ with $\lambda\mathbf{x} \in V$. Then $\mathbf{x} + \lambda\mathbf{x} \in \mathbf{x} + V \subset K$, so that $p_K(\mathbf{x}) < 1$. Conversely, if $p_K(\mathbf{x}) < 1$, then $\mathbf{x} \in \lambda K$ for some $\lambda \in (0,1)$, which yields

$$\mathbf{x} \in \mathbf{x} + (1-\lambda)K \subset \lambda K + (1-\lambda)K = K.$$

Notice that $(1-\lambda)K$ is a neighborhood of the origin. Since A contains no interior point of K it follows that $f_0(\mathbf{x}) = 1 \le p_K(\mathbf{x})$ for all $\mathbf{x} \in A$. If $\mathbf{x} \in A$ and $\lambda > 0$, then $f_0(\lambda\mathbf{x}) \le p_K(\lambda\mathbf{x})$, while for $\lambda \le 0$ we have $f_0(\lambda\mathbf{x}) \le 0 \le p_K(\lambda\mathbf{x})$. Consequently, $f_0 \le p_K$ on E_0. By Theorem B.1.1, f_0 has a linear extension f to E such that $f \le p_K$. Put $H = \{\mathbf{x}; f(\mathbf{x}) = 1\}$. Then H is a hyperplane. Since $|f(\mathbf{x})| \le p_K(\mathbf{x}) < 1$ for x in K, it follows that f is bounded on a neighborhood of 0 and thus continuous. Therefore H is a closed hyperplane and it is clear that $H \supset A$ and $H \cap K = \emptyset$. ■

B.2.2 Corollary *If U is a nonempty open convex set and F is a linear subspace such that $F \cap U = \emptyset$, then there exists a continuous linear functional f such that $f(\mathbf{x}) = 0$ if $\mathbf{x} \in F$ and $f(\mathbf{x}) > 0$ if $\mathbf{x} \in U$.*

In order to prove a strict separation result we need the following lemma of independent interest:

B.2.3 Lemma *Suppose that K_1 and K_2 are two nonempty convex sets in a real linear topological space E with $K_1 \cap K_2 = \emptyset$. If one of them is open, then there exists a closed hyperplane separating K_1 from K_2.*

Proof. If K_1 is open, then the set

$$U = K_1 - K_2 = \bigcup_{k_2 \in K_2} (K_1 - k_2)$$

is open. Since K_1 and K_2 are convex, U is convex too. Moreover, $\mathbf{0} \notin U$ since $K_1 \cap K_2 = \emptyset$. By Corollary B.2.2 there exists a continuous linear functional f

such that $f(\mathbf{x}) > 0$ on U. Therefore $f(\mathbf{x}) > f(\mathbf{y})$ for all $\mathbf{x} \in K_1$ and all $\mathbf{y} \in K_2$. Letting

$$\alpha = \inf\{f(\mathbf{x}) : \mathbf{x} \in K_1\},$$

one can show immediately that K_1 and K_2 are separated by the closed hyperplane $H = \{\mathbf{x} : f(\mathbf{x}) = \alpha\}$. The proof is complete. ■

B.2.4 Theorem (Strong separation theorem) *Let K_1 and K_2 be two nonempty convex sets in a real locally convex Hausdorff space E such that $K_1 \cap K_2 = \emptyset$. If K_1 is compact and K_2 is closed, then there exists a closed hyperplane strictly separating K_1 from K_2.*

Particularly, if K is a closed convex set in a locally convex space E and $\mathbf{x} \in E$ is not in K, then there exists a functional $f \in E^*$ such that

$$f(\mathbf{x}) > \sup\{f(\mathbf{y}) : \mathbf{y} \in K\}.$$

Proof. By our hypothesis, there exists an open convex neighborhood W of the origin such that $(K_1+W) \cap (K_2+W) = \emptyset$. This follows directly by using *reductio ad absurdum.* Since the sets $K_1 + W$ and $K_2 + W$ are convex and open, from Lemma B.2.3 we infer the existence of a separating hyperplane H. A moment's reflection shows that H separates strictly K_1 from K_2. This completes the proof. ■

The *closed convex hull* of a subset A of a locally convex space E is the smallest closed convex set $\overline{\operatorname{conv}}(A)$ containing A (that is, the intersection of all closed convex sets containing A). From Theorem B.2.4 we can infer the following result on the support of closed convex sets:

B.2.5 Corollary *If A is a nonempty subset of a real locally convex Hausdorff space E, then the closed convex hull $\overline{\operatorname{conv}}(A)$ is the intersection of all the closed half-spaces containing A. Equivalently,*

$$\overline{\operatorname{conv}}(A) = \bigcap_{f \in E'} \{\mathbf{x} : f(\mathbf{x}) \le \sup_{\mathbf{y} \in A} f(\mathbf{y})\}.$$

This corollary implies:

B.2.6 Corollary *In a real locally convex Hausdorff space E, the closed convex sets and the weakly closed convex sets are the same.*

Finally it is worth mentioning a non-topological version of the separation results above, which is important in optimization theory.

Given a set A in a linear space E, a point a of A is said to be a *core point* if for every $\mathbf{v} \in E$, $\mathbf{v} \ne \mathbf{a}$, there exists an $\varepsilon > 0$ such that $\mathbf{a} + \delta\mathbf{v} \in A$ for every δ with $|\delta| < \varepsilon$.

B.2.7 Theorem *Let K and M be two nonempty convex sets in a real linear space E. If K contains core points and M contains no core point of K, then K and M can be separated by a hyperplane.*

The details can be easily filled out by adapting the argument given in the topological case.

B.3 The Krein–Milman Theorem

Theorem 2.3.5 showed that every compact convex set in $\mathbb{R}^N$ is the convex hull of its extreme points. This result can be extended to a very general setting.

B.3.1 Theorem *Let E be a locally convex Hausdorff space and K be a nonempty compact convex subset of E. If U is an open convex subset of K such that* $\operatorname{ext} K \subset U$, *then $U = K$.*

Proof. Suppose that $U \neq K$ and consider the family $\mathcal{U}$ of all open convex sets in K which are not equal to K. By Zorn's lemma, each set $U \in \mathcal{U}$ is contained in a maximal element V of $\mathcal{U}$. For each $\mathbf{x} \in K$ and $t \in [0,1]$, let $\varphi_{\mathbf{x},t} \colon K \to K$ be the continuous map defined by $\varphi_{\mathbf{x},t}(\mathbf{y}) = t\mathbf{y} + (1-t)\mathbf{x}$. Assuming $\mathbf{x} \in V$ and $t \in [0,1)$, we shall show that $\varphi_{\mathbf{x},t}^{-1}(V)$ is an open convex set which contains V properly, hence $\varphi_{x,t}^{-1}(V) = K$. In fact, this is clear when $t = 0$. If $t \in (0,1)$, then $\varphi_{\mathbf{x},t}$ is a homeomorphism and $\varphi_{\mathbf{x},t}^{-1}(V)$ is an open convex set in K. Moreover,

$$\varphi_{\mathbf{x},t}(\overline{V}) \subset V,$$

which yields $\overline{V} \subset \varphi_{\mathbf{x},t}^{-1}(V)$, hence $\varphi_{\mathbf{x},t}^{-1}(V) = K$ by the maximality of V. Therefore $\varphi_{\mathbf{x},t}(K) \subset V$. For any open convex set W in K the intersection $V \cap W$ is also open and convex, and the maximality of V yields that either $V \cup W = V$ or $V \cup W = K$. In conclusion $K \backslash V$ is precisely a singleton $\{\mathbf{e}\}$. But such a point is necessarily an extreme point of K, which is a contradiction. ∎

B.3.2 Theorem *Theorem* (Krein–Milman theorem) *Let K be a nonempty compact convex subset of a locally convex Hausdorff space E. Then K is the closed convex hull of* $\operatorname{ext} K$.

Proof. By Theorem B.2.4, the set $L = \overline{\operatorname{conv}}(\operatorname{ext} K)$ is the intersection of all open convex sets containing L. If U is an open subset of K and $U \supset L$, then $U \supset \operatorname{ext} K$. Hence $U = K$ and $L = K$. ∎

The above proof of the Krein–Milman theorem yields the existence of extreme points as a consequence of the formula $K = \overline{\operatorname{conv}}(\operatorname{ext} K)$. However, this fact can be checked directly. Call a subset A of K *extremal* if it is closed, nonempty and verifies the following property:

$$\mathbf{x}, \mathbf{y} \in K \ \text{ and } \ (1-\lambda)\mathbf{x} + \lambda\mathbf{y} \in A \ \text{ for some } \lambda \in (0,1) \implies \mathbf{x}, \mathbf{y} \in A.$$

By Zorn's lemma we can choose a minimal extremal subset, say S. We show that S is a singleton (which yields an extreme point of K). In fact, if S contains more than one point, the separation Theorem B.2.4 proves the existence of a functional $f \in E^*$ which is not constant on S. But in this case the set

$$S_0 = \{\mathbf{x} \in S : f(\mathbf{x}) = \sup_{\mathbf{y} \in S} f(\mathbf{y})\}$$

will contradict the minimality of S. Now the formula $K = \overline{\mathrm{conv}}(\operatorname{ext} K)$ can easily be proved by noticing that the inclusion $\overline{\mathrm{conv}}(\operatorname{ext} K) \subset K$ cannot be strict.

B.3.3 Remark *The Krein–Milman theorem provides a unifying tool for many important results in mathematics. We mention here Bernstein's integral representation (1.13) of the completely monotonic functions, Bochner's integral representation of positive definite functions, and Lyapunov's convexity theorem of the range of nonatomic countably additive vector measure $\mu : \Sigma \to \mathbb{R}^N$. Details are available in the book of B. Simon [450].*

Theorem B.3.1 also yields the following result:

B.3.4 Corollary (Bauer maximum principle) *Suppose that K is a non-empty compact convex set as in Theorem B.3.1. Then every proper upper semicontinuous proper convex function $f: K \to [-\infty, \infty)$ attains its supremum at an extreme point.*

Proof. Since f is upper semicontinuous, the family of sets

$$U_n = \{\mathbf{x} \in K : f(\mathbf{x}) < n\}, \quad n \in \mathbb{N},$$

provides an open covering of K, so $K = U_n$ for some n, which shows that f is bounded above. Put $M = \sup\{f(\mathbf{x}) : \mathbf{x} \in K\}$. If f does not attain its supremum at a point of $\operatorname{ext} K$, then $U = \{\mathbf{x} \in K : f(\mathbf{x}) < M\}$ is an open convex set containing $\operatorname{ext} K$. By Theorem B.3.1 we conclude that $U = K$, which is a contradiction. ∎

It is interesting to note the following converse to Theorem B.3.1:

B.3.5 Theorem (D. P. Milman) *Suppose that K is a compact convex set (in a locally convex Hausdorff space E) and C is a subset of K such that K is the closed convex hull of C. Then the extreme points of K are contained in the closure of C.*

Coming back to Theorem 2.3.5, the fact that every point $\mathbf{x}$ of a compact convex set K in $\mathbb{R}^N$ is a convex combination of extreme points of K,

$$\mathbf{x} = \sum_{k=1}^{m} \lambda_k \mathbf{x}_k,$$

can be reformulated as an integral representation,

$$f(\mathbf{x}) = \sum_{k=1}^{m} \lambda_k f(\mathbf{x}_k) = \int_{\operatorname{ext} K} f \, \mathrm{d}\mu \tag{B.1}$$

for all $f \in (\mathbb{R}^n)^*$. Here $\mu = \sum_{k=1}^{m} \lambda_k \delta_{\mathbf{x}_k}$ is a convex combination of Dirac measures $\delta_{\mathbf{x}_k}$ and thus μ itself is a Borel probability measure on $\operatorname{ext} K$.

The integral representation (B.1) can be extended to all Borel probability measures μ on a compact convex set K (in a locally convex Hausdorff space E). We shall need some definitions.

Given a Borel probability measure μ on K, and a Borel subset $S \subset K$, we say that μ is *concentrated* on S if $\mu(K \backslash S) = 0$. For example, a Dirac measure $\delta_{\mathbf{x}}$ is concentrated on $\mathbf{x}$.

A point $\mathbf{x} \in K$ is said to be the *barycenter* of μ provided that

$$f(\mathbf{x}) = \int_K f \, \mathrm{d}\mu \quad \text{for all } f \in E^*.$$

Since the functionals separate the points of E, the point $\mathbf{x}$ is uniquely determined by μ. With this preparation, we can reformulate the Krein–Milman theorem as follows:

B.3.6 Theorem *Every point of a compact convex subset K (of a locally convex Hausdorff space E) is the barycenter of a Borel probability measure on K that is supported by the closure of the extreme points of K.*

H. Bauer pointed out that the extremal points of K are precisely the points $\mathbf{x} \in K$ for which the only Borel probability measure μ which admits $\mathbf{x}$ as a barycenter is the Dirac measure $\delta_{\mathbf{x}}$. See [398, p. 6]. This fact together with Theorem B.3.4 yields D. P. Milman's aforementioned converse of the Krein–Milman theorem. For an alternative argument see [128, pp. 103–104].

Theorem B.3.4 led G. Choquet [106] to his theory on integral representation for elements of a closed convex cone.

A Banach space E is said to have the Krein–Milman property (abbreviated, KMP) if every closed convex subset of E is the closed convex hull of its extreme points. Reflexive Banach spaces (like L^p-spaces with $1 < p < \infty$) and separable conjugate Banach spaces (like ℓ^1) have KMP. However, the general structure of Banach spaces with KMP is unknown.

Appendix C

Elementary Symmetric Functions

The *elementary symmetric functions* of n variables are defined by

$$\begin{aligned}
e_0(x_1, x_2, \ldots, x_n) &= 1 \\
e_1(x_1, x_2, \ldots, x_n) &= x_1 + x_2 + \cdots + x_n \\
e_2(x_1, x_2, \ldots, x_n) &= \sum_{1 \leq i < j \leq n} x_i x_j \\
&\vdots \\
e_n(x_1, x_2, \ldots, x_n) &= x_1 x_2 \cdots x_n.
\end{aligned}$$

The different e_k being of different degrees, they are not comparable. However, they are connected by nonlinear inequalities. To state them, it is more convenient to consider their averages,

$$E_k(x_1, x_2, \ldots, x_n) = e_k(x_1, x_2, \ldots, x_n) / \binom{n}{k}$$

and to write E_k for $E_k(x_1, x_2, \ldots, x_n)$ in order to avoid excessively long formulas.

Sections C.1 and C.2 are devoted to Newton's inequalities. While the original results due to I. Newton [350] and C. Maclaurin [292] can be proved using just Rolle's theorem, the inequalities of higher order are generated by Sylvester's discriminant families (whose existence implies the semi-algebraic character of the set of all real polynomials with all roots real). Several other inequalities discovered independently by H. F. Bohnenblust, M. Marcus and L. Lopes make the objective of Section C.3. The highlights of the very recent theory of symmetric polynomial majorization are mentioned in Section C.4.

Last, but not least, let us mention that the theory of elementary symmetric polynomials is generalized by that of hyperbolic polynomials (briefly exposed in Section 4.6).

C. P. Niculescu and L.-E. Persson, *Convex Functions and Their Applications*,
CMS Books in Mathematics, https://doi.org/10.1007/978-3-319-78337-6

C.1 Newton's Inequalities

The simplest set of inequalities relating the elementary symmetric functions was discovered by I. Newton [350] and C. Maclaurin [292]:

C.1.1 Theorem *Let $\mathcal{F}$ be an n-tuple of positive numbers. Then:*

$$E_k^2(\mathcal{F}) > E_{k-1}(\mathcal{F}) \cdot E_{k+1}(\mathcal{F}), \quad 1 \leq k \leq n-1 \tag{N}$$

and

$$E_1(\mathcal{F}) > E_2^{1/2}(\mathcal{F}) > \cdots > E_n^{1/n}(\mathcal{F}) \tag{M}$$

unless all entries of $\mathcal{F}$ coincide.

Actually Newton's inequalities (N) work for n-tuples of real, not necessarily positive, elements. An analytic proof along Maclaurin's ideas will be presented below. In Section C.2 we shall indicate an alternative argument, based on mathematical induction, which yields more Newton type inequalities, in an interpolative scheme.

The inequalities (M) can be deduced from (N) since

$$(E_0E_2)(E_1E_3)^2(E_2E_4)^3 \cdots (E_{k-1}E_{k+1})^k < E_1^2E_2^4E_3^6 \cdots E_k^{2k}$$

gives $E_{k+1}^k < E_k^{k+1}$ or, equivalently,

$$E_k^{1/k} > E_{k+1}^{1/(k+1)}.$$

Among the inequalities noticed above, the most notable is of course the *AM-GM* inequality:

$$\left(\frac{x_1 + x_2 + \cdots + x_n}{n}\right)^n \geq x_1x_2 \cdots x_n$$

for all $x_1, x_2, \ldots, x_n \geq 0$. A hundred years after C. Maclaurin, A.-L. Cauchy [99] gave his beautiful inductive argument. Notice that the *AM-GM* inequality was known to Euclid [149] in the special case where $n = 2$.

C.1.2 Remark *Newton's inequalities were intended to solve the problem of counting the number of imaginary roots of an algebraic equation. In Chapter 2 of Part 2 of* Arithmetica Universalis, *entitled* De Forma Æquationis, *Newton made (without proof) the following statement:* Given an equation with real coefficients,

$$a_0x^n + a_1x^{n-1} + \cdots + a_n = 0 \quad (a_0 \neq 0),$$

the number of its imaginary roots cannot be less than the number of changes of sign that occur in the sequence

$$a_0^2, \left(\frac{a_1}{\binom{n}{1}}\right)^2 - \frac{a_2}{\binom{n}{2}} \cdot \frac{a_0}{\binom{n}{0}}, \ldots, \left(\frac{a_{n-1}}{\binom{n}{n-1}}\right)^2 - \frac{a_n}{\binom{n}{n}} \cdot \frac{a_{n-2}}{\binom{n}{n-2}}, a_n^2.$$

Accordingly, if all the roots are real, then all the entries in the above sequence must be positive (a fact which yields Newton's inequalities).

Trying to understand which was Newton's argument, C. Maclaurin [292] gave a direct proof of the inequalities (N) and (M), but the Newton counting problem remained open until 1865, when J. Sylvester [468, 469] succeeded in proving a remarkable general result.

Quite unexpectedly, it is the real algebraic geometry (not analysis) which gives us the best understanding of Newton's inequalities. The basic fact (discovered by J. Sylvester) concerns the *semi-algebraic character* of the set of all real polynomials with all roots real.

C.1.3 Theorem (J. Sylvester) *For each natural number $n \geq 2$ there exists a set of at most $n-1$ polynomials with integer coefficients,*

$$R_{n,1}(x_1, \ldots, x_n), \ldots, R_{n,k(n)}(x_1, \ldots, x_n), \tag{R_n}$$

such that the monic real polynomials of order n,

$$P(x) = x^n + a_1 x^{n-1} + \cdots + a_n,$$

which have only real roots are precisely those for which

$$R_{n,1}(a_1, \ldots, a_n) \geq 0, \ldots, R_{n,k(n)}(a_1, \ldots, a_n) \geq 0.$$

The above result can be seen as a generalization of the well-known fact that *the roots of a quadratic polynomial $x^2 + a_1 x + a_2$ are real if and only if its discriminant*

$$D_2(1, a_1, a_2) = a_1^2 - 4a_2 \tag{D_2}$$

is positive.

Theorem C.1.3 is built on Sturm's method of counting real roots, taking into account that only the leading coefficients enter into play. It turns out that they are nothing but the principal subresultant coefficients (with convenient signs added), which are determinants extracted from the Sylvester matrix.

A set $(R_{n,k})_k^{k(n)}$ as in Theorem C.1.3 will be called a *Sylvester family* (of order n).

In Sylvester's approach, $R_{n,1}(a_1, \ldots, a_n)$ equals the *discriminant* D_n of the polynomial $P(x) = x^n + a_1 x^{n-1} + \cdots + a_n$, that is,

$$D_n = D_n(1, a_1, \ldots, a_n) = \prod_{1 \leq i < j \leq n} (x_i - x_j)^2,$$

where $x_1, \ldots, x_n$ are the roots of $P(x)$; D_n is a symmetric and homogeneous (of degree $n^2 - n$) polynomial in $\mathbb{Z}[x_1, \ldots, x_n]$. For details, see [41]. Unfortunately, at present no compact formula for D_n is known. According to [433], the number of nonzero coefficients in the expression for the discriminant increases rapidly with the degree; e.g., D_9 has 26095 terms.

For $n \in \{2, 3\}$ one can indicate Sylvester families consisting of just a single polynomial, the corresponding discriminant. An inspection of the argument given by L. Euler to solve in radicals the quartic equations allows us to write down a Sylvester family for $n = 4$. See the paper by C. P. Niculescu [352].

C.1.4 Remark *Given a Sylvester family for $n = N$, we can easily indicate such a family for each $n \in \{1, \ldots, N\}$; the trick is to replace a $P(x)$ of degree n by $x^{N-n}P(x)$, which is of degree N.*

Also, any Sylvester family $(R_{n,k})_{k=1}^{k(n)}$ (for some $n \geq 2$) allows us to decide which monic real polynomial $P(x) = x^n + a_1x^{n-1} + \cdots + a_n$ has only positive roots. A set of (necessary and) sufficient conditions consists of the following inequalities:

$$-a_1 \geq 0, \ldots, (-1)^n a_n \geq 0$$

and

$$R_{n,1}(a_1, \ldots, a_n) \geq 0, \ldots, R_{n,k(n)}(a_1, \ldots, a_n) \geq 0.$$

In fact, under the above circumstances, $x < 0$ yields $P(x) \neq 0$.

The proof of the Newton inequalities (N) presented in the book of G. H. Hardy, J. E. Littlewood and G. Pólya [209] follows Maclaurin's argument. The basic ingredient is the following lemma, a consequence of repeated application of Rolle's theorem, which we give here under the formulation of J. Sylvester [469]:

C.1.5 Lemma *If*

$$F(x, y) = c_0x^n + c_1x^{n-1}y + \cdots + c_ny^n$$

is a homogeneous function of the n-th degree in x and y, which has all its roots x/y real, then the same is true for all nonidentical 0 *equations*

$$\frac{\partial^{i+j} F}{\partial x^i \partial y^j} = 0,$$

obtained from it by partial differentiation with respect to x and y. Further, if $\mathcal{E} = 0$ is one of these equations, and it has a multiple root α, then α is also a root, of multiplicity one higher, of the equation from which $\mathcal{E}$ is derived by differentiation.

Any polynomial of the n-th degree, with real roots, can be represented as

$$E_0x^n - \binom{n}{1}E_1x^{n-1} + \binom{n}{2}E_1x^{n-2} - \cdots + (-1)^n E_n$$

and we shall apply Lemma C.1.5 to the associated homogeneous polynomial

$$F(x, y) = E_0x^n - \binom{n}{1}E_1x^{n-1}y + \binom{n}{2}E_1x^{n-2}y^2 - \cdots + (-1)^n E_ny^n.$$

Considering the case of the derivatives

$$\frac{\partial^{n-2}F}{\partial x^k \partial y^{n-2-k}} \quad (\text{for } k = 0, \dots, n-2),$$

we arrive at the fact that all the quadratic polynomials

$$E_{k-1}x^2 - 2E_k xy + E_{k+1}y^2$$

for $k = 0, \dots, n-2$ also have real roots. Consequently, the Newton inequalities express precisely this fact in the language of discriminants. That is why we shall refer to (N) as the *quadratic Newton inequalities.* An alternative argument for obtaining these inequalities is offered by R. P. Stanley [454], Theorem 2.

Stopping a step ahead, we get what S. Rosset [427] called the *cubic Newton inequalities*:

$$6E_k E_{k+1} E_{k+2} E_{k+3} + 3E_{k+1}^2 E_{k+2}^2 \geq 4E_k E_{k+2}^3 + E_k^2 E_{k+3}^2 + 4E_{k+1}^3 E_{k+3} \tag{N$_3$}$$

for $k = 0, \dots, n-3$. They are motivated by the well-known fact that a cubic real polynomial

$$x^3 + a_1 x^2 + a_2 x + a_3$$

has only real roots if and only if its discriminant

$$\begin{aligned} D_3 &= D_3(1, a_1, a_2, a_3) \\ &= 18a_1 a_2 a_3 + a_1^2 a_2^2 - 27a_3^2 - 4a_2^3 - 4a_1^3 a_3 \end{aligned}$$

is positive. Consequently, the equation

$$E_k x^3 - 3E_{k+1}x^2 y + 3E_{k+2}xy^2 - E_{k+3}y^3 = 0$$

has all its roots x/y real if and only if (N_3) holds.

S. Rosset [427] derived the inequalities (N_3) by an inductive argument and noticed that they are strictly stronger than (N). In fact, (N_3) can be rewritten as

$$4(E_{k+1}E_{k+3} - E_{k+2}^2)(E_k E_{k+2} - E_{k+1}^2) \geq (E_{k+1}E_{k+2} - E_k E_{k+3})^2$$

which yields (N).

As concerns the Newton inequalities (N_n) of order $n \geq 2$ (when applied to strings of $m \geq n$ elements), they consist of at most $n-1$ sets of relations, the first one being

$$D_n\left(1, (-1)^1 \binom{n}{1} \frac{E_{k+1}}{E_k}, (-1)^2 \binom{n}{2} \frac{E_{k+2}}{E_k}, \dots, (-1)^n \binom{n}{n} \frac{E_{k+n}}{E_k}\right) \geq 0$$

for $k \in \{0, \dots, m-n\}$.

Notice that each of these inequalities is homogeneous (for example, the last one consists of terms of weight $n^2 - n$) and the sum of all coefficients in the left-hand side is 0.

C.2 More Newton Inequalities

Our argument will yield a bit more, precisely the log-concavity of the functions $E_k\colon k \to E_k(\mathcal{F})$.

C.2.1 Theorem *Suppose that $\alpha, \beta \in \mathbb{R}_+$ and $j, k \in \mathbb{N}$ are numbers such that*

$$\alpha + \beta = 1 \quad \text{and} \quad j\alpha + k\beta \in \{0, \ldots, n\}.$$

Then

$$E_{j\alpha+k\beta}(\mathcal{F}) \geq E_j^{\alpha}(\mathcal{F}) \cdot E_k^{\beta}(\mathcal{F}),$$

for every n-tuple $\mathcal{F}$ of positive real numbers. Moreover, equality occurs if and only if all entries of $\mathcal{F}$ are equal.

The proof will be done by induction on the length of $\mathcal{F}$, after some preliminaries.

According to Rolle's theorem, if all roots of a polynomial $P \in \mathbb{R}[X]$ are real (respectively, real and distinct), then the same is true for its derivative P'. Given an n-tuple $\mathcal{F} = (x_1, \ldots, x_n)$, we shall attach to it the polynomial

$$P_{\mathcal{F}}(x) = (x - x_1) \cdots (x - x_n) = \sum_{k=0}^{n} (-1)^k \binom{n}{k} E_k(x_1, \ldots, x_n) x^{n-k}.$$

The $(n-1)$-tuple $\mathcal{F}' = \{y_1, \ldots, y_{n-1}\}$ consisting of all roots of the derivative of $P_{\mathcal{F}}(x)$ will be called the *derived n-tuple* of $\mathcal{F}$. Because

$$(x - y_1) \cdots (x - y_{n-1}) = \sum_{k=0}^{n-1} (-1)^k \binom{n-1}{k} E_k(y_1, \ldots, y_{n-1}) x^{n-k}$$

and

$$\begin{aligned}
(x - y_1) \cdots (x - y_{n-1}) &= \frac{1}{n} \cdot \frac{dP_{\mathcal{F}}}{dx}(x) \\
&= \sum_{k=0}^{n} (-1)^k \frac{n-k}{n} \binom{n}{k} E_k(x_1, \ldots, x_n) x^{n-k-1} \\
&= \sum_{k=0}^{n-1} (-1)^k \binom{n-1}{k} E_k(x_1, \ldots, x_n) x^{n-1-k}
\end{aligned}$$

we are led to the following result, which enables us to reduce the number of variables when dealing with symmetric functions.

C.2.2 Lemma $E_j(\mathcal{F}) = E_j(\mathcal{F}')$ *for every $j \in \{0, \ldots, |\mathcal{F}| - 1\}$.*

Another simple but useful fact is the following:

C.2.3 Lemma *Suppose that $\mathcal{F}$ is an n-tuple of real numbers and $0 \notin \mathcal{F}$. Put $\mathcal{F}^{-1} = \{1/a \mid a \in \mathcal{F}\}$. Then*

$$E_j(\mathcal{F}^{-1}) = E_{n-j}(\mathcal{F})/E_n(\mathcal{F})$$

for every $j \in \{0, \ldots, n\}$.

Proof of Theorem C.2.1. For $|\mathcal{F}| = 2$ we have to prove just one inequality, namely, $x_1 x_2 \le (x_1 + x_2)^2/4$, which is clearly valid for every $x_1, x_2 \in \mathbb{R}$; the equality occurs if and only if $x_1 = x_2$.

Suppose now that the assertion of Theorem B.2.1 holds for all k-tuples with $k \le n-1$. Let $\mathcal{F}$ be an n-tuple of positive numbers ($n \ge 3$), let $j, k \in \mathbb{N}$, and $\alpha, \beta \in \mathbb{R}_+ \backslash \{0\}$ be numbers such that

$$\alpha + \beta = 1 \quad \text{and} \quad j\alpha + k\beta \in \{0, \ldots, n\}.$$

According to Lemma C.2.2 (and our inductive hypothesis), we have

$$E_{j\alpha + k\beta}(\mathcal{F}) \ge E_j^{\alpha}(\mathcal{F}) \cdot E_k^{\beta}(\mathcal{F}),$$

except for the case where $j < k = n$ or $k < j = n$. Suppose, for example, that $j < k = n$; then necessarily $j\alpha + n\beta < n$. We have to show that

$$E_{j\alpha + n\beta}(\mathcal{F}) \ge E_j^{\alpha}(\mathcal{F}) \cdot E_n^{\beta}(\mathcal{F}).$$

If $0 \in \mathcal{F}$, then $E_n(\mathcal{F}) = 0$, and the inequality is clear; the equality occurs if and only if $E_{j\alpha+n\beta}(\mathcal{F}') = E_{j\alpha+n\beta}(\mathcal{F}) = 0$, that is (according to our inductive hypothesis), when all entries of $\mathcal{F}$ coincide.

If $0 \notin \mathcal{F}$, then by Lemma C.2.3 we have to prove that

$$E_{n-j\alpha-n\beta}(\mathcal{F}^{-1}) \ge E_{n-j}^{\alpha}(\mathcal{F}^{-1}),$$

or, equivalently (see Lemma C.2.2), $E_{n-j\alpha-n\beta}((\mathcal{F}^{-1})') \ge E_{n-j}^{\alpha}((\mathcal{F}^{-1})')$, which is true by our hypothesis. The proof is complete. ∎

Notice that the argument above covers Newton's inequalities even for n-tuples of real (not necessarily positive) elements.

The general problem of comparing monomials in $E_1, \ldots, E_n$ was completely solved by G. H. Hardy, J. E. Littlewood and G. Pólya in [209, Theorem 77, p. 64]:

C.2.4 Theorem *Let $\alpha_1, \ldots, \alpha_n, \beta_1, \ldots, \beta_n$ be positive numbers. Then*

$$E_1^{\alpha_1}(\mathcal{F}) \cdots E_n^{\alpha_n}(\mathcal{F}) \le E_1^{\beta_1}(\mathcal{F}) \cdots E_n^{\beta_n}(\mathcal{F})$$

for every n-tuple $\mathcal{F}$ of positive numbers if and only if

$$\alpha_m + 2\alpha_{m+1} + \cdots + (n-m+1)\alpha_n \ge \beta_m + 2\beta_{m+1} + \cdots + (n-m+1)\beta_n$$

for $1 \le m \le n$, with equality when $m = 1$.

C.3 Some Results of Bohnenblust, Marcus, and Lopes

The property of strict concavity of the geometric mean can be extended in the framework of elementary symmetric functions by asserting the strict concavity of all functions $e_r^{1/r}$ with $r \geq 2$. This fact, due independently to H. F. Bohnenblust and M. Marcus and L. Lopes [302], is a consequence of their subadditivity, since they are positively homogeneous. See Lemma 3.4.1.

C.3.1 Theorem *The sum of two n-tuples of strictly positive numbers $\mathcal{F} = \{x_1, \ldots, x_n\}$ and $\mathcal{G} = \{y_1, \ldots, y_n\}$ is defined by the formula $\mathcal{F} + \mathcal{G} = \{x_1 + y_1, \ldots, x_n + y_n\}$. Then*

$$e_r(\mathcal{F} + \mathcal{G})^{1/r} \geq e_r(\mathcal{F})^{1/r} + e_r(\mathcal{G})^{1/r}$$

for every $r = 1, \ldots, n$. The inequality is strict unless $r = 1$ or there exists $\lambda > 0$ such that $\mathcal{F} = \lambda\mathcal{G}$.

The proof combines a special case of Minkowski's inequality with the following lemma:

C.3.2 Lemma *Under the hypotheses of Theorem C.3.1, for $r = 1, \ldots, n$ and n-tuples of positive numbers not all zero, we have*

$$\frac{e_r(\mathcal{F} + \mathcal{G})}{e_{r-1}(\mathcal{F} + \mathcal{G})} \geq \frac{e_r(\mathcal{F})}{e_{r-1}(\mathcal{F})} + \frac{e_r(\mathcal{G})}{e_{r-1}(\mathcal{G})}.$$

The inequality is strict unless $r = 1$ or there exists a $\lambda > 0$ such that $\mathcal{F} = \lambda\mathcal{G}$.

Proof of Lemma C.3.2. For $r = 1$ the inequality is actually an equality. For $r = 2$, we have to look at the following identity:

$$\frac{e_2(\mathcal{F} + \mathcal{G})}{e_1(\mathcal{F} + \mathcal{G})} - \frac{e_2(\mathcal{F})}{e_1(\mathcal{F})} - \frac{e_2(\mathcal{G})}{e_1(\mathcal{G})} = \frac{\sum_{k=1}^n (x_k \sum_{j=1}^n y_j - y_k \sum_{j=1}^n x_j)^2}{2e_1(\mathcal{F} + \mathcal{G})\, e_1(\mathcal{F})\, e_1(\mathcal{G})}.$$

Assume now that $r > 2$. For an *n-tuple* $\mathcal{H} = \{z_1, \ldots, z_n\}$ we shall denote $\mathcal{H}_{\hat{k}} = \{z_1, \ldots, \widehat{z_k}, \ldots, z_n\}$, where the cap indicates omission. Then:

$$\sum_{k=1}^n x_k e_{r-1}(\mathcal{F}_{\hat{k}}) = r e_r(\mathcal{F}) \tag{C.1}$$

$$x_k e_{r-1}(\mathcal{F}_{\hat{k}}) + e_r(\mathcal{F}_{\hat{k}}) = e_r(\mathcal{F}). \tag{C.2}$$

Summing on k in (C.2) we obtain

$$n e_r(\mathcal{F}) = \sum_{k=1}^n x_k e_{r-1}(\mathcal{F}_{\hat{k}}) + \sum_{k=1}^n e_r(\mathcal{F}_{\hat{k}}),$$

and thus from (C.1) we infer that $\sum_{k=1}^{n} e_r(\mathcal{F}_{\hat{k}}) = (n-r)e_r(\mathcal{F})$. Since

$$e_r(\mathcal{F}) - e_r(\mathcal{F}_{\hat{k}}) = x_k e_{r-1}(\mathcal{F}_{\hat{k}})$$
$$= x_k e_{r-1}(\mathcal{F}) - x_k^2 e_{r-2}(\mathcal{F}_{\hat{k}})$$

we obtain

$$re_r(\mathcal{F}) = \sum_{k=1}^{n} x_k e_{r-1}(\mathcal{F}) - \sum_{k=1}^{n} x_k^2 e_{r-2}(\mathcal{F}_{\hat{k}})$$

and thus

$$\frac{e_r(\mathcal{F})}{e_{r-1}(\mathcal{F})} = \frac{1}{r}\Big[\sum_{k=1}^{n} x_k - \sum_{k=1}^{n} \frac{x_k^2\, e_{r-2}(\mathcal{F}_{\hat{k}})}{e_{r-1}(\mathcal{F})}\Big]$$
$$= \frac{1}{r}\Big[\sum_{k=1}^{n} x_k - \sum_{k=1}^{n} \frac{x_k^2}{x_k + e_{r-1}(\mathcal{F}_{\hat{k}})/e_{r-2}(\mathcal{F}_{\hat{k}})}\Big].$$

Therefore

$$\Delta = \frac{e_r(\mathcal{F}+\mathcal{G})}{e_{r-1}(\mathcal{F}+\mathcal{G})} - \frac{e_r(\mathcal{F})}{e_{r-1}(\mathcal{F})} - \frac{e_r(\mathcal{G})}{e_{r-1}(\mathcal{G})}$$
$$= \frac{1}{r}\sum_{k=1}^{n}\Big[\frac{x_k^2}{x_k + f_{r-1}(\mathcal{F}_{\hat{k}})} + \frac{y_k^2}{y_k + f_{r-1}(\mathcal{G}_{\hat{k}})} - \frac{(x_k+y_k)^2}{x_k + y_k + f_{r-1}((\mathcal{F}+\mathcal{G})_{\hat{k}})}\Big],$$

where $f_s(\mathcal{F}) = e_s(\mathcal{F})/e_{s-1}(\mathcal{F})$.

The proof ends by induction. Assume that the statement of the lemma is true for $r-1$, that is,

$$f_{r-1}((\mathcal{F}+\mathcal{G})_{\hat{k}}) > f_{r-1}(\mathcal{F}_{\hat{k}}) + f_{r-1}(\mathcal{G}_{\hat{k}}), \tag{C.3}$$

unless $\mathcal{F}_{\hat{k}}$ and $\mathcal{G}_{\hat{k}}$ are proportional (when equality holds). Then

$$\Delta > \frac{1}{r}\sum_{k=1}^{n}\Big[\frac{x_k^2}{x_k + f_{r-1}(\mathcal{F}_{\hat{k}})} + \frac{y_k^2}{y_k + f_{r-1}(\mathcal{G}_{\hat{k}})}$$
$$- \frac{(x_k+y_k)^2}{x_k + y_k + f_{r-1}(\mathcal{F}_{\hat{k}}) + f_{r-1}(\mathcal{G}_{\hat{k}})}\Big]$$
$$= \frac{1}{r}\sum_{k=1}^{n} \frac{[x_k f_{r-1}(\mathcal{G}_{\hat{k}}) - y_k f_{r-1}(\mathcal{F}_{\hat{k}})]^2}{[x_k + f_{r-1}(\mathcal{F}_{\hat{k}})][y_k + f_{r-1}(\mathcal{G}_{\hat{k}})][x_k + y_k + f_{r-1}(\mathcal{F}_{\hat{k}}) + f_{r-1}(\mathcal{G}_{\hat{k}})]}$$

provided that at least one of the inequalities (C.3) is strict. Thus (C.3) holds also for $r-1$ replaced by r and the proof is complete. ■

An alternative proof of Lemma C.3.2, based on the properties of perspective functions, was found by E. K. Godunova [183].

Proof of Theorem C.3.1. In fact, by Minkowski's inequality for $p = 0$ (see Example 3.1.5) and Lemma C.3.2, we have

$$
\begin{aligned}
e_r(\mathcal{F}+\mathcal{G})^{1/r} &= \Big[\frac{e_r(\mathcal{F}+\mathcal{G})}{e_{r-1}(\mathcal{F}+\mathcal{G})} \cdot \frac{e_{r-1}(\mathcal{F}+\mathcal{G})}{e_{r-2}(\mathcal{F}+\mathcal{G})} \cdots \frac{e_1(\mathcal{F}+\mathcal{G})}{e_0(\mathcal{F}+\mathcal{G})}\Big]^{1/r} \\
&\geq \Big\{\Big[\frac{e_r(\mathcal{F})}{e_{r-1}(\mathcal{F})} + \frac{e_r(\mathcal{G})}{e_{r-1}(\mathcal{G})}\Big] \cdots \Big[\frac{e_1(\mathcal{F})}{e_0(\mathcal{F})} + \frac{e_1(\mathcal{G})}{e_0(\mathcal{G})}\Big]\Big\}^{1/r} \\
&\geq \Big(\prod_{k=1}^{r} \frac{e_k(\mathcal{F})}{e_{k-1}(\mathcal{F})}\Big)^{1/r} + \Big(\prod_{k=1}^{r} \frac{e_k(\mathcal{G})}{e_{k-1}(\mathcal{G})}\Big)^{1/r} = e_r(\mathcal{F})^{1/r} + e_r(\mathcal{G})^{1/r}.
\end{aligned}
$$

The problem of equality is left to the reader. ∎

Theorem C.3.1 has important consequences for positive matrices $A \in \mathrm{M}_n(\mathbb{C})$, $A = (a_{ij})_{i,j=1}^n$. In this case all eigenvalues $\lambda_1(A), \ldots, \lambda_n(A)$ are positive and the symmetric elementary functions of them can be easily computed via the Cauchy–Binet formulae:

$$
\begin{gathered}
\sum_{k=1}^{n} \lambda_k(A) = \sum_{k=1}^{n} a_{kk} \\
\sum_{i<j} \lambda_i(A)\lambda_j(A) = \det \begin{pmatrix} a_{11} & a_{12} \\ a_{21} & a_{22} \end{pmatrix} + \cdots + \det \begin{pmatrix} a_{n-1\,n-1} & a_{n-1\,n} \\ a_{n\,n-1} & a_{n\,n} \end{pmatrix} \\
\vdots \\
\prod_{k=1}^{n} \lambda_k(A) = \det(a_{ij})_{i,j=1}^n.
\end{gathered}
$$

In particular, we recover the result of Example 3.1.5: If A and B are positive matrices, and $\alpha \in (0,1)$, then

$$
\begin{aligned}
\big(\det((1-\alpha)A + \alpha B)\big)^{1/n} &\geq (1-\alpha)(\det A)^{1/n} + \alpha(\det B)^{1/n} \\
&\geq (\det A)^{(1-\alpha)/n}(\det B)^{\alpha/n}.
\end{aligned}
$$

Newton's inequalities (of any order) have equivalent formulations in terms of positive matrices (and their principal minors). We shall recall here the analogue of the *AM-GM* inequality: *If A is a strictly positive matrix in*$\mathrm{M}_n(\mathbb{R})$, *then*

$$
\Big(\frac{\operatorname{trace} A}{n}\Big)^n > \det A,
$$

unless A is a multiple of the unit matrix I.

C.3.3 Remark *In differential geometry, the higher order mean curvatures are defined as the elementary symmetric functions of the principal curvatures. In fact, if S is a hypersurface in $\mathbb{R}^n$ and p is a point of S, one considers the Gauss* *map, $g\colon p \to N(p)$, whose differential at p is diagonalized by the principal curvature directions at p,*

$$dg_p(e_j) = -k_j e_j \quad \text{for } j = 1, \ldots, n.$$

Then the j-th-order mean curvatures H_j are given by

$$\prod_{k=1}^{n-1}(1 + tk_j) = \sum_{j=0}^{n-1}\binom{n-1}{j} H_j t^j.$$

See R. Osserman [384] *for details. It would be interesting to explore the applications of various inequalities of convexity to this area.*

C.4 Symmetric Polynomial Majorization

Motivated by an optimization problem in solid-body mechanics, M. Bîrsan, P. Neff and J. Lankeit [58] have considered the following variant of majorization for vectors in $\mathbb{R}^n_{++}$ $(N \geq 2)$. Precisely, if $x, y \in \mathbb{R}^n_{++}$, we say that x is less than or equal to y with respect to the *symmetric polynomial majorization* (that is, $x \prec_{SPM} y$) if

$$e_k(x) \leq e_k(y) \quad \text{for all } k \in \{1, ..., n-1\}$$

and

$$e_n(x) = e_n(y).$$

Accordingly, a function $f : (0, \infty) \to \mathbb{R}$ is called $\prec_{SPM}$- *convex* if

$$\sum_{k=1}^{n} f(x_k) \leq \sum_{k=1}^{n} f(y_k)$$

whenever $x \prec_{SPM} y$. The simplest examples of such functions are

$$\pm 1,\ x,\ \pm \log x,\ 1/x,$$

as well as all linear combinations $a + bx + c\log x + d/x$ with $a, c \in \mathbb{R}$ and $b, d \geq 0$. Other examples are provided by the family of functions

$$L_t(x) = \log \frac{x+t}{1+t} - \frac{\log x}{1+t^2},$$

where $t > 0$ is a parameter. Indeed,

$$\begin{aligned}\sum_{k=1}^{n} L_t(x_k) &= \log \frac{(x_1+t)\cdots(x_n+t)}{(1+t)^n} - \frac{\log(x_1\cdots x_n)}{1+t^2} \\ &= \log \sum_{k=1}^{n} \left(t^{n-k} e_k(x_1, \ldots, x_n)\right) - n\log(1+t) - \frac{\log e_n(x_1, \ldots, x_n)}{1+t^2},\end{aligned}$$

which makes clear the implication

$$x \prec_{SPM} y \text{ implies } \sum_{k=1}^{n} L_t(x_k) \leq \sum_{k=1}^{n} L_t(y_k).$$

As was noticed by M. Šilhavý [446], the examples listed above generate *all* $\prec_{SPM}$- convex functions. For example, so is the case of squared logarithms, which combines the $\prec_{SPM}$- convexity of the functions L_t and the following representation formula:

$$\log^2 x = 2\int_0^\infty \left[\log \frac{x+t}{1+t} - \frac{\log x}{1+t^2}\right] \frac{dt}{t}.$$

Indeed, this formula takes place for $x = 1$ and the two sides have the same derivative for $x > 0$.

Appendix D

Second-Order Differentiability of Convex Functions

The famous result of A. D. Alexandrov [9] on the second differentiability of convex functions defined on $\mathbb{R}^N$ has received a great deal of attention during the last two decades, motivated by its connection to mass transportation problems. The proof we present here follows the argument of M. G. Crandall, H. Ishii and P.-L. Lions [118] (which in turn were inspired by a paper of F. Mignot [320]). A basic ingredient in their approach is Rademacher's theorem on the almost differentiability of Lipschitz functions defined on an Euclidean space. Its proof makes the objective of Section D.1.

There are known many other proofs of Alexandrov's theorem but all make use of the machinery of hard analysis. See, for example, G. Alberti and L. Ambrosio [7], G. Bianchi, A. Colesanti and C. Pucci [54] (which includes also some historical comments) and L. C. Evans and R. Gariepy [152] to cite just a few.

D.1 Rademacher's Theorem

The almost everywhere differentiability of a convex function defined on a convex open subset of $\mathbb{R}^N$ is the particular case of the following basic result in nonlinear analysis:

D.1.1 Theorem (Rademacher's theorem) *Every locally Lipschitz function f acting on Euclidean spaces is almost everywhere differentiable.*

Proof. Since a vector-valued function $f\colon \mathbb{R}^N \to \mathbb{R}^m$ is differentiable at a point if and only if all of its components are differentiable at that point, we may restrict ourselves to the case of real-valued functions. Also, since differentiability is a

C. P. Niculescu and L.-E. Persson, *Convex Functions and Their Applications*, CMS Books in Mathematics, https://doi.org/10.1007/978-3-319-78337-6

local property, we may as well assume that f is Lipschitz. See Kirszbraun's Theorem 3.7.4. The remainder of the proof will be done in three steps.

Step 1. Fix arbitrarily a vector $\mathbf{v} \in \mathbb{R}^N$, $\|\mathbf{v}\| = 1$. We shall show that the directional derivative $f'(\mathbf{x}; \mathbf{v})$ exists for almost every $\mathbf{x} \in \mathbb{R}^N$. In fact, for each $\mathbf{x} \in \mathbb{R}^N$ consider the limits

$$\underline{D}f(\mathbf{x}; \mathbf{v}) = \liminf_{t \to 0} \frac{f(\mathbf{x} + t\mathbf{v}) - f(\mathbf{x})}{t}$$

and

$$\overline{D}f(\mathbf{x}; \mathbf{v}) = \limsup_{t \to 0} \frac{f(\mathbf{x} + t\mathbf{v}) - f(\mathbf{x})}{t},$$

which exist in $\overline{\mathbb{R}}$. The set

$$E_{\mathbf{v}} = \{\mathbf{x} \in \mathbb{R}^n : \underline{D}f(\mathbf{x}; \mathbf{v}) < \overline{D}f(\mathbf{x}; \mathbf{v})\}$$

equals the set where the directional derivative $f'(\mathbf{x}; \mathbf{v})$ does not exist. As in the proof of Theorem 3.7.3 we may conclude that $E_{\mathbf{v}}$ is Lebesgue measurable. We shall show that $E_{\mathbf{v}}$ is actually Lebesgue negligible. In fact, by Lebesgue's theory on the differentiability of absolutely continuous functions (see [107], or [214]), we infer that the functions

$$g(t) = f(\mathbf{x} + t\mathbf{v}), \quad t \in \mathbb{R},$$

are differentiable almost everywhere. This implies that the Lebesgue measure of the intersection of E_v with any line L is Lebesgue negligible. Then, by Fubini's theorem, we conclude that $E_{\mathbf{v}}$ is itself Lebesgue negligible.

Step 2. According to the discussion above we know that

$$\nabla f(\mathbf{x}) = \Big(\frac{\partial f}{\partial x_1}(\mathbf{x}), \ldots, \frac{\partial f}{\partial x_n}(\mathbf{x})\Big)$$

exists almost everywhere. We shall show that

$$f'(\mathbf{x}; \mathbf{v}) = \langle \mathbf{v}, \nabla f(\mathbf{x}) \rangle$$

for almost every $\mathbf{x} \in \mathbb{R}^N$. In fact, for an arbitrary fixed $\varphi \in C_c^\infty(\mathbb{R}^n)$ we have

$$\int_{\mathbb{R}^n} \Big[\frac{f(\mathbf{x} + t\mathbf{v}) - f(\mathbf{x})}{t}\Big] \varphi(\mathbf{x}) \, \mathrm{d}\mathbf{x} = -\int_{\mathbb{R}^n} f(\mathbf{x}) \Big[\frac{\varphi(\mathbf{x}) - \varphi(\mathbf{x} - t\mathbf{v})}{t}\Big] \, \mathrm{d}\mathbf{x}.$$

Since

$$\Big|\frac{f(\mathbf{x} + t\mathbf{v}) - f(\mathbf{x})}{t}\Big| \leq \mathrm{Lip}(f),$$

we can apply the dominated convergence theorem to get

$$\int_{\mathbb{R}^N} f'(\mathbf{x}; \mathbf{v}) \varphi(\mathbf{x}) \, \mathrm{d}\mathbf{x} = -\int_{\mathbb{R}^N} f(\mathbf{x}) \varphi'(\mathbf{x}; \mathbf{v}) \, \mathrm{d}\mathbf{x}.$$

By taking into account Fubini's theorem and the absolute continuity of f on lines we can continue as follows:

$$
\begin{aligned}
-\int_{\mathbb{R}^N} f(\mathbf{x})\varphi'(\mathbf{x};\mathbf{v})\,\mathrm{d}\mathbf{x} &= -\sum_{k=1}^{N} v_k \int_{\mathbb{R}^N} f(\mathbf{x})\frac{\partial \varphi}{\partial x_k}(\mathbf{x})\,\mathrm{d}\mathbf{x} \\
&= \sum_{k=1}^{N} v_k \int_{\mathbb{R}^N} \frac{\partial f}{\partial x_k}(\mathbf{x})\varphi(\mathbf{x})\,\mathrm{d}\mathbf{x} = \int_{\mathbb{R}^N} \langle \mathbf{v}, \nabla f(\mathbf{x})\rangle \varphi(\mathbf{x})\,\mathrm{d}\mathbf{x}
\end{aligned}
$$

and this leads us to the formula $f'(\mathbf{x};\mathbf{v}) = \langle \mathbf{v}, \nabla f(\mathbf{x})\rangle$, as φ was arbitrarily fixed.

Step 3. Consider now a countable family $(\mathbf{u}_i)_i$ of unit vectors, which is dense in the unit sphere of $\mathbb{R}^N$. By the above reasoning we infer that the complement of each of the sets

$$
A_i = \{\mathbf{x} \in \mathbb{R}^N : Df(\mathbf{x};\mathbf{u}_i) \text{ and } \nabla f(\mathbf{x}) \text{ exist and } f'(\mathbf{x};\mathbf{u}_i) = \langle \mathbf{u}_i, \nabla f(\mathbf{x})\rangle\}
$$

is Lebesgue negligible, and thus the same is true for the complement of

$$
A = \bigcap_{i=1}^{\infty} A_i.
$$

We shall show that f is differentiable at all points of A. This will be done by considering the function

$$
R(\mathbf{x},\mathbf{v},t) = \frac{f(\mathbf{x}+t\mathbf{v}) - f(\mathbf{x})}{t} - \langle \mathbf{v}, \nabla f(\mathbf{x})\rangle,
$$

for $\mathbf{x} \in A$, $\mathbf{v} \in \mathbb{R}^N$, $\|\mathbf{v}\| = 1$, and $t \in \mathbb{R}\backslash\{0\}$. Since

$$
\begin{aligned}
|R(\mathbf{x},\mathbf{v},t) - R(\mathbf{x},\mathbf{v}',t)| &\leq \operatorname{Lip}(f)\cdot\|\mathbf{v}-\mathbf{v}'\| + \|\nabla f(\mathbf{x})\|\cdot\|\mathbf{v}-\mathbf{v}'\| \\
&\leq (\sqrt{N}+1)\operatorname{Lip}(f)\cdot\|\mathbf{v}-\mathbf{v}'\|,
\end{aligned}
$$

the function $R(\mathbf{x},\mathbf{v},t)$ is Lipschitz in $\mathbf{v}$. Suppose there are given a point $\mathbf{a} \in A$ and a number $\varepsilon > 0$. Since the unit sphere of $\mathbb{R}^N$ is compact and the family $(\mathbf{u}_i)_i$ is a dense subset, we can choose a natural number n such that

$$
\inf_{i\in\{0,\dots,n\}} \|\mathbf{v} - u_i\| < \frac{\varepsilon}{2(\sqrt{N}+1)\operatorname{Lip}(f)}
$$

for all $\mathbf{v}$ with $\|\mathbf{v}\| = 1$. By the definition of A, there exists a $\delta > 0$ such that

$$
|R(\mathbf{a},\mathbf{u}_i,t)| < \frac{\varepsilon}{2}
$$

for all $i \in \{0,\dots,N\}$ and all $|t| < \delta$. Then

$$
\begin{aligned}
|R(\mathbf{a},\mathbf{v},t)| &< \inf_{i\in\{0,\dots,n\}} \left(|R(\mathbf{a},\mathbf{u}_i,t)| + |R(\mathbf{a},\mathbf{u}_i,t) - R(\mathbf{a},\mathbf{v},t)|\right) \\
&\leq \frac{\varepsilon}{2} + (\sqrt{N}+1)\operatorname{Lip}(f)\cdot\frac{\varepsilon}{2(\sqrt{N}+1)\operatorname{Lip}(f)} = \varepsilon,
\end{aligned}
$$

for all $\mathbf{v}$ in the unit sphere of $\mathbb{R}^N$ and all t with $|t| < \delta$. This assures the differentiability of f at $\mathbf{a}$. ■

F. Mignot [320] proved the following generalization of Rademacher's theorem: *Every Lipschitz function from a separable Hilbert space to a Hilbert space is Gâteaux differentiable at densely many points.* However, as shown in the case of the function

$$F\colon L^2[0,1] \to \mathbb{R}, \quad F(f) = \left(\int_0^1 (f^+(t))^2 \, \mathrm{d}t \right)^{1/2},$$

the set of points of Fréchet differentiability may be empty. The hypothesis on separability is essential for the validity of Mignot's result. A counterexample is provided by the projection of $\ell^2(A)$ (for an uncountable index set A) onto the cone of positive elements.

D.2 Alexandrov's Theorem

We pass now to the problem of almost everywhere second differentiability of convex functions. In order to simplify the exposition we shall make constant use of Landau's little-o symbol, precisely,

$$f = o(g) \text{ for } \mathbf{x} \to 0 \text{ if and only if } f = hg \text{ with } \lim_{\mathbf{x} \to 0} h(\mathbf{x}) = 0.$$

D.2.1 Theorem (A. D. Alexandrov [9]) *Every convex function f from $\mathbb{R}^N$ to $\mathbb{R}$ is twice differentiable almost everywhere in the following sense: f is twice differentiable at $\mathbf{a}$, with Alexandrov Hessian $\nabla^2 f(\mathbf{a})$ in $\mathrm{Sym}^+(N, \mathbb{R})$, if $\nabla f(\mathbf{a})$ exists, and if for every $\varepsilon > 0$ there exists $\delta > 0$ such that*

$$\|\mathbf{x} - \mathbf{a}\| < \delta \text{ implies } \sup_{\mathbf{y} \in \partial f(\mathbf{x})} \|\mathbf{y} - \nabla f(\mathbf{a}) - \nabla^2 f(\mathbf{a})(\mathbf{x} - \mathbf{a})\| \leq \varepsilon \|\mathbf{x} - \mathbf{a}\|.$$

Moreover, if $\mathbf{a}$ is such a point, then

$$\lim_{\mathbf{h} \to 0} \frac{f(\mathbf{a} + \mathbf{h}) - f(\mathbf{a}) - \langle \nabla f(\mathbf{a}), \mathbf{h} \rangle - \frac{1}{2} \langle \nabla^2 f(\mathbf{a})\mathbf{h}, \mathbf{h} \rangle}{\|\mathbf{h}\|^2} = 0.$$

Proof. By Theorem 3.3.1, the domain of the subdifferential ∂f is the whole space $\mathbb{R}^N$, while Proposition 3.6.9 shows that $\partial f(\mathbf{x}) = \{\mathrm{d}f(\mathbf{x})\}$ for all $\mathbf{x}$ in

$$X_1 = \{\mathbf{x} \in \mathbb{R}^N : f \text{ is differentiable at } \mathbf{x}\},$$

which is a set whose complement is a negligible set. We shall show that for almost all $\mathbf{x}$ in $\mathbb{R}^N$ there exists a matrix A in $\mathrm{Sym}^+(n, \mathbb{R})$ such that

$$\mathrm{d}f(\mathbf{y}) = \mathrm{d}f(\mathbf{x}) + A(\mathbf{y} - \mathbf{x}) + o(\|\mathbf{y} - \mathbf{x}\|) \quad \text{for all } \mathbf{y} \in X_1. \tag{D.1}$$

We need the fact that $J = (I + \partial f)^{-1}$ is a nonexpansive map of $\mathbb{R}^N$ into itself. See Corollary 3.3.11. This yields a new set,

$$X_2 = \{\mathbf{x} : J \text{ is differentiable at } \mathbf{x} \text{ and } \mathrm{d}J(\mathbf{x}) \text{ is nonsingular}\},$$

whose complement is also a negligible set. In fact, by Rademacher's theorem, J is differentiable almost everywhere. Since J is Lipschitz, we may apply the area formula (see L. C. Evans and R. F. Gariepy [152, Theorem 3.3.2, p. 96]) to get

$$\int_B |\det(\mathrm{d}J(\mathbf{x}))|\,\mathrm{d}\mathbf{x} = \int_{\mathbb{R}^N} \#(B \cap J^{-1}(\mathbf{y}))\,\mathrm{d}\mathbf{y} \quad \text{for all Borel sets } B \text{ in } \mathbb{R}^N,$$

where $\#$ is the counting measure. By this formula (and the fact that J is onto) we infer that the complementary set of

$$\{\mathbf{x} : J \text{ is differentiable at } \mathbf{x} \text{ and } \mathrm{d}J(\mathbf{x}) \text{ is nonsingular}\}$$

is a negligible set. On the other hand, any Lipschitz function maps negligible sets into negligible sets. See [429], Lemma 7.25. Hence X_2 is indeed a set whose complementary set is negligible. We shall show that the formula (D.1) works for all $\mathbf{x}$ in $X_3 = X_1 \cap X_2$ (which is a set with negligible complementary set). Our argument is based on the following fact concerning the solvability of nonlinear equations in $\mathbb{R}^N$: If $F\colon \overline{B}_\delta(0) \to \mathbb{R}^N$ is continuous, $0 < \varepsilon < \delta$ and $\|F(\mathbf{x}) - \mathbf{x}\| < \varepsilon$ for all $\mathbf{x} \in \mathbb{R}^N$ with $\|\mathbf{x}\| = \delta$, then $F(B_\delta(\mathbf{0})) \supset B_{\delta-\varepsilon}(\mathbf{0})$. See W. Rudin [429, Lemma 7.23] for a proof based on the Brouwer fixed-point theorem. By the definition of J,

$$\mathrm{d}f(J(\mathbf{x})) = \mathbf{x} - J(\mathbf{x})$$

for all $\mathbf{x}$ with $J(\mathbf{x}) \in X_3$ $(\subset X_1)$. Suppose that $J(\mathbf{x}) + \tilde{\mathbf{y}} \in X_1$, where $\tilde{\mathbf{y}}$ is small. Since J is Lipschitz and $dJ(\mathbf{x})$ is nonsingular, if $\tilde{y}$ is sufficiently small, then there exists $\tilde{\mathbf{x}}$ such that $J(\mathbf{x} + \tilde{\mathbf{x}}) = J(\mathbf{x}) + \tilde{\mathbf{y}}$. Moreover, we may choose $\tilde{\mathbf{x}}$ to verify $\|\tilde{\mathbf{x}}\| \leq C\|\tilde{\mathbf{y}}\|$ for some constant $C > 0$. Use a remark above (on the solvability of nonlinear equations) and the fact that $J(\mathbf{x}+\mathbf{h}) = J(\mathbf{x}) + \mathrm{d}J(\mathbf{x})\mathbf{h} + o(\|\mathbf{h}\|)$. Since J is nonexpansive, we also have $\|\tilde{\mathbf{y}}\| \leq \|\tilde{\mathbf{x}}\|$, hence $\|\tilde{\mathbf{x}}\|$ and $\|\tilde{\mathbf{y}}\|$ are comparable. Then

$$\begin{aligned}\mathrm{d}f(J(x) + \tilde{y}) &= \mathrm{d}f(J(\mathbf{x} + \tilde{\mathbf{x}})) = \mathbf{x} + \tilde{\mathbf{x}} - J(\mathbf{x} + \tilde{\mathbf{x}}) \\ &= df(J(\mathbf{x})) + (\mathrm{I} - \mathrm{d}J(\mathbf{x}))\tilde{\mathbf{x}} + o(\|\tilde{\mathbf{x}}\|).\end{aligned}$$

Due to the relation $J(\mathbf{x}) + \tilde{\mathbf{y}} = J(\mathbf{x}) + \mathrm{d}J(\mathbf{x})\tilde{\mathbf{x}} + o(\|\tilde{\mathbf{x}}\|)$ and the comparability of $\|\tilde{\mathbf{x}}\|$ and $\|\tilde{\mathbf{y}}\|$, we have

$$\tilde{\mathbf{x}} = (\mathrm{d}J(\mathbf{x}))^{-1}\tilde{\mathbf{y}} + o(\|\tilde{\mathbf{y}}\|).$$

Hence $\mathrm{d}(\mathrm{d}f)(J(\mathbf{x}))$ exists and equals $(\mathrm{d}J(\mathbf{x}))^{-1} - \mathrm{I}$. It remains to prove that

$$f(J(x) + \tilde{\mathbf{y}}) = f(J(\mathbf{x})) + \mathrm{d}f(J(\mathbf{x}))\tilde{\mathbf{y}} + \frac{1}{2}\langle((\mathrm{d}J(\mathbf{x}))^{-1} - \mathrm{I})\tilde{\mathbf{y}}, \tilde{\mathbf{y}}\rangle + o(\|\tilde{\mathbf{y}}\|^2)$$

for $J(\mathbf{x}) \in X_3$. Letting

$$R(\tilde{\mathbf{y}}) = f(J(\mathbf{x}) + \tilde{\mathbf{y}}) - f(J(\mathbf{x})) - \mathrm{d}f(J(\mathbf{x}))\tilde{\mathbf{y}} - \frac{1}{2}\langle((\mathrm{d}J(\mathbf{x}))^{-1} - \mathrm{I})\tilde{\mathbf{y}}, \tilde{\mathbf{y}}\rangle,$$

we get a locally Lipschitz function R such that $R(\mathbf{0}) = 0$, and for almost all small $\tilde{y}$,

$$dR(\tilde{\mathbf{y}}) = o(\|\tilde{\mathbf{y}}\|).$$

By the mean value theorem we conclude that $R(\tilde{\mathbf{y}}) = o(\|\tilde{\mathbf{y}}\|^2)$ and the proof is complete. ■

The result of Theorem D.2.1 can easily be extended to conclude that every proper convex function $f\colon \mathbb{R}^N \to \mathbb{R} \cup \{\infty\}$ is twice differentiable almost everywhere on its effective domain. Alexandrov's theorem has important applications to convex geometric analysis and partial differential equations. Both Theorems D.1.1 and D.2.1 remain valid in the more general framework of semiconvex functions. A function f defined on a convex set in $\mathbb{R}^N$ is said to be *semiconvex* if $f + \lambda\|\cdot\|^2$ is a convex function for some $\lambda > 0$.

In Mignot's approach [320], Alexandrov's theorem appears as a consequence of the following differentiability property of monotone maps:

D.2.2 Theorem (*Differentiability of monotone maps*). *Let u be a maximal monotone map on $\mathbb{R}^n$ and let D be the set of points x such that $u(x)$ is a singleton. Then u is differentiable at almost every $a \in D$, that is, there exists an $n \times n$ matrix $\nabla u(a)$ such that*

$$\lim_{\substack{\mathbf{x}\to\mathbf{a} \\ \mathbf{y}\in u(\mathbf{x})}} \frac{\mathbf{y} - u(\mathbf{a}) - \nabla u(\mathbf{a})(\mathbf{x}-\mathbf{a})}{\|\mathbf{x}-\mathbf{a}\|} = 0.$$

In fact, if we apply this result to the subdifferential of a convex function $f\colon \mathbb{R}^n \to \mathbb{R}$, we obtain that for almost every $\mathbf{a} \in \mathbb{R}^n$ where ∂f is a singleton (that is, where f is differentiable), there exists a matrix $\nabla^2 f(\mathbf{a})$ such that

$$\lim_{\substack{\mathbf{x}\to\mathbf{a} \\ \mathbf{y}\in \partial f(\mathbf{x})}} \frac{\mathbf{y} - \nabla f(\mathbf{a}) - \nabla^2 f(\mathbf{a})(\mathbf{x}-\mathbf{a})}{\|\mathbf{x}-\mathbf{a}\|} = 0. \tag{D.2}$$

If (D.2) holds, then $A = \nabla^2 f(\mathbf{a})$ proves to be the Alexandrov Hessian of f at $\mathbf{a}$. To show this, it suffices to restrict ourselves to the case where $\mathbf{a} = 0$, $f(\mathbf{a}) = 0$, and $\nabla f(\mathbf{a}) = 0$. We shall prove that $\varphi(\mathbf{h}) = f(\mathbf{h}) - \frac{1}{2}\langle A\mathbf{h}, \mathbf{h}\rangle$ verifies $\lim_{\mathbf{h}\to 0} \varphi(\mathbf{h})/\|\mathbf{h}\|^2 = 0$. In fact, fixing an $\mathbf{h} \neq \mathbf{0}$, by the nonsmooth version of the mean value theorem (see Exercise 4, Section 3.6) we get a point y in the segment joining $\mathbf{0}$ to $\mathbf{h}$, and a $\mathbf{p} \in \partial\varphi(\mathbf{y})$ such that $\varphi(\mathbf{h}) - \varphi(\mathbf{0}) = \langle \mathbf{h}, \mathbf{p}\rangle$. Then

$$\varphi(\mathbf{h}) = \langle \mathbf{q} - A\mathbf{y}, \mathbf{h}\rangle$$

for some vector $\mathbf{q} \in \partial f(\mathbf{y})$. According to (D.2), $\lim_{\mathbf{h}\to 0} \|\mathbf{q} - A(\mathbf{y})\|/\|\mathbf{h}\| = 0$, which yields $\lim_{\mathbf{h}\to \mathbf{0}} \varphi(\mathbf{h})/\|\mathbf{h}\|^2 = 0$.

Appendix E

The Variational Approach of PDE

The aim of this appendix is to illustrate a number of problems in partial differential equations (PDE) which can be solved by seeking a global minimum of suitable convex functionals. This idea goes back to Fermat's rule for finding points of extrema.

E.1 The Minimum of Convex Functionals

An important criterion for the existence of a global minimum is the following variant of Weierstrass' extreme value theorem:

E.1.1 Theorem *Let C be a closed convex set in a reflexive Banach space V and let $J\colon C \to \mathbb{R}$ be a function which fulfills the following two conditions:*

(a) *J is weakly lower semicontinuous, that is,*

$$\mathbf{u}_n \to \mathbf{u} \text{ weakly in } V \text{ implies } J(\mathbf{u}) \leq \liminf_{n\to\infty} J(\mathbf{u}_n);$$

(b) *Either C is bounded or $\lim_{\|\mathbf{u}\|\to\infty} J(\mathbf{u}) = \infty$.*

Then J admits at least one global minimum. If, moreover, J is quasiconvex, the points of global minimum constitute a convex set.

Proof. Put

$$m = \inf_{\mathbf{u}\in C} J(\mathbf{u}).$$

Clearly, $m < \infty$, and there exists a sequence $(\mathbf{u}_n)_n$ of elements in C such that $J(\mathbf{u}_n) \to m$. By our hypothesis (b), the sequence $(\mathbf{u}_n)_n$ is bounded, so by Theorem B.1.8, we may assume (replacing $(\mathbf{u}_n)_n$ by a subsequence if necessary) that it is also weakly converging to an element $\mathbf{u}$ in C. Here we used the fact that C is weakly closed (which is a consequence of Corollary B.2.6). Then

$$m \leq J(\mathbf{u}) \leq \liminf_{n\to\infty} J(\mathbf{u}_n) = m,$$

C. P. Niculescu and L.-E. Persson, *Convex Functions and Their Applications*,
CMS Books in Mathematics, https://doi.org/10.1007/978-3-319-78337-6

and thus $\mathbf{u}$ is a global minimum. The remainder of the proof is left to the reader as an exercise. ∎

Under the presence of convexity and differentiability one can prove much more powerful results.

E.1.2 Theorem *Let V be a reflexive Banach space V and let $J: V \to \mathbb{R}$ be a convex functional with the following properties:*

(a) *J is Gâteaux differentiable;*

(b) $\lim_{\|\mathbf{u}\| \to \infty} J(\mathbf{u}) = \infty$.

Then J admits at least one global minimum and the points of global minimum are precisely the solutions $\mathbf{u}$ of the so-called Euler–Lagrange equation,

$$J'(\mathbf{u}) = 0.$$

If, moreover, J is strictly convex, then there is a unique global minimum.

Proof. First notice that J is weakly lower semicontinuous. In fact, by Proposition 3.6.9,

$$J(\mathbf{u}_n) \geq J(\mathbf{u}) + J'(\mathbf{u}; \mathbf{u}_n - \mathbf{u})$$

for all n, so if $\mathbf{u}_n \xrightarrow{w} \mathbf{u}$, then $J'(\mathbf{u}; \mathbf{u}_n - \mathbf{u}) = J'(\mathbf{u})(\mathbf{u}_n - \mathbf{u}) \to 0$ and we infer that

$$\liminf_{n \to \infty} J(\mathbf{u}_n) \geq J(\mathbf{u}).$$

Hence, according to Theorem E.1.1, J admits a global minimum. If $\mathbf{u}$ is a global minimizer, then for each $\mathbf{v} \in V$ there is a $\delta > 0$ such that

$$\frac{J(\mathbf{u} + \varepsilon \mathbf{v}) - J(\mathbf{u})}{\varepsilon} \geq 0, \quad \text{whenever } |\varepsilon| < \delta.$$

This yields $J'(\mathbf{u}; \mathbf{v}) \geq 0$. Replacing $\mathbf{v}$ by $-\mathbf{v}$, we obtain

$$-J'(\mathbf{u}; \mathbf{v}) = J'(\mathbf{u}; -\mathbf{v}) \geq 0,$$

and thus $J'(\mathbf{u}; \mathbf{v}) = 0$. Conversely, if $J'(\mathbf{u}; \mathbf{v}) = 0$ for all $\mathbf{v} \in V$, then by Proposition 3.6.9 we get

$$J(\mathbf{v}) \geq J(\mathbf{u}) + J'(\mathbf{u}, \mathbf{v} - \mathbf{u}) = J(\mathbf{u}),$$

that is, $\mathbf{u}$ is a global minimizer. ∎

For example, Theorem E.1.2 applies to the convex functionals of the form

$$J(\mathbf{u}) = \frac{1}{2} \|\mathbf{u} - \mathbf{w}\|^2 + \varphi(\mathbf{u}), \quad \mathbf{u} \in V,$$

where V is an L^p-space with $p \in (1, \infty)$, or a Sobolev space; $\mathbf{w}$ is an arbitrary fixed element of V; and $\varphi: V \to \mathbb{R}$ is a weakly lower semicontinuous convex function. These functionals play an important role in connection with the existence of solutions of partial differential equations.

The following consequence of Theorem E.1.2 is borrowed from the book of J. Céa [101]:

E.1.3 Corollary *Let Ω be a nonempty open set in $\mathbb{R}^N$ and let $p > 1$. Consider a function $g \in C^1(\mathbb{R})$ which verifies the following properties:*

(a) $g(0) = 0$ and $g(t) \geq \alpha|t|^p$ for a suitable constant $\alpha > 0$;

(b) The derivative g' is increasing and $|g'(t)| \leq \beta|t|^{p-1}$ for a suitable constant $\beta > 0$.

Then the linear space $V = L^p(\Omega) \cap L^2(\Omega)$ is reflexive when endowed with the norm

$$\|u\|_V = \|u\|_{L^p} + \|u\|_{L^2},$$

and for all $f \in L^2(\Omega)$ the functional

$$J(u) = \int_\Omega g(u(x))\,\mathrm{d}x + \frac{1}{2}\int_\Omega |u(x)|^2\,\mathrm{d}x - \int_\Omega f(x)u(x)\,\mathrm{d}x, \quad u \in V$$

is convex and Gâteaux differentiable with

$$J'(u;v) = \int_\Omega g'(u(\mathbf{x}))v(\mathbf{x})\,\mathrm{d}\mathbf{x} + \int_\Omega u(\mathbf{x})v(\mathbf{x})\,\mathrm{d}\mathbf{x} - \int_\Omega f(\mathbf{x})v(\mathbf{x})\,\mathrm{d}\mathbf{x}.$$

Moreover, J admits a unique global minimizer $\bar{u}$, which is the solution of the Euler–Lagrange equation

$$J'(u) = 0.$$

Proof. V is a closed subspace of $L^2(\Omega)$ and thus it is a reflexive space. Then notice that

$$\begin{aligned} |g(t)| &= |g(t) - g(0)| \\ &= \left|\int_0^t g'(s)\,\mathrm{d}s\right| \leq \frac{\beta}{p}|t|^p, \end{aligned}$$

from which it follows easily that J is well defined. Letting

$$J_1(u) = \int_\Omega g(u(x))\,\mathrm{d}x,$$

by Lagrange's mean value theorem,

$$\begin{aligned} J_1(u+tv) &= \int_\Omega g(u(\mathbf{x}) + tv(\mathbf{x}))\,\mathrm{d}\mathbf{x} \\ &= \int_\Omega g(u(\mathbf{x}))\,\mathrm{d}\mathbf{x} + t\int_\Omega g'(u(\mathbf{x}) + \tau(\mathbf{x})v(\mathbf{x}))v(\mathbf{x})\,\mathrm{d}\mathbf{x}, \end{aligned}$$

where $0 < \tau(\mathbf{x}) < t$ for all $\mathbf{x}$, provided that $t > 0$. Then

$$\frac{J_1(u+tv) - J_1(u)}{t} = \int_\Omega g'(u(\mathbf{x}) + \tau(\mathbf{x})v(\mathbf{x}))v(\mathbf{x})\,\mathrm{d}\mathbf{x},$$

and letting $t \to 0+$ we get the desired formula for $J'(u;v)$.

Again by Lagrange's mean value theorem, and the fact that g' is increasing, we have

$$\begin{aligned} J_1(v) &= J_1(u) + \int_\Omega g'(u(\mathbf{x}) + \tau(\mathbf{x})(v(\mathbf{x}) - u(\mathbf{x}))) \cdot (v(\mathbf{x}) - u(\mathbf{x}))\, \mathrm{d}\mathbf{x} \\ &\geq J_1(u) + \int_\Omega g'(u(\mathbf{x})) \cdot (v(\mathbf{x}) - u(\mathbf{x}))\, \mathrm{d}\mathbf{x} \\ &= J_1(u) + J_1'(u, v - u), \end{aligned}$$

which shows that J_1 is convex. Then the functional J is the sum of a convex function and a strictly convex function. Finally,

$$\begin{aligned} J(u) &\geq \alpha \int_\Omega |u(\mathbf{x})|^p\, \mathrm{d}\mathbf{x} + \frac{1}{2}\int_\Omega |u(\mathbf{x})|^2\, \mathrm{d}\mathbf{x} - \left|\int_\Omega f(\mathbf{x})u(\mathbf{x})\, \mathrm{d}\mathbf{x}\right| \\ &\geq \alpha \|u\|_{L^p}^p + \frac{1}{2}\|u\|_{L^2}^2 - \|f\|_{L^2}\|u\|_{L^2}, \end{aligned}$$

from which it follows that

$$\lim_{\|u\|_V \to \infty} J(u) = \infty,$$

and the conclusion follows from Theorem E.1.2. ■

The result of Corollary E.1.3 extends (with obvious changes) to the case where V is defined as the space of all $u \in L^2(\Omega)$ such that $Au \in L^p(\Omega)$ for a given linear differential operator A. Also, we can consider finitely many functions g_k (verifying the conditions (a) and (b) of Corollary E.1.3) for different exponents $p_k > 1$) and finitely many linear differential operators A_k. In that case we shall deal with the functional

$$J(u) = \sum_{k=1}^m \int_\Omega g_k(A_k u)\, \mathrm{d}x + \frac{1}{2}\int_\Omega |u|^2\, \mathrm{d}x - \int_\Omega f u\, \mathrm{d}x,$$

defined on $V = \bigcap_{k=1}^m \left\{u \in L^2(\Omega) : A_k u \in L^{p_k}(\Omega)\right\}$; V is a reflexive Banach space when endowed with the norm

$$\|u\|_V = \sum_{k=1}^m \|A_k u\|_{L^{p_k}} + \|u\|_{L^2}.$$

E.2 Preliminaries on Sobolev Spaces

Some basic results on Sobolev spaces are recalled here for the convenience of the reader. The details are available from many sources, for example, R. A. Adams and J. J. F. Fournier and W. P. Ziemer [499]. Let Ω be a bounded open set in $\mathbb{R}^N$ with Lipschitz boundary $\partial\Omega$, and let m and p be two strictly positive integers. The Sobolev space $W^{m,p}(\Omega)$ consists of all functions $u \in L^p(\Omega)$ which admit weak derivatives $D^\alpha u$ in $L^p(\Omega)$, for all multi-indices α with $|\alpha| \le m$. This means the existence of functions $v_\alpha \in L^p(\Omega)$ such that

$$\int_\Omega v_\alpha \cdot \varphi \, dx = (-1)^{|\alpha|} \int_\Omega u \cdot D^\alpha \varphi \, dx \tag{E.1}$$

for all φ in the space $C_c^\infty(\Omega)$ and all α with $|\alpha| \le m$. Due to the denseness of $C_c^\infty(\Omega)$ in $L^p(\Omega)$, the functions v_α are uniquely defined by (E.1), and they are usually denoted as $D^\alpha u$. One can easily prove that $W^{m,p}(\Omega)$ is a separable Banach space when endowed with the norm $\|\cdot\|_{W^{m,p}}$ defined by the formula

$$\langle u, v\rangle_{W^{m,p}} = \Big(\sum_{|\alpha|\le m} \|D^\alpha u\|_{L^p}^p \Big)^{1/p}.$$

Notice that $C^m(\overline{\Omega})$ is a dense subspace of $W^{m,p}(\Omega)$.

It is usual to denote $W^{m,p}(\Omega)$ by $H^m(\Omega)$ for $p = 2$; this is a Hilbert space relative to the norm $\|\cdot\|_{H^m}$, associated to the scalar product

$$\langle u, v\rangle_{H^m} = \sum_{|\alpha|\le m} \int_\Omega D^\alpha u \cdot D^\alpha v \, dx.$$

E.2.1 Theorem (The trace theorem) *There is a continuous linear operator*

$$\gamma = (\gamma_0, \dots, \gamma_{m-1})\colon W^{m,p}(\Omega) \to L^p(\partial\Omega)^{m-1}$$

such that

$$\gamma_0(u) = u|_{\partial\Omega}, \quad \gamma_1(u) = \frac{\partial u}{\partial n}, \dots, \gamma_{m-1}(u) = \frac{\partial^{m-1} u}{\partial n^{m-1}}$$

for all u in $C^m(\overline{\Omega})$.

The closure of $C_c^\infty(\Omega)$ in $W^{m,p}(\Omega)$ is the Sobolev space $W_0^{m,p}(\Omega)$ (denoted $H_0^m(\Omega)$ when $p = 2$). This space coincides with the kernel of the trace operator γ, indicated in Theorem E.2.1.

On $H_0^1(\Omega)$, the norm $\|\cdot\|_{H^1}$ can be replaced by an equivalent norm,

$$\|u\|_{H_0^1} = \left(\int_\Omega \|\nabla u\|^2 \, dx \right)^{1/2}.$$

In fact, there exists a constant $c > 0$ such that

$$\|u\|_{H_0^1} \le \|u\|_{H^1} \le c\|u\|_{H_0^1} \quad \text{for all } u \in H_0^1(\Omega).$$

This is a consequence of a basic inequality in partial differential equations.

E.2.2 Theorem (Poincaré's inequality) *If Ω is a bounded open subset of $\mathbb{R}^N$, then for every $p \in [1, \infty)$ there exists a constant $C > 0$ (depending only on Ω and p) such that*

$$\|u\|_{L^p} \le C\left(\int_\Omega \|\nabla u\|^p \, dx \right)^{1/p}$$

for all $u \in W_0^{1,p}(\Omega)$.

Proof. Since $C_c^\infty(\Omega)$ is dense into $W_0^{1,p}(\Omega)$, it suffices to prove Poincaré's inequality for functions $u \in C_c^\infty(\Omega) \subset C_c^\infty(\mathbb{R}^N)$. The fact that Ω is bounded yields two real numbers a and b such that

$$\Omega \subset \{\mathbf{x} = (\mathbf{x}', x_N) \in \mathbb{R}^{N-1} \times \mathbb{R} : a \leq x_N \leq b\}.$$

We have

$$u(x', x_N) = \int_a^{x_N} \frac{\partial u}{\partial x_N}(x', t)\, \mathrm{d}t,$$

and an application of the Rogers–Hölder inequality gives us

$$\begin{aligned} |u(\mathbf{x}', x_N)|^p &\leq (x_N - a)^{p/q} \int_a^{x_N} \left|\frac{\partial u}{\partial x_N}(\mathbf{x}', t)\right|^p \mathrm{d}t \\ &\leq (x_N - a)^{p/q} \int_{\mathbb{R}} \left|\frac{\partial u}{\partial x_N}(\mathbf{x}', t)\right|^p \mathrm{d}t. \end{aligned}$$

Then

$$\int_{\mathbb{R}^{N-1}} |u(\mathbf{x}', t)|^p \, \mathrm{d}\mathbf{x}' \leq (x_N - a)^{p/q} \int_{\mathbb{R}^N} \left|\frac{\partial u}{\partial x_N}(\mathbf{x})\right|^p \mathrm{d}\mathbf{x},$$

which leads to

$$\int_{\mathbb{R}^N} |u(\mathbf{x})|^p \, \mathrm{d}\mathbf{x} = \int_a^b \int_{\mathbb{R}^{N-1}} |u(\mathbf{x}', t)|^2 \, \mathrm{d}\mathbf{x}' \leq \frac{(b-a)^p}{p} \int_{\mathbb{R}^N} \left|\frac{\partial u}{\partial x_N}(\mathbf{x})\right|^p \mathrm{d}\mathbf{x}$$

and now the assertion of Theorem E.2.2 is clear. ■

By Poincaré's inequality, the inclusion $W_0^{m,p}(\Omega) \subset W^{m,p}(\Omega)$ is strict whenever Ω is bounded. Notice that $W_0^{m,p}(\mathbb{R}^N) = W^{m,p}(\mathbb{R}^N)$, due to the possibility to approximate (via mollification) the functions in $W^{m,p}(\mathbb{R}^N)$ by functions in $C_c^\infty(\mathbb{R}^N)$.

E.3 Applications to Elliptic Boundary-Value Problems

In what follows we shall illustrate the role of the variational methods in solving some problems in partial differential equations. More advanced applications may be found in books like those by B. Dacorogna [119], [120] and E. Giusti [182].

E.3.1 Dirichlet Problems. Let Ω be a bounded open set in $\mathbb{R}^N$ and let $f \in C(\overline{\Omega})$. A function $u \in C^2(\Omega) \cap C(\overline{\Omega})$ is said to be a classical solution of the Dirichlet problem

$$\begin{cases} -\Delta u + u = f & \text{in } \Omega \\ u = 0 & \text{on } \partial\Omega, \end{cases} \tag{E.2}$$

provided that it satisfies the equation and the boundary condition pointwise. If u is a classical solution to this problem, then the equation $-\Delta u + u = f$ is equivalent to

$$\int_\Omega (-\Delta u + u) \cdot v \, d\mathbf{x} = \int_\Omega f \cdot v \, d\mathbf{x} \quad \text{for all } v \in H_0^1(\Omega).$$

By Green's formula,

$$\int_\Omega (-\Delta u + u) \cdot v \, d\mathbf{x} = -\int_{\partial\Omega} \frac{\partial u}{\partial n} \cdot v \, d\mathbf{x} + \int_\Omega u \cdot v \, d\mathbf{x} + \sum_{k=1}^N \int_\Omega \frac{\partial u}{\partial x_k} \cdot \frac{\partial v}{\partial x_k} \, d\mathbf{x},$$

so that we arrive at the following restatement of (E.2):

$$\sum_{k=1}^N \int_\Omega \frac{\partial u}{\partial x_k} \cdot \frac{\partial v}{\partial x_k} \, d\mathbf{x} + \int_\Omega u \cdot v \, d\mathbf{x} = \int_\Omega f \cdot v \, d\mathbf{x} \tag{E.3}$$

for all $v \in C_c^\infty(\Omega)$. It turns out that (E.3) makes sense for $u \in H_0^1(\Omega)$ and $f \in L^2(\Omega)$. We shall say that a function $u \in H_0^1(\Omega)$ is a *weak solution* for the Dirichlet problem (E.2) with $f \in L^2(\Omega)$ if it satisfies (E.3) for all $v \in H_0^1(\Omega)$. The existence and uniqueness of the weak solution for the Dirichlet problem (E.2) follows from Theorem E.1.2, applied to the functional

$$J(u) = \frac{1}{2} \|u\|_{H_0^1}^2 - \langle f, u \rangle_{L^2}, \quad u \in H_0^1(\Omega).$$

In fact, this functional is strictly convex and twice Gâteaux differentiable, with

$$J'(u; v) = \langle u, v \rangle_{H_0^1} - \langle f, v \rangle_{L^2}$$
$$J''(u; v, w) = \langle w, v \rangle_{H_0^1}.$$

According to Theorem E.1.2, the unique point of global minimum of J is the unique solution of the equation

$$J'(u; v) = 0 \quad \text{for all } v \in H_0^1(\Omega),$$

and clearly, the latter is equivalent to (E.3).

E.3.2 Neumann Problems. Let Ω be a bounded open set in $\mathbb{R}^N$ (with Lipschitz boundary) and let $f \in C(\overline{\Omega})$. A function $u \in C^2(\Omega) \cap C^1(\overline{\Omega})$ is said to be a classical solution of the Neumann problem

$$\begin{cases} -\Delta u + u = f & \text{in } \Omega \\ \dfrac{\partial u}{\partial n} = 0 & \text{on } \partial\Omega, \end{cases} \tag{E.4}$$

provided that it satisfies the equation and the boundary condition pointwise. If u is a classical solution to this problem, then the equation $-\Delta u + u = f$ is equivalent to

$$\int_\Omega (-\Delta u + u) \cdot v \, d\mathbf{x} = \int_\Omega f \cdot v \, d\mathbf{x} \quad \text{for all } v \in H^1(\Omega),$$

and thus with

$$\sum_{k=1}^{N}\int_{\Omega}\frac{\partial u}{\partial x_k}\cdot\frac{\partial v}{\partial x_k}\,d\mathbf{x}+\int_{\Omega}u\cdot v\,d\mathbf{x}=\int_{\Omega}f\cdot v\,d\mathbf{x}\quad\text{for all } v\in H^1(\Omega),\tag{E.5}$$

taking into account Green's formula and the boundary condition $\frac{\partial u}{\partial n}=0$ on $\partial\Omega$. As in the case of Dirichlet problem, we can introduce a concept of a weak solution for the Neumann problem (E.4) with $f\in L^2(\Omega)$. We shall say that a function $u\in H^1(\Omega)$ is a *weak solution* for the problem (E.4) if it satisfies (E.5) for all $v\in H^1(\Omega)$. The existence and uniqueness of the weak solution for the Neumann problem follows from Theorem E.1.2, applied to the functional

$$J(u)=\frac{1}{2}\,\|u\|_{H^1}^2-\langle f,u\rangle_{L^2},\quad u\in H^1(\Omega).$$

The details are similar to the above case of Dirichlet problem.

Corollary E.1.3 (and its generalization to finite families of functions g_K) allows us to prove the existence and uniqueness of the weak solutions even for some Neumann problems such as

$$\begin{cases}-\Delta u+u+u^3=f & \text{in }\Omega\\ \dfrac{\partial u}{\partial n}=0 & \text{on }\partial\Omega,\end{cases}\tag{E.6}$$

where $f\in L^2(\Omega)$. This corresponds to the case where

$$\begin{gathered}g_1(t)=\cdots=g_N(t)=t^2/2,\quad g_{N+1}(t)=t^4/4,\\ A_k u=\partial u/\partial x_k\ \text{for } k=1,\ldots,N,\quad A_{N+1}u=u,\\ p_1=\cdots=p_N=2,\ p_{N+1}=4,\end{gathered}$$

and

$$J(u)=\frac{1}{2}\,\|u\|_{H^1}^2+\frac{1}{4}\,\|u\|_{L^4}^4-\langle f,u\rangle_{L^2},\quad u\in V=H^1(\Omega)\cap L^4(\Omega).$$

According to Corollary E.1.3, there is a unique global minimum of J and this is done by the equation

$$J'(u;v)=0\quad\text{for all } v\in V,$$

that is, by

$$\sum_{k=1}^{N}\int_{\Omega}\frac{\partial u}{\partial x_k}\cdot\frac{\partial v}{\partial x_k}\,d\mathbf{x}+\int_{\Omega}u\cdot v\,d\mathbf{x}+\int_{\Omega}u^3\cdot v\,d\mathbf{x}=\int_{\Omega}f\cdot v\,d\mathbf{x}$$

for all $v\in V$. Notice that the latter equation represents the weak form of (E.6). The conditions under which the weak solutions provide classical solutions are discussed in textbooks like that by M. Renardy and R. C. Rogers [416].

E.4 The Galerkin Method

It is important to give here an idea how the global minimum of convex functionals can be determined via numerical algorithms. For this, consider a reflexive real Banach space V, with Schauder basis $(\mathbf{e}_k)_k$. This means that every $\mathbf{u} \in V$ admits a unique representation

$$\mathbf{u} = \sum_{k=1}^{\infty} c_k \mathbf{e}_k$$

with $c_k \in \mathbb{R}$, the convergence being in the norm topology. As a consequence, for each $n \in N$ there is a linear projection

$$P_n \colon V \to V, \quad P_n\mathbf{u} = \sum_{k=1}^{n} c_k \mathbf{e}_k.$$

Since $P_n\mathbf{u} \to \mathbf{u}$ for every $\mathbf{u}$, the Banach–Steinhaus theorem in functional analysis assures that $\sup \|P_n\| < \infty$. Consider a functional $J \colon V \to \mathbb{R}$ which is twice Gâteaux differentiable and for each $\mathbf{u} \in V$ there exist $\nabla J(\mathbf{u}) \in V^*$ and $H(\mathbf{u}) \in L(V, V^*)$ such that

$$J'(\mathbf{u}; \mathbf{v}) = \langle \nabla J(\mathbf{u}), \mathbf{v} \rangle$$
$$J''(\mathbf{u}; \mathbf{v}, \mathbf{w}) = \langle H(\mathbf{u})\mathbf{v}, \mathbf{w} \rangle$$

for all $\mathbf{u}, \mathbf{v}, \mathbf{w} \in V$. In addition, we assume that $H(\mathbf{u})$ satisfies estimates of the form:

$$\begin{cases} |\langle H(\mathbf{u})\mathbf{v}, \mathbf{w} \rangle| \leq M\|\mathbf{v}\| \, \|\mathbf{w}\| \\ \langle H(\mathbf{u})\mathbf{v}, \mathbf{v} \rangle \geq \alpha\|\mathbf{v}\|^2 \end{cases} \tag{E.7}$$

for all $\mathbf{u}, \mathbf{v}, \mathbf{w} \in V$. Here M and α are positive constants. By Taylor's formula, J is strictly convex and $\lim_{\|\mathbf{u}\| \to \infty} J(\mathbf{u}) = \infty$. Then, by Theorem E.1.2, J is lower semicontinuous and admits a unique global minimum. In the Galerkin method , the global minimum $\mathbf{u}$ of J is found by a finite-dimensional approximation process. More precisely, one considers the restriction of J to $V_n = \text{span}\{\mathbf{e}_1, \ldots, \mathbf{e}_n\}$ and one computes the global minimum $\mathbf{u}_n$ of this restriction by solving the equation

$$\langle \nabla J(\mathbf{u}_n), \mathbf{v} \rangle = 0 \quad \text{for all } \mathbf{v} \in V_n.$$

The existence of $\mathbf{u}_n$ follows again from Theorem E.1.2. Remarkably, these minimum points approximate the global minimum $\mathbf{u}$ in the following strong way:

E.4.1 Theorem *We have*

$$\lim_{n \to \infty} \|\mathbf{u}_n - \mathbf{u}\| = 0.$$

Proof. Letting $v_n = P_n\mathbf{u}$, we know that $\mathbf{v}_n \to \mathbf{u}$. By Taylor's formula, for each n there is a $\lambda_n \in (0,1)$ such that

$$J(\mathbf{v}_n) = J(\mathbf{u}) + \langle \nabla J(\mathbf{u}), \mathbf{v}_n - \mathbf{u}\rangle + \frac{1}{2}\langle H(\mathbf{u} + \lambda_n(\mathbf{v}_n - \mathbf{u}))(\mathbf{v}_n - \mathbf{u}), \mathbf{v}_n - \mathbf{u}\rangle.$$

Combining this with the first estimate in (E.7), we get $J(\mathbf{v}_n) \to J(\mathbf{u})$. By the choice of $\mathbf{u}_n$, it yields that

$$J(\mathbf{u}) \le J(\mathbf{u}_n) \le J(\mathbf{v}_n),$$

so that $J(\mathbf{u}_n) \to J(\mathbf{u})$ too. Also, $\sup J(\mathbf{u}_n) < \infty$. Since $\lim_{\|\mathbf{u}\|\to\infty} J(\mathbf{u}) = \infty$, we deduce that the sequence $(\mathbf{u}_n)_n$ is norm bounded. According to Theorem B.1.6, it follows that $(\mathbf{u}_n)_n$ has a weak converging subsequence, say $\mathbf{u}_{k(n)} \stackrel{w}{\to} \mathbf{u}'$. Since J is lower semicontinuous, we have

$$J(\mathbf{u}') \le \liminf_{n\to\infty} J(\mathbf{u}_{k(n)}) \le J(\mathbf{u}),$$

from which it follows that $\mathbf{u}' = \mathbf{u}$ and $\mathbf{u}_n \stackrel{w}{\to} \mathbf{u}$. Again by Taylor's formula, for each n there is $\mu_n \in (0,1)$ such that

$$J(\mathbf{u}_n) = J(\mathbf{u}) + \langle \nabla J(\mathbf{u}), \mathbf{u}_n - \mathbf{u}\rangle + \frac{1}{2}\langle H(\mathbf{u} + \mu_n(\mathbf{u}_n - \mathbf{u}))(\mathbf{u}_n - \mathbf{u}), \mathbf{u}_n - \mathbf{u}\rangle.$$

This relation, when combined with the second estimate in (E.7), leads to

$$\frac{2}{\alpha}\|\mathbf{u}_n - \mathbf{u}\|^2 \le |J(\mathbf{u}_n) - J(\mathbf{u})| + |\langle \nabla J(\mathbf{u}), \mathbf{u}_n - \mathbf{u}\rangle|$$

and the conclusion of the theorem is now obvious. ■

References

[1] N. H. Abel, *Untersuchungen über die Reihe* $1+\frac{m}{1}x+\frac{m(m-1)}{1\cdot 2}x^2+\frac{m(m-1)(m-2)}{1\cdot 2\cdot 3}x^3+\cdots$, J. Reine Angew. Math. **1** (1826), 311–339. See also *Œuvres complètes de N. H. Abel*, t. I, 66–92, Christiania, 1839; available online at http://archive.org/stream/oeuvrescomplte01abel

[2] S. Abramovich, G. Jameson and G. Sinnamon, *Refining Jensen's Inequality*, Bull. Math. Soc. Sci. Math. Roumanie **47** (95) (2004), no. 1–2, 3–14.

[3] J. Aczél, *The notion of mean values*, Norske Vid. Selsk. Forh., Trondhjem, **19** (1947), 83–86.

[4] J. Aczél, *A generalization of the notion of convex functions*, Norske Vid. Selsk. Forhdl., Trondhjem, **19** (1947), 87–90.

[5] J. Aczél and J. Dhombres, *Functional equations in several variables*, Encyclopedia of Mathematics and its Applications, Vol. **31**, Cambridge University Press, Cambridge, 1989.

[6] R. A. Adams and J. J. F. Fournier, *Sobolev Spaces*, Second Edition, Pure and Applied Mathematics, Vol. **140**, Elsevier Academic Press, Amsterdam, 2003.

[7] G. Alberti and L. Ambrosio, *A geometrical approach to monotone functions in* $\mathbb{R}^n$, Math. Z. **230** (1999), no. 2, 259–316.

[8] J. M. Aldaz, *A stability version of Hölder's inequality*, J. Math. Anal. Appl. **343** (2008), no. 2, 842 852.

[9] A. D. Alexandrov, *Almost everywhere existence of the second differential of a convex function and some properties of convex surfaces connected to it*, Leningrad State Univ. Ann., Math. Ser. **6** (1939), 3–35. (Russian)

[10] E. M. Alfsen, *Compact convex sets and boundary integrals*, Ergebnisse der Mathematik und ihrer Grenzgebiete, Band **57**, Springer-Verlag, New York - Heidelberg, 1971.

C. P. Niculescu and L.-E. Persson, *Convex Functions and Their Applications*,
CMS Books in Mathematics, https://doi.org/10.1007/978-3-319-78337-6

[11] Y. Al-Manasrah and F. Kittaneh, *A generalization of two refined Young inequalities*, Positivity **19** (2015), no. 4, 757–768.

[12] H. W. Alt, *Linear functional analysis. An application-oriented introduction.* Translated from the German edition 1992, Springer-Verlag, London, 2016.

[13] H. Alzer, *On an integral inequality*, Anal. Numér. Théor. Approx. **18** (1989), 101–103.

[14] H. Alzer, *The inequality of Ky Fan and related results*, Acta Appl. Math. **38** (1995), no. 3, 305–354.

[15] L. Ambrosio, G. Da Prato and A. Mennucci, *Introduction to Measure Theory and Integration*, Scuola Normale Superiore, Pisa, 2011.

[16] G. D. Anderson, M. K. Vamanamurthy and M. Vuorinen, *Generalized convexity and inequalities*, J. Math. Anal. Appl. **335** (2007), no. 2, 1294–1308.

[17] T. Ando, *Majorization, doubly stochastic matrices, and comparison of eigenvalues*, Linear Algebra Appl. **118** (1989), 163–248

[18] G. Andrews, R. Askey and R. Roy, *Special Functions*, Encyclopedia of Mathematics and its Applications, Vol. **71**, Cambridge University Press, Cambridge, 1999.

[19] B. C. Arnold, *Majorization and the Lorenz Order: A Brief Introduction*, Springer-Verlag, Lecture Notes in Statistics, Vol. **43**, 1987.

[20] E. Artin, *The gamma function.* English translation of German original, *Einführung in die Theorie der Gammafunktion* (Verlag B. G. Teubner, Leipzig, 1931). Dover Books on Mathematics, 2015.

[21] S. Artstein-Avidan and V. Milman, *The concept of duality in convex analysis, and the characterization of the Legendre transform*, Ann. of Math. **169** (2009), 661–674.

[22] W. Arveson and R. Kadison, *Diagonals of self-adjoint operators.* In vol. *Operator theory, operator algebras andapplications*, Contemp. Math. **414** (D. Han, P. Jorgensen and D. Larson, eds.), pp. 247–263. Amer. Math. Soc. Publ. Providence, R. I., 2006

[23] E. Asplund, *Fréchet differentiability of convex functions*, Acta Math. **121** (1968), 31–47.

[24] M. F. Atiyah, *Angular momentum, convex polyhedra and algebraic geometry*, Proc. Edinburgh Math. Soc. **26** (1983), no. 2, 121–138.

[25] G. Aumann, *Konvexe Funktionen und die Induktion bei Ungleichungen zwischen Mittelwerten*, Bayer. Akad. Wiss. Math.-Natur. Kl. S.-B., 1933, 403–415.

[26] M. Avriel, W. E. Diewert, S. Schaible and I. Zang I., *Generalized Concavity*, Classics in Applied Mathematics **63**, SIAM, Society for Industrial and Applied Mathematics, Philadelphia, 2010.

[27] D. Azagra and C. Mudarra, *Global approximation of convex functions by differentiable convex functions on Banach spaces*, J. Convex Anal. **22** (2015), no. 4, 1197–1205.

[28] K. Ball, *Logarithmically concave functions and sections of convex sets in* $\mathbb{R}^n$, Studia Math. **88** (1988), 69–84.

[29] V. Barbu and T. Precupanu, *Convexity and optimization in Banach spaces*, Fourth Edition, Springer Monographs in Mathematics, Springer-Verlag, Dordrecht, 2012.

[30] A. Baricz, *Generalized Bessel functions of the first kind*, Lecture Notes in Mathematics **1994**, Springer-Verlag, Berlin, 2010.

[31] F. Barthe, *Inégalités de Brascamp–Lieb et convexité*, C. R. Acad. Sci. Paris, Sér. I Math. **324** (1997), no. 8, 885–887.

[32] F. Barthe, *Optimal Young's inequality and its converse: a simple proof*, Geom. Funct. Anal. **8** (1998), 234–242.

[33] S. Barza and C. P. Niculescu, *Integral inequalities for concave functions*, Publ. Math. Debrecen **68** (2006), no. 1–2, 139–142.

[34] H. H. Bauschke and P. L. Combettes, *Convex Analysis and Monotone Operator Theory in Hilbert Spaces*, Second Edition, CMS Books in Mathematics, Springer International Publishing AG, 2017.

[35] H. H. Bauschke, O. Güler, A. S. Lewis and H. S. Sendov, *Hyperbolic Polynomials and Convex Analysis*, Canad. J. Math. **53** (2001), no. 3, 470–488.

[36] B. Beauzamy, *Introduction to Banach spaces and their geometry,* North-Holland Mathematics Studies, Vol. **68**, North Holland Publishing Co., Amsterdam-NewYork-Oxford, 1982.

[37] E. F. Beckenbach and R. Bellman, *Inequalities,* Second revised printing. Ergebnisse der Mathematik und ihrer Grenzgebiete, Neue Folge, Band **30**, Springer-Verlag, New York, 1965.

[38] R. Bellman, *Introduction to matrix analysis*, Second Edition, Society for Industrial and Applied Mathematics, Philadelphia, 1997.

[39] M. Bencze, C. P. Niculescu and F. Popovici, *Popoviciu's inequality for functions of several variables.* J. Math. Anal. Appl. **365** (2010), no. 1, 399–409.

[40] J. Bendat and S. Sherman, *Monotone and convex operator functions*, Trans. Amer. Math. Soc. **79** (1955), 58–71.

[41] R. Benedetti and J.-J. Risler, *Real algebraic and semi-algebraic sets*, Actualités Mathématiques, Hermann, Editeurs des Sciences et des Arts, Paris, 1990.

[42] C. Bennett and R. Sharpley, *Interpolation of Operators*, Pure and Applied Mathematics, Vol. **129**, Academic Press, Boston, 1988.

[43] G. Berkhan, *Zur projektivischen Behandlung der Dreiecksgeometrie*, Arch. Math. Phys. (3) **11** (1907), 1–31.

[44] J. Bernoulli, *Positiones Arithmeticae de seriebus infinitas, earumque summa finita*, Basileae, 1689, *Opera* 1, 375–402.

[45] M. Berger, *Convexity*, Amer. Math. Monthly **97** (1990), no. 8, 650–678.

[46] L. Berwald, *Verallgemeinerung eines Mittelwertsatzes von J. Favard, für positive konkave Funktionen*, Acta Math. **79** (1947), 17–37.

[47] M. Bessenyei and Z. Páles, *Higher order generalizations of Hadamard's inequality*, Publ. Math. Debrecen **61** (2002), no. 3–4, 623–643.

[48] M. Bessenyei and Z. Páles, *Hadamard-type inequalities for generalized convex functions,* Math. Inequal. Appl. **6** (2003), no. 3, 379–392.

[49] R. Bhatia, *Matrix analysis*, Graduate Texts in Mathematics **169**, Springer-Verlag, New York, 1997.

[50] R. Bhatia, *Linear algebra to quantum cohomology: The story of Alfred Horn's inequalities*, Amer. Math. Monthly **108** (2001), no. 4, 289–318.

[51] R. Bhatia, *Notes on functional analysis*, Texts and Readings in Mathematics, **50**, Hindustan Book Agency, New Dehli, 2009.

[52] R. Bhatia, *Positive Definite Matrices*, Princeton University Press, Princeton, 2015.

[53] R. Bhatia and C. Davis, *A better bound on the variance*, Amer. Math. Monthly **107** (2000), no. 4, 353–357.

[54] G. Bianchi, A. Colesanti and C. Pucci, *On the second differentiability of convex surfaces*, Geom. Dedicata **60** (1996), no. 1, 39–48.

[55] G. Birkhoff, *Tres observaciones sobre el algebra lineal*, Univ. Nac. Tucumán, Revista Ser. A **5** (1946), 147–150.

[56] Z. W. Birnbaum and W. Orlicz, *Über die Verallgemeinerung des Begriffes der zueinander konjugierten Potenzen*, Studia Math. **3** (1931), 1–67.

[57] E. Bishop and K. de Leeuw, *The representation of linear functionals by measures on sets of extreme points*, Ann. Inst. Fourier (Grenoble), **9** (1959), 305–331.

[58] M. Bîrsan, P. Neff, and J. Lankeit, *Sum of squared logarithms – an inequality relating positive definite matrices and their matrix logarithm.* J. Inequal. Appl. **2013** (2013), Paper no. 168, 16 pages.

[59] H. Blumberg, *On convex functions,* Trans. Amer. Math. Soc. **20** (1919), no. 1, 40-44.

[60] V. I. Bogachev, *Measure Theory,* Vols. I, II, Springer-Verlag, Berlin, 2007.

[61] H. F. Bohnenblust, S. Karlin and L. S. Shapley, *Games with continuous, convex pay-off.* In vol. *Contributions to the Theory of Games* (H.W. Kuhn and A.W. Tucker eds.), 181–192, Annals of Mathematics Studies **24**, Princeton University Press, 1950.

[62] H. Bohr and J. Mollerup, *Laerebog i Mathematisk Analyse*, vol. III, Kopenhagen, 1922.

[63] J. Borcea, *Equilibrium points of logarithmic potentials induced by positive charge distributions.* I. *Generalized de Bruijn-Springer relations*, Trans. Amer. Math. Soc. **359** (2007), no. 7, 3209–3237.

[64] Ch. Borell, *Convex set functions in d-space*, Period. Math. Hungar. **6** (1975), no. 2, 111–136.

[65] Ch. Borell, *Diffusion equations and geometric inequalities*, Potential Anal. **12** (2000), no. 1, 49–71.

[66] L. Borisov, P. Neff, S. Sra and Ch. Thiel, *The sum of squared logarithms inequality in arbitrary dimensions*, Linear Algebra Appl. **528** (2017), no. 1, 124–146.

[67] D. Borwein, J. M. Borwein, G. Fee and R. Girgensohn, *Refined convexity and special cases of the Blaschke-Santalo inequality*, Math. Inequal. Appl. **4** (2001), no. 4, 631–638.

[68] J. M. Borwein, *Continuity and differentiability properties of convex operators*, Proc. London Math. Soc. **44** (1982), no. 3, 420–444.

[69] J. M. Borwein, *Proximality and Chebyshev sets*, Optim. Lett. **1** (2007), no. 1, 21–32.

[70] J. M. Borwein, D. Bailey, and R. Girgensohn, *Experimentation in Mathematics*, A K Peters, Natick, MA, 2004.

[71] J. M. Borwein and P. B. Borwein, *The Way of All Means,* Amer. Math. Monthly **94** (1987), no. 6, 519–522.

[72] J. M. Borwein and A. S. Lewis, *Convex analysis and nonlinear optimization. Theory and examples,* Springer Science & Business Media, 2010.

[73] J. M. Borwein and J. Vanderwerff, *Convex functions: constructions, characterizations and counterexamples*, Encyclopedia of Mathematics and Its Applications **109,** Cambridge University Press, Cambridge, 2010.

[74] J. M. Borwein and Q. J. Zhu, *Techniques of variational analysis*, Canadian Mathematical Society Books in Maths, Vol. **20**, Springer-Verlag, New York, 2005.

[75] S. Boyd and L. Vandenberghe, *Convex Optimization*, Cambridge University Press, Cambridge, 2004.

[76] H. J. Brascamp and E. H. Lieb, *Best constants in Young's inequality, its converse and its generalization to more than three functions*, Adv. Math. **20** (1976), no. 2, 151–173.

[77] Y. Brenier, *Polar factorization and monotone rearrangement of vector-valued functions*, Comm. Pure Appl. Math. **44** (1991), no. 4, 375–417.

[78] Y. Brenier, *Hidden convexity in some nonlinear PDEs from geometry and physics*, J. Convex Anal. **17** (2010), no. 3–4, 945–959.

[79] J. L. Brenner and B. C. Carlson, *Homogeneous mean values: weights and asymptotics*, J. Math. Anal. Appl. **123** (1987), no. 1, 265–280.

[80] H. Brezis, *Opérateurs maximaux monotones et semigroups de contractions dans les espaces de Hilbert*, North-Holland Math. Stud. 5, North Holland, Amsterdam, 1973.

[81] P. S. Bullen, *The inequalities of Rado and Popoviciu*, Publ. Elektrotehn. Fak. Ser. Mat. Fiz. No. **332** (1970), 23–33.

[82] P. S. Bullen, *Error estimates for some elementary quadrature rules*, Publ. Elektrotehn. Fak. Ser. Mat. Fiz. No. **602–633** (1978), 97–103.

[83] P. S. Bullen, D. S. Mitrinović and P. M. Vasić, *Means and Their Inequalities*, D. Reidel Publishing Company, Dordrecht, 1988.

[84] V. Bouniakowsky, *Sur quelques inegalités concernant les intégrales aux différences finies*, Mem. Acad. Sci. St. Petersbourg, **1** (1859), no. 9, 1–18.

[85] Y. D. Burago and V. A. Zalgaller, *Geometric Inequalities*, Grundlehren der Mathematischen Wissenschaften **285**, Springer-Verlag, Berlin, 1988.

[86] F. Burk, *The Geometric, Logarithmic and Arithmetic Mean Inequality,* Amer. Math. Monthly **94** (1987), no. 6, 527–528.

[87] Ch. L. Byrne, *A First Course in Optimization*, CRC Press, Boca Raton, FL, 2015.

[88] A. Cambini and L. Martein, *Generalized convexity and optimization. Theory and applications*, Lecture Notes in Economics and Mathematical Systems **616**, Springer-Verlag, Berlin, 2009.

[89] P. Cannarsa and C. Sinestrari, *Semiconcave functions, Hamilton–Jacobi equations, and optimal control*, Progress in Nonlinear Differential Equations and their Applications **58**, Birkhäuser, Boston, MA, 2004.

[90] G. T. Cargo, *Comparable means and generalized convexity*, J. Math. Anal. Appl. **12** (1965), 387–392.

[91] T. Carleman, *Sur les fonctions quasi-analitiques.* In *Proc. 5th Scand. Math. Congress*, Helsingfors, Finland, 1923, 181–196.

[92] E. A. Carlen, *Trace inequalities and quantum entropy: an introductory course*, Contemporary Math. **529**, 73–140, Amer. Math. Soc., Providence, RI, 2010.

[93] B. C. Carlson, *Algorithms involving arithmetic and geometric means*, Amer. Math. Monthly **78** (1971), no. 5, 496–505.

[94] N. L. Carothers, *A short course on Banach space theory*, London Mathematical Society Students Texts **64,** Cambridge University Press, Cambridge, 2004.

[95] E. Cartan, *Leçons sur la géométrie des espaces de Riemann*, 2ème éd., Gauthiers-Villars, Paris, 1946.

[96] P. Cartier, J. M. G. Fell and P.-A. Meyer, *Comparaison des mesures portées par un ensemble convexe compact,* Bull. Soc. Math. France **92** (1964), 435–445.

[97] D. I. Cartwright and M. J. Field, *A refinement of the arithmetic mean-geometric mean inequality*, Proc. Amer. Math. Soc. **71** (1978), no. 1, 36–38.

[98] J. W. S. Cassels, *Measures of the non-convexity of sets and the Shapley-Folkman-Starr theorem*, Mathematical Proceedings of the Cambridge Philosophical Society, Vol. **78**, no. 3, 433–436, Cambridge University Press, 1975.

[99] A.-L. Cauchy, *Cours d'analyse de l'Ecole Royale Polytechnique*, 1ère partie, *Analyse algébrique*, Paris, 1821. See also, *Œuvres complètes*, IIe série, VII.

[100] A. Causa and F. Raciti, *A purely geometric approach to the problem of computing the projection of a point on a simplex*, J. Optim. Theory Appl. **156** (2013), no. 2, 524–528.

[101] J. Céa, *Optimisation. Théorie et Algorithmes*, Dunod, Paris, 1971.

[102] S. H. Chang, *On the distribution of the characteristic values and singular values of linear integral equations,* Trans. Amer. Math. Soc. **67** (1949), 351–369.

[103] G. Chiti, *Rearrangements of functions and convergence in Orlicz spaces.* Applicable Anal. **9** (1979), no. 1, 23–27.

[104] K.-M. Chong, *Some extensions of a theorem of Hardy, Littlewood and Pólya and their applications*, Can. J. Math. **26** (1974), 1321–1340.

[105] G. Choquet, *Le théorème de représentation intégrale dans les ensembles convexes compacts*, Ann. Inst. Fourier (Grenoble) **10** (1960), 333–344.

[106] G. Choquet, *Les cônes convexes faiblement complets dans l'Analyse*, Proc. Intern. Congr. Mathematicians, Stockholm (1962), 317–330.

[107] A. D. R. Choudary and C. P. Niculescu, *Real analysis on intervals*, Springer, New Delhi-Heidelberg-New York-Dordrecht-London, 2014.

[108] A. Čižmešija and J. E. Pečarić, *Mixed means and Hardy's inequality*, Math. Inequal. Appl. **1** (1998), no. 4, 491–506.

[109] A. Čižmešija, J. E. Pečarić and L.-E. Persson, *On strengthened Hardy and Pólya–Knopp's inequalities*, J. Approx. Theory **125** (2003), no. 3, 74–84.

[110] F. H. Clarke, *Optimization and nonsmooth analysis*, Second Edition, Classics in Applied Mathematics, Vol. **5**, Society for Industrial and Applied Mathematics (SIAM), Philadelphia, PA, 1990.

[111] J. A. Clarkson, *Uniformly convex spaces*, Trans. Amer. Math. Soc. **40** (1936), no. 3, 396–414.

[112] R. Coleman, *Calculus on normed vector spaces*, Springer-Verlag, New York, 2012.

[113] A. Colesanti, *Brunn–Minkowski inequalities for variational functionals and related problems*, Adv. Math. **194** (2005), 105–140

[114] P. L. Combettes, *Perspective functions: Properties, constructions, and examples*, Set-Valued and Variational Analysis, pp. 1–18, 2017.

[115] J. B. Conway, *A course in functional analysis*, Second Edition, Springer-Verlag, Berlin, 1997.

[116] D. Cordero-Erausquin, *Some applications of mass transport to Gaussian-type inequalities*, Arch. Ration. Mech. Anal. **161** (2002), no. 3, 257–269.

[117] D. Cordero-Erausquin, R. J. McCann and M. Schmuckenschläger, *A Riemannian interpolation inequality à la Borell, Brascamp and Lieb*, Invent. Math. **146** (2001), no. 2, 219–257.

[118] M. G. Crandall, H. Ishii and P.-L. Lions, *User's guide to viscosity solutions of second order partial differential equations*, Bull. Amer. Math. Soc. **27** (1992), no. 1, 1–67.

[119] B. Dacorogna, *Direct Methods in the Calculus of Variations*, Second Edition, Springer-Verlag, New York, 2008.

[120] B. Dacorogna, *Introduction to the Calculus of Variations,* Third Edition, World Scientific Publishing, 2014.

[121] S. Dancs and B. Uhrin, *On a class of integral inequalities and their measure-theoretic consequences*, J. Math. Anal. Appl. **74** (1980), no. 2, 388–400.

[122] L. Danzer, B. Grunbaum, and V. Klee, *Helly's theorem and its relatives.* In: *Convexity*, Proc. Symposia Pure Math., Vol. **VII**, 101–180, Amer. Math. Soc., Providence, RI., 1963.

[123] J. Dattorro, *Convex Optimization and Euclidean Distance Geometry,* Meboo Publishing, USA, 2005.

[124] C. Davis, *All convex invariant functions of hermitian matrices.* Arch. Math. **8** (1957), no. 4, 276–278.

[125] E. B. Davies, *The structure and ideal theory of the pre-dual of a Banach lattice*, Trans. Amer. Math. Soc. **131** (1968), 544–555.

[126] E. B. Davies, *Heat Kernels and Spectral Theory,* Pure and Applied Mathematics, Vol. **X**, Cambridge University Press, Cambridge, 1989.

[127] K. R. Davidson and A. P. Donsig, *Real Analysis with Real Applications,* Prentice Hall, Upper Saddle River, N.J., 2002.

[128] M. M. Day, *Normed linear spaces*, Third Edition, Springer-Verlag, Berlin, 1973.

[129] M. de Carvalho, *Mean, what do you Mean?*, Amer. Statist. **70** (2016), no. 3, 764–776.

[130] J. B. Diaz and F. T. Metcalf, *Stronger forms of a class of inequalities of G. Pólya-G. Szegö, and L. V. Kantorovich*, Bull. Amer. Math. Soc. **69** (1963), no. 3, 415–418.

[131] J. Dieudonné, *Foundations of modern analysis*, Academic Press, New York-London, 1960.

[132] J. Diestel, H. Jarchow and A. Tonge, *Absolutely Summing Operators,* Cambridge Studies in Advanced Mathematics, Vol. **43**, Cambridge University Press, Cambridge, 1995.

[133] S. S. Dragomir, *On Hadamard's inequality for the convex mappings defined on a ball in the space and applications*, Math. Inequal. Appl. **3** (2000), no. 2, 177–187.

[134] S. S. Dragomir and N. M. Ionescu, *Some converse of Jensen's inequality and applications*, Rev. Anal. Numér. Théor. Approx. **23** (1994), no. 1, 71–78.

[135] S. S. Dragomir, C. E. M. Pearce and J. E. Pečarić, *On Jessen's and related inequalities for isotonic sublinear functionals*, Acta Sci. Math. (Szeged) **61** (1995), no. 1–4, 373–382.

[136] S. Dubuc, *Critères de convexité et inégalités intégrales*, Ann. Inst. Fourier Grenoble **27** (1977), no. 1, 135–165.

[137] J. Duncan and C. M. McGregor, *Carleman's inequality*, Amer. Math. Monthly **110** (2003), no. 5, 424–431.

[138] D. E. Dutkay, C. P. Niculescu and F. Popovici, *A short proof of Stirling's formula*, Amer. Math. Monthly **120** (2013), no. 8, 733–736.

[139] G. Duvaut and J. L. Lions, *Inequalities in Mechanics and Physics*, Grundlehren der mathematischen Wissenschaften **219**, Springer-Verlag, Berlin-New York, 1976.

[140] A. Ebadian, I. Nikoufar and M. E. Gordji, *Perspectives of matrix convex functions*, Proc. Natl. Acad. Sci. U.S.A. **108** (2011), no. 18, 7313–7314.

[141] J. Eells and B. Fuglede, *Harmonic maps between Riemannian polyhedra*, Cambridge University Press, 2001.

[142] E. G. Effros, *A matrix convexity approach to some celebrated quantum inequalities*, Proc. Natl. Acad. Sci. U.S.A. **106** (2009), no. 4, 1006–1008.

[143] E. G. Effros and F. Hansen, *Non-commutative perspectives*, Ann. Funct. Anal. **5** (2014), no. 2, 74–79.

[144] H. G. Eggleston, *Convexity,* Cambridge Tracts in Mathematics and Mathematical Physics, No. **47**, Cambridge University Press, New York, 1958.

[145] I. Ekeland and R. Temam, *Convex Analysis and Variational Problems*, North-Holland Publishing Company, Amsterdam-Oxford, 1976.

[146] N. Elezović and J. E. Pečarić, *Differential and integral F-means and applications to digamma function*, Math. Inequal. Appl., **3** (2000), no. 2, 189–196.

[147] P. Embrechts and M. Hofert, *A note on generalized inverses*, Math. Methods Oper. Res. **77** (2013), no. 3, 423–432.

[148] A. Engel, *Problem-solving strategies*, Problem Books in Mathematics, Springer-Verlag, New York, 1998.

[149] Euclid, *The thirteen books of Euclid's Elements* (translated by Sir Thomas Heath, Cambridge, 1908).

[150] L.C. Evans, *Weak Convergence Methods for Nonlinear Partial Differential Equations*, CBMS Regional Conference Series in Mathematics **74**, American Mathematical Society, Providence, RI, 1990.

[151] L. C. Evans, *Partial differential equations*, Second Edition, Graduate Studies in Mathematics, Vol. **19**, American Mathematical Society, Providence, 2010.

[152] L. C. Evans and R. F. Gariepy, *Measure theory and fine properties of functions*, Revised Edition, Textbooks in Mathematics, CRC Press, Boca Raton–New York-London-Tokyo, 2015.

[153] M. Egozcue, G. L. Fuentes and W. K. Wong, *On some covariance inequalities for monotonic and non-monotonic functions*, J. Inequal. Pure Appl. Math. (JIPAM) **10** (2009), no. 3, Article 75, 7 pages.

[154] K. Fan, *On a theorem of Weyl concerning eigenvalues of linear transformations* I, Proc. Nat. Acad. Sci. U.S.A. **35** (1949), no. 11, 652–655.

[155] K. Fan, *On a theorem of Weyl concerning eigenvalues of linear transformations* II, Proc. Nat. Acad. Sci. U.S.A. **36** (1950), no. 1, 31–35.

[156] K. Fan, *A minimax inequality and applications*, in vol. *Inequalities* III (O. Shisha Editor), pp. 103–113, Academic Press, New York, 1972.

[157] J. Farkas, *Über die Theorie der Einfachen Ungleichungen*, J. Reine Angew. Math. **124** (1902), 1–27

[158] J. Favard, *Sur les valeurs moyennes,* Bull. Sci. Math. **57** (1933), 54–64.

[159] H. Federer, *Geometric measure theory*, Classics in Mathematics, Springer-Verlag, New York, 2014.

[160] W. Fenchel, *On conjugate convex functions,* Canadian J. Math. **1** (1949), 73–77.

[161] W. Fenchel, *Convex cones, sets and functions* (mimeographed lecture notes). Princeton University Press, Princeton, 1951.

[162] B. de Finetti, *Sul concetto di media*, Giornale dell' Instituto Italiano degli Attuari **2** (1931), 369–396.

[163] A. M. Fink, *Kolmogorov-Landau inequalities for monotone functions*, J. Math. Anal. Appl. **90** (1982), no. 1, 251–258.

[164] A. M. Fink, *A best possible Hadamard inequality*, Math. Inequal. Appl. **1** (1998), no. 2, 223–230.

[165] P. Fischer and J. A. R. Holbrook, *Balayage defined by the nonnegative convex functions*, Proc. Amer. Math. Soc. **79** (1980), 445–448.

[166] A. Florea and C. P. Niculescu, *A note on Ostrowski's inequality.* J. Ineq. Appl. 2005 (2005), no. 5, 459–468.

[167] A. Florea and C. P. Niculescu, *A Hermite-Hadamard inequality for convex-concave symmetric functions*, Bull. Math. Soc. Sci. Math. Roumanie **50** (98) (2007), no. 2, 149–156.

[168] J. Franklin, *Methods of mathematical economics. Linear and nonlinear programming, fixed-point theorems.* Classics in applied mathematics Vol. **37**, SIAM, 2002.

[169] J. Franklin, *Mathematical Methods of Economics*, Amer. Math. Monthly **90** (1983), no. 4, 229–244.

[170] L. Fuchs, *A new proof of an inequality of Hardy, Littlewood and Pólya*, Mat. Tidsskr. B., 1947, 53–54.

[171] W. Fulton, *Eigenvalues, invariant factors, highest weights and Schubert calculus.* Bull. Amer. Math. Soc. **37** (2000), no. 3, 209–249.

[172] D. Gale, V. Klee and R. T. Rockafellar, *Convex functions on convex polytopes*, Proc. Amer. Math. Soc. **19** (1968), 867–873.

[173] L. Galvani, *Sullefunzioni converse di una o due variabili definite in aggregate qualunque,* Rend. Circ. Mat. Palermo **41** (1916), 103–134.

[174] W. Gangbo and R. J. McCann, *The geometry of optimal transportation*, Acta Math. **177** (1996), no. 2, 113–161.

[175] R. J. Gardner, *The Brunn-Minkowski inequality: A survey with proofs.* Preprint, 2001.

[176] R. J. Gardner, *The Brunn-Minkowski inequality*, Bull. Amer. Math. Soc. **39** (2002), no. 3, 355–405.

[177] L. Gårding, *Linear hyperbolic differential equations with constant coefficients*, Acta Math. **85** (1951), 2–62.

[178] L. Gårding, *An inequality for hyperbolic polynomials*, J. Math. Mech. **8** (1959), no. 6, 957–965.

[179] M. Gidea and C.P. Niculescu, *A brief account on Lagrange's identity*, The Math. Intelligencer **34** (2012), no. 3, 55–61.

[180] M. Gieraltowska-Kedzierska and F. S. van Vleck, *Fréchet vs. Gâteaux differentiability of Lipschitzian functions*, Proc. Amer. Math. Soc. **114** (1992), no. 4, 905–907.

[181] P. M. Gill, C. E. M. Pearce, and J. Pečarić, *Hadamard's inequality for r-convex functions*, J. Math. Anal. Appl. **215** (1997), no. 2, 461–470.

[182] E. Giusti, *Direct Methods in the Calculus of Variations,* World Scientific Publishing, 2003.

[183] E. K. Godunova, *Convexity of composed functions and its application to the proof of inequalities*, Mat. Zametki **1** (1967), 495–500. (Russian)

[184] J. Gondzio, *Interior point methods 25 years later*, European J. Oper. Res. **218** (2012), no. 3, 587–601.

[185] C.E. Gounaris and C.A. Floudas, *Convexity of products of univariate functions and convexification transformations for geometric programming*, J. Optim. Theory Appl. **138** (2008), no. 3, 407–427.

[186] I. S. Gradshteyn and I. M. Ryzhik, *Table of integrals, series and products,* Seventh edition. Elsevier Academic Press, Amsterdam, 2007. (CD-ROM version)

[187] H. J. Greenberg and W. P. Pierskalla, *A review of quasi-convex functions,* Operations Research, **19** (1971), no. 7, 1553–1570.

[188] D. Gronau and J. Matkowski, *Geometrical convexity and generalizations of the Bohr-Mollerup theorem on the Gamma function*, Math. Panon. **4** (1993), no. 2, 153–160.

[189] D. Gronau and J. Matkowski, *Geometrically convex solutions of certain difference equations and generalized Bohr-Mollerup type theorems*, Results in Math. **26** (1994), no. 3–4, 290–297.

[190] P. M. Gruber, *Aspects of convexity and its applications,* Expo. Math. **2** (1984), no. 1, 47–83.

[191] P. M. Gruber and J. M. Willis (eds), *Convexity and its Applications,* Birkhäuser Verlag, Basel, 1983.

[192] P. M. Gruber and J. M. Willis (eds), *Handbook of convex geometry*, Vol. A, B, North-Holland, Amsterdam, 1993.

[193] G. Grüss, *Über das Maximum des absoluten Betrages von* $\frac{1}{b-a}\int_a^b f(x)g(x)dx - \frac{1}{b-a}\int_a^b f(x)dx\frac{1}{b-a}\int_a^b g(x)dx$, Math. Z. **39** (1935), no. 1, 215–226.

[194] K. Gustafson, *The Toeplitz-Hausdorff theorem for linear operators*, Proc. Amer. Math. Soc. **25** (1970), 203–204.

[195] J. Hadamard, *Étude sur les propriétés des fonctions entières et en particulier d'une fonction considerée par Riemann,* J. Math. Pures Appl. **58** (1893), 171–215.

[196] G. Hamel, *Eine Basis aller Zahlen und die unstetigen Lösungen der Funktionalgleichung* $f(x+y) = f(x) + f(y)$, Math. Ann. **60** (1905), no. 3, 459–462.

[197] P. C. Hammer, *The midpoint method of numerical integration*, Math. Mag. **31** (1958), no. 4, 193–195.

[198] O. Hanner, *On the uniform convexity of* L^p *and* l^p, Ark. Mat. **3** (1956), no. 3, 239–244.

[199] F. Hansen, *The fast track to Löwner's theorem*, Linear Algebra Appl. **438**, (2013), no. 11, 4557–4571.

[200] F. Hansen, *Regular operator mappings and multivariate geometric means*, Linear Algebra Appl. **461** (2014), no. 1, 123–138.

[201] F. Hansen and G. K. Pedersen, *Jensen's inequality for operators and Löwner's theorem*, Math. Ann. **258** (1982), no. 3, 229–241.

[202] F. Hansen and G. K. Pedersen, *Jensen's operator inequality*, Bull. London Math. Soc. **35** (2003), no. 4, 553–564.

[203] F. Hansen, J. E. Pečarić and I. Perić, *Jensen's operator inequality and its converses*, Math. Scand. **100** (2007), 61–73.

[204] A. Hantoute, M. A. López, and C. Zălinescu, *Subdifferential calculus rules in convex analysis: a unifying approach via pointwise supremum functions*, SIAM J. Opim., **19** (2008), no. 2, 863–882.

[205] G. H. Hardy, *Notes on some points in the integral calculus,* LX. *An inequality between integrals,* Messenger Math. **54** (1925), 150–156.

[206] G. H. Hardy, *Notes on some points in the integral calculus,* LXIV. *Further inequalities between integrals,* Messenger Math. **57** (1928), 12–16.

[207] G. H. Hardy, *A note on two inequalities*, J. London Math. Soc. **11** (1936), 167–170.

[208] G. H. Hardy, J. E. Littlewood and G. Pólya, *Some simple inequalities satisfied by convex functions*, Messenger Math. **58** (1929), 145–152.

[209] G. H. Hardy, J. E. Littlewood and G. Pólya, *Inequalities,* Second Edition, Cambridge University Press, 1952. Reprinted 1988.

[210] H. Heinig and L. Maligranda, *Weighted inequalities for monotone and concave functions*, Studia Math. **116** (1995), no. 2, 133–165.

[211] J. W. Helton and V. Vinnikov, *Linear matrix inequality representation of sets*, Comm. Pure Appl. Math. **60** (2007), no. 5, 654–674.

[212] R. Henstock and A. M. Macbeath, *On the measure of sum sets* I. *The theorems of Brunn, Minkowski and Lusternik*, Proc. London Math. Soc. **3** (1953), 182–194.

[213] Ch. Hermite, *Sur deux limites d'une intégrale définie*, Mathesis **3** (1883), 82.

[214] E. Hewitt and K. Stromberg, *Real and abstract analysis*, Springer-Verlag, Berlin-Heidelberg-New York, 1965.

[215] J.-B. Hiriart-Urruty, *Ensembles de Tchebychev vs ensembles convexes: l'état de la situation vu via l'analyse non lisse*, Ann. Sci. Math. Québec, **22** (1998), no. 1, 47–62.

[216] J.-B. Hiriart-Urruty, *La lutte pour les inégalités*, Revue de la filière mathématiques RMS (ex-Revue Mathématiques Spéciales) **122** (2011–2012), no. 2, 12–18.

[217] J.-B. Hiriart-Urruty and C. Lemaréchal, *Convex Analysis and Minimization Algorithms.* Springer-Verlag, New York, 1993.

[218] J.-B. Hiriart-Urruty and C. Lemaréchal. *Fundamentals of convex analysis.* Corrected Second Printing 2004, Springer-Verlag, Berlin-Heidelberg, 2012.

[219] A. J. Hoffman and H. W. Wielandt, *The variation of the spectrum of a normal matrix*, Duke Math. J. **20** (1953), 37–40.

[220] A. Horn, *Doubly stochastic matrices and the diagonal of a rotation matrix*, Amer. J. Math. **76** (1954), no. 3, 620–630.

[221] A. Horn, *On the eigenvalues of a matrix with prescribed singular values*, Proc. Amer. Math. Soc. **5** (1954), no. 1, 4–7.

[222] A. Horn, *Eigenvalues of Sums of Hermitian matrices*, Pacific J. Math. **12** (1962), no. 1, 225–241.

[223] R. A. Horn and C. R. Johnson, *Matrix analysis.* Corrected reprint of the 1985 original. Cambridge University Press, Cambridge, 1990.

[224] H. Hornich, *Eine Ungleichung für Vektorlängen*, Math. Z. **48** (1942), 268–274.

[225] O. Hölder, *Über einen Mittelwertsatz*, Nachr. Ges. Wiss. Goettingen, 1889, 38–47.

[226] L. Hörmander, *Sur la fonction d'appui des ensembles convexes dans une espace localement convexe*, Ark. Mat. **3** (1955), no. 2, 181–186.

[227] L. Hörmander, *Notions of Convexity*, Reprint of the 1994 edition. Modern Birkhäuser Classics, Birkhäuser, Boston, 2007.

[228] K. S. K. Iyengar, *Note on an inequality*, Math. Student **6** (1938), 75–76.

[229] R. C. James, *Weakly compact sets*, Trans. Amer. Math. Soc. **113** (1964), 129–140.

[230] J. L. W. V. Jensen, *Om konvexe Funktioner og Uligheder mellem Middelvaerdier,* Nyt. Tidsskr. Math. **16B** (1905), 49–69.

[231] J. L. W. V. Jensen, *Sur les fonctions convexes et les inegalités entre les valeurs moyennes*, Acta Math. **30** (1906), 175–193.

[232] B. Jessen, *Bemaerkinger om konvekse Funktioner og Uligheder imellem Middelvaerdier* I, Mat. Tidsskrift B 1931, 17–28.

[233] K. Jichang, *Some extensions and refinements of Minc-Sathre inequality,* Math. Gazette **83** (1999), Note 83.17, 123–127.

[234] M. Johansson, L.-E. Persson and A. Wedestig, *Carleman's inequality-history, proofs and some new generalizations*, J. Inequal. Pure Appl. Math. (JIPAM) **4** (2003), no. 3, article 53, 19 pp.

[235] W. B. Johnson and J. Lindenstrauss, *Basic concepts in the geometry of Banach spaces.* Handbook of the geometry of Banach spaces, Vol. I, 599–670, North-Holland, Amsterdam, 2001.

[236] B. Josefson, *Weak sequential convergence in the dual of a Banach space does not imply norm convergence*, Ark. Mat. **13** (1975), 79–89.

[237] J. Jost, *Nonpositive curvature: geometric and analytic aspects*, Lectures in Mathematics ETH Zürich, Birkhäuser Verlag, Basel, 1997.

[238] K. Jörgens, *Linear Integral Operators,* Surveys and Reference Works in Mathematics, **7**, Pitman, Boston-London-Melbourne, 1982.

[239] I. R. Kachurovskii, *On monotone operators and convex functionals*, Uspekhi Mat. Nauk **15** (1960), no. 4, 213–215.

[240] S. Kaijser, L.-E. Persson and A. Öberg, *On Carleman and Knopp's inequalities*, J. Approx. Theory **117** (2002), no. 1, 140–151.

[241] J. Karamata, *Sur une inégalité rélative aux fonctions convexes*, Publ. Math. Univ. Belgrade **1** (1932), no. 1, 145–147.

[242] S. Karlin and A. Novikoff, *Generalized convex inequalities.* Pac. J. Math. **13** (1963), 1251–1279.

[243] N. D. Kazarinoff, *Analytic Inequalities,* Holt, Rinehart and Winston, New York, 1961.

[244] B. Kawohl, *When are superharmonic functions concave? Applications to the St. Venant torsion problem and to thefundamental mode of the clamped membrane*, Z. Angew. Math. Mech. **64** (1984), no. 5, 364–366.

[245] B. Kawohl, *Rearrangements and Convexity of Level Sets in PDE*, Lecture Notes in Mathematics **1150**, Springer-Verlag, Berlin, 1985.

[246] K. Kedlaya, *Proof of a mixed arithmetic-mean, geometric-mean inequality*, Amer. Math. Monthly **101** (1994), no. 4, 355–357.

[247] J. L. Kelley, *General Topology,* Courier Dover Publications, 2017.

[248] S. Kerov, *Interlacing measures*. In *Kirillov's Seminar on Representation Theory*, Amer. Math. Soc. Transl. Ser. 2, vol. **181**, Amer. Math. Society, Providence, RI (1998), 35–83.

[249] L. G. Khanin, *Problem* M 1083, Kvant, **18** (1988), No. 1, p. 35 and Kvant **18** (1988), no. 5, p. 35.

[250] C. H. Kimberling, *Some corollaries to an integral inequality*, Amer. Math. Monthly **81** (1974), no. 3, 269–270.

[251] M. S. Klamkin, *Inequalities for inscribed and circumscribed polygons*, Amer. Math. Monthly **87** (1980), no. 6, 469–473.

[252] V. Klee, *Some new results on smoothness and rotundity in normed linear spaces*, Math. Ann. **139** (1959), no. 1, 51–63.

[253] V. Klee, *Convexity of Chebyshev sets*, Math. Ann. **142** (1961), no. 3, 292–304.

[254] A. A. Klyachko, *Stable bundles, representation theory and Hermitian operators*, Selecta Math. **4** (1998), no. 3, 419–445.

[255] K. Knopp, *Über Reihen mit positiven Gliedern*, J. London Math. Soc. **3** (1928), no. 1, 205–211.

[256] A. Knutson and T. Tao, *The honeycomb model of $GL_n(\mathbb{C})$ tensor products I: Proof of the saturation conjecture.* J. Amer. Math. Soc. **12** (1999), no. 4, 1055–1090.

[257] A. Kolmogorov, *Sur la notion de la moyenne*, Atti della Academia Nazionale dei Lincei, **12** (1930), 323–343; also available in *Selected Works of A. N. Kolmogorov*, Vol. 1: *Mathematics and Mechanics*, 144–146, Dordrecht, Kluwer, 1991.

[258] B. Kostant, *On convexity, the Weyl group and the Iwasawa decomposition*, Ann. Sci. École Norm. Sup. **6** (1973) no. 4, 413–455.

[259] M. A. Krasnosel'skii and Ya. B. Rutickii, *Convex functions and Orlicz spaces*, P. Nordhoff, Groningen, 1961.

[260] J. L. Krivine, *Théorèmes de factorisation dans les espaces réticulés*, Séminaire Analyse fonctionnelle (dit" Maurey-Schwartz"), 1973–74, Exposés 22–23, Ecole Polytechnique, Paris.

[261] F. Kubo and T. Ando, *Means of positive linear operators*, Math. Ann. **246** (1980), no. 3, 205–224.

[262] A. Kufner, *Weighted inequalities and spectral problems*, Banach J. Math. Anal. **4** (2010), no. 1, 116–121

[263] A. Kufner, L. Maligranda and L.-E. Persson, *The prehistory of the Hardy inequality*, Amer. Math. Monthly **113** (2006), no. 8, 715–732.

[264] A. Kufner, L. Maligranda and L.-E. Persson, *The Hardy inequality – About its history and some related results*, Vydavatelski Servis Publishing House, Pilzen, 2007.

[265] A. Kufner, L.-E. Persson and N. Samko, *Weighted Inequalities of Hardy type,* Second Edition, World Scientific, New Jersey-London-Singapore, 2017.

[266] J. L. Lagrange, *Solutions analytiques de quelques problémes sur les pyramides triangulaires.* Nouveaux Mémoirs de l'Académie Royale de Berlin, 1773; see *Oeuvres* de Lagrange, vol. **3**, 661–692, Gauthier-Villars, Paris, 1867.

[267] J. L. Lagrange: *Sur une nouvelle proprieté du centre de gravité*, Nouveaux Mémoirs de l'Académie Royale de Berlin, 1783; see *Oeuvres* de Lagrange, vol. **5**, 535–540, Gauthier-Villars, Paris, 1870.

[268] J. Lamperti, *On the isometries of certain function-spaces*, Pacific J. Math. **8** (1958), 459–466.

[269] R. Lang, *A note on the measurability of convex sets*, Arch. Math. (Basel) **47** (1986), no. 1, 90–92.

[270] S. Lang, *Analysis* I, Addison-Wesley Publ. Co., Reading, Massachusetts, 1968.

[271] S. Lang, *Analysis* II, Addison-Wesley Publ. Co., Reading, Massachusetts, 1969.

[272] J. M. Lasry and P.-L. Lions, *A remark on regularization in Hilbert spaces*, Israel J. Math. **55** (1986), no. 3, 257–266.

[273] P.-J. Laurent, *Approximation et optimisation*, Hermann, Paris, 1972.

[274] J. D. Lawson and Y. Lim, *The geometric mean, matrices, metrics and more*, Amer. Math. Monthly **108** (2001), no. 9, 797–812.

[275] J. Lawson and Y. Lim, *The least squares mean of positive Hilbert-Schmidt operators*, J. Math. Anal. Appl. **403** (2013), 365–375.

[276] S. Lehmich, P. Neff, and J. Lankeit, *On the convexity of the function $c \to f(\det c)$* on positive-definite matrices, Math. Mech. Solids **19** (2014), 369–375.

[277] V. I. Levin and S. B. Stečkin, *Inequalities*, Amer. Math. Soc. Transl. **14** (1960), 1–22.

[278] A. S. Lewis, *Convex analysis on the Hermitian matrices*, SIAM J. Optim. **6** (1996), 164–177.

[279] C.-K. Li and R. Mathias, *The Lidskii-Mirsky-Wielandt theorem - additive and multiplicative versions*, Numer. Math. **81** (1999), no. 3, 377–413.

[280] X. Li, R. N. Mohapatra, and R. S. Rodriguez, *Grüss-type inequalities*, J. Math. Anal. Appl. **267** (2002), no. 2, 434–443.

[281] V. B. Lidskii, *The proper values of the sum and product of symmetric matrices*, Dokl. Akad. Nauk S.S.S.R. **74** (1950), 769–772.

[282] E. H. Lieb and M. Loss, *Analysis*, Second Edition, Graduate Studies in Mathematics **14**, American Mathematical Society, Providence, R. I., 2001.

[283] T.-C. Lim, *Some L^p inequalities and their applications to fixed point theorems of uniformly Lipschitzian mappings.* In: Nonlinear functional analysis and its applications, Part 2, 119–125, Proc. Sympos. Pure Math. **45**, Amer. Math. Soc., Providence, RI, 1986.

[284] T.-P. Lin, *The power mean and the logarithmic mean*, Amer. Math. Monthly **81** (1974), 879–883.

[285] G. Lindblad, *Expectations and entropy inequalities for finite quantum systems*, Comm. Math. Phys. **39** (1974), 111–119.

[286] J. Lindenstrauss and A. Pełczyński, *Absolutely summing operators in $\mathcal{L}_p$-spaces and their applications*, Studia Math. **29** (1968), no. 3, 275–326.

[287] J. Lindenstrauss and L. Tzafriri, *Classical Banach spaces* I: *sequence spaces.* Springer-Verlag, Berlin-Heidelberg-New York, 1996.

[288] J. Lindenstrauss and L. Tzafriri, *Classical Banach spaces* II: *function spaces.* Springer Science & Business Media, 2013.

[289] G. L. Litvinov, V. P. Maslov, and G. B. Shpiz, *Idempotent functional analysis: an algebraic approach* (available as an arXiv preprint at https://arxiv.org/pdf/math/0009128.pdf).

[290] L. L. Liu and Yi Wang, *On the log-convexity of combinatorial sequences*, Adv. Appl. Math. **39** (2007), no. 4, 453–476.

[291] A. Lyapunov, *Nouvelle forme du théorème sur la limite de probabilité*, Memoires de l'Acad. de St.-Petersburg, **12** (1901), no. 5.

[292] C. Maclaurin, *A second letter to Martin Folkes, Esq.; concerning the roots of equations, with the demonstration of other rules in algebra,* Phil. Transactions, **36** (1729), 59–96.

[293] J. R. Magnus and H. Neudecker, *Matrix differential calculus with applications in statistics and econometrics*, Third Edition, John Wiley & Sons, New York, 2007.

[294] S. M. Malamud, *Some complements to the Jensen and Chebyshev inequality and a problem of W. Walter*, Proc. Amer. Math. Soc. **129** (2001), no. 9, 2671–2678.

[295] S. M. Malamud, *A converse to the Jensen inequality, its matrix extensions and inequalities for minors and eigenvalues*, Linear Algebra Appl. **322** (2001), no. 1–3, 19–41.

[296] S. M. Malamud, *An analogue of the Poincaré separation theorem for normal matrices, and the Gauss–Lucas theorem*, Funct. Anal. Appl. **37** (2003), no. 3, 232–235.

[297] S. M. Malamud, *Inverse spectral problem for normal matrices and the Gauss–Lucas theorem*, Trans. Amer. Math. Soc. **357** (2005), no. 10, 4043–4064. Circulated also as a preprint on ArXiv math.CV/0304158/6 July 2003.

[298] L. Maligranda, *Concavity and convexity of rearrangements*, Comment. Math. (Prace Mat.) **32** (1992), 85–90.

[299] L. Maligranda, *Why Hölder's inequality should be called Rogers' inequality*?, Math. Inequal. Appl., **1** (1998), no. 1, 69–83.

[300] L. Maligranda, J. E. Pečarić and L.-E. Persson, *On some inequalities of the Grüss-Barnes and Borell type*, J. Math. Anal. Appl. **187** (1994), no. 1, 306–323.

[301] L. Maligranda, J. E. Pečarić and L.-E. Persson, *Weighted Favard and Berwald inequalities*, J. Math. Anal. Appl. **190** (1995), no. 1, 248–262.

[302] M. Marcus and J. Lopes, *Inequalities for symmetric functions and Hermitian matrices,* Canad. J. Math. **9** (1957), 305–312.

[303] P. Maréchal, *On the convexity of the multiplicative potential and penalty functions and related topics*, Math. Program. Ser. A **89** (2001), 505–516

[304] P. Maréchal, *On a functional operation generating convex functions.* I. Duality. J. Optim. Theory Appl. **126** (2005), no. 1, 175–189.

[305] A. W. Marshall, I. Olkin and B. Arnold, *Inequalities: Theory of majorization and its applications*, Second Edition, Springer Series in Statistics, Springer, New York, 2011.

[306] M. Matić, C. E. M. Pearce, and J. E. Pečarić, *Improvements of some bounds on entropy measures in information theory*, Math. Inequal. Appl. **1** (1998), no. 2, 295–304.

[307] M. Matić and J. E. Pečarić, *Some companion inequalities to Jensen's inequality*, Math. Inequal. Appl. **3** (2000), no. 3, 355–368.

[308] M. Matić, J. E. Pečarić and A. Vukelić, *On generalization of Bullen-Simpson's inequality*, Rocky Mountain J. Math. **35** (2005), no. 5, 1727–1754.

[309] J. Matkowski, *Affine and convex functions with respect to the logarithmic mean*, Colloq. Math. **95** (2003), no. 2, 217–230.

[310] J. Matkowski and J. Rätz, *Convexity of power functions with respect to symmetric homogeneous means*, In vol. *General Inequalities***7** (Oberwolfach, 1995), 231–247, Birkhäuser, Basel, 1997.

[311] J. Matkowski and J. Rätz, *Convex functions with respect to an arbitrary mean*, In vol. *General Inequalities***7** (Oberwolfach, 1995), 249–258, Birkhäuser, Basel, 1997.

[312] M. E. Mayes, *Functions which parametrize means*, Amer. Math. Monthly **90** (1983), no. 10, 677–683.

[313] S. Mazur and S. Ulam, *Sur les transformations isométriques d'espaces vectoriels normés*, C. R. Acad. Sci. Paris **194** (1932), 946–948.

[314] R. J. McCann, *Existence and uniqueness of monotone measure-preserving maps*, Duke Math. J. **80** (1995), no. 2, 309–323.

[315] R. J. McCann, *A convexity principle for interacting gases*, Adv. Math. **128** (1997), no. 1, 153–179.

[316] J. McKay, *On computing discriminants*, Amer. Math. Monthly, **94** (1987), no. 6, 523–527.

[317] D. McLaughlin, R. Poliquin, J. Vanderwerff, and V. Zizler, *Second-order Gâteaux differentiable bump functions and approximations in Banach spaces*, Canad. J. Math. **45** (1993), no. 3, 612–625.

[318] M. Merkle, *Completely monotone functions - a digest*. In vol.. *Analytic Number Theory, Approximation Theory, and Special Functions* (G. V. Milovanović and M. Th. Rassias Eds.), 347–364, Springer, New York, 2014.

[319] C. D. Meyer, *Matrix Analysis and Applied Linear Algebra*, SIAM, 2000.

[320] F. Mignot, *Contrôle dans les inéquations variationelles elliptiques*, J. Funct. Anal. **22** (1976), no. 2, 130–185.

[321] M. V. Mihai and F.-C. Mitroi-Symeonidis, *New extensions of Popoviciu's inequality*, Mediterr. J. Math. **13** (2016), no. 5, 3121–3133.

[322] M. V. Mihai and C. P. Niculescu, *A simple proof of the Jensen type inequality of Fink and Jodeit*, Mediterr. J. Math. 13 (2016), no. 1, 119–126.

[323] M. Mihailescu and C. P. Niculescu, *An extension of the Hermite-Hadamard inequality through subharmonic functions,* Glasg. Math. J. **49** (2007), no. 3, 509–514.

[324] K. S. Miller and S. G. Samko, *Completely monotonic functions*, Integral Transform Spec. Funct. **12** (2001), no. 4, 389–402.

[325] H. Minkowski, *Theorie der Konvexen Körper, Insbesondere Begründung ihres Oberflächenbegriffs,* Gesammelte Abhandlungen II, Leipzig, 1911.

[326] G. Minty, *Monotone (nonlinear operators) in a Hilbert space*, Duke Math. J. **29** (1962), no. 3, 341–346.

[327] D. S. Mitrinović (in cooperation with P. M. Vasić), *Analytic Inequalities*, Grundlehren der mathematischen Wissenschaften **165,** Springer-Verlag, Berlin, 1970.

[328] D. S. Mitrinović and I. B. Lacković, *Hermite and convexity*, Aequat. Math. **28** (1985), no. 3, 229–232.

[329] D. S. Mitrinović, J. E. Pečarić and A. M. Fink, *Inequalities involving functions and their integrals andderivatives*, Kluwer Academic Publishers, Dordrecht, 1991.

[330] D. S. Mitrinović, J. E. Pečarić and A. M. Fink, *Classical and new inequalities in analysis*, Kluwer Academic Publishers, Dordrecht, 1993.

[331] D. S. Mitrinović and P. M. Vasić, *History, variations and generalizations of the Čebyšev inequality and the question of some priorities*, Univ. Beograd, Publ. Elektrotehn. Fak. Ser. Mat. Fiz. No. 461–497 (1974), 1–30.

[332] C.-F. Mitroi and C. P. Niculescu, *An Extension of Young's Inequality*, Abstr. Appl. Anal. **2011** (2011), Article ID 162049, 18 pp.

[333] B. Mond and J. E. Pečarić, *Convex inequalities in Hilbert space*, Houston J. Math. **19** (1993), no. 3, 405–420.

[334] B. Mond and J. E. Pečarić, *A mixed means inequality*, Austral. Math. Soc. Gaz. **23** (1996), no. 2, 67–70.

[335] P. Montel, *Sur les functions convexes et les fonctions sousharmoniques*, Journal de Math., 9ème série, **7** (1928), 29–60.

[336] M. H. Moore, *A convex matrix function*, Amer. Math. Monthly **80** (1973), no. 4, 408–409.

[337] J. J. Moreau, *Fonctions convexes en dualité*, Faculté des Sciences de Montpellier, Séminaires de Mathématiques Université de Montpellier, Montpellier, 1962 (multigraphié, 18 pp.)

[338] J. J. Moreau, *Fonctions convexes duales et points proximaux dans un espace hilbertien*, C. R. Acad. Sci. Paris Sér. A Math. **255** (1962), 2897–2899.

[339] J. J. Moreau, *Fonctionelles convexes.* Notes polycopiées, College de France, 1966.

[340] T. S. Motzkin, *The arithmetic-geometric inequality.* In: *Inequalities* (O. Shisha, ed.), 205–224, Academic Press, New York, 1967.

[341] R. F. Muirhead, *Some methods applicable to identities and inequalities of symmetric algebraic functions of n letters,* Proc. Edinburgh Math. Soc. **21** (1903), 144–157.

[342] J. Munkres, *Topology*, Second Edition, Prentice Hall, 2000.

[343] I. Namioka and R. R. Phelps, *Banach spaces which are Asplund spaces*, Duke Math. J. **42** (1975), no. 4, 735–750.

[344] J. Nash, *Non-cooperative games*, Ann. Math. **54** (1951), 286–295.

[345] T. Needham, *A visual explanation of Jensen's Inequality*, Amer. Math. Monthly **100** (1993), no. 8, 768–771.

[346] E. Neuman, *The weighted logarithmic mean*, J. Math. Anal. Appl. **188** (1994), no. 3, 885–900.

[347] E. Neuman and J. Sándor, *On the Ky Fan inequality and related inequalities* I, Math. Inequal. Appl. **5** (2002), no. 1, 49–56.

[348] E. Neuman, *Stolarski means in several variables*, JIPAM, J. Inequal. Pure Appl. Math. **6** (2005), no. 2, Article 30.

[349] J. von Neumann, *Some matrix-inequalities and metrization of matric-space*, Tomsk University Rev. **1** (1937), 286–300. In *Collected Works*, Vol. IV, Pergamon, Oxford, 1962, pp. 205–218.

[350] I. Newton, *Arithmetica universalis: sive de compositione et resolutione arithmetica liber,* 1707.

[351] C. P. Niculescu, *Convexity according to the geometric mean*, Math. Inequal. Appl. **3** (2000), no. 2, 155–167.

[352] C. P. Niculescu, *A new look at Newton's inequalities,* JIPAM, J. Inequal. Pure Appl. Math. **1** (2000), no. 2, article no. 17.

[353] C. P. Niculescu, *A multiplicative mean value and its applications.* In Vol. *Inequality Theory and Applications* (Y. J. Cho, S. S. Dragomir and J. Kim, Eds.), Vol. **1**, 243–255, Nova Science Publishers, Huntington, New York, 2001.

[354] C. P. Niculescu, *An extension of Chebyshev's inequality and its connection with Jensen's inequality,* J. Inequal. Appl. **6** (2001), no. 4, 451–462.

[355] C. P. Niculescu, *A note on the Hermite-Hadamard Inequality*, Math. Gazette, July 2001, 48–50.

[356] C. P. Niculescu, *Choquet theory for signed measures,* Math. Inequal. Appl. **5** (2002), no. 3, 479–489.

[357] C. P. Niculescu, *The Hermite-Hadamard inequality for functions of several variables,* Math. Inequal. Appl. **5** (2002), no. 4, 619–623.

[358] C. P. Niculescu, *Convexity according to means,* Math. Inequal. Appl. **6** (2003), no. 3, 571–579.

[359] C. P. Niculescu, *An extension of the Mazur–Ulam theorem.* In vol.: *Global Analysis and Applied Mathematics* (K. Tas, D. Krupka, O. Krupkova and D. Baleanu, eds.), 248–256, AIP Conf. Proc. **729**, Amer. Inst. Phys., Melville, New York, 2004.

[360] C. P. Niculescu, *Interpolating Newton's Inequalities*, Bull. Math. Soc. Sci. Math. Roumanie **47** (95), 2004, no. 1–2, 67–83.

[361] C. P. Niculescu, *The Hermite-Hadamard inequality for convex functions on a global NPC space*, J. Math. Anal. Appl. **356** (2009), no. 1, 295–301.

[362] C. P. Niculescu, *The integral version of Popoviciu's inequality*, J. Math. Inequal. **3** (2009), no. 3, 323–328.

[363] C. P. Niculescu, *On a result of G. Bennett*, Bull. Math. Soc. Sci. Math. Roumanie **54** (102), 2011, no. 3, 261–267.

[364] C. P. Niculescu, *The Hermite-Hadamard inequality for log-convex functions*, Nonlinear Anal. **75** (2012), no. 2, 662–669.

[365] C. P. Niculescu and J. E. Pečarić, *The equivalence of Chebyshev's inequality with the Hermite-Hadamard inequality*, Math. Rep. Buchar. **12** (62), 2010, no. 2, 145–156.

[366] C. P. Niculescu and L.-E. Persson, *Old and New on the Hermite-Hadamard Inequality*, Real Anal. Exchange, **29** (2003/2004), no. 2, 663–685.

[367] C. P. Niculescu and L.-E. Persson, *Convex Functions and their Applications. A Contemporary Approach*, First Edition, CMS Books in Mathematics vol. **23**, Springer-Verlag, New York, 2006.

[368] C. P. Niculescu and F. Popovici, *A note on the Denjoy-Bourbaki theorem*, Real Anal. Exchange **29** (2003/2004), no. 2, 639–646.

[369] C. P. Niculescu and F. Popovici, *A refinement of Popoviciu's inequality*, Bull. Math. Soc. Sci. Math. Roumanie **49** (97), 2006, no. 3, 285–290.

[370] C. P. Niculescu and F. Popovici, *The extension of majorization inequalities within the framework of relative convexity,* JIPAM, J. Inequal. Pure Appl. Math. **7** (2006), no. 1, Article 27, 6 pp.

[371] C. P. Niculescu and G. Prăjitură, *The integral version of Popoviciu's inequality on real line*, Ann. Acad. Rom. Sci., Math. Appl. **8** (2016), no. 1, 56–67.

[372] C. P. Niculescu and I. Rovenţa, *An extension of Chebyshev's algebraic inequality*, Math. Rep. Buchar. **15** (65) (2013), no. 1, 91–95.

[373] C. P. Niculescu and I. Rovenţa, *An approach of majorization in spaces with a curved geometry*, J. Math. Anal. Appl. **411** (2014), no. 1, 119–128.

[374] C. P. Niculescu and I. Rovenţa, *Relative convexity and its applications*, Aequat. Math. **89** (2015), no. 5, 1389–1400.

[375] C. P. Niculescu and I. Rovenţa, *Relative Schur convexity on global NPC spaces*, Math. Inequal. Appl. **18** (2015), no. 3, 1111–1119.

[376] C. P. Niculescu and I. Rovenţa, *Hardy-Littlewood-Pólya theorem of majorization in the framework of generalized convexity*, Carpathian J. Math. **33** (2017), no. 1, 87–95.

[377] C. P. Niculescu and C. I. Spiridon, *New Jensen-type inequalities*. J. Math. Anal. Appl. **401** (2013), no. 1, 343–348.

[378] C. P. Niculescu and M. M. Stănescu, *The Steffensen-Popoviciu measures in the context of quasiconvex functions,* J. Math. Inequal. **11** (2017), no. 2, 469–483.

[379] C. P. Niculescu and M. M. Stănescu, *A Note on Abel's Partial Summation Formula,* Aequat. Math. **91** (2017), no. 6, 1009–1024.

[380] C. P. Niculescu and H. Stephan, *Lagrange's barycentric identity from an analytic viewpoint*, Bull. Math. Soc. Sci. Math. Roumanie **56** (104), 2013, no. 4, 487–496.

[381] L. Nikolova, L.-E. Persson and T. Zachariades, *On Clarkson's inequality, type and cotype of Edmunds–Triebellogarithmic spaces*, Arch. Math. **80** (2003), no. 2, 165–176.

[382] A. Nissenzweig, *$w^\star$-sequential convergence*, Israel J. Math. **22** (1975), no. 3–4, 266–272.

[383] B. Opic and A. Kufner, *Hardy Type Inequalities*, Longman, Harlow, 1990.

[384] R. Osserman, *Curvature in the`eighties*, Amer. Math. Monthly **97** (1990), no. 8, 731–756.

[385] A. M. Ostrowski, *Sur quelques applications des fonctions convexes et concaves au sens de I. Schur,* J. Math. Pures Appl. **31** (9) (1952), 253–292.

[386] C. E. M. Pearce and S. S. Dragomir, *Selected topics on Hermite-Hadamard Inequality and Applications*, Victoria University, Melbourne, 2000. Available at http://rgmia.org/monographs/hermite_hadamard.html

[387] J. E. Pečarić, *A simple proof of the Jensen-Steffensen inequality*, Amer. Math. Monthly **91** (1984), no. 3, 195–196.

[388] J. E. Pečarić, F. Proschan and Y. C. Tong, *Convex functions, partial orderings and statistical applications*, Mathematics in science and engineering **187**, Academic Press, Boston, 1992.

[389] J. E. Pečarić and V. Šimić, *Stolarsky–Tobey mean in n variables*, Math. Inequal. Appl. **2** (1999), no. 3, 325–341.

[390] J. E. Pečarić and K. B. Stolarsky, *Carleman's inequality: history and new generalizations*, Aequat. Math. **61** (2001), no. 1–2, 49–62.

[391] J. Peetre and L.-E. Persson, *General Beckenbach's inequality with applications.* In: Function Spaces, Differential Operators and Nonlinear Analysis, Pitman Research Notes in Math., Ser. **121**, 1989, 125–139.

[392] J.-P. Penot, *Glimpses upon quasiconvex analysis*, ESAIM: Proc., **20** (2007) 170–194

[393] R. Pereira, *Differentiators and the geometry of polynomials*, J. Math. Anal. Appl. **285** (2003), no. 1, 336–348.

[394] L.-E. Persson, *Some elementary inequalities in 'connection with X^p-spaces*, pp. 367–376, Publ. Bulgarian Acad. Sci., 1988.

[395] L.-E. Persson, *Lecture Notes*, Collège de France, November 2015. Research report 1, Department of Engineering Sciences and Mathematics, Luleå University of Technology, 2017, 69 pp.

[396] L.-E. Persson and N. Samko, *What should have happened if Hardy had discovered this?*, J. Inequal. Appl. 2012, no. 29, 11 pp.

[397] R. R. Phelps, *Convex functions, Monotone Operators, and Differentiability,* Second Edition, Lecture Notes in Math. No. 1364, Springer-Verlag, Berlin, 1993.

[398] R. R. Phelps, *Lectures on Choquet's theorem*, Second Edition, Lecture Notes in Mathematics **1757**, Springer-Verlag, Berlin, 2001.

[399] A. O. Pittenger, *The logarithmic mean in n variables*, Amer. Math. Monthly **92** (1985), no. 2, 99–104.

[400] H. Poincaré, *Sur les equations aux dérivées partielles de la physique mathématique*, Amer. J. Math. **12** (1890), no. 3, 211–294.

[401] G. Pólya, *Remark on Weyl's note "Inequalities between the two kinds of eigenvalues of a linear transformation"* , Proc. Nat. Acad. Sci. U.S.A. **36** (1950), 49–51.

[402] G. Pólya, *Mathematics and plausible reasoning,* Vols. I and II, Princeton University Press, Princeton, New Jersey, 1954.

[403] G. Pólya and G. Szegö, *Aufgaben und Lehrsätze aus Analysis*, Vols. I & II, Springer-Verlag, 1925. English edition, Springer-Verlag, 1972.

[404] B.T. Polyak, *Existence theorems and convergence of minimizing sequences in extremum problems with restrictions*, Soviet Math. Dokl. **7** (1966), 72–75. (Russian)

[405] T. Popoviciu, *Sur les equations algebriques ayant toutes leurs racines reelles*, Mathematica (Cluj) **9** (1935), 129–145.

[406] T. Popoviciu, *Notes sur les fonctions convexes d'ordre superieurIX*, Bull. Math. Soc. Roumaine Sci. **43** (1941), 85–141.

[407] T. Popoviciu, *Les fonctions convexes,* Actualités Sci. Ind. no. 992, Hermann, Paris, 1944.

[408] T. Popoviciu, *Sur certaines inégalités qui caractérisent les fonctions convexes*, An. Şti. "Al. I. Cuza", Iasi, Secţia Mat., **11** (1965), 155–164.

[409] B. H. Pourciau, *Modern Multiplier Rules*, Amer. Math. Monthly **87** (1980), no. 6, 433–452.

[410] W. Pusz and S. L. Woronowicz, *Functional calculus for sesquilinear forms and the purification map*, Rep. Mathematical Phys. **8** (1975), no. 2, 159–170.

[411] S. T. Rachev and L. Rüschendorf, *Mass Transportation Problems*: Volume I: *Theory*. Probability and its Applications, Springer-Verlag, Berlin, 1998.

[412] R. Rado, *An inequality*, J. London Math. Soc., **27** (1952), 1–6.

[413] Q. I. Rahman and G. Schmeisser, *Analytic theory of polynomials*, London Mathematical Society Monographs. New Series, vol. **26**, The Clarendon Press, Oxford University Press, Oxford, 2002.

[414] M. M. Rao and Z. D. Ren, *Theory of Orlicz Spaces*, Monographs and textbooks in Pure and Applied Mathematics **146**, Marcel Dekker, New York, 1991.

[415] M. Rădulescu and S. Rădulescu, *Generalization of Dobrusin's inequalities and applications*, J. Math. Anal. Appl. **204** (1996), no. 3, 631–645.

[416] M. Renardy and R. C. Rogers, *An introduction to partial differential equations*, Texts in Applied Mathematics **13**, Springer-Verlag, New York, 1993.

[417] J. Renegar, *Hyperbolic Programs, and Their Derivative Relaxations*, J. Found. Comput. Math. **6** (2006), no. 1, 59–79.

[418] Yu. G. Reshetnyak, *Generalized derivatives and differentiability almost everywhere*, Mat. Sb. **75** (1968), no. 3, 323–334.

[419] F. Riesz, *Über lineare Funktionalgleichungen.* Acta Math. **41** (1918), 71–98.

[420] A. W. Roberts and D. E. Varberg, *Convex functions*, Pure and Applied Mathematics **57**, Academic Press, New York - London, 1973.

[421] R. T. Rockafellar, *Convex analysis*, Princeton Mathematical Series **28**, Princeton University Press, Princeton, New Jersey, 1970.

[422] R. T. Rockafellar and R. J. -B. Wets, *Variational Analysis*, Corrected 3rd printing, Springer-Verlag, Berlin, 2009.

[423] G. Rodé, *Eine abstrakte Version des Satzes von Hahn–Banach*, Arch. Math. (Basel) **31** (1978/1979), no. 5, 474–481.

[424] L. J. Rogers, *An extension of a certain theorem in inequalities*, Messenger Math. **17** (1888), 145–150.

[425] D. Romik, *Explicit formulas for hook walks on continual Young diagrams*, Adv. Appl. Math. **32** (2004), no. 4, 625–654.

[426] P. Roselli and M. Willem, *A convexity inequality*, Amer. Math. Monthly **109** (2002), no. 1, 64–70.

[427] S. Rosset, *Normalized symmetric functions, Newton's inequalities and a new set of stronger inequalities*, Amer. Math. Monthly **96** (1989), no. 9, 815–820.

[428] H. Rubin and O. Wesler, *A note on convexity in Euclidean n-space*, Proc. Amer. Math. Soc. **9** (1958), 522–523.

[429] W. Rudin, *Real and complex analysis*, Third Edition, McGraw–Hill Book Co., New York, 1987.

[430] L. Rüschendorf, *Monge–Kantorovich transportation problem and optimal couplings*, Jahresber. Deutsch. Math.-Verein. **109** (2007), 113–137.

[431] L. Rüschendorf, *Dependence, Risk Bounds, Optimal Allocations and Portfolios,* Springer Series in Operations Research and Financial Engineering, Springer-Verlag, Berlin, 2013.

[432] S. Saks, *Sur un théorème de M. Montel,* C. R. Acad. Sci. Paris, **187** (1928), 276–277.

[433] T. Sasaki, Y. Kanada and S. Watanabe, *Calculation of discriminants of higher degree equations*, Tokyo J. Math., **4** (1981), 493–503.

[434] H. H. Schaefer, *Banach lattices and positive operators*, Springer, Berlin, 1974.

[435] H. H. Schaefer and M. P. Wolff, *Topological vector spaces*, Second Edition, Graduate Texts in Mathematics **3**, Springer Verlag, New York, 1999.

[436] R. Schneider, *Convex Bodies: The Brunn-Minkowski Theory,* Second Expanded Edition, Encyclopedia of Mathematics and its Applications **151**, Cambridge University Press, Cambridge, 2014.

[437] I. Schur, *Über die charakteristischen Wurzeln einer linearen Substitution mit einer Anwendung auf die Theorie der Integralgleichungen*, Math. Ann. **66** (1909), no. 4, 488–510.

[438] I. Schur, *Bemerkungen zur Theorie der beschränken Bilinearformen mit unendlich vielen Veränderlichen,* J. Reine Angew. Math. **140** (1911), 1–28

[439] I. Schur, *Über eine Klasse von Mittelbildungen mit Anwendungdie Determinanten,* Theorie Sitzungsber. Berlin. Math. Gesellschaft **22** (1923), 9–20.

[440] H. A. Schwarz, *Über ein Flächen kleinsten Flächeninhalts betreffendes Problem der Variationsrechnung*, Acta Societatis Scientiarum Fennicae **XV** (1888), 318–362.

[441] D. Serre, *Weyl and Lidskii Inequalities for General Hyperbolic Polynomials*, Chin. Ann. Math. B **30** (2009), no. 6, 785–802

[442] S. Sherman, *On a theorem of Hardy, Littlewood, Pólya, and Blackwell*, Proc. Natl. Acad. Sci. USA, **37** (1951), 826–831; Errata, ibid. **38** (1952), 382.

[443] R. E. Showalter, *Monotone operators in Banach space and nonlinear partial differential equations*, Mathematical Surveys and Monographs **49**, American Mathematical Society, Providence, RI, 1997.

[444] W. Sierpiński, *Sur la question de la mesurabilité de la base de M. Hamel*, Fund. Math. **1** (1920), 105–111.

[445] W. Sierpiński, *Sur les fonctions convexes mesurables*, Fund. Math., **1** (1920), 125–129.

[446] M. Šilhavý, *A functional inequality related to analytic continuation.* Preprint No. 37-2015, Institute of Mathematics of the AS ČR. Available at: https://www.researchgate.net/profile/Miroslav_Silhavy

[447] M. Šilhavý, *The convexity of* $C \to h(\det C)$, Technische Mechanik **35** (2015), no. 1, 60–61.

[448] B. Simon, *The statistical mechanics of lattice gases*, Vol. 1, Princeton Series in Physics, Princeton University Press, Princeton, 1993.

[449] B. Simon, *Trace ideals and their applications*, Second Edition, Mathematical Surveys and Monographs **120**, American Mathematical Society, 2005.

[450] B. Simon, *Convexity. An Analytic Viewpoint*, Cambridge Tracts in Mathematics **187**, Cambridge University Press, Cambridge, 2011.

[451] S. Sra, *Inequalities via symmetric polynomial majorization*, Preprint, September 18, 2015.

[452] J. Šremr, *Absolutely continuous functions of two variables in the sense of Carathéodory*, Electronic Journal of Differential Equations, **2010** (2010), no. 154, 11 pp.

[453] A. J. Stam, *Some inequalities satisfied by the quantities of information of Fisher and Shannon*, Information and Control **2** (1959), 101–112.

[454] R. P. Stanley, *Log-concave and unimodal sequences in algebra, combinatorics, and geometry*, Ann. New York Acad. Sei. **576** (1989), no. 1, 500–535.

[455] R. Starr, *Quasi-equilibria in markets with nonconvex preferences,* Econometrica: journal of the Econometric Society **37** (1969), no. 1, 25–38.

[456] J. M. Steele, *The Cauchy-Schwarz master class. An introduction to the art of mathematical inequalities*, Cambridge University Press, Cambridge, 2004.

[457] J. F. Steffensen, *On certain inequalities between mean values, and their application to actuarial problems,* Scandinavian Actuarial Journal 1918, no. 1, 82–97.

[458] J. F. Steffensen, *On certain inequalities and methods of approximation*, J. Inst. Actuaries, **51** (1919), no. 3, 274–297.

[459] J. F. Steffensen, *On a generalization of certain inequalities by Tchebycheff and Jensen*, Scandinavian Actuarial Journal, 1925, no. 3–4, 137–147.

[460] J. Stoer and C. Witzgall, *Convexity and optimization in finite dimensions* I, Die Grundlehren der Mathematischen Wissenschaften **163**, Springer Verlag, New York - Berlin, 1970.

[461] K. B. Stolarsky, *Generalizations of the logarithmic mean*, Math. Mag. **48** (1975), no. 2, 87–92.

[462] K. B. Stolarsky, *The power and generalized logarithmic means*, Amer. Math. Monthly **87** (1980), no. 7, 545–548

[463] O. Stolz, *Grunzüge der Differential und Integralrechnung*, Vol. 1, Teubner, Leipzig, 1893.

[464] Ph. D. Straffin, *Game Theory and Strategy*, The Mathematical Association of America, 1993.

[465] S. Strătilă and L. Zsidó, *Lectures on von Neumann Algebras*, Editura Academiei Bucharest and Abacus Press, Tunbridge Wells, Kent, England, 1979.

[466] K. T. Sturm, Probability measures on metric spaces of nonpositive curvature. In vol.: *Heat kernels and analysis on manifolds, graphs, and metric spaces* (Pascal Auscher et al. editors). Lecture notes from a quarter program on heat kernels, random walks, and analysis on manifolds and graphs, April 16–July 13, 2002, Paris, France. Contemp. Math. 338 (2003), 357–390.

[467] K. Sundaresan and S. Swaminathan, *Geometry and nonlinear analysis in Banach spaces,* Lecture Notes in Mathematics **1131**, Springer-Verlag, Berlin, 2006.

[468] J. Sylvester, *On Newton's Rule for the discovery of imaginary roots of equations.* See: *The Collected Mathematical Papers of James Joseph Sylvester*, Vol. II (1854–1873), 493–494, Cambridge University Press, 1908.

[469] J. Sylvester, *On an elementary proof and generalization of Sir Isaac Newton's hitherto undemonstrated rule for discovery of imaginary roots.* See: *The Collected Mathematical Papers of James Joseph Sylvester*, Vol. II (1854–1873), 498–513, Cambridge University Press, 1908.

[470] M. Tomić, *Théorème de Gauss relatif au centre de gravité et son application*, Bull. Soc. Math. Phys. Serbie **1** (1949), 31–40.

[471] Y. L. Tong, *Probability inequalities in multivariate distributions*, Academic Press, New York, 1980.

[472] E. Torgersen, *Comparison of Statistical Experiments*, Encyclopedia of Mathematics and its Applications **36**, Cambridge University Press, Cambridge, 1991.

[473] T. Trif, *Convexity of the gamma function with respect to Hölder means.* In: *Inequality Theory and Applications*, Vol. **3** (Y. J. Cho, J. K. Kim, and S. S. Dragomir, eds.), Nova Science Publishers, Inc., Hauppage, New York, 2003, 189–195.

[474] J. Tropp, *From joint convexity of quantum relative entropy to a concavity theorem of Lieb*, Proc. Amer. Math. Soc. **140** (2012), no. 5, 1757–1760.

[475] J. Väisälä, *A proof of the Mazur–Ulam theorem*, Amer. Math. Monthly **110** (2003), no. 7, 633–635.

[476] L. Veselý, *Jensen's integral inequality in locally convex spaces*, Ann. Acad. Rom. Sci. Math. Appl. **9** (2017), no. 2, 136–144.

[477] C. Villani, *Topics in optimal transportation*, Graduate Studies in Mathematics, Vol. **58**, American Mathematical Society, Providence, RI, 2003.

[478] C. Villani, *Optimal transport. Old and new*, Grundlehren der mathematischen Wissenschaften **338**, Springer Verlag, Berlin, 2009.

[479] A. Vogt, *Maps which preserve equality of distance*, Studia Math. **45** (1973), 43–48.

[480] M. Volle, J.-B. Hiriart-Urruty and C. Zălinescu, *When some variational properties force convexity*, ESAIM: Control Optim. Calc. Var. **19** (2013), no. 3, 701–709.

[481] R. Výborný, *The Hadamard three-circles theorems for partial differential equations*, Bull. Amer. Math. Soc. **80** (1974), no. 1, 81–84.

[482] H.-T. Wang, *Convex functions and Fourier coefficients*, Proc. Amer. Math. Soc. **94** (1985), no. 4, 641–646.

[483] R. Webster, *Convexity*, Oxford Science Publications, The Clarendon Press, Oxford University Press, New York, 1994.

[484] H. Weyl, *Das asymtotische Verteilungsgesetz der Eigenwerte lineare partieller Differentialgleichungen*, Math. Ann. **71** (1912), no. 4, 441–479.

[485] H. Weyl, *Inequalities between two kinds of eigenvalues of a linear transformation*, Proc. Natl. Acad. Sci. U.S.A. **35** (1949), no. 7, 408–411.

[486] E. T. Whittaker and G.N. Watson, *A course on Modern Analysis.* Reprint of the fourth (1927) edition, Cambridge Mathematical Library, Cambridge University Press, Cambridge, 1996.

[487] D. V. Widder, *The Laplace transform*, Princeton Mathematical Series **6**, Second printing, Princeton University Press, Princeton, 1946.

[488] H. Wielandt, *An extremum property of sums of eigenvalues*, Proc. Amer. Math. Soc. **6** (1955), no. 1, 106–110.

[489] M. Willem, *Functional analysis: Fundamentals and applications*, Cornerstones, Birkhäuser, New York, 2013.

[490] G. J. Woeginger, *When Cauchy and Hölder met Minkowski: A tour through well-known inequalities*, Math. Mag. **82** (2009), no. 3, 202–207.

[491] R. K. P. Zia, E. F. Redish and S. R. McKay, *Making Sense of the Legendre Transform*, American Journal of Physics **77** (2009), 614–625.

[492] L. C. Young, *Lectures on the calculus of variations*, Chelsea, New York, 1980.

[493] W. H. Young, *On classes of summable functions and their Fourier series*, Proc. Roy. Soc. London, Ser. A **87** (1912), 225–229.

[494] A. J. Yudine, *Solution of two problems on the theory of partially ordered spaces*, Dokl. Akad. Nauk SSSR **23** (1939), 418–422.

[495] T. Zamfirescu, *The curvature of most convex surfaces vanishes almost everywhere*, Math. Z. **174** (1980), no. 2, 135–139.

[496] T. Zamfirescu, *Nonexistence of curvature in most points of most convex surfaces*, Math. Ann. **252** (1980), no. 3, 217–219.

[497] T. Zamfirescu, *Curvature properties of typical convex surfaces*, Pacific J. Math. **131** (1988), no. 1, 191–207.

[498] C. Zalinescu, *Convex analysis in general vector spaces*, World Scientific Publishing Co., River Edge, NJ., 2002.

[499] W. P. Ziemer, *Weakly differentiable functions, Sobolev spaces and functions of bounded variation*, Graduate Texts in Mathematics **120**, Springer-Verlag, New York, 1989.

[500] A. Zygmund, *Trigonometric Series,* Vols. I, II, Second Edition, Cambridge University Press, New york, 1959.

Index

A
Alexandrov Hessian, 181, 364

B
Banach lattice, 96
barycenter, 42, 75, 161, 251, 310, 350
barycentric coordinates, 74
Brenier map, 299

C
center of mass, 75
combination
  affine, 72
  convex, 73
cone
  convex, 75
  dual, 95
  future light, 75
  generating, 96
  hyperbolicity, 214
  monotone positive, 75
  normal, 173
  positive, 95
  proper, 95
  recession, 271
  tangent, 173
continuous functional calculus, 105
convex body, 175
core point, 345
Courant–Fischer minimax principle, 203
covariance, 46

D
derivative
  directional, 141
  lower/upper, 28
  lower/upper second symmetric, 29
  weak, 373

E
effective domain, 117
ellipsoid, 77
entropy
  differential, 44
  quantum relative, 230
  von Neumann, 230
epigraph, 115
expectation, 41, 161
extreme point, 90

F
formula
  Abel's partial summation, 57
  Hopf–Lax, 292
  Stirling's, 60
  Taylor's, 156
  Tolland–Singer, 262
function
  (Fréchet) differentiable, 147
  $\prec_{SPM}$- convex, 361
  absolutely continuous, 35
  affine, 1, 108
  Asplund, 270
  closed, 118
  closure, 121
  Cobb–Douglas, 134
  coercive, 166
  complementary cumulative distribution, 224
  completely monotonic, 21
  concave, 1, 108, 333
  conjugate, 255

C. P. Niculescu and L.-E. Persson, *Convex Functions and Their Applications*,
CMS Books in Mathematics, https://doi.org/10.1007/978-3-319-78337-6

convex, 1, 107, 117, 234, 333
convex conjugate, 38
convex hull, 121
cost, 268
cumulative distribution, 42
distance, 108
gamma, 20
Gâteaux differentiable, 144, 145
gauge, 132
Hamiltonian, 290
indicator, 117
integral sine, 338
invariant under permutations, 189
Lagrangian, 273, 290
Leontief, 134
Lobacevski's, 336
log-concave, 20
log-convex, 20, 108
log-determinant, 110
logarithmic integral, 336
lower envelope, 309
lower semicontinuous, 39, 118
midpoint convex, 5
multiplicatively convex, 336
nonexpansive, 126
O(N)-invariant, 228
operator convex, 235
operator monotone, 236
Orlicz, 39
perspective, 135
piecewise linear, 55
positively homogeneous, 130, 341
proper, 38, 117
proper concave, 117
quasiconcave, 4, 108, 117
quasiconvex, 4, 108, 116
recession, 264
Schur-convex, 197
semiconcave, 297
semiconvex, 366
separately convex, 114
slope, 31
strictly concave, 1, 108, 331
strictly convex, 1, 107, 331
strictly Schur-convex, 197
strongly concave, 108
strongly convex, 31, 107
subadditive, 130, 339
sublinear, 130, 339
supercoercive, 169
superquadratic, 67
support, 261
symmetric-decreasing rearrangement, 225
twice Gâteaux differentiable, 155
upper envelope, 309
upper semicontinuous, 118

G

Galerkin method, 375
generalized inverse, 66
generic property, 70
gradient, 145

H

half-space, 89
Hessian matrix, 155
Hilbert cube, 145
hull
affine, 73
convex, 73
hyperplane, 89

I

inequality
Abel–Steffensen, 57
AM–GM, 7
weighted AM–GM, 8
Apéry's, 308
Bellman's, 193
Berezin's, 239
Bernoulli's, 16
Berwald's, 19
Borell–Brascamp–Lieb, 176
Brunn–Minkowski, 175
Carleman's, 54
Cauchy–Bunyakovsky–Schwarz, 13, 15, 102
Chebyshev's algebraic, 50
Chebyshev's probabilistic, 46
Clarkson's, 86

Dunkl–Williams, 88
Fenchel–Young, 39, 256
Fink's, 193
Gibbs', 19
Grüss, 48
Hadamard determinant, 196, 240
Hanner's, 139
Hardy's, 52, 54
Hardy–Littlewood–Pólya, 186
Hardy–Littlewood–Pólya–Fuchs, 194
Hermite–Hadamard, 60, 321, 326
Hilbert's, 18
Hlawka's, 59
integral *AM–GM*, 48
Iyengar's, 70
Jensen's, 7, 32, 43, 47, 48, 116, 140, 163
Jensen's operator, 237
Jensen–Steffensen, 57, 58, 311
Kantorovich's, 32
Klein's, 241
Ky Fan minimax, 279
Ky Fan's, 205
Lidskii–Wielandt, 206
Lyapunov's, 19
Maclaurin's, 352
Massera-Schäffer, 87
Minkowski's, 14, 18, 137, 138, 159
mixed arithmetic–geometric, 67
Newton's, 352
Ostrowski's, 70
Pólya–Knopp, 54
Peierls, 239
Peierls–Bogoliubov, 240
Pinsker's, 19
Poincaré's, 18, 371
Popoviciu's, 56, 69
Prékopa–Leindler, 178
rearrangement, 59
Rogers–Hölder, 13, 15, 17, 18, 139
Schur's, 195
Schweitzer's, 33
Steffensen's, 307
subgradient, 123
three chords, 25
von Neumann's, 206
Weyl's, 191, 204
Young's, 11, 12, 16, 17
infimal convolution, 283

K
Karush–Kuhn–Tucker conditions, 274, 278
Kullback–Leibler divergence, 19, 238

L
lemma
accessibility, 78, 81
Aczel's, 333
line segment, 71
Lipschitz constant, 70
Lorenz curve, 225

M
majorization
Choquet order, $\prec_{Ch}$, 314
doubly stochastic, $\prec_{\mathrm{ds}}$, 226
Hardy-Littlewood-Pólya $\prec_{HLP}$, 186
Sherman, $\prec_{Sh}$, 218
symmetric polynomial majorization, 359
unitary doubly stochastic, $\prec_{\mathrm{uds}}$, 226
weak, $\prec_{HLPw}$, 185
map
diagonal, 99
duality, 340
eigenvalues, 202
monotone, 84
matrix
covariance, 165
doubly stochastic, 91
Markov, 93
permutation, 91
positive (positive semi-definite), 76
strictly positive (positive definite), 77
maximum principle, 171, 347
mean
arithmetic, 41, 330

geometric, 242, 328
harmonic, 328
identric, 330
logarithmic, 330
power (Hölder's), 10, 48, 330
quasi-arithmetic, 331
symmetric, 328
weighted arithmetic, 321
weighted geometric, 321
weighted identric, 320
weighted logarithmic, 321
mean value, 41
measure
Borel probability, 342
Dirichlet, 321
dual Steffensen–Popoviciu, 308
Gauss, 179
Hermite–Hadamard, 322
log-concave, 179
M_p-concave, 179
push-forward, 162
Steffensen–Popoviciu, 301
strong Steffensen–Popoviciu, 303, 305
Minkowski addition, 71
mollifier, 181
Monge–Ampère equation, 299
Moreau–Yosida approximation, 284

N

Nash equilibrium, 281
norm
Hilbert–Schmidt, 76
nuclear, 253
operator, 76
nuclear operator, 253
numerical range, 102

O

order
convex, 194
increasing convex, 194
orthogonal projection, 83

P

point
of convexity, 122
polyhedron, 94
polynomial
homogeneous, 212
hyperbolic, 212
polytope, 74
Pompeiu–Hausdorff distance, 79
programming problem
convex, 272
linear, 272

R

random variable, 41
continuous, 45
normal, 46
relative boundary, 80
relative interior, 80
resultant, 42, 161

S

saddle point, 273
seminorm, 130
separation
of points, 345
of sets, 89
sequence
concave, 11
convex, 11
log-concave, 24
log-convex, 24
set
affine, 72
Chebyshev, 85
convex, 71
extremal, 348
invariant under permutations, 189
level, 39, 116
$O(N)$-*i*nvariant, 228
of best approximation, 85
polar, 94
proximinal, 85
set-valued map
cyclically monotone, 128
maximal monotone, 125
monotone, 125
simplex, 74
singular value, 99

Slater's condition, 276
solution
classical, 373, 374
weak, 373, 374
space
dual, 342
finite measure, 41
ordered linear, 95
probability, 41
reflexive, 342
regularly ordered, 100
smooth, 151
strictly convex, 86
uniformly convex, 85
spectral decomposition, 98
standard deviation, 45
subdifferential, 123, 125
subgradient, 36
support
of a convex function, 122
of a convex set, 131
supporting hyperplane, 90
supporting line, 36

T

T-transformation, 187
theorem
Alexandrov's, 181, 366
Artin's, 23
Banach–Alaoglu, 341
basic separation, 90
bipolar, 94
Bohr–Mollerup, 21
Brenier's, 299
Brouwer's, 279
Bunt's, 85
Carathéodory's, 74
Choquet–Bishop–de Leeuw, 319
convex mean value, 40
Davis', 228
Eberlein–Šmulyan, 342
Farkas' alternative, 93, 276
Fenchel–Moreau duality, 266
Galvani's, 31
Gauss–Lucas, 73
Gordan's alternative, 93
Hahn–Banach, 339, 341
Helly's, 82
Hermite–Hadamard, 317
Hiriart-Urruty's, 263
Josephson–Nissenzweig, 150
Kirzsbraun's, 126
KKM, 279
Krein–Milman, 348
Lim's, 137
Mazur's, 343
mean value, 147
Mignot's, 366
Minkowski's, 90
Montel's, 335
Radon's, 81
Riesz–Kakutani, 342
Schur-Horn, 195
separation, 89
Shapley–Folkman, 78
Sherman's, 219
Stolz, 25
strong separation, 90, 345
support, 94
Sylvester's, 157
Toeplitz–Hausdorff, 102
Tomić–Weyl, 187
Tonelli's, 182
trace, 371
Weyl's perturbation, 205
trace, 101, 253
trace metric, 242
transform
Cayley, 126
Laplace, 24
Legendre–Fenchel, 38, 255

V

variance, 45

W

weak topology, 341
weak-star topology, 341
Weyl's monotonicity principle, 203

Printed in the United States
By Bookmasters